ERRATUM

The following is a correction to volume 84 in AGU's Geophysical Monograph Series *Solar System Plasmas in Space and Time*, J. L. Burch and J. H. Waite, Jr., Editors.

In the paper "Multi-Spacecraft Study of a Substorm Growth and Expansion Phase Features Using a Time-Evolving Field Model" by D. N. Baker et al., the caption to Figure 7 on page 107 was omitted and reads as follows:

Fig. 7. (a) A projection of field lines onto the equatorial plane in the modified magnetic field model discussed in the text. Two sets of field lines are shown, one set emanating from 59° magnetic latitude and the other from 61° magnetic latitude. The heavy dashed radial line shows the substorm onset meridian (21 MLT) at 0111 UT. Panels (b) and (c) show the subsequent longitudinal spread of the substorm disturbance as seen by the several high-altitude spacecraft [adapted from Baker et al., 1993].

Solar System Plasmas in Space and Time

Geophysical Monograph Series

Including

IUGG Volumes

Maurice Ewing Volumes

Mineral Physics Volumes

GEOPHYSICAL MONOGRAPH SERIES

Geophysical Monograph Volumes

1 Antarctica in the International Geophysical Year *A. P. Crary, L. M. Gould, E. O. Hulburt, Hugh Odishaw, and Waldo E. Smith (Eds.)*

2 Geophysics and the IGY *Hugh Odishaw and Stanley Ruttenberg (Eds.)*

3 Atmospheric Chemistry of Chlorine and Sulfur Compounds *James P. Lodge, Jr. (Ed.)*

4 Contemporary Geodesy *Charles A. Whitten and Kenneth H. Drummond (Eds.)*

5 Physics of Precipitation *Helmut Weickmann (Ed.)*

6 The Crust of the Pacific Basin *Gordon A. Macdonald and Hisashi Kuno (Eds.)*

7 Antarctica Research: The Matthew Fontaine Maury Memorial Symposium *H. Wexler, M. J. Rubin, and J. E. Caskey, Jr. (Eds.)*

8 Terrestrial Heat Flow *William H. K. Lee (Ed.)*

9 Gravity Anomalies: Unsurveyed Areas *Hyman Orlin (Ed.)*

10 The Earth Beneath the Continents: A Volume of Geophysical Studies in Honor of Merle A. Tuve *John S. Steinhart and T. Jefferson Smith (Eds.)*

11 Isotope Techniques in the Hydrologic Cycle *Glenn E. Stout (Ed.)*

12 The Crust and Upper Mantle of the Pacific Area *Leon Knopoff, Charles L. Drake, and Pembroke J. Hart (Eds.)*

13 The Earth's Crust and Upper Mantle *Pembroke J. Hart (Ed.)*

14 The Structure and Physical Properties of the Earth's Crust *John G. Heacock (Ed.)*

15 The Use of Artificial Satellites for Geodesy *Soren W. Henricksen, Armando Mancini, and Bernard H. Chovitz (Eds.)*

16 Flow and Fracture of Rocks *H. C. Heard, I. Y. Borg, N. L. Carter, and C. B. Raleigh (Eds.)*

17 Man-Made Lakes: Their Problems and Environmental Effects *William C. Ackermann, Gilbert F. White, and E. B. Worthington (Eds.)*

18 The Upper Atmosphere in Motion: A Selection of Papers With Annotation *C. O. Hines and Colleagues*

19 The Geophysics of the Pacific Ocean Basin and Its Margin: A Volume in Honor of George P. Woollard *George H. Sutton, Murli H. Manghnani, and Ralph Moberly (Eds.)*

20 The Earth's Crust: Its Nature and Physical Properties *John C. Heacock (Ed.)*

21 Quantitative Modeling of Magnetospheric Processes *W. P. Olson (Ed.)*

22 Derivation, Meaning, and Use of Geomagnetic Indices *P. N. Mayaud*

23 The Tectonic and Geologic Evolution of Southeast Asian Seas and Islands *Dennis E. Hayes (Ed.)*

24 Mechanical Behavior of Crustal Rocks: The Handin Volume *N. L. Carter, M. Friedman, J. M. Logan, and D. W. Stearns (Eds.)*

25 Physics of Auroral Arc Formation *S.-I. Akasofu and J. R. Kan (Eds.)*

26 Heterogeneous Atmospheric Chemistry *David R. Schryer (Ed.)*

27 The Tectonic and Geologic Evolution of Southeast Asian Seas and Islands: Part 2 *Dennis E. Hayes (Ed.)*

28 Magnetospheric Currents *Thomas A. Potemra (Ed.)*

29 Climate Processes and Climate Sensitivity (Maurice Ewing Volume 5) *James E. Hansen and Taro Takahashi (Eds.)*

30 Magnetic Reconnection in Space and Laboratory Plasmas *Edward W. Hones, Jr. (Ed.)*

31 Point Defects in Minerals (Mineral Physics Volume 1) *Robert N. Schock (Ed.)*

32 The Carbon Cycle and Atmospheric CO_2: Natural Variations Archean to Present *E. T. Sundquist and W. S. Broecker (Eds.)*

33 Greenland Ice Core: Geophysics, Geochemistry, and the Environment *C. C. Langway, Jr., H. Oeschger, and W. Dansgaard (Eds.)*

34 Collisionless Shocks in the Heliosphere: A Tutorial Review *Robert G. Stone and Bruce T. Tsurutani (Eds.)*

35 Collisionless Shocks in the Heliosphere: Reviews of Current Research *Bruce T. Tsurutani and Robert G. Stone (Eds.)*

36 Mineral and Rock Deformation: Laboratory Studies —The Paterson Volume *B. E. Hobbs and H. C. Heard (Eds.)*

37 Earthquake Source Mechanics (Maurice Ewing Volume 6) *Shamita Das, John Boatwright, and Christopher H. Scholz (Eds.)*

38 Ion Acceleration in the Magnetosphere and Ionosphere *Tom Chang (Ed.)*

39 High Pressure Research in Mineral Physics (Mineral Physics Volume 2) *Murli H. Manghnani and Yasuhiko Syono (Eds.)*

40 Gondwana Six: Structure Tectonics, and Geophysics *Gary D. McKenzie (Ed.)*

41 Gondwana Six: Stratigraphy, Sedimentology, and Paleontology *Garry D. McKenzie (Ed.)*

42 Flow and Transport Through Unsaturated Fractured Rock *Daniel D. Evans and Thomas J. Nicholson (Eds.)*

43 **Seamounts, Islands, and Atolls** *Barbara H. Keating, Patricia Fryer, Rodey Batiza, and George W. Boehlert (Eds.)*

44 **Modeling Magnetospheric Plasma** *T. E. Moore and J. H. Waite, Jr. (Eds.)*

45 **Perovskite: A Structure of Great Interest to Geophysics and Materials Science** *Alexandra Navrotsky and Donald J. Weidner (Eds.)*

46 **Structure and Dynamics of Earth's Deep Interior (IUGG Volume 1)** *D. E. Smylie and Raymond Hide (Eds.)*

47 **Hydrological Regimes and Their Subsurface Thermal Effects (IUGG Volume 2)** *Alan E. Beck, Grant Garven, and Lajos Stegena (Eds.)*

48 **Origin and Evolution of Sedimentary Basins and Their Energy and Mineral Resources (IUGG Volume 3)** *Raymond A. Price (Ed.)*

49 **Slow Deformation and Transmission of Stress in the Earth (IUGG Volume 4)** *Steven C. Cohen and Petr Vaníček (Eds.)*

50 **Deep Structure and Past Kinematics of Accreted Terranes (IUGG Volume 5)** *John W. Hillhouse (Ed.)*

51 **Properties and Processes of Earth's Lower Crust (IUGG Volume 6)** *Robert F. Mereu, Stephan Mueller, and David M. Fountain (Eds.)*

52 **Understanding Climate Change (IUGG Volume 7)** *Andre L. Berger, Robert E. Dickinson, and J. Kidson (Eds.)*

53 **Plasma Waves and Istabilities at Comets and in Magnetospheres** *Bruce T. Tsurutani and Hiroshi Oya (Eds.)*

54 **Solar System Plasma Physics** *J. H. Waite, Jr., J. L. Burch, and R. L. Moore (Eds.)*

55 **Aspects of Climate Variability in the Pacific and Western Americas** *David H. Peterson (Ed.)*

56 **The Brittle-Ductile Transition in Rocks** *A. G. Duba, W. B. Durham, J. W. Handin, and H. F. Wang (Eds.)*

57 **Evolution of Mid Ocean Ridges (IUGG Volume 8)** *John M. Sinton (Ed.)*

58 **Physics of Magnetic Flux Ropes** *C. T. Russell, E. R. Priest, and L. C. Lee (Eds.)*

59 **Variations in Earth Rotation (IUGG Volume 9)** *Dennis D. McCarthy and Williams E. Carter (Eds.)*

60 **Quo Vadimus Geophysics for the Next Generation (IUGG Volume 10)** *George D. Garland and John R. Apel (Eds.)*

61 **Cometary Plasma Processes** *Alan D. Johnstone (Ed.)*

62 **Modeling Magnetospheric Plasma Processes** *Gordon K. Wilson (Ed.)*

63 **Marine Particles Analysis and Characterization** *David C. Hurd and Derek W. Spencer (Eds.)*

64 **Magnetospheric Substorms** *Joseph R. Kan, Thomas A. Potemra, Susumu Kokubun, and Takesi Iijima (Eds.)*

65 **Explosion Source Phenomenology** *Steven R. Taylor, Howard J. Patton, and Paul G. Richards (Eds.)*

66 **Venus and Mars: Atmospheres, Ionospheres, and Solar Wind Interactions** *Janet G. Luhmann, Mariella Tatrallyay, and Robert O. Pepin (Eds.)*

67 **High-Pressure Research: Application to Earth and Planetary Sciences (Mineral Physics Volume 3)** *Yasuhiko Syono and Murli H. Manghnani (Eds.)*

68 **Microwave Remote Sensing of Sea Ice** *Frank Carsey, Roger Barry, Josefino Comiso, D. Andrew Rothrock, Robert Shuchman, W. Terry Tucker, Wilford Weeks, and Dale Winebrenner*

69 **Sea Level Changes: Determination and Effects (IUGG Volume 11)** *P. L. Woodworth, D. T. Pugh, J. G. DeRonde, R. G. Warrick, and J. Hannah*

70 **Synthesis of Results from Scientific Drilling in the Indian Ocean** *Robert A. Duncan, David K. Rea, Robert B. Kidd, Ulrich von Rad, and Jeffrey K. Weissel (Eds.)*

71 **Mantle Flow and Melt Generation at Mid-Ocean Ridges** *Jason Phipps Morgan, Donna K. Blackman, and John M. Sinton (Eds.)*

72 **Dynamics of Earth's Deep Interior and Earth Rotation (IUGG Volume 12)** *Jean-Louis Le Mouël, D.E. Smylie, and Thomas Herring (Eds.)*

73 **Environmental Effects on Spacecraft Positioning and Trajectories (IUGG Volume 13)** *A. Vallance Jones (Ed.)*

74 **Evolution of the Earth and Planets (IUGG Volume 14)** *E. Takahashi, Raymond Jeanloz, and David Rubie (Eds.)*

75 **Interactions Between Global Climate Subsystems: The Legacy of Hann (IUGG Volume 15)** *G. A. McBean and M. Hantel (Eds.)*

76 **Relating Geophysical Structures and Processes: The Jeffreys Volume (IUGG Volume 16)** *K. Aki and R. Dmowska (Eds.)*

77 **The Mesozoic Pacific: Geology, Tectonics and Volcanism—A Volume in Memory of Sy Schlanger** *Malcolm S. Pringle, William W. Sager, William V. Sliter, and Seth Stein (Eds.)*

78 **Climate Change in Continental Isotopic Records** *P. K. Swart, K. C. Lohmann, J. McKenzie, and S. Savin (Eds.)*

79 **The Tornado: Its Structure, Dynamics, Prediction, and Hazards** *C. Church, D. Burgess, C. Doswell, R. Davies-Jones (Eds.)*

80 **Auroral Plasma Dynamics** *R. L. Lysak (Ed.)*

81 **Solar Wind Sources of Magnetospheric Ultra-Low Frequency Waves** *M. J. Engebretson, K. Takahashi, and M. Scholer (Eds.)*

82 **Gravimetry and Space Techniques Applied to Geodynamics and Ocean Dynamics** *Bob E. Schutz, Allen Anderson, Claude Froidevaux, and Michael Parke (Eds.)*

83 **Nonlinear Dynamics and Predictability of Geophysical Phenomena** *William I. Newman, Andrei Gabrielov, and Donald L. Turcotte (Eds.)*

Maurice Ewing Volumes

1 **Island Arcs, Deep Sea Trenches, and Back-Arc Basins** *Manik Talwani and Walter C. Pitman III (Eds.)*

2 **Deep Drilling Results in the Atlantic Ocean: Ocean Crust** *Manik Talwani, Christopher G. Harrison, and Dennis E. Hayes (Eds.)*

3 **Deep Drilling Results in the Atlantic Ocean: Continental Margins and Paleoenvironment** *Manik Talwani, William Hay, and William B. F. Ryan (Eds.)*

4 **Earthquake Prediction—An International Review** *David W. Simpson and Paul G. Richards (Eds.)*

5 **Climate Processes and Climate Sensitivity** *James E. Hansen and Taro Takahashi (Eds.)*

6 **Earthquake Source Mechanics** *Shamita Das, John Boatwright, and Christopher H. Scholz (Eds.)*

IUGG Volumes

1 **Structure and Dynamics of Earth's Deep Interior** *D. E. Smylie and Raymond Hide (Eds.)*

2 **Hydrological Regimes and Their Subsurface Thermal Effects** *Alan E. Beck, Grant Garven, and Lajos Stegena (Eds.)*

3 **Origin and Evolution of Sedimentary Basins and Their Energy and Mineral Resources** *Raymond A. Price (Ed.)*

4 **Slow Deformation and Transmission of Stress in the Earth** *Steven C. Cohen and Petr Vaníček (Eds.)*

5 **Deep Structure and Past Kinematics of Accreted Terranes** *John W. Hillhouse (Ed.)*

6 **Properties and Processes of Earth's Lower Crust** *Robert F. Mereu, Stephan Mueller, and David M. Fountain (Eds.)*

7 **Understanding Climate Change** *Andre L. Berger, Robert E. Dickinson, and J. Kidson (Eds.)*

8 **Evolution of Mid Ocean Ridges** *John M. Sinton (Ed.)*

9 **Variations in Earth Rotation** *Dennis D. McCarthy and William E. Carter (Eds.)*

10 **Quo Vadimus Geophysics for the Next Generation** *George D. Garland and John R. Apel (Eds.)*

11 **Sea Level Changes: Determinations and Effects** *Philip L. Woodworth, David T. Pugh, John G. DeRonde, Richard G. Warrick, and John Hannah (Eds.)*

12 **Dynamics of Earth's Deep Interior and Earth Rotation** *Jean-Louis Le Mouël, D.E. Smylie, and Thomas Herring (Eds.)*

13 **Environmental Effects on Spacecraft Positioning and Trajectories** *A. Vallance Jones (Ed.)*

14 **Evolution of the Earth and Planets** *E. Takahashi, Raymond Jeanloz, and David Rubie (Eds.)*

15 **Interactions Between Global Climate Subsystems: The Legacy of Hann** *G. A. McBean and M. Hantel (Eds.)*

16 **Relating Geophysical Structures and Processes: The Jeffreys Volume** *K. Aki and R. Dmowska (Eds.)*

17 **Gravimetry and Space Techniques Applied to Geodynamics and Ocean Dynamics** *Bob E. Schutz, Allen Anderson, Claude Froidevaux, and Michael Parke (Eds.)*

Mineral Physics Volumes

1 **Point Defects in Minerals** *Robert N. Schock (Ed.)*

2 **High Pressure Research in Mineral Physics** *Murli H. Manghnani and Yasuhiko Syona (Eds.)*

3 **High Pressure Research: Application to Earth and Planetary Sciences** *Yasuhiko Syono and Murli H. Manghnani (Eds.)*

Geophysical Monograph 84

Solar System Plasmas in Space and Time

J. L. Burch
J. H. Waite, Jr.
Editors

American Geophysical Union

Published under the aegis of the AGU Books Board.

Library of Congress Cataloging-in-Publication Data

Solar system plasmas in space and time / J. L. Burch, and J. H. Waite, Jr., editors.
     p.   cm. — (Geophysical monograph ; 84)
    Includes bibliographical references.
    ISBN 0-87590-041-0
    1. Solar wind.   2. Space plasmas.   3. Magnetosphere.   I. Burch, J. L. (James L.)   II. Waite, Jr., J. H. (Jack H.)   III. Series.
QB529.S625   1994
523.2—dc20
                                                                          94-25481
                                                                           CIP

ISSN 0065-8448
ISBN 0-87590-041-0

Copyright 1994 by the American Geophysical Union, 2000 Florida Avenue, NW, Washington, DC 20009, U.S.A.

Figures, tables, and short excerpts may be reprinted in scientific books and journals if the source is properly cited.

    Authorization to photocopy items for internal or personal use, or the internal or personal use of specific clients, is granted by the American Geophysical Union for libraries and other users registered with the Copyright Clearance Center (CCC) Transactional Reporting Service, provided that the base fee of $1.00 per copy plus $0.10 per page is paid directly to CCC, 222 Rosewood Dr., Danvers, MA 01923. 0065-8448/94/$01.00+0.20.
    This consent does not extend to other kinds of copying, such as copying for creating new collective works or for resale. The reproduction of multiple copies and the use of full articles or the use of extracts, including figures and tables, for commercial purposes requires permission from AGU.

Printed in the United States of America.

CONTENTS

Preface
J. L. Burch, J. H. Waite, Jr., and W. S. Lewis   xiii

Field Structures

Heating of X-ray Bright Points and Other Coronal Structures
E. R. Priest   1

Solar Flare Hard X-ray Angular Distribution and Its Time Evolution
Peng Li   15

The Experimental Signatures of Coronal Heating Mechanisms
Peter J. Cargill   21

Space-time Structure of Langmuir Turbulence in the Lower Solar Corona
M. V. Goldman and D. L. Newman   33

Development of Fractal Structure in the Solar Wind and Distribution of Magnetic Field in the Photosphere
A. V. Milovanov and L. M. Zelenyi   43

Evolution of the Interplanetary Magnetic Field
David J. McComas   53

The Solar Flare Myth in Solar-Terrestrial Physics
J. T. Gosling   65

Generation and Nonlinear Evolution of Oblique Magnetosonic Waves: Application to Foreshock and Comets
N. Omidi, H. Karimabadi, D. Krauss-Varban, and K. Killen   71

Magnetospheric and Solar Wind Studies with Co-orbiting Spacecraft
C. T. Russell   85

Multi-Spacecraft Study of a Substorm Growth and Expansion Phase Features Using a Time-Evolving Field Model
D. N. Baker, T. I. Pulkkinen, R. L. McPherron, and C. R. Clauer   101

Ion Anisotropy-Driven Waves in the Earth's Magnetosheath and Plasma Depletion Layer
Richard E. Denton, Brian J. Anderson, Stephen A. Fuselier, S. Peter Gary, and Mary K. Hudson   111

Multidimensional Fourier Analysis of a Whistler Pulse Excited by a Loop Antenna
C. L. Rousculp, J. M. Urrutia, and R. L. Stenzel   121

CONTENTS

Thermal Magnetic Fluctuations in Maxwellian and Non-Maxwellian Plasmas at Whistler and Electron Cyclotron Harmonic Frequencies
R. L. Stenzel, G. Golubyatnikov, and J. M. Urrutia   125

Magnetic Dipole Antennas in Moving Plasmas: a Laboratory Simulation
J. M. Urrutia, C. L. Rousculp, and R. L. Stenzel   129

Particles and Flow Structures

Interaction Between Global MHD and Kinetic Processes in the Magnetotail
G. Ganguli, H. Romero, and J. Fedder   135

The Structure and Dynamics of the Plasma Sheet During the Galileo Earth-1 Flyby
G. D. Reeves, T. A. Fritz, R. D. Belian, R. W. McEntire, D. J. Williams, E. C. Roelof, M. G. Kivelson, and B. Wilken   149

Initial Observations of the Medium Distance Magnetotail Plasma by GEOTAIL: Cold Ion Beams
T. Mukai, M. Hirahara, S. Machida, Y. Saito, T. Terasawa, and A. Nishida   155

Temporal Evolution and Spatial Dispersion of Ion Conics: Evidence for a Polar Cusp Heating Wall
D. J. Knudsen, B. A. Whalen, T. Abe, and A. Yau   163

Temporal and Spatial Signatures in the Injection of Magnetosheath Plasma into the Cusp/Cleft
R. M. Winglee, J. D. Menietti, W. K. Peterson, J. L. Burch, and J. H. Waite, Jr.   171

The Location of Magnetopause Reconnection for Northward and Southward Interplanetary Magnetic Field
T. G. Onsager and S. A. Fuselier   183

The Shape and Size of Convection Cells in the Jovian Magnetosphere
T. W. Hill   199

Structure of the Venus Tail
O. Vaisberg, V. Smirnov, A. O. Fedorov, L. Avanov, F. Dunjushkin, J. G. Luhmann, and C. T. Russell   207

Ion Scattering and Acceleration by Low Frequency Waves in the Cometary Environment
H. Karimabadi, N. Omidi, and S. P. Gary   221

Axisymmetric Modeling of Cometary Mass Loading on an Adaptively Refined Grid: Hydrodynamic Results
Tamas J. Gombosi and Kenneth G. Powell   237

CONTENTS

Missions and Strategies

First High-Resolution Measurements by the Freja Satellite
R. Lundin, L. Eliasson, O. Norberg, G. Marklund, L. R. Zanetti, B. A. Whalen, B. Holback, J. S. Murphree, G. Haerendel, M. Boehm, and G. Paschmann 247

The Inner Magnetosphere Imager Mission
D. L. Gallagher 265

Imaging of Magnetospheric Dynamics Using Low Energy Neutral Atom Detection
H. O. Funsten, D. J. McComas, K. R. Moore, E. E. Scime, and M. F. Thomsen 275

The NASA High Energy Solar Physics (HESP) Mission for the Next Solar Maximum
R. P. Lin, B. R. Dennis, R. Ramaty, A. G. Emslie, R. Canfield, and G. Doschek 283

Base of Upper Yosemite Falls, Yosemite National Park, California, c. 1950. Photograph by Ansel Adams. Copyright © 1994 by the Trustees of The Ansel Adams Publishing Rights Trust. All Rights Reserved.

PREFACE

Solar system plasmas are highly structured and dynamic and are characterized by great variability in both space and time. The variations in their spatial distribution and temporal evolution occur on a variety of scales, ranging from kilometers (ion gyroradius) to hundreds of thousands of kilometers (coronal mass ejections) and from microseconds (electron plasma frequency) to years (solar sunspot cycle). Space plasma physicists seeking to understand the complex plasma phenomena that occur at the Sun, in the solar wind, and in the magnetospheres and ionospheres of the Earth and other solar system bodies thus face twin challenges. First, they must distinguish variations that are spatial in nature from those that are temporal. The heavy reliance in past investigations on singlepoint in situ measurements has significantly limited their ability to do this. Second, space physicists must elucidate the interrelationships among micro-, meso-, and macroscale plasma phenomena, relationships that organize the various solar system plasmas into a single heliospheric plasma system embedded in the interstellar medium. Here, too, experimental limitations have constrained the development of a global picture of solar system plasmas. However, new technologies promise a significant advance in our understanding of the interconnectedness of solar system plasmas.

The two challenges just mentioned, which give direction to much of the forefront research being conducted in space plasma physics, provide the thematic focus for the papers collected in this volume. The research reported here is intended to be broadly representative of work undertaken in the various subdisciplines of space physics to investigate the spatial and temporal structure of space plasmas and to relate local and global plasma processes. Terrestrial magnetospheric physics is represented by nine papers, solar physics by four, heliospheric physics by three, and cometary and planetary physics by five. In addition, the volume includes three laboratory plasma studies and four papers describing experimental techniques and missions.

The perspective offered by this collection is thus a multidisciplinary one, and it is inevitable that it is. For the phenomena under investigation will, in the final analysis, only yield their secrets to an approach that draws on and synthesizes the insights acquired by each of the subdisciplines that constitute the larger field of space physics. To underscore the multidisciplinary character of this collection (and of the enterprise of space plasma physics), we have chosen to organize the papers in three sections according not to subdiscipline but to three broad themes: electric and magnetic field structures, particles and flow structures, and missions and strategies. Each section is devoted to the development of one of these themes from a variety of disciplinary perspectives.

Field Structures. The papers in this section deal mainly with the behavior of electric and magnetic fields in space plasmas. The first four papers focus on plasma processes—reconnection, nanoflares, Langmuir turbulence—occurring at the Sun. Priest reviews possible mechanisms for the heating of coronal holes and coronal loops and proposes a new three-phase "converging flux model" for the production of X-ray bright points, while Li reports the results of a statistical study of the angular distribution of hard X-ray emissions in solar flares and its temporal evolution. Cargill presents theoretical considerations relevant to the identification of coronal radiative signatures that could provide clues to the mechanisms responsible for coronal heating and computes the spectral line profiles for one such mechanism (nanoflare heating). Goldman and Newman describe a theoretical model for the spatial and temporal structure of Langmuir turbulence excited by electron beams associated with Type III radio emission events in the lower solar corona.

The three papers that follow offer a wider-angle view of events occurring at the Sun, which are discussed in the context of their relationship to structures and processes in the solar wind. Milovanov and Zelenyi present a model of the fractal structure of solar wind turbulence and examine the relationship of fractal solar wind properties to nonlinear processes in the photosphere and convection zone. McComas et al. relate local, in situ plasma measurements made in the interplanetary medium to global images of the solar corona to draw conclusions about the effects of coronal mass ejections and coronal disconnection events on the evolution of the interplanetary magnetic field. In the "Solar Flare Myth in Solar-Terrestrial Physics" Gosling reviews observational evidence that calls into question the standard paradigm according to which solar flares are the cause of disturbances in the near-Earth space environment. In the new paradigm that Gosling proposes, events such as geomagnetic storms result from shocks driven by coronal mass ejections instead of from flare activity.

With the next four papers, the emphasis shifts from solar processes and their manifestations in the solar wind to the interaction between the solar wind and the Earth, other planets, and comets. Omidi et al. use linear theory and hybrid simulations to study the formation of magnetosonic waves (shocklets) upstream of planetary bowshocks and at comets and their subsequent nonlinear evolution. Russell reviews the lessons learned from the ISEE mission about the role of measurements by multiple spacecraft at varying separations in investigating a variety of shock and wave phenomena occurring in the solar wind, at the bow shock, and in the magnetosheath and in understanding such structures as flux transfer events and the tail current sheet. Baker et al. examine the Earth's magnetotail current sheet during the growth and expansion phase of a large substorm. Using magnetic field data from four spacecraft in the near-tail region and auroral images from two other spacecraft, together with ground-based observations, these authors successfully separate spatial and temporal effects in the substorm. Denton et al. consider recent studies of AMPTE observations of low-frequency (ion cyclotron and mirror mode) waves in the plasma depletion layer and the magnetosheath and show that the observed wave spectral types, which occur at different plasma betas, are related to distance from the magnetopause.

The final three papers in this section present the results of related laboratory plasma experiments performed by a group of researchers at UCLA—Rousculp et al. (a multidimensional Fourier analysis of whistler pulses), Stenzel et al. (measurements of thermal magnetic fluctuations at whistler and cyclotron harmonic frequencies), and Urrutia et al. (effects of the perturbation of a stationary plasma by a pulsed magnetic dipole). These results obtained in these laboratory studies are relevant to active space plasma experiments.

Particles and Flow Structures. The papers in the second section focus on plasma dynamics in the terrestrial and planetary magnetospheres and in cometary plasma environments. The first three papers examine magnetotail dynamics. Ganguli et al. present results of a modeling study of the structure and nonlinear dynamics of the plasma sheet boundary layer (PSBL). Although not yet complete, their model in its present form illuminates certain aspects of the coupling between the global dynamics of the solar wind-magnetosphere interaction and microscale dissipative processes (e.g., resistivity, viscosity) occurring in the stressed magnetotail. The structure and dynamics of the plasma sheet are also the subject of the paper by Reeves et al., who analyze particle flux data from Galileo and a geosynchronous spacecraft to study spatial and temporal variability in the plasma sheet during a substorm growth phase. The third paper, by Mukai et al., discusses cold ion beams observed by Geotail in the magnetotail lobe during a substorm. The observed ions (H^+ and O^+) presumably originated in the dayside polar ionosphere and were convected tailward to the mid-tail region. Temporal variations in their flow velocity may have been caused by the passage of a plasmoid.

The next three papers concern plasma populations and processes occurring in the cusp/cleft region of the Earth's magnetosphere or at the subsolar magnetopause. Knudsen et al. analyze EXOS-D measurements of cusp ions of ionospheric origin and find that the observed ion conic pitch angle behavior could be explained in terms of the acceleration of the ions in a wall-like heating region at the equatorward boundary of the cusp. The observed ions are subsequently convected northward. While Knudsen et al. are concerned with the upflow of ions from the ionosphere into the magnetosphere, Winglee et al. consider the injection of ions from the magnetosheath into the cusp/cleft and the interaction of these ions with those of ionospheric origin. To resolve the spatial and temporal aspects of this interaction, in which there is a direct transfer of energy from the injected magnetosheath ions to the upwelling ionospheric ions, the authors employ mesoscale particle simulations of magnetosheath plasma injection and compare the simulation results with DE-1 particle observations. One of the key solar wind/magnetosphere coupling processes responsible for the injection of solar wind (magnetosheath) plasma into the magnetosphere is reconnection. Onsager and Fuselier use AMPTE CCE measurements of magnetosheath He^{++} in the low-latitude boundary layer to estimate the location of magnetopause reconnection sites. They consider two cases, one for southward IMF and one for northward IMF, and find that reconnection occurred within about 6 R_E of the subsolar magnetopause and equatorward of the cusp.

The last four papers in this section examine plasma dynamics in the Jovian magnetosphere, the induced magnetotail of Venus, and in the interaction of cometary atmospheres with the solar wind. Hill offers a theoretical discussion of the morphology and size of mesoscale convection cells assumed to exist within the global pattern of corotating convection in the Jovian magnetosphere. Depending on the convection model used, cell scale sizes ranging from 0.2 R_J to 2.0 R_J can be calculated. Vaisberg et al. discuss the various plasma populations observed in the induced Venusian magnetotail, which is formed by the mass loading of the solar wind in its interaction with the

Venus ionosphere. Solar wind mass loading, such as occurs at Venus (during solar maximum), is also the fundamental process in the interaction of the solar wind with comets; this interaction is the subject of two modeling studies presented in this volume. Karimabadi et al. use a hybrid code to simulate the instabilities associated occurring in the solar wind/coma interaction and offer a theoretical explanation for the particle scattering and velocity diffusion observed in the simulation. Gombosi and Powell describe a new numerical model used to investigate certain global aspects of the mass loading of the solar wind by water group ions from the coma of a Halley-type comet. Though still under development, this model appears to overcome some of the limitations of earlier global models of the solar wind/comet interaction (e.g., low resolution, inability to encompass the entire cometary atmosphere).

Missions and Strategies. Resolving space-time ambiguities and understanding the interrelatedness of local and global plasma phenomena require appropriate measurement strategies and experimental techniques. These are the subject of the third section of this volume, which begins with an overview by Lundin et al. of the joint Swedish-German Freja mission. The goals of the Freja mission are to explore the lower portion of the auroral acceleration region and to investigate energization processes responsible for the outflow of ionospheric plasma into the magnetosphere. While still only a single-point measurement platform, the Freja spacecraft carries improved instrumentation (compared with that carried by earlier spacecraft in similar orbits) that makes possible plasma measurements with the high spatial and temporal resolution needed to resolve, for example, the fine structure of auroral arcs.

To develop a truly comprehensive picture of magnetospheric (or solar system) plasma processes and regions and their interrelationships, it is necessary to complement single- and multipoint in situ measurements with global imaging. The use of global imaging in magnetospheric investigations is the subject of the papers by Gallagher and by Funsten et al. Gallagher outlines the Inner Magnetospheric Imager (IMI) mission, which would place into an elliptical polar orbit a spacecraft equipped with a suite of imagers (energetic neutral atom, far and extreme UV, and X-ray). IMI would image the ring current, plasmasphere, inner plasma sheet, aurora, and geocorona, offering macroscopic views of these regions and their relationship to one another. The paper by Funsten et al. discusses the potential offered by one imaging technique, low-energy neutral atom (LENA) detection, for studying the dynamical behavior of the plasma sheet during quiet and storm times. Tests of a prototype LENA imager have produced encouraging results.

In the final paper in this section, Lin et al. describe the High Energy Solar Physics (HESP) mission, a multiple-spacecraft mission designed to investigate processes involved in the energy release and particle acceleration in solar flares. The primary HESP instrument is a highenergy imaging spectrometer that will provide high-resolution spectrometric images of energetic flare photons over an energy range from soft x-rays to gamma rays. This instrument, along with context instruments, can be carried on Lightsats placed into equatorial orbit by Pegasus launch vehicles. Satellite observations will be complemented by ground-based observations.

This volume was inspired and motivated by the stimulating presentations and discussions at the 1993 Yosemite Conference, which was sponsored by Southwest Research Institute, the National Aeronautics and Space Administration, and the National Science Foundation and held in Yosemite National Park, California, in early February 1993. The four-day conference brought together over 75 representatives of the various subdisciplines within space plasma physics to discuss in an interdisciplinary setting problems related to the investigation of the spatial and temporal structure of space plasmas in the magnetospheres and ionospheres of the Earth and other planets, the interplanetary medium, and the solar atmosphere.

J. L. Burch
J. H. Waite, Jr.
W. S. Lewis
Southwest Research Institute
San Antonio, Texas

Heating of X-ray Bright Points and Other Coronal Structures

by

E R Priest

Mathematical Sciences Dept, The University, St Andrews KY16 9SS, UK

A brief summary is given of the observations of X-ray bright points in the solar corona and of their relation to cancelling magnetic features in the underlying photosphere. A new *Converging Flux Model* for bright points is proposed with three phases. In the *Preinteraction Phase* photospheric magnetic fragments of opposite polarity are unconnected and approach one another. In the *Interaction Phase* the X-ray bright point is created by coronal reconnection. Finally, in the *Cancellation Phase* the cancelling magnetic feature is produced by photospheric reconnection. It is suggested that coronal reconnection driven by footpoint motions may represent an elementary heating event that is heating all coronal loops and not just bright points.

1. INTRODUCTION

The solar corona is a beautiful and complex environment, heated and structured by the magnetic field and having a temperature of a few million degrees that is five hundred times hotter than the solar surface. It has a three-fold structure of coronal holes, coronal loops and x-ray bright points, which was revealed by soft x-ray pictures from rocket flights and Skylab (Figure 1). In these pictures, the active regions above sunspot groups are rather fuzzy and the bright points are mainly just unresolved points of emission. However, recently, the remarkable NIXT photographs of Leon Golub at a temperature of 2-3 x 10^6K and with a five-times better resolution of 500km have shown that active regions consist of many fine-scale loops and bright points appear to include several interacting loops (Figure 2), as was previously glimpsed by Skylab [Sheeley and Golub, 1979].

1.1 Heating Mechanisms

Coronal magnetic field lines (where the plasma beta ($\beta = p2\mu/B^2$) is much less than unity) are anchored in the dense photosphere, where $\beta >> 1$ and their feet move around due to a variety of motions. Granulation has a speed of 0.25 - 2km s^{-1} and a lifetime of 8 minutes or less, while supergranulation has a speed of 0.3 km s^{-1} and a lifetime of a day or two. The corona evolves in response to these footpoint motions in a way that depends on their time-scale τ. If $\tau < 10\tau_A$, where $\tau_A = L/V_A$ is the time for an Alfven wave to travel at speed V_A along the length of a coronal field line, then the effect of the rapid motions is for waves to propagate upwards. For active-region values of magnetic field (B = 100G), density (n = $10^{15}m^{-3}$) and length (L = 100Mm) this condition becomes $\tau < 2$ min. If, on the other hand, $\tau > 10\tau_A$, then the coronal field tries to evolve slowly through a series of force-free equilibria. We take $10\tau_A$ rather than τ_A as a typical value since it takes several Alfvén travel times before an Alfvén wave damps away by, say, phase mixing.

2 HEATING OF X-RAY BRIGHT POINTS

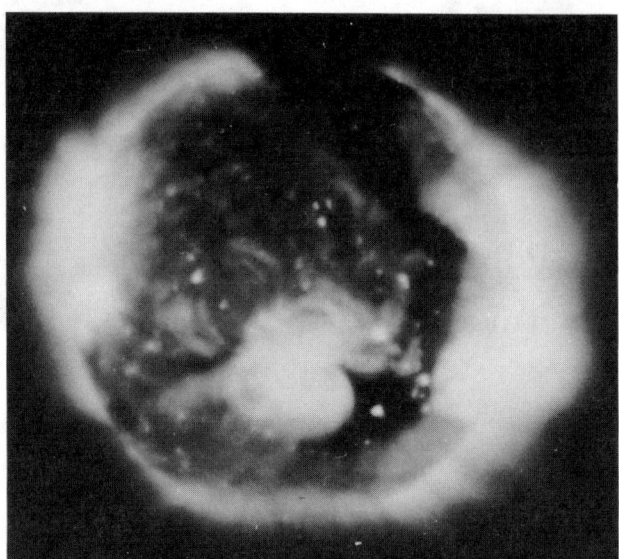

Fig 1. Soft x-ray picture of the corona. (Courtesy D. Webb)

Energy is injected from the photosphere as a Poynting flux

$$E\frac{B}{\mu} \approx v\frac{B^2}{\mu} \qquad (1.1)$$

which is about 10^4Wm^{-2} for typical values of v = 0.1 km s^{-1} and B = 100G, say. This is greater than the required energy to heat the corona, which has been estimated to be 600Wm^{-2} for coronal holes, 300Wm^{-2} for quiet regions and 5000Wm^{-2} for active regions [Withbroe and Noyes, 1977]. Thus the energy flux from the photosphere is sufficient but how is it dissipated magnetically? Below we summarise some possible mechanisms for coronal holes (§1.2) and coronal loops (§3) and put forward in detail a new mechanism for bright points which may also be occurring in coronal loops (§2).

1.2 *Coronal Holes*

At the moment the most attractive means of heating large-scale magnetically open regions of the corona is by Alfven waves. An efficient means of dissipating them is by phase mixing [Heyvaerts and Priest, 1983; Sakurai, 1985; Cally, 1991]. The basic idea is very simple and is due to the inhomogeneous nature of the coronal magnetic field. For instance, consider a field of the form $B_o(x)\hat{\mathbf{z}}$ with straight vertical field lines and perturb them with a motion $v(x,z,t)\hat{\mathbf{y}}$ in the y-direction which produces an extra field $B_1(x,z,t)\hat{\mathbf{y}}$. The MHD equations then reduce to a wave equation

$$\frac{\partial^2 v}{\partial t^2} = v_A^2(x)\frac{\partial^2 v}{\partial z^2} \qquad (1.2)$$

with solution

$$v \sim \exp i[\omega t - k_z(x)z]. \qquad (1.3)$$

Thus, if motions of frequency ω are excited at the base ($z = 0$), the wavelength $k_z(x) = \omega/v_A(x)$ varies with x. In otherwords, there is a continuous spectrum of wavelengths. The waves propagate along different field lines with different wavelengths and so their phases become mixed in space. In particular, by differentiating the above solution (1.2) it can be seen that the horizontal gradients are of the form

$$\frac{\partial v}{\partial x} = \frac{dk_z}{dx} z \, v \qquad (1.4)$$

and so they grow with height z. Steep gradients are generated and dissipation comes into play at a height of a few wavelengths. Laminar phase mixing is a very efficient means of dissipating the waves, but it may be noted that the dissipation may be enhanced by small-scale Kelvin-Helmholtz and tearing mode instabilities within the waves, as described in detail by Browning and Priest [1984]. Also a similar process of phase mixing exists for the other polarization when the motions are in the x-direction but the resulting equation is much more complicated to analyse [Goedbloed 1983, Goossens 1991].

1.3 *X-ray Bright Points*

X-ray bright points (BP) were discovered by Vaiana et al [1970] from rocket images and studied with Skylab by Golub et al [1974, 1976, 1977], where they have the form of diffuse clouds of diameter 20Mm which grow at 1km

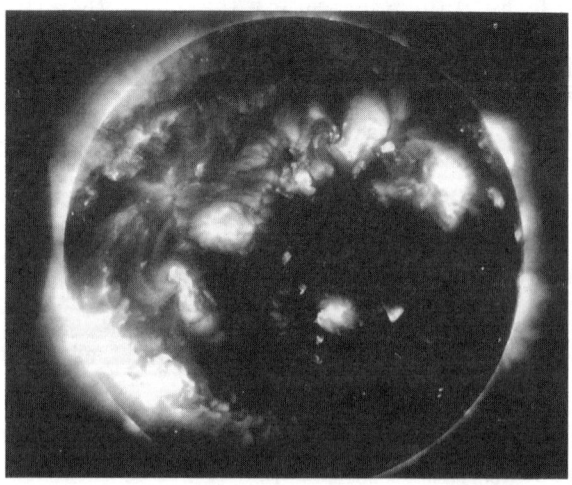

Fig 2. NIXT coronal picture. (Courtesy L. Golub)

s^{-1} and have a bright core of size 3Mm. Their height above the photosphere is not yet known in detail. They are uniformly distributed over the solar surface at solar minimum, when 200 may be present at one time. The number increases as their lifetime decreases. They are situated above pairs of opposite polarity magnetic fragments in the photosphere. Two types of such positive-negative pairs are observed. *Ephemeral active regions* (ER) are small emerging magnetic bipoles of flux $2\text{-}3 \times 10^{11}$Wb ($2\text{-}3 \times 10^9$Mx) with random orientations [Harvey and Martin, 1973, 1975; Martin and Harvey, 1979]. *Cancelling magnetic features* (CMF Figure 3) are opposite polarity fragments that are cancelling [Martin et al, 1983; Martin, 1986, 1988, 1990a, b]. Furthermore, Helium 10830 *dark points* (DP) are a very useful proxy for the 30% largest BP [Harvey et al, 1975; Golub et al, 1989]. The lifetimes are 2-48 hrs for BPs (with a mean of 8 hrs), 1-2 days for ERs, a few hrs for DPs, 1-36 hrs for CMFs. The sizes are 20Mm for BPs, < 30Mm for ERs, 7-20Mm for DPs, 2-20Mm for CMFs. The number born per day is estimated to be 1500 for BPs and 100 for ERs, while the number of CMFs is three or four times that of BPs.

It was natural to assume BPs represent emerging flux [Heyvaerts et al, 1977; Tur and Priest, 1976; Forbes and Priest, 1984]. This was the standard explanation, but the number of BPs is out of phase with the solar cycle, while the number of ERs is in phase. This mystery was solved by Karen Harvey [1984, 1985] who found that one-third of DPs overlie ERs, but two-thirds overlie CMFs, so she suggested that most BPs are due to chance encounters of opposite polarity fragments in the network or of emerging flux with the network. Cancellations depend on the amount of mixed polarity areas, which decreases by a factor of six from solar minimum to solar maximum, while the number of BPs also decreases by the same factor, which explains the anti-correlation of BPs with the solar cycle.

Two other important observational features are of note. Hermans and Martin [1986] found that in half of the cases a tiny filament forms and erupts during the process of cancellation. Also, Strong et al [1992] have discovered from the Japanese Yohkoh satellite that, when a BP flares, long neighbouring loops brighten up.

So, what is happening during a CMF? The obvious explanation is that it represents the simple submergence of a flux tube as its feet come together, but how can this explain the coronal brightening above? Also, the BP tends to start before cancellation [Harvey, 1985] and no chromospheric fibrils are seen to join the fragments - instead, they turn away from each other [Martin, 1986]. Furthermore, the magnetic fragments are initially widely separated and unconnected, so a key point of a model is to include the effect of the ambient magnetic field to which the fragments are initially connected.

2. A CONVERGING FLUX MODEL FOR BRIGHT POINTS

Together with Clare Parnell and Sara Martin, we are proposing a Cancelling Flux Model (Priest *et al*, 1994), which has three phases (Figure 4). In the Pre-interaction Phase a pair of oppositely directed magnetic fragments in the photosphere are unconnected and approach one another. They are separated by a channel of overlying flux, which is squeezed by the approach until a null point forms in the photosphere (Figure 4(ii)). In the Interaction Phase the null point moves upwards from the photosphere to the corona and coronal reconnection creates an x-ray bright point, whose structure consists of two newly reconnected and heated flux tubes, one a small loop linking the fragments and the other a large

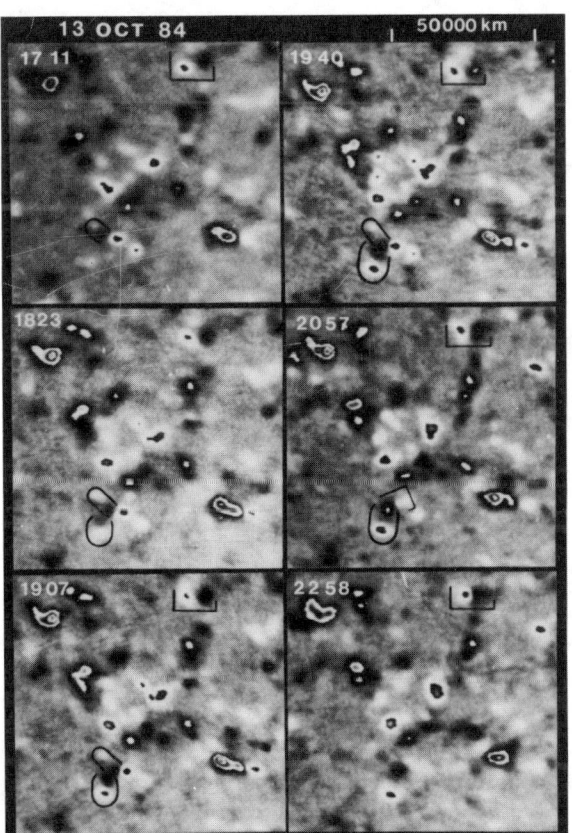

Fig 3. An example of an ER (surrounded by an oval) and a CMF (surrounded by rectangle). (Courtesy S. Martin)

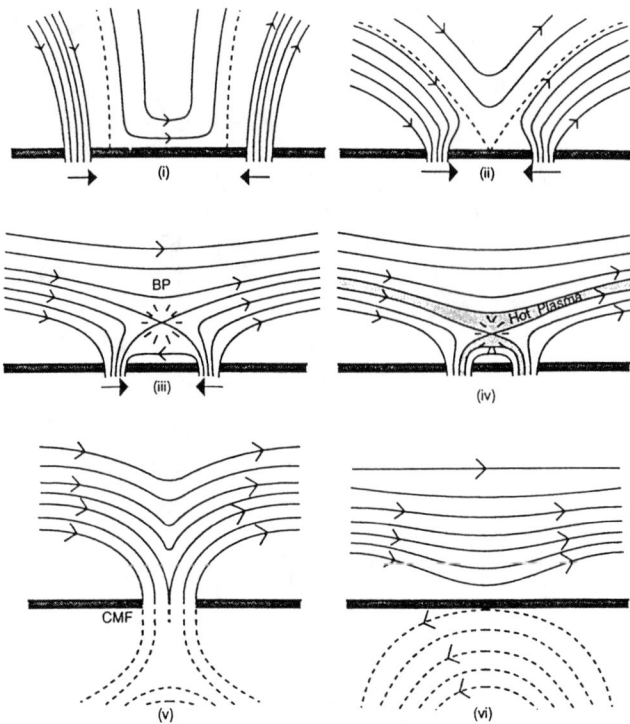

Fig 4. The magnetic field lines for the Converging Flux Model, showing: (i) and (ii) the Pre-interaction Phase; (iii) and (iv) the Interaction Phase; (v)-(vi) the Cancellation Phase.

loop (as seen in NIXT and Yohkoh images) linking to distant locations. Towards the end of the Interaction Phase the reconnection point moves back down from the corona to the photosphere. In the Cancellation Phase the fragments come into contact and cancel by photospheric reconnection. In the special case when the initial fragments are equal in magnitude the final state consists of two disconnected fields, one above the photosphere and one below.

A simple way of modelling the above processes is to represent the sources by poles of flux $\pm f$ at locations $z = \pm a$, where $z = x + iy$ as shown in Figure 5. The field components (B_x, B_y) due to such sources together with a uniform horizontal ambient field (B_o) may be written

$$B_y + iB_x = \frac{if/\pi}{z-a} - \frac{if/\pi}{z+a} + iB_o$$
$$= iB_o \frac{z^2 - b^2}{z^2 - a^2} \qquad (2.1)$$

where

$$b = (a^2 - ad)^{\frac{1}{2}} \qquad (2.2)$$

is the half-width of the channel and

$$d = \frac{2f}{\pi B_o} \qquad (2.3)$$

we refer to as the "interaction distance".

In the Pre-interaction Phase, the poles are assumed to approach at a speed much slower than the Alfven speed

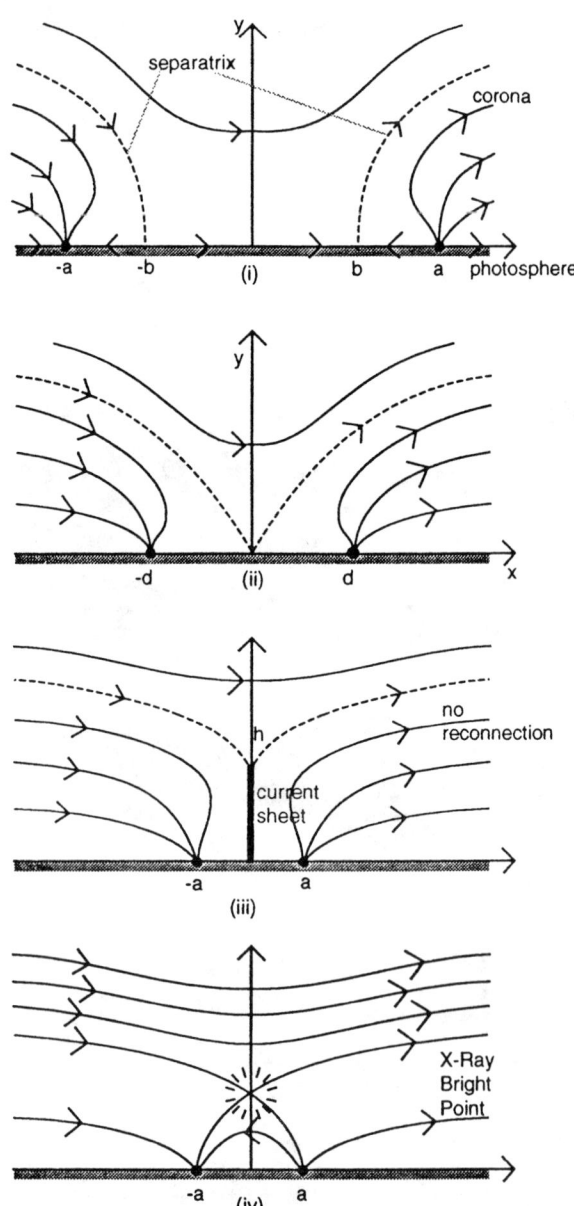

Fig 5. Model of ((i) and (ii)) Pre-interaction and ((iii) and (iv)) Interaction Phases.

and to make the overlying field evolve through a series of potential states given by (2.1). As the source position (a) decreases, the half width (b) of the channel decreases while its flux is conserved, until at the interaction distance (a=d) the null point forms at the origin (Figure 5b). It is a second-order null point with

$$B_y + iB_x \approx -iB_o \frac{z^2}{d^2}$$

so that the field components are

$$(B_x, B_y) = \frac{B_o}{d^2}(-x^2 + y^2, 2xy) \quad (2.4)$$

When a<d, we have the Interaction Phase with reconnection driven at the null point by the motion of the sources. Such reconnection would probably be in the flux pile-up regime [Priest and Forbes, 1986] and may be impulsive and bursty [Priest, 1986; Lee and Fu, 1986] as the diffusion region goes unstable to secondary tearing, which could explain the rapid time-variations observed in bright points [Sheeley and Golub, 1979]. We plan to perform a numerical experiment on such reconnection, but a simple model is to suppose the evolution through potential states continues. In equation (2.1) b^2 now becomes negative and so the field vanishes at an X-point on the y-axis at a height

$$|b| = (ad - a^2), \quad (2.5)$$

which rises as a decreases to a maximum of $\frac{1}{2}d$ when $a = \frac{1}{2}d$ and then decreases to zero as the sources approach the origin.

If instead there is no reconnection, the topology is preserved and a current sheet forms so that the magnetic energy (W) exceeds the energy (W_o) of the potential state by an amount that can be released by reconnection to give the bright point. The field is given by

$$B_y + iB_x = \frac{iB_o(z^2 + h^2)^{\frac{1}{2}} z}{z^2 - a^2}, \quad (2.6)$$

which tends to a uniform field iB_o at infinity and has a cut (the current sheet) stretching along the y-axis from the origin to $z = ih$. The condition that the flux above the sheet is preserved gives the length of the sheet as

$$h = (d^2 - a^2)^{\frac{1}{2}}, \quad (2.7)$$

which increases from zero to d as a decreases from d to zero.

The energy (W) is given by

$$2\mu W = \int B^2 dV = \int \mathbf{B}.\nabla \times \mathbf{A} \ dV \quad (2.8)$$

$$= \int \mathbf{A}.\nabla \times \mathbf{B} + \nabla.(\mathbf{A} \times \mathbf{B}) \ dV \quad (2.9)$$

since $\mathbf{B} = \nabla \times \mathbf{A}$. We assume $\mathbf{A}$ vanishes on the current sheet, the only place where $\nabla \times \mathbf{B} = \mu \mathbf{j}$ is nonzero, and so the first term vanishes. The second term may be transformed by the divergence theorem to give

$$2\mu W = \int \mathbf{A} \times \mathbf{B}.\mathbf{dS}$$

or, in our two-dimensional geometry,

$$2\mu W = -2 \int_o^\infty (AB_x)_{y=o} \ dx \quad (2.10)$$

The potential field ($\mathbf{B}_o$) with the same normal field (B_y) and therefore flux function on the base ($y = 0$) has energy

$$2\mu W = -2 \int_o^\infty (AB_{ox})_{y=o} \ dx,$$

and so by subtracting from (2.10) we obtain the stored energy in excess of potential as

$$2\mu W_f = 4 \int_o^\infty (AB_{sx})_{y=o} \ dx. \quad (2.11)$$

Here

$$\mathbf{B} - \mathbf{B}_o = \mathbf{B}_s + \mathbf{B}_{si} \quad (2.12)$$

where $\mathbf{B}_s$ is the field due to the sheet and $\mathbf{B}_{si}$ due to its image, so that $B_{sx} = B_{six}$ on $y = 0$.

However, the contribution due to the sheet is

$$B_{sx} = \int_o^h \frac{\mu I(y)}{2\pi(x^2 + y^2)^{\frac{1}{2}}} \frac{y}{(x^2 + y^2)^{\frac{1}{2}}} \ dy \quad (2.13)$$

where

$$I(y) = \frac{2B_{ys}}{\mu} = \frac{2B_o}{\mu} \frac{(h^2 - y^2)^{\frac{1}{2}} y}{y^2 + a^2} \quad (2.14)$$

is the current in the sheet. Thus, after integrating over x, (2.11) reduces to

$$W_f = \frac{B_o^2 d^3}{2\mu} \int_o^h \frac{2(h^2 - y^2)^{\frac{1}{2}} y}{d(y^2 + a^2)} \left(\frac{\pi}{2} - tan^{-1}\frac{a}{y}\right) dy \quad (2.15)$$

The factor outside the integral represents the energy in a cube of side d, while the integral is a dimensionless factor that depends on a/d. The resulting variation of W_f with a/d and d is shown in Figure 6 for $B_0 = 10G$. It represents an estimate (an upper limit) of the energy

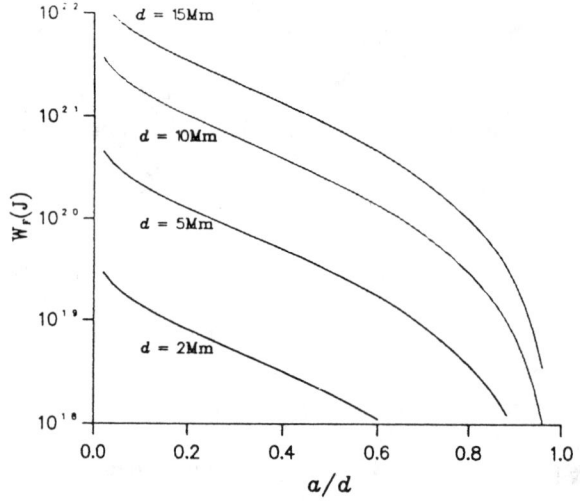

Fig 6. The stored energy (W_f) as a function of source position (a) for several values of the interaction distance (d).

released by the reconnection. Now, the energy released in a bright point is between 3.10^{20} and 3.10^{21} J (3.10^{27}-3.10^{28} erg) and so W_f is of this order for $d = 5$-10Mm.

Several points may be noted about the model. First of all, the duration of the bright point is roughly the duration of the interaction phase, namely d/v_o, where v_o is the speed of aproach. Putting $d = 7.5$Mm and $v_o = 0.5$km s^{-1}, we obtain a duration of 3.10^4sec or 8 hours, as required.

Another point is that so far we have been describing the interaction between fragments of equal and opposite magnitude (Figure 7a), when coronal reconnection in the interaction phase creates the bright point, photospheric reconnection in the cancellation phase creates the CMF and the final state has no local flux sources. When instead the fragments have unequal magnitudes the configuration is asymmetric and flux of one polarity is left behind in the final state (Figure 7b). Furthermore, interaction between a strong bipole (such as an ephemeral region) and a weak unipolar element (such

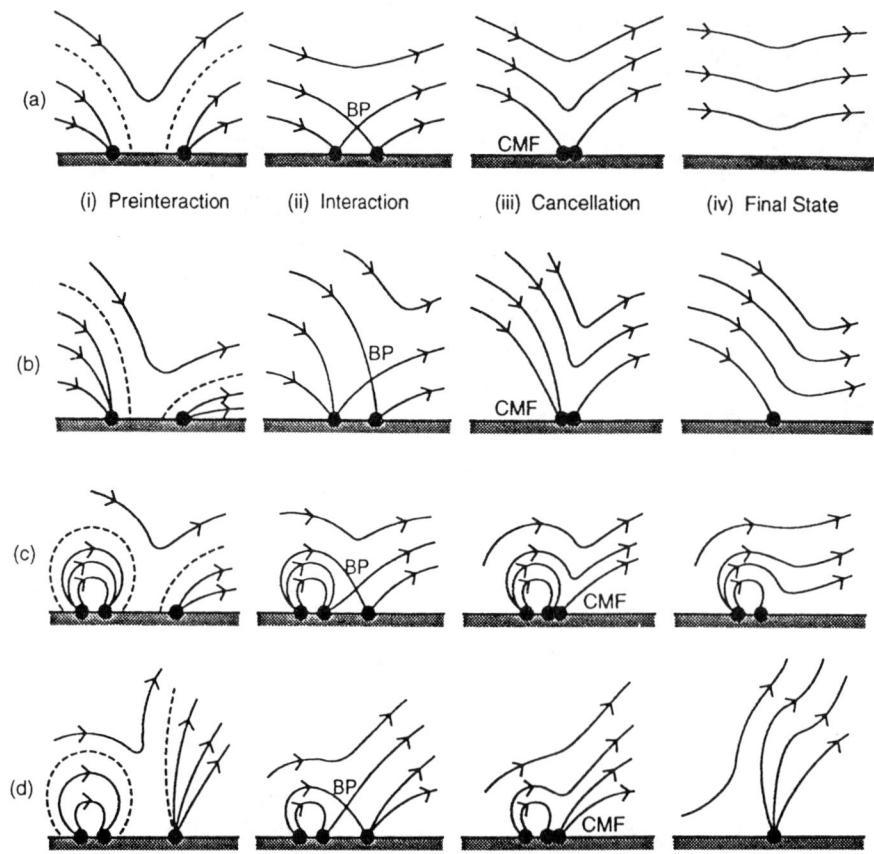

Fig 7. Interactions between: (a) equal and opposite poles, (b) unequal poles, (c) a strong bipole and a weak pole, and (d) a weak bipole and a strong pole.

as an intracell fragment) leads to the evolution sketched in Figure 7c, whereas when the bipole is weak a single flux source remains behind at the end (Figure 7d).

In addition, it may be noted that the formation of a filament during the Cancellation Phase is a natural consequence of our model (Figure 8), either by compression and cooling during the late phase of cancellation or by plasmoid creation during the early stages. The dominant magnetic component is out of the plane of Figure 8 and so leads to a filament bisecting the line joining the fragment centres. In Figure 8a the plasma collects in the dip above and between the cancelling features and cools by, say, radiative instability, especially since the pressure is enhanced by the convergence. In Figure 8b, a new X-point forms at the photosphere below the coronal X-point and so creates a magnetic island within which the plasma is thermally isolated and cools to prominence temperatures. Finally, although we are suggesting coronal reconnection as a mechanism for heating the low-lying complex loop systems seen in the NIXT photographs, the question arises: how are the much larger overlying loops heated, such as the closed parts of helmet streamers extending outwards by a solar radius? One possibility is that they are heated by conduction from the much hotter loops below them, provided they are connected magnetically. Another possibility is the waves that are thought to heat coronal holes.

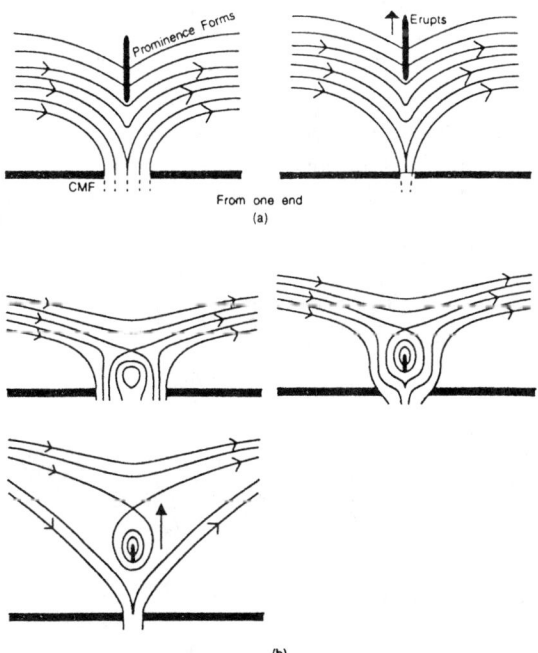

Fig 8. The formation of a filament by two mechanisms during the Cancellation Phase.

3. CORONAL LOOPS AND ARCADES

Slow footpoint motions imply that the corona tries to evolve through a series of equilibria, but often such equilibria are not smooth and instead contain singularities (current sheets), where dissipation can occur and reconnection can take place in the way we have modelled above for bright points. There are several ways of forming such current sheets. One is due to braiding by random footpoint motions, as Parker[1972] proposed in his topological dissipation idea [Berger 1991; Rosner and Knoblock, 1982; Van Ballegooijen, 1985]. Numerical experiments by Mikic et al [1989] have confirmed that current concentrations are formed which grow in time. A second way of forming current sheets is by the collapse of an X-point, or in three dimensions, a separator (para 3.1). A third way of is by shearing a region with separatrix surfaces (para 3.2). Furthermore, the resulting highly complex and dynamic state with many small current sheets may be modelled by using MHD turbulence ideas (para 3.3).

3.1 *X-point Collapse*

The collapse of an X-point due to motions of the distant sources of magnetic field was first proposed by Dungey [1953], and a self-similar solution was discovered by Imshennik and Syrovatsky [1967]. More recently, Craig and McClymont [1991] have included resistivity for small oscillations about an X-type equilibrium. Green [1965] and Somov and Syrovatsky [1976] used complex variable theory to deduce the resulting equilibrium state with a current sheet that may develop from an X-type field. Thus, if one starts with a potential field in two dimensions having components $(B_x, B_y) = (y, x)$, they may be written in the form

$$B_y + iB_x = z. \qquad (3.1)$$

Any analytic function of z gives a potential field and so one possibility for the potential field with a current sheet that grows from (3.1) is (Green, 1965)

$$B_y + iB_x = \left(z^2 + L^2\right)^{\frac{1}{2}}, \qquad (3.2)$$

which has a cut in the complex plane (a current sheet) stretching from $z = -iL$ to $z = iL$ and Y-type null points at the ends of the sheet (middle of Figure 9). However, Somov and Syrovatsky (1976) suggested another possibility, namely

$$B_y + iB_x = \frac{z^2 + a^2}{(z^2 + L^2)^{\frac{1}{2}}}, \qquad (3.3)$$

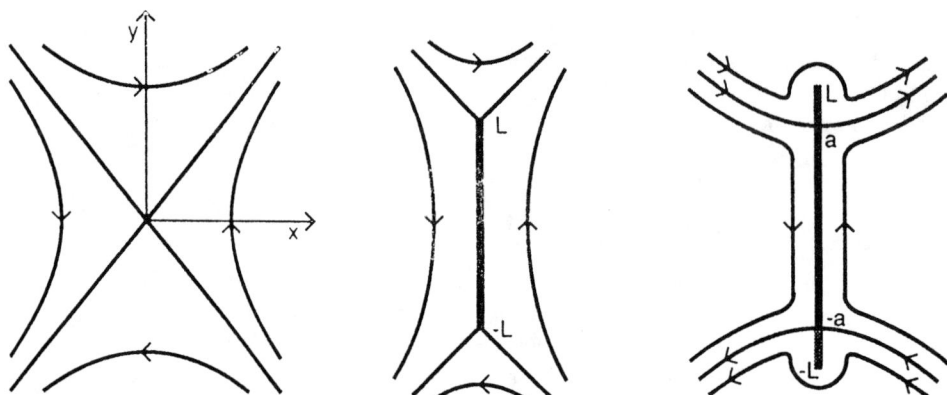

Fig 9. X-point collapse to give a current sheet

which has regions of reversed current between $z = \pm ia$ and $z = \pm iL$ and singularities at the ends of $z = \pm iL$ of the sheet where the field becomes infinite (right of Figure 9). Thus the question arises: does the X-type field (3.1) collapse to the form (3.2), as has been commonly assumed, or to (3.3) or to some other state?

We have recently made progress on this question (Priest, Titov and Rickard, 1994). The above solutions (3.2) and (3.3) represent a slow evolution through a series of equilibria, but we have recently discovered some new nonlinear self-similar compressible solutions for the dynamic time-dependent formation of a current sheet in the situation when

$$c_s << v << v_A \qquad (3.4)$$

so that the flow speed lies between the sound and Alfven speeds. We solve the equation of motion

$$\rho \frac{d\mathbf{v}}{dt} = \mathbf{j} \times \mathbf{B} \qquad (3.5)$$

with no pressure gradient and the ideal induction equation

$$\frac{\partial \mathbf{B}}{\partial t} = \nabla \times (\mathbf{v} \times \mathbf{B}). \qquad (3.6)$$

To lowest order, we find a self-similar collapse through current-free states surrounding a growing current sheet. The motion of the field lines determines the flow speed perpendicular to the field lines, while the flow ($v_\parallel$) along the moving field lines is determined by a balance between the coriolis and centrifugal forces associated with the rotation of the field lines so that the acceleration is perpendicular to the magnetic field. The resulting field lines in the first quadrant are shown in Figure 10.

In dimensionless variables the current sheet stretches from $-2\sqrt{t}$ to $+2\sqrt{t}$ along the x-axis. As the sheet grows in length the magnetic dissipation increases and it swallows up half of the magnetic flux ahead of it, so creating a transverse y-component of field threading the sheet. The other half piles up ahead of the sheet and creates a region of reversed current near the ends. The individ-

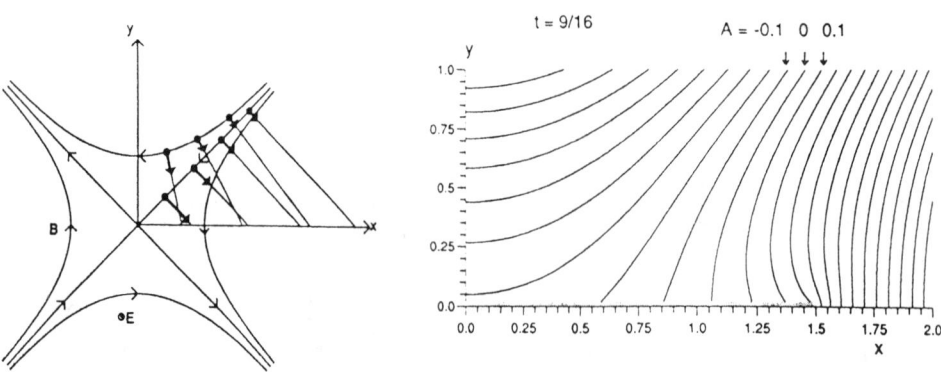

Fig 10. Magnetic field lines at t=0 and t=9/16

ual plasma elements converge on the x-axis during the collapse.

The solution may be written in an elegant form in terms of $z = x + iy$ as

$$B_y + iB_x = -\frac{[z/2 + \sqrt{(z^2/4 - t)}]^2}{2\sqrt{(z^2/4 - t)}}. \quad (3.7)$$

It may be derived as follows. First of all, expand the variables in powers of the Alfven Mach number $\epsilon(<< 1)$, so that

$$\mathbf{v} = \epsilon \mathbf{v}_o + \epsilon^2 \mathbf{v}_1 + ..., \mathbf{B} = \mathbf{B}_o + \epsilon \mathbf{B}_1 + ... \quad (3.8)$$

To zeroth order (3.5) becomes

$$\mathbf{j}_o = \mathbf{O} \quad (3.9)$$

and to first order

$$\rho_o \frac{d\mathbf{v}_o}{dt} = \mathbf{j}_1 \times \mathbf{B}_o \quad (3.10)$$

Thus (3.9) implies an evolution through potential states ($\mathbf{B}_o$), while (3.10) implies that

$$\frac{d\mathbf{v}_o}{dt} \cdot \mathbf{B}_o = 0 \quad (3.11)$$

so that the acceleration must be perpendicular to $\mathbf{B}_o$.

Next we write the field in terms of a flux function A_o,

$$(B_{ox}, B_{oy}) = \left(\frac{\partial A}{\partial y}, -\frac{\partial A}{\partial x}\right) \quad (3.12)$$

where (3.9) and (3.6) imply that

$$\nabla^2 A = 0 \text{ and } \frac{\partial A}{\partial t} + \mathbf{v}_o \cdot \nabla A = 0 \quad (3.13)$$

so that A is frozen to the plasma. We introduce then the conjugate harmonic function Φ and assume that it too is frozen to the plasma so that

$$\frac{\partial \Phi}{\partial t} + \mathbf{v}_o \cdot \nabla \Phi = 0. \quad (3.14)$$

Thus a plasma element preserves the values of A and Φ as it moves.

Then we write

$$f = A + i\Phi \quad (3.15)$$

and

$$z = x + iy, \quad (3.16)$$

so the object is to determine $z(f, t)$ and its inverse $f(z, t)$. In this formalism (3.11) may be written (Priest, Titov and Rickard, 1994)

$$\frac{\partial^2 z}{\partial t^2} / \frac{\partial z}{\partial f} - \overline{\frac{\partial^2 z}{\partial t^2}} / \overline{\frac{\partial z}{\partial f}} = 0,$$

where the bar denotes a complex conjugate, which implies that

$$\frac{\partial^2 z}{\partial t^2} = \chi(t) \frac{\partial z}{\partial f}, \quad (3.17)$$

where $\chi(t)$ is an arbitrary real function. We now for simplicity assume $\chi \equiv 0$ for acceleration-free flow, so that the solution of (3.17) becomes

$$z = z_o(f) + v_o(f) t \quad (3.18)$$

where $z_o(f)$ is the initial position of a plasma element and $v_o(f)$ its velocity.

Suppose now the initial flux function is $A_o = \frac{1}{2}(x^2 - y^2)$ giving $f_o = \frac{1}{2}z_o^2$ and so $z_o = \sqrt{2f}$. Suppose also the initial velocity is $v_o = 1/\sqrt{2f}$ (so that the electric field equals 1 initially). Then (3.18) becomes

$$z = z_o + \frac{t}{z_o}, \quad (3.19)$$

which may be rewritten to give z_o as

$$z_o = \frac{1}{2}z + (\tfrac{1}{4}z^2 - t)^{\frac{1}{2}}.$$

However, the plasma preserves its value of $f(= f_o = \frac{1}{2}z_o^2)$ as it moves and so

$$f(z, t) = \frac{1}{2}[\tfrac{1}{2}z + (\tfrac{1}{4}z^2 - t)^{\frac{1}{2}}]^2, \quad (3.20)$$

which is the solution we have been seeking. The resulting magnetic field components are (3.7), namely

$$B_y + iB_x = -\frac{\partial f}{\partial z} = \frac{[\tfrac{1}{2}z + (\tfrac{1}{4}z^2 - t)^{\frac{1}{2}}]^2}{2(\tfrac{1}{4}z^2 - t)^{\frac{1}{2}}}, \quad (3.21)$$

while the velocity components are

$$v_x + iv_y = \frac{\partial z}{\partial t} = \frac{1}{\tfrac{1}{2}z + (\tfrac{1}{4}z^2 - t)^{\frac{1}{2}}} \quad (3.22)$$

and the density and electric field are

$$\rho = \frac{\rho_o}{|\partial z/\partial z_o|^2} = \frac{\rho_o}{|1 - t[\tfrac{1}{2}z + (\tfrac{1}{4}z^2 - t)^{\frac{1}{2}}]^{-2}|^2} \quad (3.23)$$

and

$$E = -Re\frac{\partial f}{\partial t} = \tfrac{1}{2}Re[1 + \tfrac{1}{2}z(\tfrac{1}{4}z^2 - t)^{-\frac{1}{2}}]. \quad (3.24)$$

This is an elegant formalism with the equations being (at least to the author!) almost as beautiful as Yohkoh or NIXT pictures, but the resulting expressions for, say, $B_x(x, y, t)$ are rather nasty. In this simplest of a whole family of solutions, the individual plasma elements converge on the x-axis and form a current sheet of length $4\sqrt{t}$. As the magnetic field collapses, the sheet grows in length and the magnetic dissipation increases. The end of the sheet moves with speed $1/\sqrt{t}$: it swallows up half of the magnetic flux and causes the remainder to pile up in a region of reversed current.

3.2 Current Sheets Near Separatrices

An X-point is structurally unstable in the sense that it splits into a pair of Y-points joined by a current sheet under the action of converging motions. However (Figure 11), shearing motions can instead produce current sheets all along the separatrices, the field lines which link to the X-point [Low and Wolfson, 1988; Vekstein, Priest and Amari, 1990; Vekstein and Priest 1992]. In a 2.5-dimensional field its Cartesian components are, in terms of the flux function (A),

$$(B_x, B_y, B_z) = \left(\frac{\partial A}{\partial y}, -\frac{\partial A}{\partial x}, B_z(A)\right) \quad (3.25)$$

and the force-free equation ($\mathbf{j} \times \mathbf{B} = \mathbf{0}$) reduces to the Grad-Shafranov equation

$$\nabla^2 A + B_z \frac{dB_z}{dA} = 0. \quad (3.26)$$

If the footpoint positions $\xi_z(X)$ at the photosphere are imposed, the toroidal field $(B_z(A))$ is given from the equations of a field line by

$$B_z = \frac{d(A)}{V(A)}, \quad (3.27)$$

where d is the difference in footpoint displacement between the ends of the field and $V(A) = \int ds/B_p$ is a property of the poloidal field. $B_z(A)$ is constant along a given field line but it may be very different on field lines just above and below the separatrix - thus the whole separatrix must become a current sheet.

Finn and Lau [1991] suggested that, in response to shearing, the X-point would just close up slightly, but in general this cannot remain in equilibrium since B_p tends to zero as one approaches the X-point from any direction and so the separatrix cannot support a jump in magnetic pressure across it associated with the jump in B_z. The answer is to use a cusp [Vekstein and Priest, 1992] because it has the property that B_p tends to zero from one side and to constants from the other two sides, so there is a jump in B_p^2 across the separatrix which can balance the jump in B_z^2.

Consider the simplest case where there is shearing present only in the region, (I) say, below the X-point so that in the regions (II) and (III) to either side $B_z \equiv 0$ and the field is potential with $\nabla^2 A = 0$.

In (I) near the cusp there is a self-similar solution

$$A = r^\alpha f(\xi), \quad (3.28)$$

where

$$\xi = \frac{\theta}{r^\beta}, \quad (3.29)$$

so the separatrix ($A = 0$, say) is $\xi = 1$, say; in other words, it is not a straight line but a curve $\theta = r^\beta$. Then the field components are by differentiating (3.28)

$$B_r = \frac{1}{r}\frac{\partial A}{\partial \theta} = r^{\alpha-1-\beta} f'(\xi)$$

$$B_\theta = -\frac{\partial A}{\partial r} = -\alpha r^{\alpha-1} f(\xi) + \beta r^{\alpha-1} f'(\xi)\xi.$$

The equilibrium equation (3.26), namely

$$\nabla^2 A = -B_z \frac{dB_z}{dA},$$

must have the right-hand side of the form $-\epsilon A^{-n}$, where substitution of (3.28) gives

$$n = \frac{2\beta + 2 - \alpha}{\alpha} \quad (3.30)$$

and to lowest order the function $f(\xi)$ is given by

$$f'' = -\epsilon f^{-n}. \quad (3.31)$$

In region II the field is potential and an appropriate form for A is

$$A = B_0 r \sin\theta + b r^p \sin p(\theta - \pi) \quad (3.32)$$

which has $A = 0$ on $\theta = \pi$, the vertical arm of the separatrix, as required. As far as region II is concerned the curved part of the separatrix ($A = 0$) is given from (3.32) by

$$\theta = \frac{b}{B_0} r^{p-1} \sin p\pi$$

and so, by comparing with the form $\theta = r^\beta$ in region I, we see that

$$p = 1 + \beta. \quad (3.33)$$

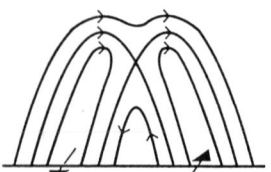
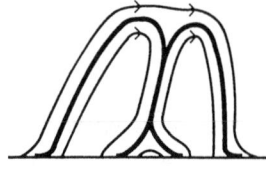

Fig 11. Current sheets created by shearing

Finally, magnetic pressure balance across the separatrix separating regions I and II gives

$$B_{z0}^2 + cr^{2(a-1-\beta)} = B_0^2 + 2pbB_0 \ r^\beta \ \cos \ p\pi$$

so that in order that the variations with r should agree we need

$$\alpha = 1 + \frac{3\beta}{2}. \quad (3.34)$$

3.3 Self-Consistent Model for Heating by MHD Turbulence

Many coronal heating mechanisms, such as braiding and current sheet formation or resistive instabilities or waves, all lead to a state of MHD turbulence, so an important question on which we have recently made progress is: how can we analyse it? Heyvaerts and Priest [1984] made a start by adapting Taylor's relaxation theory to the coronal environment, in which the field lines thread the boundary rather than being parallel to it.

In Taylor's model the global magnetic helicity is conserved, but in our model footpoint motions make the coronal field evolve through a series of linear force-free fields, satisfying

$$\nabla \times \mathbf{B} = \alpha_0 \mathbf{B}. \quad (3.35)$$

The footpoint connections are not preserved, but instead the constant α_0 is determined from the evolution of the *magnetic helicity*.

$$K = \int \mathbf{A}.\mathbf{B} \ dV \quad (3.36)$$

where $\mathbf{B} = \nabla \times \mathbf{A}$, due to the injection by boundary motions according to

$$\frac{dK}{dt} = \int (\mathbf{A}.\mathbf{v})\mathbf{B}.\mathbf{dS}. \quad (3.37)$$

The resulting heating flux is of the form

$$F_H - \frac{B^2 v}{\mu} \frac{\tau_d}{\tau_0} \quad (3.38)$$

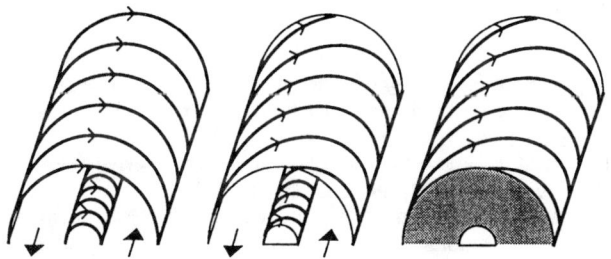

Fig 12. The scenario for Taylor-Heyvaerts-Priest relaxation

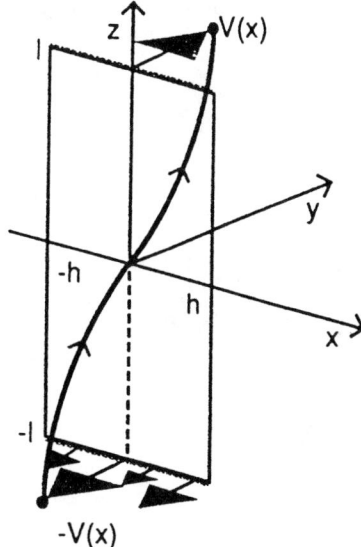

Fig 13. Nomenclature for model of turbulent heating.

where τ_d is the dissipation time and τ_0 the time-scale for footpoint motions.

Although many mechanisms produce a turbulent state, they are incomplete in the sense that there is a free parameter present, such as τ_d in (3.38) or a correlation time or a relaxation time. In other words, they don't determine the heating flux (F_H) in terms of photospheric motions alone. Heyvaerts and Priest [1993] have therefore begun a new approach in which they assume photospheric motions inject energy into the corona and maintain it in a turbulent state with a turbulent magnetic diffusivity (η^*) and viscosity (ν^*). There are two parts to their theory. First of all, they calculate the global MHD state driven by boundary motions, which gives F_H in terms of ν^*. Secondly, they invoke cascade theories of MHD turbulence to determine the ν^* and η^* that result from (F_H). In other words the circle is completed and (F_H) is determined independently of ν^* and η^*. They apply their general philosophy to a simple example of one-dimensional random photospheric motions producing a two-dimensional coronal magnetic field.

Suppose the boundary motions are $\pm V(x)\hat{\mathbf{y}}$ at $z = \pm \ell$ and produce motions $v(x,z)\hat{\mathbf{y}}$ and field $B_0\hat{\mathbf{z}} + \mathbf{B}_y(x,z)\hat{\mathbf{y}}$ within the volume between $z = -\ell$ and $z = \ell$. Then the steady MHD equations of motion and induction reduce simply to

$$0 = B_0 \frac{\partial B_y}{\partial z} + \nu^* \nabla^2 v,$$

$$0 = B_0 \frac{\partial B_y}{\partial z} + \eta^* \nabla^2 B_y.$$

The solutions may be found and the resulting Poynting

energy flux through the boundary is, in contrast to the earlier form (3.38) found by Heyvaerts and Priest (1984),

$$F_H = \frac{B_0^2 v_{AO}}{\mu} \sum_0^\infty \frac{V_n^2 H}{\eta} \left(1 + 2\lambda_n^2/\sqrt{1+4\lambda_n^2}\right)$$

$$\frac{\sinh\left(\sqrt{1+4\lambda_n^2}/H\right) + \sinh(1/H)}{\cosh\left(\sqrt{1+4\lambda_n^2}/H\right) - \cosh(1/H)} \quad (3.39)$$

where V_n are the Fourier coefficients of $V(x)$ and

$$H = \frac{\sqrt{\eta^* \nu^*}}{\ell v_{AO}}.$$

For the second step, invoking Pouquet theory gives $\nu^* = \eta^*$ and

$$F_H = \frac{27\nu^{*2}\pi^3}{2h^3/\ell^2} \frac{B_0^2 v_{AO}}{\mu \ell^2 v_{AO}^2} \quad (3.40)$$

so that equating (3.39) and (3.40) gives a single expression for ν^*. They find typically for a quiet-region loop that a density 2×10^{16} m^{-3} and a magnetic field of 30-50G produces a heating (F_H) of 2.4-5.5 $\times 10^2$ Wm^{-2} and a turbulent velocity of 24-33 km s^{-1}, whereas values of 5×10^{16} m^{-3} and 100G for an active-region loop give 2×10^3 Wm^{-2} for the heating and 40 km s^{-1} for the turbulent velocities. Given the limitations of the model, these reasonable values are very encouraging.

4. CONCLUSION

The solar corona is likely to be heated by a variety of mechanisms. My current view is that coronal holes are probably heated by phase mixing of propagating waves, while coronal loops may well be heated by many small current sheets. In response to footpoint motions the highly complex coronal magnetic field is likely to form many current sheets. These may collapse and grow by the mechanism described in Section 3.1 or may form in response to shearing (§3.2). The resulting highly turbulent state may be described in a self-consistent manner in terms of MHD turbulence theory.

In this paper we have also put forward a new Converging Flux Model for X-ray bright points. It has: a Pre-interaction Phase, in which the photospheric magnetic fragments are not yet connected magnetically; an Interaction Phase, where coronal reconnection creates an X-ray bright point; and a Cancellation Phase, where photospheric reconnection produces a Cancelling Magnetic Feature. Indeed, coronal reconnection driven in this manner by footpoint motion may well represent an "elementary heating event" that is heating not just bright points but also all coronal loops, at least low-lying ones.

Acknowledgements. I am most grateful to Jim Burch and Hunter Waite for organising such a valuable conference and to them and the UK Science and Engineering Research Council for financial support.

REFERENCES

Berger, M., Magnetic heating of the solar corona, in *Advances in Solar System MHD* (eds. E.R. Priest and A.W. Hood), Cambridge, p241-256, 1991.

Browning, P.K. and Priest, E.R., Kelvin-Helmholtz instability of phase mixed Alfven waves, *Astron. Astrophys.* **131**, 283-290, 1984.

Cally, P., Phase mixing of surface waves: a new interpretation, *J Plasma Phys.* **45**, 453, 1991.

Craig, I.J.D. and McClymont, A.N., Dynamic magnetic reconnection at an X-type neutral point, *Astrophys. J.* **371**, L41, 1991.

Dungey, J.W., Conditions for the occurrence of electrical discharges in astrophysical systems, *Phil. Mag.* **44**, 725, 1953.

Finn, J. and Lau, Y-T, *Phys. Fluids* **B3**, 2675, 1991.

Forbes, T.G. and Priest, E.R., Numerical simulation of reconnection in an emerging flux region, *Solar Phys.* **94**, 315-340, 1984.

Goedbloed, J.P., Lecture Notes on Ideal MHD, Rijnhuizen Report 83-145, 1983.

Golub, L., Krieger, A.S., Silk, J., Timothy, A. and Vaiana G., Solar x-ray bright points, *Astrophys. J.* **189**, L93 - L97, 1974.

Golub, L., Krieger, A.S. and Vaiana, G.S., Distribution of lifetimes for coronal soft x-ray bright points, *Solar Phys.* **49**, 79-116, 1976a.

Golub, L., Krieger, A.S., Harvey, J. and Vaiana, G., Magnetic properties of x-ray bright points, *Solar Phys.* **53**, 111-122, 1977.

Golub, L., Harvey, K., Herant, M. and Webb, D., X-ray bright points and He 10830 dark points, *Solar Phys.* **124**, 211-217, 1989.

Goossens, M., MHD waves and wave heating in non-uniform plasmas, in *Advances in Solar System MHD* (ed E.R. Priest and A.W. Hood) Cambridge, p137-172, 1991.

Green, R.M., Models of annihilation and reconnection of magnetic fields *IAU Symp.* **2**, 398, 1965.

Harvey, K.L., Solar cycle variation of ephemeral active regions, *Proc. 4th European Meeting on Solar Phys.* ESA SP 220, 235-236, 1984.

Harvey, K.L., The relationship between coronal bright points as seen in He 10830 and the evolution of the photospheric network magnetic fields, *Aust J Phys.* **38**, 875-883, 1985.

Harvey, K.L. and Martin, S.F., Ephemeral active regions, *Solar Phys.* **32**, 389-402, 1973.

Harvey, K.L., Harvey, J.W. and Martin, S.F., Ephemeral active regions in 1970 and 1973, *Solar Phys.* **40**, 87-102, 1975.

Heyvaerts, J. and Priest, E.R., Coronal heating by phase-

Heyvaerts, J. and Priest, E.R., Coronal heating by phase mixed Alfven waves, *Astron. Astrophys.* **117**, 220-234, 1983.

Heyvaerts, J. and Priest, E.R., Coronal Heating by reconnection in DC current systems. A theory based on Taylor's hypothesis, *Astron. Astrophys.* **137**, 63-78, 1984.

Heyvaerts, J. and Priest, E.R., A self-consistent turbulent model for solar coronal heating, *Astrophys. J.* **390**, 297-308, 1993.

Heyvaerts, J., Priest, E.R. and Rust, D.M., An emerging flux model for the solar flare phenomenon, *Astrophys. J.* **216**, 123-137, 1977.

Imshennik, V.S. and Syrovatsky, S.I., Two-dimensional flow of an ideally conducting gas in the vicinity of the zero line of a magnetic field, *Sov. Phys. JETP* **25**, 656-664, 1967.

Lee, L.C. and Fu, Z.F., Multiple x-line reconnection, 1, A criterion for the transition from a single x-line to a multiple x-line reconnection, *J Geophys Res.* **91**, 6807, 1986.

Low, B.C. and Wolfson, R., Spontaneous formation of electric current sheets and the origin of solar flares, *Astrophys. J.* **324**, 574-581, 1988.

Martin, S.F., *Coronal and Prominence Plasmas* (ed. A. Poland) NASA CP 2442, p73, 1986.

Martin, S.F., The identification and interaction of network, intranetwork and ephemeral region magnetic fields, *Solar Phys.* **117**, 243-259, 1988.

Martin, S.F., *IAU Symp.* **138**, 130, 1990a.

Martin, S.F., Elementary bipoles of active regions and ephemeral active regions, *Mem S A It.* **61**, 293-315, 1990b.

Martin, S.F., and Harvey, K.L., Ephemeral regions during solar minimum, *Solar Phys.* **64**, 93-108, 1979.

Martin, S.F., Livi, S.H.B. and Wang, J., The cancellation of magnetic flux. II In a decaying active region, *Aust. J. Phys.* **38**, 929-959, 1985.

Mikic, Z., Schnack, D.D. and Van Hoven, G., Creation of current filaments in the solar corona, *Astrophys J.* **338**, 1148-1157, 1989.

Parker, E.N., Topological dissipation and the small-scale fields in turbulent gases, *Astrophys. J.* **174**, 499-510, 1972.

Priest, E.R., Magnetic reconnection on the Sun, *Mit. Astron. Ges.* **65**, 41-51, 1986.

Priest, E.R. and Forbes, T.G., New models for fast steady state magnetic reconnection, *J. Geophys. Res.* **91**, 5579-5588, 1986.

Priest, E R, Parnell, C E and Martin, S F, A converging flux model of an X-ray bright point and an associated cancelling magnetic feature, *Astrophys. J.*, in press, 1994.

Priest, E R, Titov, V S and Rickard, G K, The formation of magnetic singularities by nonlinear time-dependent collapse of an X-type magnetic field, *Phil. Trans. Roy. Soc. Lond.*, submitted, 1994.

Rosner, R. and Knobloch, E., On perturbations of magnetic field configurations, *Astrophys. J.* **262**, 349-357, 1982.

Sakurai, T., Phase mixing of Alfven waves, in *Th. Probs in High Resoln Solar Phys.* (ed. H. Schmidt), 263-267, 1985.

Sheeley, N.R. and Golub, L., Rapid changes in the fine structure of a coronal bright point and a small coronal active region, *Solar Phys.* **63**, 119-126, 1979.

Somov, B.V. and Syrovatsky, S.I., Hydrodynamic plasma flow in a strong magnetic field, *Proc. Lebedev. Phys. Inst.* **74**, 13-71, Consultants Bureau, New York, 1976.

Strong, K., Harvey, K., Hirayama, T., Nitta, N., Shimizu, T. and Tsuneta, S., Observations of the variability of coronal bright points by the soft x-ray telescope on Yohkoh, *Pub. Astron. Soc. Jap.*, 1992, in press,

Tur, T.J. and Priest, E.R., The formation of current sheets during the emergence of new magnetic flux from below the photosphere, *Solar Phys.* **48**, 89-100, 1976.

Vaiana, G.S., Krieger, A.S., Van Speybroeck, L.P. and Zehnpfennig, T., *Bull. Am. Phys. Soc.* **15**, 611, 1970.

Van Ballegooijen, A., Electric currents in the solar corona and the existence of magnetostatic equilibrium, *Astrophys. J.* **298**, 421-430, 1985.

Vekstein, G.E., Priest, E.R. and Amari, T., Formation of current sheets in force-free magnetic fields, *Astron. Astrophys.* **243**, 492-500, 1990.

Vekstein, G.E. and Priest, E.R., Magnetohydrostatic equilibria and cusp formation at an X-type neutral line by footpoint shearing, *Astron. Astrophys* **384**, 333-340, 1992.

Withbroe, G.L. and Noyes, R.W., Mass and energy flow in the solar chromosphere and corona, *Ann. Rev. Astron. Astrophys.* **15**, 363-387, 1977.

Solar Flare Hard X-ray Angular Distribution and Its Time Evolution

Peng Li

Space Sciences Laboratory, University of California, Berkeley, CA 94720

We have identified 72 large solar flares (peak counting rates >1000 count s^{-1}) observed by Hard X-ray Burst Spectroscopy (HXRBS) on-board the Solar Maximum Mission (SMM). Using a database of these flares, we have studied the hard X-ray (50-850 keV) spectral center-to-limb variation and its evolution with time. The major results are the following: (1) During the rise phase, the center-to-limb spectral variation is small, with a hardness of $\Delta\delta=0.02\pm0.25$; (2) During the peak phase, the center-to-limb variation is $\Delta\delta=0.13\pm0.13$; (3) During the decay phase, the center-to-limb variation changes to softening. The softness is relatively larger with $\Delta\delta=-0.25\pm0.21$; (4) The spectral index variation $d(\Delta\delta)=-0.38\pm0.25$ (1.5σ) at most; (5) The linear least-squares fits to the spectral center-to-limb variations do not have slopes that significantly differ from zero during all those three phases; (6) The spectral distributions of center events and limb events are shown to be not different by using the Kolmogorov-Smirnov two-sample test; (7) The fraction of events detected near the limb is marginally consistent with that expected from isotropically emitting flares. These results suggest that there is no statistically significant evidence for hard X-ray directivity during the rise, peak, and decay phases of solar flares. The hard X-ray angular distribution at those energies is almost isotropic during all those phases.

1. INTRODUCTION

Determination of the hard X-ray source directivity in solar flares is important for inferring the electron angular distribution. Furthermore, hard X-ray directivity measurements can place important constraints on solar flare electron models (e.g., beam model, trap model, and trap plus precipitation model). Although a single spacecraft is not sufficient to measure the hard X-ray directivity for a given flare, single spacecraft observations may be used to study the mean X-ray angular distribution on a statistical basis. By assuming flares have same geometry and size distribution, any variations with view angle can be attributed to flare hard X-ray anisotropy. Using these statistical results, together with some modeling efforts, the electron angular distribution can be determined. The corresponding directivity can also be estimated.

Solar System Plasmas in Space and Time
Geophysical Monograph 84
Copyright 1994 by the American Geophysical Union.

The statistical method has been used by several researchers in the past two decades. The studies of Ohki [1969], Pinter [1969], and Kane [1974] used data from OGO-1, OGO-3, and OGO-5, and found that at energies >20 keV, the flare frequency is consistent with an isotropic radiation pattern. The work by Datlowe et al. [1977], based on observations of 148 flares detected by OSO-7, also found that >20 keV hard X-ray emission is isotropic. However, Vestrand et al. [1987] have analyzed 146 (with 87 spectra determined) flares at energies >300 keV observed by GRS on board the SMM and have found that flare energetic emission at those energies is anisotropic, based on significant center-to-limb spectral hardening and increase in the frequency of detection. It was also found that all the flares observed above 10 MeV are located near the limb, indicating an even larger anisotropy. However, the statistics for these >10 MeV flares are poor (only 10 events).

Up until now, the spectral analyses were performed either over the entire event or during the flare peak phase in all existing statistical directivity measurements. However, the questions of whether the directivity is time-dependent and how does it evolve with time, have not been investigated yet. The answer to this question should provide physical insight to the flare energetic photon angular distribution and its time evolution. With this information, the angular distribution of energetic electrons can be inferred through appropriate modeling efforts. The deduced electron angular distribution in this manner is crucially important for a better understanding of electron acceleration and transport mechanisms. Moreover, the answer to this question can tell us when is the best time to measure a significant directivity if there is any. It can be seen that the study of this problem is pressing and important. In this paper, we therefore study the temporal evolution of hard X-ray anisotropy using the statistical method. We will use the SMM HXRBS observation as our database.

2. SAMPLE DATA SELECTION

Compton backscattering adds complications to the flare hard X-ray directivity measurement and its interpretation. Bai and Ramaty [1978] have studied the effect of Compton backscattering on the hard X-ray anisotropy. They found that this effect could contribute to a significant level to "wash out" or "shield" the hard X-ray anisotropy, particularly at low energies around or below 30 keV. The importance of this effect diminishes as photon energy increases. The fact that most of the statistical measurements around 1970 s yielded isotropic photon distribution seems to support the importance of Compton backscattering to the hard X-ray anisotropy, due to their ≤ 30 keV lower energy thresholds in observations. In addition to Compton backscattering, flare thermal bremsstrahlung may be important (see, e.g., Li and Emslie [1990]) for large events at energies ~30 keV and also serves to "wash out" the possible emission anisotropy, because thermal bremsstrahlung radiates isotropically. It can be seen that we need to use hard X-ray observations whose lower energy threshold is higher than 30 keV to measure the hard X-ray directivity. The "shielding" due to Compton backscattering and thermal bremsstrahlung emission are expected to be reduced at higher energies (>30 keV) and unambiguous measurements can be made.

The HXRBS lower energy threshold drifted from ~25 keV at the beginning of the mission (1980) to ~52 keV at the end of the mission (1989) due to the gain shift [Dennis et al., 1991]. Between 1980 and 1988, the lower energy threshold drifting was small, typically in the range of ~ 25-35 keV. Starting from December 1988, the lower energy threshold drifted to ~52 keV. It can be seen that the hard X-ray observation from 1980-1988 contained a certain amount of contributions from both Compton backscattered photons and thermal bremsstrahlung photons at low energies. The observations during this time period may not be suitable for studying the hard X-ray anisotropy through statistical methods, since the anisotropy could be shielded by those two physical processes. Nevertheless, the observations obtained in 1989 did not have significant thermal bremsstrahlung due to the significantly increased lower energy threshold. Furthermore, the Compton backscattering is reduced at energies > 50 keV, according to Bai and Ramaty [1978]. Therefore, the observation of 1989 serves better for the purpose of the directivity study. Since the cycle 22 peaks at 1989, the flare events of this year should be good enough for performing a meaningful statistical measurement.

To study the time evolution of flare spectra, we need to study those events whose count rates are large enough so that statistical uncertainties are small, yielding the best determination of the photon spectrum. Initially, we identified about 150 events with peak rate larger than 1000 count s^{-1}. However, among these events, some had no corresponding Hα flare coincident in time, while others exhibited clear pulse pile-up. To reduce uncertainties, flares with either of these problems were removed from the initial selection. Finally, a total of 72 "good" events were identified. For simplicity, we assumed a single power-law form to fit the spectrum, yielding a best determined spectral index δ by minimizing χ^2.

3. RESULTS

3.1. Spectral Distribution and Its Time Evolution

We show spectral center-to-limb variation and spectral distribution in Figure 1 for different phases of hard X-ray solar flares. We first discuss the results for the rise phase (left column of Figure 1). At this early phase, the spectral index is largely scattered, ranging from δ = 3 to δ = 9. The mean spectral index is $\bar{\delta}$ = 5.25 ± 0.18. We did a linear least-squares fit for the data, with a best fit described by δ = -0.0016 (± 0.0085)θ + 5.33 (± 0.47). From this we know that the slope of the linear fit is small, indicating an increase in view angle results in a slight decrease in spectral index. However, the uncertainty on the slope is large and the center-to-limb hardening is not significant.

To further examine whether the observed radiation pattern is anisotropic or isotropic during the rise phase, we classify the flares into two groups. The first group are those with |θ| < 60°, which we call center events. The second group are those with |θ| ≥ 60°, which we call limb events. This events classification scheme is the same as that employed by Vestrand et al. [1987] in their statistical study. Looking at the middle and lower panels of Figure 1 for the

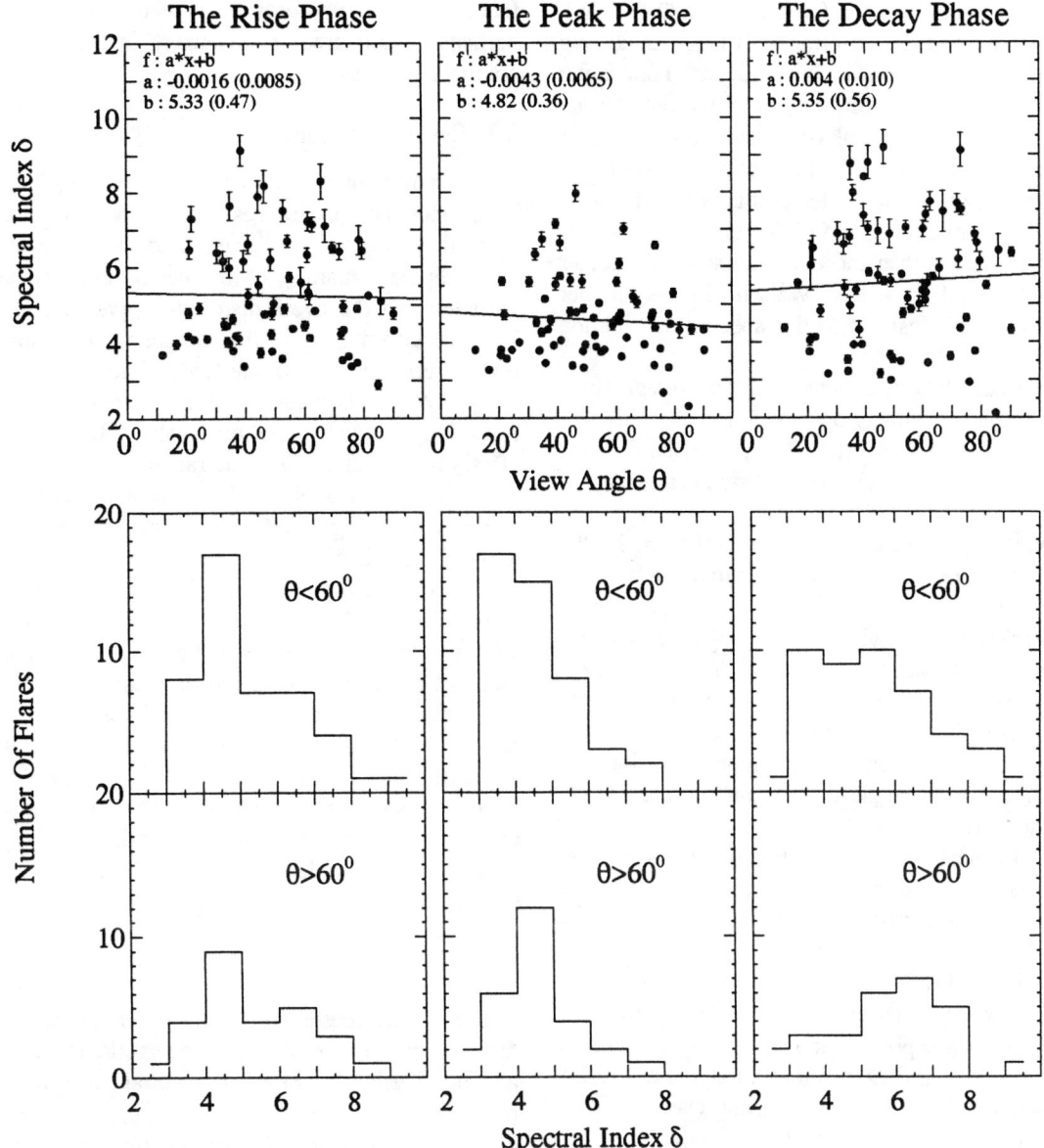

Fig. 1. Spectral index center-to-limb variation (upper panels), spectral distribution of center events (middle panels), and spectral distribution of limb events (lower panels), for rise, peak, and decay phases of solar flares.

rise phase, we see that the difference between these two distributions is small. They have roughly similar profiles and peak around $\delta \approx 4 \sim 5$.

We can also calculate the center-to-limb variation quantitatively. For the 45 center events, the average spectral index is $\delta_c = 5.25 \pm 0.18$, while for the 27 limb events, the average spectral index is $\delta_l = 5.23 \pm 0.17$. Defining the spectral hardness $\Delta\delta$ as the difference between the average center and limb spectral indices, $\Delta\delta = \delta_c - \delta_l = 0.02$ (± 0.25). The absolute value of $\Delta\delta$ is very small. The analysis of GRS observations for the energy range 0.3 - 1 MeV yielded $\Delta\delta = 0.37 \pm 0.11$ at a 3σ level [see, i.e., *Vestrand et al.*, 1987], which is 18 times larger than HXRBS hardness. This calculation, together with the analyses of Figure 1 rise phase results, shows that there is no clear evidence for anisotropy during the rise phase.

During the peak phase (middle column of Figure 1), the range of spectral indices has decreased to $\delta = 2 \sim 8$ (upper panel). The mean spectral index is $\bar{\delta} = 4.60 \pm 0.09$ and is smaller than the average for the rise phase. A linear least-squares fit yields δ = -0.0043 (± 0.0065) θ + 4.82 (± 0.36). The slope is small and the uncertainty is large, indicating the center-to-limb variation is insignificant. For center events (middle panel), most of the spectral indices are distributed over δ ≈ 3 ~ 6. The peak occurs near δ ≈ 3 ~ 4. For the limb events (lower panel), the peak occurs at δ ≈ 4 ~ 5. Since the distribution of the center events is broader, it is hard to judge whether distribution of the limb events is generally softer or harder than the distribution of the center events through visual inspection. We will discuss another more conclusive statistical test to this spectral distribution later this section.

Quantitative calculations indicate that the average spectral index for center events is $\delta_c = 4.65 \pm 0.09$, while for limb events it is $\delta_l = 4.52 \pm 0.10$. The spectral hardness is $\Delta\delta = \delta_c - \delta_l = 0.13 (\pm 0.13)$, which is still small and insignificant.

During the decay phase (right column of Figure 1), the spectral indices range from as low as ~2, to as high as ~9, with a mean spectral index $\bar{\delta} = 5.51 \pm 0.19$ (upper panel). The linear least-squares fit yields δ = 0.004 (± 0.010) θ + 5.35 (± 0.56). In contrast to the results of the rise and peak phases, the slope of the linear fit is positive for the decay phase. This means that the spectral index gets softer as the viewing angle increases, However, the slope of the linear fit is the same order as in the peak phase and the uncertainty associated with it is still large.

The shift in the peak of the distribution of limb events relative to the distribution of center events is relatively clear compared with that in the rise and peak phases (middle and lower panels). The distribution of center events is broader compared with the distribution of limb events. The mean spectral index for the center events is found to be δ_c= 5.48 (± 0.19), and the mean spectral index for the limb events is found to be δ_l = 5.73 ± 0.09, resulting in a softness of $\Delta\delta = \delta_c - \delta_l = -0.25 \pm 0.21$, a 1.2σ significant level. The evidence for anisotropy is not significant.

3.2. Kolmogorov-Smirnov Test

Although the visual inspection shows some similarity between spectral distributions of center and limb events, it will be more conclusive if we examine those two distributions through some statistical test. We use the Kolmogorov-Smirnov two-sample test to check whether those two distributions originate from the same distribution (or they are the same). The answer is affirmative. We found that the center events distribution and limb events distribution for all phases are the same at a 99% confidence level (α=0.01). If the radiation pattern were strongly anisotropic, we should observe a clear difference in those two distributions. The fact that we did not observe this feature indicates a small or perhaps none degree of anisotropy, in fact consistent with complete isotropy.

3.3. Events Distributions

The nature of the hard X-ray source radiation pattern determines the appearances of the flare spectral index center-to-limb variation and event distribution. Coversely, from the observed signatures of the spectral index center-to-limb variation and the event distribution, we can, in principle, infer the information on the energetic photon source radiation pattern. Using a Monte Carlo simulation, the expected number of events located at the limb can be calculated by assuming a certain radiation pattern. Vestrand and Ghosh [1987] have found that if the radiation pattern is isotropic, then about 29.5% of the flares should have heliocentric angle (or view angle in this paper) such that sin θ > 0.9 (or θ > 64.20); this is $N_l = 0.295N \pm 0.456 N^{1/2}$. Since our database is composed of 72 events, N_l can be found as N_l = 21 ± 4 (or $17 \leq N_l \leq 25$). Among those 72 events, we observed 18 events with sin θ > 0.9, which is still in the acceptable range of N_l. This analysis indicates that the event distribution is consistent with the isotropic distribution.

The events distribution, by its nature, is the same for all the flare phases. The isotropic radiation pattern inferred from events distribution studies agrees with those inferred from the spectral center-to-limb variation during all the phases of flares. This also suggests that the hard X-ray angular distribution does not significantly vary with time.

DISCUSSION AND CONCLUSIONS

The results presented in this paper indicate interesting features concerning the flare energetic radiation and its angular distribution. First, the spectral center-to-limb variation from the HXRBS observations is significantly smaller than the one from the GRS observations. Second, the spectral index distribution of center events and limb events are the same for all flare phases, in contrast to the GRS observations in which those two distributions were significantly different. Third, the inferred photon source angular distribution does not vary with time significantly and is almost isotropic. The results we found are consistent with all the stereoscopic directivity measurements to date [see, e.g., *Kane et al.*, 1988; *Li et al.*, 1993]. What can we learn from those results and their comparison with previous studies? To find out the nature of flare energetic photons and energetic electrons, we explore some plausible physical scenarios to explain our results and GRS results in a self-consistent manner.

The GRS results indicated that the energetic photon (>300 keV) source is anisotropic with a center-to-limb hardening of 0.37 ± 0.11 and a deduced directivity of 2~4. Although the directionality of hard X-ray bremsstrahlung is dependent on the photon energy, it should not make much difference in the 50-300 keV energy range [see, e.g., *Dermer and Ramaty* 1986]. Some other effects must cause the differences between HXRBS results and GRS results. Vestrand et al. [1987] had invoked an anisotropic pancake electron angular distribution (peaks along the direction perpendicular to the magnetic field line at the flare loop footpoints) and a downward peaked Gaussian angular distribution to explain their observed significant anisotropy. If, at the principal site of hard X-ray emission, the electron angular distribution does not vary with energy significantly over the energy range of 0.05 - 1MeV, one possible interpretation is that Compton backscattering is still significant (much more than expected from Bai and Ramaty [1978]) at energies of 50-300 keV, so that no significant anisotropy can be measured at energies > 50 keV. If the electron angular distribution does vary with energy significantly, one may expect that the anisotropy of low energy electrons is much smaller than high energy electrons. In other words, the anisotropy increases with energy. Since electron spectrum decays as a power-law, the high energy electrons produce insignificant amount of photons at low energies. It follows that the near isotropic electrons produce isotropic photons at low energies, while highly anisotropic electrons produce anisotropic photons at high energies, therefore, in agreement with the both HXRBS and GRS observations. However, whether this variation with energy is due to acceleration or energy transport remains to be further studied.

To our knowledge, theoretical studies on energetic photon source directivity have not been able to include the time effect [e.g., *Bai and Ramaty*, 1978; *Petrosian*, 1985; *Dermer and Ramaty*, 1986; *McTiernan and Petrosian*, 1991]. Moreover, all models assumed that the magnetic field line is perpendicular to the photosphere at the footpoints of the flare loop. This assumption may not be true since the flare magnetic structures are likely sheared or twisted, particularly during the early phase of the flare. The magnetic shearing or twisting may play an important role in determining the energetic photon bremsstrahlung radiation angular distribution. Moreover, our results suggest that Bai and Ramaty [1978] could underestimate the Compton backscattering at energies > 50 keV. Further theoretical studies including all those physical processes and their comparison with the results demonstrated in this paper and others should provide promising results to understand the time evolution of flare energetic photon source angular distribution and the nature of flare energetic electrons.

We have studied, for the first time, the time evolution of solar flare hard X-ray (50 ~ 850 keV) angular distribution, using 72 large flares (peak counting rate > 1000 counts s^{-1}) observed by HXRBS during solar cycle 22 peak 1989. The major results are as follows: (1) during the rise phase, the spectral index exhibits a small center-to-limb hardening, with a hardness $\Delta \delta = 0.02$ (± 0.25) at a 0.1σ level, and a linear least-squares fit of $\delta = -0.0016$ (± 0.0085)θ + 5.33 (± 0.47); (2) during the peak phase, the spectral hardness is $\Delta \delta = 0.13$ (± 0.13) significant at the 1σ level. The linear least-squares fit is $\delta = -0.0043$ (± 0.0065)θ + 4.82 (± 0.36); (3) during the decay phase, the spectral index exhibits center-to-limb softening, with softness $\Delta \delta = -0.25$ (± 0.21), significant at the 1.2 σ level. The linear least-squares fit is $\delta = 0.004$ (± 0.010)θ + 5.35 (± 0.56); (4) the center events and limb events spectral distributions are found to be the same during the rise, peak and decay phases, using the Kolmogorov-Smirnov two-sample test; (5) among these 72 flares, 18 were located at $\sin\theta > 0.9$ (or $\theta > 64.20°$), which is 25% of the total, slightly less than the fraction 29.5% expected from an isotropic radiation source, but still within the statistical errors.

In summary, there is no statistically significant evidences in favor of hard X-ray anisotropy at energies >50 keV at all times during solar flare bursts.

Acknowledgments. I am grateful to Kevin Hurley and Gordon Emslie for their stimulating discussions. I thank Kim Tolbert for her help with the DCP and HXRBS data. This work was supported at UCB by JPL contract 958056 and NASA grant NAG5-935.

REFERENCES

Bai, T., and R. Ramaty, Backscatter, Anisotropy, and polarization of solar hard X-rays, *Astrophys. J.*, 210, 705, 1978.

Datlowe, D. W., S. L. O' Dell, L. E. Peterson and M. J. Elcan, An upper limit to the anisotropy of solar heard X-ray emission, *Astrophys. J.*, 212, 561, 1977.

Dennis, B. R., et al., NASA Technical Memorandum 4332, p5, 1991.

Dermer, C., and R. Ramaty, Directionality of bremsstrahlung from relativistic electrons in solar flares, *Astrophys. J.*, 301, 962, 1986.

Kane, S. R., in *IAU Symposium 57, Coronal Disturbances*, Ed. G. Newkirk, Jr. (Porodrechtii Reidel), Op, 105, 1974.

Kane, S. R., E. E. Fenimore, R. W. Klebesadel, and J. G. Laros, Directivity of 100 keV- 1 MeV photon sources in solar flares, *Astrophys. J.*, 326, 1017, 1988.

Li, P., and A. G. Emslie, Comparison of thermal and nonthermal hard X-ray emission in electron-heated solar flares, *Solar Phys.*, 129, 113, 1990.

Li, P., K. C. Hurley, C. Barat, M. Niel, R. Talon, and V. Kurt, Directivity of 100-500 keV solar flare hard X-ray emission, submitted to *Astrophys. J.*, 1993.

McTiernan, J. M., and V. Petrosian, Center-to-limb variations of the characteristics of solar flare hard X-ray and gamma-ray emission, *Astrophys. J.*, 379, 381, 1991.

Ohki, L. E., Directivity of solar hard X-ray bursts, Solar Phys., 7, 260, 1969.

Petrosian, V., Directivity of bremsstrahlung radiation from relativistic beams and the gamma rays from solar flares, *Astrophys. J.*, 299, 987, 1985.

Pinter, S., Longitudinal distribution of X-bremsstrahlung on solar disc, *Solar Phys.*, 8, 142, 1969.

Vestrand, W. T., and A. Ghosh, Simulations of stereoscopic solar flare observations, *Proc. 20th Internat. Comic Ray Conf. (Moscow)*, 1987.

Vestrand, W. T., D. J. Forrest, E. L. Chupp, E. Rieger, and G. H. Share, The directivity of high-energy emission from solar flares: solar Maximum Mission observations, *Astrophys. J.*, 322, 1010, 1987.

P. Li, Space Sciences Laboratory, University of California, Berkeley, Berkeley, CA 94702

The Experimental Signatures of Coronal Heating Mechanisms

PETER J. CARGILL

Science Applications International Corporation, McLean, VA

and

Beam Physics Branch, Plasma Physics Division, Naval Research Laboratory, Washington, D.C. 20375-5320

Methods by which coronal heating mechanisms can be identified are reviewed and a number of presently popular mechanisms are discussed in light of these constraints. Coronal heating by nanoflares is summarized. Assuming that the corona is comprised of many hundred small, elemental loops that are heated impulsively by nanoflares, the cooling of such loops is examined. It is shown that the emission measure of such a corona satisfies a two part power law, which at lower temperatures is in agreement with observations. Also, even though the coronal energy release may be highly filamentary, the resulting coronal plasma is rather homogeneous, filling factors of order unity being obtained. By parameterizing the bulk plasma flows arising from nanoflares, spectral line profiles can be computed and it is shown that the observed profiles can be functions of the location along a loop.

1. INTRODUCTION

The existence of a hot solar corona ($T > 10^6$ K) has been known for over 50 years, while the presence of similar features in solar-like stars has been apparent since the 1970s. The 1973-74 Skylab mission highlighted the fact that the corona is divided between open regions (coronal holes: the source of the solar wind) and closed regions (coronal loops), and also revealed the highly structured nature of the closed corona, results emphasized in recent years by the Solar Maximum Mission (SMM), YOHKOH and numerous rocket flights. Just why the corona is hot has been less readily apparent. While it is believed that the upper photosphere and chromosphere can be heated by upward travelling acoustic or fast modes magnetohydrodynamic waves, the power in these waves entering the corona is probably inadequate to account for the energy losses of the active corona [*Athay and White*, 1979]. It is thus widely believed that the corona is heated by dissipation of electric currents associated with the coronal magnetic field. The strength of this magnetic field cannot at present be determined to any accuracy, but is believed to lie somewhere between ten and a few hundred gauss.

Solar System Plasmas in Space and Time
Geophysical Monograph 84
This paper is not subject to U.S. copyright. Published in 1994 by the American Geophysical Union.

The general energy requirements of coronal heating are well known. Using spectral information, particularly in the EUV, *Withbroe and Noyes* [1977] presented a simple estimate of the energy loss from different regions of the corona. In coronal holes, the energy loss was estimated to be $\approx 10^6$ ergs cm^{-2} sec^{-1}, much of which was accounted for by the solar wind mass flux. In the closed corona, quiet (active) regions accounted for losses of $\approx 3 \times 10^5$ (10^7) ergs cm^{-2} sec^{-1}. Summed over the whole solar surface, the total losses are somewhere in the region of 10^{29} ergs sec^{-1}, a tiny fraction of the total solar radiative loss of 4×10^{33} ergs sec^{-1}. It should be noted that these numbers use data obtained in the early and mid-1970s, and are taken from a region of the spectrum that has not been re-examined in detail since then. The upcoming Solar and Heliospheric Observatory (SOHO), scheduled for launch in 1996, will carry a broad array of EUV instruments. It is imperative that the estimates of Withbroe and Noyes be verified, or updated, by SOHO.

A number of mechanisms have been proposed for this heating, but it is not the purpose of this paper to provide a detailed review of the physics of these mechanisms. That has been done both extensively and well elsewhere [e.g. *Ulmschneider et al.*, 1991; *Browning*, 1991; *Spicer and MacNiece*, 1992; *Zirker*, 1993]. Rather the aim of this paper is to examine the coronal heating problem from the perspective of the goal of the 1993 Yosemite meeting, namely to discuss how theory and observations can be closely tied together to determine the relevant spatial and temporal scales associated with coronal heating. In this aspect, the present paper is closest to the review of *Zirker* [1993], but addresses the issues from a different perspective.

In Section 2, we list the means at our disposal to identify the coronal heating mechanism. Section 3 reviews the various mechanisms in light of existing and future experimental results. Section 4 discusses the likelyhood that coronal heating needs to be understood in terms of a convolution of many small sources (Parker's nanoflares) and presents methods for comparing the theoretical predictions of such models with experimental data. A fuller description of much of Section 4 will appear elsewhere [*Cargill*, 1994: hereafter C94].

2. PRESENT AND FUTURE EXPERIMENTAL DATA BASES

As will become apparent in the next Section, it is unlikely that we will be able to resolve spatially the scales associated directly with many of the models for energy release in the corona. This is a direct manifestation of the inefficiency of dissipation in the coronal plasma, which requires small scales to become effective. We are thus forced into consideration of indirect detection of the relevant heating mechanism. By necessity, all of the detectable signatures arise from solar radiative output at the radio, H_α, EUV, and X-rays (soft and hard) wavelengths. H_α (and associated magnetograms) can provide information about motions of photospheric plasma which are widely believed to stress coronal magnetic fields, hence increasing the free energy in the corona. The following is a summary of the diagnostic tools at our disposal either at the present time, or in the forseeable future.

2.1 Emission Measure

In pre-Skylab days, as well as during the Skylab mission, a wide range of lines in UV and EUV were observed and it was possible to construct a curve showing the height-integrated emission measure ($\int n_e^2 dh$, where dh is directed along the line of sight) as a function of temperature [e.g. *Gabriel*, 1976; *Jordan*, 1976; 1980]. The curve has a minimum at around 2×10^5 K and scales as $T^{1.5}$ up to just over 10^6 K, where the good coverage of emission lines began to fall off. Below 2×10^5, the emission measure increases as the temperature decreases. (This increase at low temperatures has yet to be explained adequately despite some interesting efforts [*Antiochos and Noci*, 1986; *Woods et al.*, 1990, *Roumeliotis*, 1992], but it is unclear whether this behavior is related to the coronal heating mechanism. Rather it is possible that the cool plasma arises from different loops, which must still of course be heated, but by a distinct mechanism.) The shape of the emission measure between 2×10^5 and 10^6 K may not be a particularly good discriminator between different heating mechanisms, but a reproduction of it is a minimum requirement for any mechanism. Rather more useful may be computations or predictions of the emission measure curve above 10^6 K. SOHO will provide an excellent coverage in some of this temperature range.

2.2 Doppler Shifts and Line Broadening

The instruments on Skylab, OSO-8 and other missions observed systematic spectral line-broadening of transition region lines (such as CIV which forms at around 10^5 K) of up to 30 km/sec [e.g. *Klimchuk*, 1986; *Athay*, 1988 and references therein]. SMM observed line broadening of 60 km/sec in lines formed at approximately 4×10^6 K [*Saba and Strong*, 1991]. This may indicate a strong source of turbulence in the corona and transition region, but whether the broadening is caused by oscillations in a nearly homogeneous plasma (such as Alfven waves) or the convolution of many small, individual plasma jets is unclear at present. An interesting feature of the active region studied by Saba and Strong was that the line profile showed little change from disc to limb. However, only one active region has been studied in detail, and, although other data sets do exist, resources do not permit their study at this time. It is also expected that SOHO will give useful information about the structure of spectral lines formed at coronal temperatures. This is an important diagnostic, particularly if the line profile is in fact a function of the location of the active region on the solar disc.

Skylab also observed systematic redshifts in CIV of perhaps 10 km/sec [e.g. *Doschek et al.*, 1976]. It is unclear whether this corresponds to a continuous draining of coronal loops that have been impulsively heated, or to material that is impulsively injected into the corona from below (such as spicules), returning to the chromosphere. In either case, the corresponding amount of material is not seen going upward in blue shifts in CIV. Such flows would not appear to be a natural consequence of a steady state coronal heating mechanism.

2.3 Hard X-rays and Energetic Electrons

Hard X-rays (energies > 10 keV) are widely believed to arise when energetic electrons precipitate into the photosphere, and have been used as a diagnostic in solar flare studies for many years [e.g. *Emslie*, 1989]. It is only recently that their potential in studying coronal heating has been realized. Particularly influential was the balloon flight reported by *Lin et al.* [1984], who identified small "elemental" bursts, corresponding to the release of approximately 10^{27} ergs, generally referred to as microflares. These microflares seem to be composed of yet smaller bursts. This led *Parker* [1988] to conjecture that the elemental unit of coronal energy release was the nanoflare (10^{24} ergs), and that microflares were just the simultaneous "firing" of many nanoflares. As pointed out by *Zirker* [1993], direct detection of the nanoflare is elusive, and likely to remain so for the forseeable future. However, R. P. Lin (private communication, 1993) has indicated that the Solar Probe could perhaps detect nanoflares.

Hard X-ray bursts are widely viewed as being a signal of accelerated electrons, so if the signals in the 22 - 33 keV energy bands used by Lin et al. persist to nanoflares, the heating mechanism must be able to produce energetic electrons that precipitate into the photosphere. Future hard

X-ray observations are thus an important part of unravelling the coronal heating problem. The High Energy Solar Physics (HESP) mission is important in this regard.

Other evidence of energetic particles in the corona has come from the radio data, with the impulsive millisecond bursts of *Benz* [1987] being particularly interesting. Much of the data comes from the very impulsive phase of solar flares, but is very suggestive in that it shows closely linked sequences of intense bursts, interpreted by Benz to correspond to "elemental" units of energy release. Better observations are needed to see to what extent such phenomena persist in active regions.

2.4 Filling Factors

Since present instruments (or those to be built in the forseeable future) will probably not be able to resolve coronal fine structure with scales much less that 1 arcsecond ($\approx$ 740 km), it is of some interest to know what fraction of an unresolved image is actually at coronal temperatures. It is possible that much of the corona may be devoid of material, with the observed emission coming from a few isolated strands. The ratio of radiating volume to the total volume is defined here as the coronal filling factor (ϕ) and can be determined experimentally. Given the emission measure and an estimate of the volume in the image, the "density" can be estimated. Note that this "density" assumes that the coronal volume is filled uniformly with plasma. An independent estimate of the density can also be made from density-sensitive spectral lines [e.g. *Dere et al.*, 1979]. This density is the real density of any radiating filamentary structure. Given these two determinations of the density, the filling factor follows readily. ϕ was estimated from Skylab as being very small in the transition region but of order unity in the corona [*Dere*, 1982]. These observations must be repeated from SOHO. Estimates of ϕ can be used to shed some light on the scale of the coronal energy release: for example very small scale energy release might be expected to give $\phi \ll 1$, whereas a homogeneous release mechanism could lead to $\phi \approx 1$. We return to this later.

It can be seen that while we have many potential diagnostic tools, the fundamental problem of a lack of spatial, temporal and spectral resolution will continue to plague coronal heating studies for the forseeable future. While some of the instruments that will fly on SOHO claim arcsecond spatial resolution, one second temporal resolution and spectral resolution corresponding to Doppler shifts of 1 km/sec, these cannot in all likelihood be achieved simultaneously. Compromises are necessary, as is advance planning of observing runs such as to optimize the available resources. An excellent example of this planning is the "blue book" of the Coronal Diagnostic Spectrometer [CDS: *Harrison*, 1991].

Since it is unlikely that future missions will be able to resolve the scales of coronal energy release, it is then equally important that theories of coronal heating compute observables. Such observables could be used as guidance in mission planning (e.g. to suggest observing runs that would not otherwise be made). In the next section, we briefly summarize theories of coronal heating, indicating possible observables and in Section 4 compute some observables from one particular mechanism.

3. THEORY OF CORONAL HEATING

3.1 Alfven Waves

Given the apparent failure of acoustic waves as a coronal heating mechanism, a natural alternative is Alfven waves. Unfortunately, it is difficult to damp Alfven waves in the corona, in large part due to their incompressible nature. However, Hollweg, Davila and collaborators [e.g. *Hollweg*, 1987; *Davila*, 1987 and references therein] have addressed this issue in a series of papers, based on the realization that in a medium with a varying Alfven speed, the Alfven wave can excite a resonance and, as a result of mode conversion, be damped by resistive or viscous effects. Resistive damping does not seem to be effective [*Steinolfson and Davila*, 1993], leading to unrealistically large velocities in the resonance layer, and the preliminary results of *Offman et al.*, [1993] indicate that compressive viscosity has difficulties as well.

Alfven wave heating would appear to be consistent with the line-broadening seen by *Saba and Strong* [1991], but a steady state mechanism such as are proposed at this time would probably not lead to the transition region downflows. Also, the viscous damping mechanism has no parallel electric field associated with it, so that energetic electrons would not be an expected signature. The small dissipation scales predicted by Offman et al. would probably lead to small filling factors, but the scaling of the emission measure as a function of temperature has yet to be computed. The biggest problem with Alfven waves is their apparent inability to heat short loops. As has been pointed out by *Parker* [1988], the frequency required for the waves to be transmitted from the photosphere into these loops is higher than that at which the power spectrum peaks. Further studies of photospheric turbulence are needed to confirm this, but in the absence of mode conversion of some description in the transition zone, it appears that Alfven waves face some problems in this regard.

3.2 Velocity Filtration

Scudder [1992a,b] has made the suggestion that the corona is not heated at all, but that the observed emission is just the remnant of high energy tails of the electron and ion distribution function that exist in the lower atmosphere. Based on well known concepts from heliospheric and magnetospheric physics concerning non-Maxwellian distributions and energetic tails thereon, Scudder argues that if a Kappa distribution function is present at the base of the transition region, there are enough particles in the tail to account for the coronal energy losses. A fundamental assumption of this mechanism is that these distribution functions can be generated in the lower atmosphere. Scudder appeals to the presence of MHD turbulence in these regions, but at this time has not carried out detailed calculations of how the tails are generated. The model has some interesting features. It predicts a corona that is heated roughly uniformly over scales of a few arcseconds, with the different power losses in different regions being accounted for by different photospheric densities there. Interestingly, by generalizing the definition of the broadening of spectral lines to account for non-Maxwellian ion distributions, Scudder can account for the observed line broadening. This is an important achievement of the model. Scudder has not computed an emission measure curve at this time, and the model would not be expected to lead to significant hard X-ray signatures. It is important to realize that if the postulated non-Maxwellian tail exists, then much of what Scudder proposes must occur.

3.3 Anomalous Current Disruption

The idea that the corona may be heated due to plasma micro-instabilities of very filamented currents has been in the literature for some time [*Hinata*, 1982; *Benford*, 1983; *Sturrock*, 1989; *Sturrock et al.*, 1990]. This mechanism has two problems. One is that the conditions for these instabilities are very severe. For example, the Buneman instability requires scales of order c/ω_e, the ion acoustic and lower hybrid instabilities scales of c/ω_i etc. (where ω_e and ω_i are the electron and ion plasma frequencies respectively). For typical coronal densities, these correspond to scales of a few hundred cm or less. While it may be possible that these instabilities can lead to high temperatures, the power released in each event is tiny. To power the coronal energy loss requires millions of these events per second. The second problem relates to the difficulty of actually getting any heating at all. For most of the instabilities currently considered, the threshold is directly related to the temperature of the plasma. An instability that gives plasma heating automatically raises the threshold for any subsequent instability. It is easy to see that enormous drifts ($v \approx c$) arise rapidly if the instability is to be sustained over any meaningful time (J. Klimchuk, private communication, 1992). The problem is one of removing hot plasma from the heating site so that more cold plasma can then be heated, rather than the hot plasma being reheated more (but see the interesting effort of *Amendt and Benford* [1989] to get around this problem). What is really required is a plasma instability whose threshold is quite insensitive to the plasma temperature. Examples do exist [e.g. *Ganguli et al.*, 1988], but have not been considered in the solar context at this time.

At this stage, little has been done in the way of computing observables. However, anomalous resistivity can lead to significant electron energization [*Papadopoulos*, 1977], so that hard X-ray signatures can be expected. Localized heating will give rise to plasma hot spots, which will expand at some fraction of the local sound speed. The convolution of these expansions can lead to distinctive structures in the spectral line profiles (see Section 4.2). Finally, the anomalous heating can be incorporated in the formalism developed in

Section 4.1, so that filling factors and emission measures can be computed (C94 contains an example of this).

3.4 Current Sheet Formation, Turbulent Cascades and Taylor Relaxation

I have grouped these mechanisms together since they all involve the slow injection of energy into the corona by photospheric footpoint motions, and the subsequent rapid dissipation of the resultant electric currents. *Parker* [1972] argued that if the footpoint motions led to the coronal magnetic field having a fully three-dimensional nature, then no coronal equilibrium could exist, and current sheets would form. These current sheets could then relax by rapid reconnection, releasing of order 10^{24} ergs in a few seconds [*Petschek*, 1964; *Parker*, 1988]. [An important aspect of invoking rapid reconnection as the energy release mechanism is that there is no problem with cross-field heat transport. Newly reconnected (and heated) field lines are convected away from the reconnection site, resulting in a significant volume of plasma being heated.] This process has been contested by *Van Ballegooijen* [1985], *Zweibel and Li* [1987], *Antiochos*, [1987] and others (see *Browning*, [1991] for a good review of both this and other material discussed in this paragraph). Instead, *Van Ballegooijen* [1986] proposed that the random walk of the photospheric footpoints would ultimately lead to small scale magnetic field structures being formed, which could ultimately dissipate, a process that he referred to as a "cascade" of magnetic energy to small scales. In the opinion of this author, these two mechanisms are indistinguishable from a viewpoint of the actual phase of energy release. Small scale currents sheets will form, and dissipate rapidly. We examine the consequences of this and discuss observables in Section 4.

An alternative way to address the issue of coronal heating through photospheric motions and current sheet formation has been developed by *Heyvaerts and Priest* [1984]: (see *Browning*, [1991] for references to subsequent work). Drawing on the postulate of *Taylor* [1974] that the lowest energy state in a weakly resistive plasma is a constant-α force-free magnetic field, they argued that a stressed coronal structure would ultimately relax to such a state. Estimates of the energy released are generally quite favorable, but specific observables of this model do not appear to have been computed at this time.

4. SIGNATURES OF A NANOFLARE HEATED CORONA

The hard X-ray data of *Lin et al.* [1984] indicated that energy could be released in elements of as small as 10^{27} ergs, which was named a mircoflare in recognition of the fact that it was approximately 10^6 times smaller than the largest solar flare. These data showed intriguing evidence for finer scale structure, which led *Parker* [1988] to propose that microflares were comprised of many thousand yet smaller events, with energy $\approx 10^{24}$ ergs, the nanoflare. Each of these nanoflares would correspond to the relaxation of a current sheet formed either spontaneously [*Parker*, 1972] or due to a turbulent cascade [*Van Ballegooijen*, 1986]. Parker cited additional evidence for such small elements of energy release in the data of *Porter et al.*, [1984] from the Ultraviolet Spectrometer and Polarimeter (UVSP) instrument on SMM. *Porter et al.*, [1984] noted significant variations in the light curves of transition region lines (such as $SiIV$) on timescales of order a minute or less. Since lines such as $SiIV$ and CIV are very sentitive diagnostics of the transport of energy from corona to photosphere, Parker argues that these fluctuations were evidence of elemental bursts of energy release on the scale of a nanoflare. Other evidence for the presence of nanoflares is slender, and indeed arguments have been made that, if the coronal magnetic energy is released in elemental bursts, there is not enough power at around 10^{24} ergs to heat the corona. *Hudson* [1991] and *Lee et al.*, [1993] have extrapolated X-ray data obtained from a variety of sources, from the observed regime ($> 10^{27}$ ergs) to the nanoflare regime, and show that the amount of energy in the nanoflare domain is far too small. However, at least the *Lee et al.* [1993] analysis has the problem that one of their free parameters is the ratio of energy going into hard X-rays (probably through energetic electrons) and that going into plasma heating. It is widely believed [*Emslie*, 1989] that a significant fraction of the energy in flares appears as hard X-rays. It seems inevitable that a much smaller fraction of the energy of a nanoflare goes into hard X-rays, so that the analysis of Lee et al. cannot be used to rule out the nanoflare hypothesis.

4.1 Emission Measure and Filling Factors

To compute the signatures of the nanoflare model, we assume that the large scale (observed) coronal structures are comprised of many hundred small, elemental flux tubes. Each flux tube can be heated randomly by nanoflares, such that the total coronal energy loss requirement is satisfied. If the elemental flux tubes have a cross-section A_h, then a nanoflare will occur somewhere along this flux tube on average every τ_{nano} seconds, where

$$\tau_{nano} = \frac{Q}{A_h \Lambda}, \qquad (1)$$

Q is the average nanoflare energy and Λ is the coronal energy loss in units of ergs cm^{-2} sec^{-1}. It is likely that the timescale for energy release is of order a few seconds to a minute [*Parker*, 1988], a result easily seen from the theory of rapid reconnection [e.g. *Forbes and Priest*, 1987]. The principal observed signature from such a system arises from the cooling of these impulsively heated loops.

We have developed a model to study the cooling of many hundred elemental loops. Since the coronal magnetic field is strong (the plasma $\beta = 8\pi p/B^2$ is $\ll 1$), and cross-field thermal conduction is negligible [*Priest*, 1982], each elemental flux loop acts as an isolated mini-atmosphere. The problem then reduces to following the thermal evolution along the magnetic field lines of each loop in isolation. This is governed by the following energy balance equation:

$$\frac{\partial p}{\partial t} + V_s \frac{\partial p}{\partial s} + \frac{5}{3} p \frac{\partial V_s}{\partial s} = \frac{2}{3} \left(\frac{\partial}{\partial s} \left(\kappa_0 T^{5/2} \frac{\partial T}{\partial s} \right) - n^2 R \right), \quad (2)$$

where the first three terms represent the familiar adiabatic expansion, the first term on the right is field aligned thermal conduction (κ_0 is the coefficient of conductivity, taken as 10^{-6} in c.g.s units) and the second term is an optically thin radiative loss function. In (2), s is the coordinate along the magnetic field line and the ratio of specific heats has been taken to be 5/3. In general, we approximate R as a piecewise continuous function of temperature: $R = \chi T^\alpha$ and use new estimates of R due to *Cook et al.*, [1989]:

$$R = \begin{cases} 2 \times 10^{-23}, & T > 10^{6.5} \\ 3.5 \times 10^{-7} T^{-2.5}, & 10^6 < T < 10^{6.5} \\ 3.5 \times 10^{-22}, & 10^5 < T < 10^6 \\ 1.1 \times 10^{-27} T^{1.1}, & 10^4 < T < 10^5. \end{cases}$$

This function differs in important ways from those popularly in use at this time [e.g. *Rosner et al.*, 1978] in that the radiation at the peak of the curve (near 10^5K) is reduced significantly. These differences are due to revised estimates of the coronal abundances of heavy elements and further modifications of R are to be expected as our knowledge of these abundances is refined further.

In principle, one would like to solve (2) numerically for each elemental flux tube, as has been done by others [e.g. *Doschek et al.*, 1982; *Serio et al.*, 1992], but a quick estimate of the computing requirements to follow hundreds of loops for several hours indicates that this is not feasible at the present time. Instead, we are forced to use analytic solutions. A dimensional analysis of (2) shows that a loop will cool by conduction (radiation) on a timescales τ_c (τ_r) respectively, where:

$$\tau_c \approx 2.69 \times 10^{-10} \frac{nL^2}{T^{5/2}}, \quad \tau_r \approx \frac{3kT}{nR}, \qquad (3)$$

where $2L$ is the total loop length. For typical solar parameters, it is readily seen that loops with high (low) temperatures and low (high) densities cool predominantely by conduction (radiation). We thus argue that the loop cooling can be modelled in two distinct stages: immediately after impulsive nanoflare heating, the loop cools solely by conduction and, below some critical temperature, it cools solely by radiation. At no time are conduction and radiation operating simultaneously. The high powers of temperature that arise in (3) suggest that the transition from purely conductive to purely radiative cooling will be a rapid one. In this analysis, we do not consider the spatial structure of the loop temperature and density. The quantities T and n should then be viewed as averages over the entire loop length.

By treating the cooling processes separately, we can make use of analytic solutions of (2). In the conductive regime, it is assumed that the loop cools by subsonic evaporative cooling [*Antiochos and Sturrock*, 1978]. Here, the coronal heat flux cannot be radiated away by the chromosphere and transition region, which responds by driving ("evaporating") material into the corona. If the downward heat flux is balanced by an upward enthalpy flux, the thermal energy is constant, and the loop density is inversely proportional to the temperature, which satisfies:

$$T(t) = T_0 \left(1 + \frac{t}{\tau_{c0}}\right)^{-2/7}, \qquad (4)$$

where T_0 is the initial loop temperature and $\tau_{c0} = \tau_c(T = T_0)$. In the radiative regime, the loop temperature is given by:

$$T(t) = T_0 \left[1 - \frac{3}{2}\left(\frac{1}{2} - \alpha\right)\frac{t}{\tau_{r0}}\right]^{\frac{1}{2-\alpha}}, \qquad (5)$$

where T_0 is now the temperature at the start of the radiative phase and $\tau_{r0} = \tau_r(T = T_0)$. We have assumed that matter drains from the loop in a way such that $T \propto n^2$. This scaling was found using numerical simulations by *Serio et al.*, [1992] and (5) turns out also to be one of the family of analytic solutions for radiative, draining cooling found by *Antiochos* [1980]. We allow the loop to cool radiatively to 10^4 K, at which point matter is assumed to drain from the loop, leaving a cold, evacuated region. This draining will arise because, for a 10^4 K plasma, the gravitational scale height is approximately 600 km, much less that the characteristic loop height. Therefore, a significant body of cold plasma cannot be supported against gravity.

To follow the behavior of such an ensemble of loops, we choose the overall dimensions of the system and the number of loops to be studied. This gives A_h, and specification of Λ, Q and the time the loops are to be followed then determines how many nanoflares are needed. In this paper, we will look at a surface area of dimensions $2 \times 10^8 \times 2 \times 10^8$ cm^2, 500 elemental loops, $\Lambda = 10^7$ ergs cm^{-2} sec^{-1} and Q randomly distributed between 5×10^{23} and 2×10^{24} ergs. This implies $A_h = 8 \times 10^{13}$ cm^2, much smaller that 1 arcsecond^2. The loop system will be followed for 10^4 seconds. A description of how the results depend on these parameters can be found elsewhere (C94).

The loop temperature immediately after a nanoflare is given by:

$$T = T_0 + 1.2 \times 10^{15} \frac{Q}{n_0 A_h L}, \qquad (6)$$

where it is assumed that all of the nanoflare energy goes into plasma heating. T_0 and n_0 are the loop temperature and density prior to the nanoflare, so that the temperature at any given time can depend on the previous history of the loop. One problem with (6) is that it often predicts initial loop temperatures at which the conditions required for (4) to be valid are not satisfied, i.e. the upflows are not subsonic. This is mostly a problem when the loops are allowed to drain before being reheated. Unfortunately analytic solutions of (2) with supersonic flows are not known. We therefore assume that during this early phase, the thermal energy in the loop remains constant, and any supersonic flows are thermalized. The initial loop temperature is then given by

the temperature at which subsonic flows become valid: this is typically $> 5 \times 10^6$ K (see *Antiochos and Sturrock*, 1978 and C94 for details).

Figure 1 shows results from a model of a nanoflare heated corona. The top two panels show the distribution of loop temperatures and densities for the last 5000 seconds of the simulation. These have been obtained by binning the temperatures and densities every 20 timesteps: the vertical axes are thus arbitrary in each case. As noted above, we follow 500 loops, but in distinction from C94, we allow them to have a variety of different values of L between 10^9 and 3×10^9 cm. This more realistically models part of an active region, where loops of many different lengths are seen. The coronal temperature distribution peaks at $\approx 3 \times 10^6$ K, with a fairly broad spread on either side. The density distribution has a broad plateau from $\approx 6 \times 10^9$ to 3×10^{10} cm^{-3}. Both of these are around the commonly accepted values for the active region corona.

The lower panel shows the result from binning n^2 as a function of temperature. Since the cross-sectional area of the loops are all the same, this is just the line of sight emission measure discussed in Section 2. The curve has two distinct parts, a gentle increase with temperature up to 10^6 K and a steeper part from 10^6 up to the maximum of the distribution. These can be shown to be power laws, scaling as $T^{1.5}$ and $T^{4.75}$ respectively. It is interesting to note that incorporating many different loop lengths does not change these slopes from the case when all the loops have the same length (C94). The part of the curve with $T > 10^5$ K is in agreement with existing EUV observations [e.g. *Jordan*, 1980]. Below 10^5 K, we are unable to reproduce the increase of the emission measure with decreasing temperature (see Section 2 for a discussion of this issue). The upper (steeper) part of the curve has not to our knowledge been seen in active regions, so is a prediction of the present model. Such steep slopes have been seen in the the decay phase of flares [*Underwood et al.*, 1978]. The similarity between the computed nanoflare emission measure and the observed flare data is perhaps suggestive of a continuity of physical processes from nanoflares to flares. The slopes of these curves can be understood by recognising that the emission measure will scale as $n^2 \tau_r$. Using our form of R, scalings of $T^{1.5}$ (T^4) are to be expected below (above) 10^6 K. The former is in agreement with the simulation, but the latter less so, due mostly to the presence of loops that are still cooling conductively below the temperature maximum.

We have computed the filling factor (ϕ) in this case to be 0.67, where ϕ is defined as the ratio of the number of loops at temperatures above 10^5 K to the total loop number. (The lower cut-off here is not terribly crucial, since Figure 1a shows that most of the loops are at temperatures greater than 10^6 K). This is an important result, since it implies that for a corona where the energy release scale is much less than 1 arcsecond (and so unresolvable), the corona is largely radiating in observable wavebands.

It is possible to derive analytic estimates for ϕ by calculating the ratio of τ_{nano} to the total loop cooling time, provided some power close to unity. Experimental determination of ϕ could prove useful in constraining this ratio, since L and Λ can be readily determined from observations. Note that it is unlikely that A_h and Q can be varied independent of each other. Q can be related to the free magnetic energy in the energy release volume by:

$$Q \approx \frac{B_t^2}{8\pi} A_h L_h , \qquad (8)$$

where L_h is the length along the loop axis over which energy is released and B_t is the strength of the coronal field component lying perpendicular to the loop axis (where the notation of *Parker*, 1988 is used). If Q is held fixed while A_h is decreased, B_t becomes unrealistically big at some point, so that the energy in the nanoflare must be decreased.

4.2 Spectral Line Broadening and Doppler Shifts

As was noted in Section 2, an important diagnostic of coronal heating is the presence of broadening or Doppler shifts in spectral lines. Line broadening could be a signature of a background level of turbulence or Alfven waves, or alternatively it could arise from many small, unresolved directed plasma jets. The presence of Doppler shifts (which have not been observed at this time in active region X-ray lines) indicates bulk plasma flows. Coronal heating by small scale reconnection (or nanoflares) has the potential to produce bulk plasma flows in two ways. First, it is well known that a characteristic property of reconnection is a pair of plasma jets directed away fron the reconnection line. These jets have speeds that are typically of order the Alfven speed, based on the reconnecting field component [*Petschek*, 1964; *Forbes and Priest*, 1987]. Secondly, local plasma heating will generate a pressure enhancement that will relax by trying to expand along the local magnetic field at a fraction of the local sound speed. The experimental signatures of these mechanisms are dependent on the angle made by the magnetic field with the observer's line of sight. For example, if the dominant flow is a field-aligned expansion, and the apex of a loop system is being observed from directly above, the dominant flow will be perpendicular to the line of sight.

We have examined the spectral signatures arising from many such small jets for a model twisted magnetic field geometry. The flux loop is chosen to be cylindrical, so that the magnetic field is of the form $\mathbf{B} = (B_r, B_\theta, B_z)$, with the z axis lying along the loop, which is assumed to lie in the east-west direction. The elemental loops discussed above are assumed to be embedded in this larger scale field. The magnetic field satisfies:

$$\mathbf{B} = B_0 \left(0, \frac{r/a}{1+(r/a)^2}, \sqrt{\frac{B_{z0}^2}{B_0^2} + \frac{1}{(1+(r/a)^2)^2}} \right), \qquad (9)$$

where B_{z0} is a constant magnetic field along the axis of the loop and a is a characteristic width of the loop. This field was originally proposed as a model for loops by *Gold and Hoyle* [1960]. The loop is assumed to form a semicircle rising

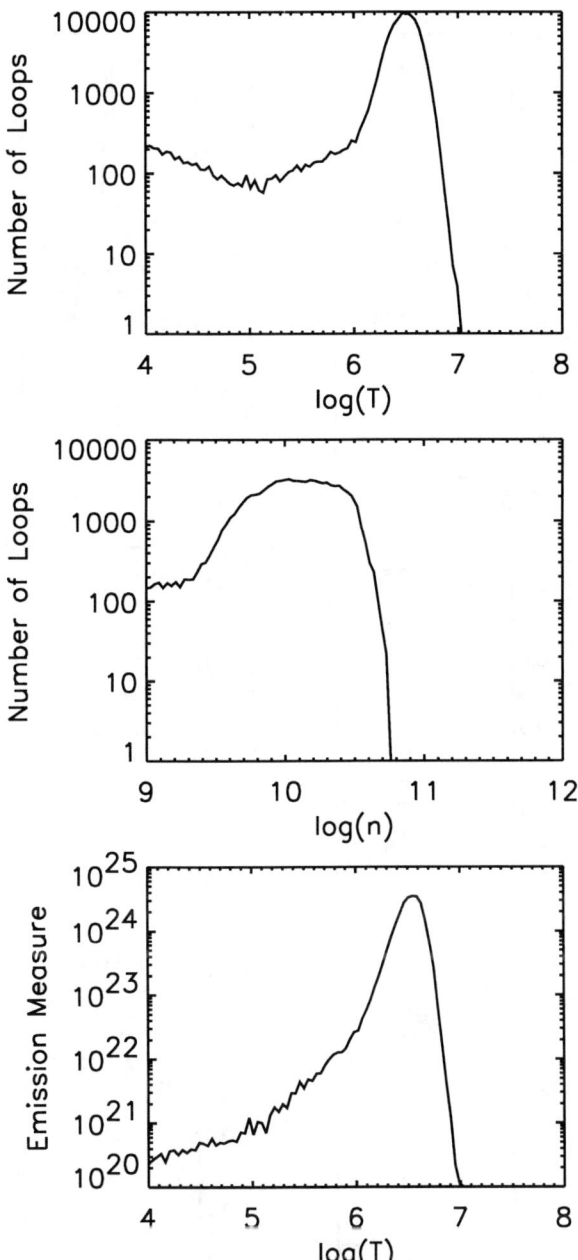

Fig. 1. From top to bottom, the distribution of loop temperatures and densities and the emission measure as a function of temperature for a coronal model consisting of 500 loops that are randomly heated and then allowed to cool. The vertical axis is arbitrary in each panel.

the radiative losses can be modelled by a single power law. Upon use of (4) and (5), it is readily shown (C94) that

$$\phi = CL\Lambda \left(\frac{A_h}{Q}\right)^{\frac{7}{m}} \left[\frac{2.5 \times 10^{-96} (3.08 \times 10^6)^\alpha}{\chi^{7/2}} \right]^{\frac{1}{m}} , \qquad (7)$$

where $m = 11/2 - \alpha$ and $C = (2-\alpha)/(1-\alpha)$. For most reasonable values of α, ϕ scales with the ratio A_h/Q by above the photosphere: although (9) is clearly not an exact equilibrium solution in this geometry, this simple model does reveal the essential physics of the problem. For reconnection jets, the velocity satisfies:

$$\mathbf{V} = V_A \left(0, \frac{B_\theta}{B}, \frac{\sqrt{B_z^2 - B_{z0}^2}}{B}\right) \quad (10)$$

and for field-aligned flows:

$$\mathbf{V} = C_s \left(0, \frac{B_\theta}{B}, \frac{B_z}{B}\right), \quad (11)$$

where C_s (V_A) are the sound and Alfven speeds respectively. The spectral line profiles due to these flows have been computed for the $OVII$ line, which has a peak formation temperature at around $2 \times 10^6 K$ and wavelength of 21.6Å. This temperature is probably rather smaller than that at which the jets are formed, but the object here is to demonstrate the method rather than to compute specific results. Note also that the spectra do not include any stationary component that would arise from plasma in loop where the flows had been thermalized. This aspect is discussed further below.

Figure 2 shows the spectra arising from 2000 reconnection jets for the case of the loop footpoints (left hand column: B_z field directed along the line of sight) and the loop apex (right hand column: B_z field normal to the line of sight). We choose $B_{z0}/B_0 = 4$ and the maximum value of r to be $2a$. Five different values of the jet speed are shown and the jets have been distributed randomly throughout the loop. The Alfven speed in the reconnecting plane is chosen as 10^3 km/sec. Figure 3 shows the corresponding spectra generated by field-aligned flows. The temperature of the jets have been distributed randomly between 5×10^5 and $3 \times 10^6 K$. The four different cases each show different spectra as the plasma velocities get larger. For the case of reconnection jets, the spectra develop an larger line-broadening as the velocities increase. The differences between the loop apex and footpoint cases for large velocities are quantatative and depend to some degree on the chosen magnetic field model. However, the wavelength shift ($\Delta\lambda/\lambda \approx 5 \times 10^{-4}$) corresponds to velocities of order 150 km/sec, well in excess of those observed by SMM [*Saba and Strong*, 1991]. The footpoint to apex variations for the case of field-aligned flows are of a qualitative nature. In any twisted loop structure with a dominant axial (B_z) field, field-aligned flows at the loop apex will lead to a general line broadening, since B_θ will vanish at the loop center, reach a maximum and then vanish again at larger distances. Thus, the field-aligned flows in the line of sight direction will be spread out over a range of values from zero to some maximum. This leads to a systematic line broadening, with values that compare reasonably well with the SMM data. However, at the footpoints, clear Doppler shifts will arise due to the dominant field in the line of sight direction.

A problem with interpreting these results is that the thermalization (or relaxation) time for the flows is not known. For example, the reconnection jets are likely to be seriously

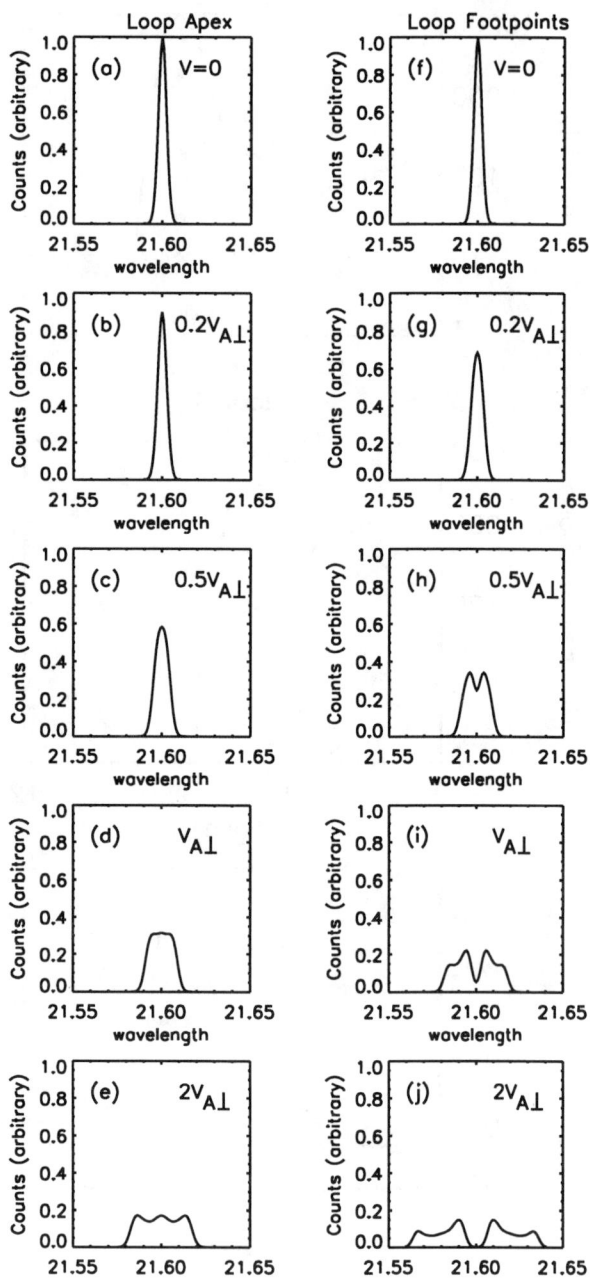

Fig. 2. The spectral line profiles of $OVII$ for reconnection jets with a number of jet speeds shown. The left (right) columns show the results at the loop apex (footpoints).

impeded by the neighboring magnetic field, and may thermalize rather rapidly, as was suggested by *Forbes* [1986]. The lifetime of the flow is then likely to be a few $\sqrt{A_h}/V_{A\perp}$. For $A_h = 10^{14}$ km^2 and $V_{A\perp} = 10^8$ km sec^{-1}, this is less that one second. In this case, the lifetime of the jet will be given by the nanoflare lifetime, perhaps 10 - 20 seconds [*Parker*, 1988]. The ratio of the line-broadened to stationary

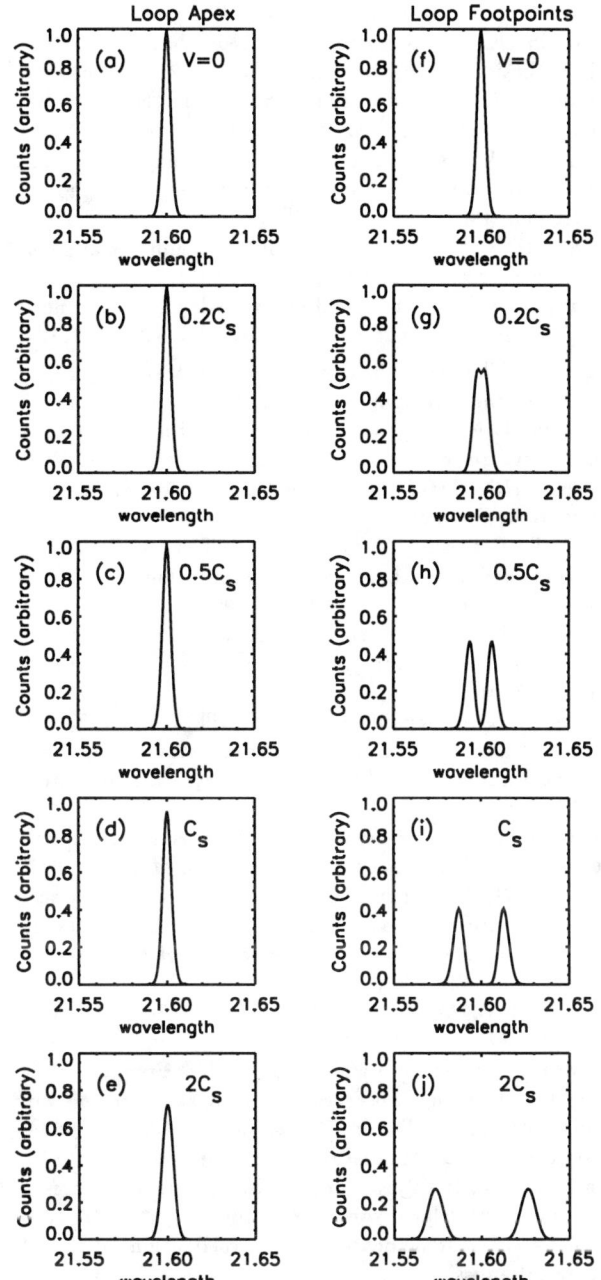

Fig. 3. The spectral line profiles of $OVII$ for field-aligned expansions. The left (right) columns show the results at the loop apex (footpoints).

Similar considerations hold for the field-aligned expansions. A pressure enhancement can relax by expansion or thermal conduction. The conductive relaxation time is given by (4), which for typical parameters ($L = 2 \times 10^9$, $n = 5 \times 10^9$ and $T = 3 \times 10^6$) is roughly 350 seconds. The expansion time is of order L/C_s, where C_s is the sound speed, typically 100 seconds. In this case, the line-broadened (or Doppler shifted) component could comprise a significant fraction of the total count rate. New data from SOHO is badly needed to sort these issues out.

5. CONCLUSIONS

While upcoming missions such as SOHO and HESP have the potential to shed more light on the coronal heating problem, we are in all likelyhood many years away from getting the observations required to settle the issue in a definitive manner. The small dissipation scales in the solar atmosphere make it difficult to see the actual energy release sites themselves, so that we are reduced to attempting to see whether the different proposed mechanisms lead to any unique coronal radiative signatures, or quantities that can be deduced from these signatures. It is therefore incumbent on theorists to compute signatures that are of relevance to present and future observations. The present paper has given some very simple examples of how such quantities can be evaluated. Many of the coronal heating mechanisms do not involve steady dissipation over a large volume, but require many small, localized dissipation sites scattered randomly throughout the corona. Unfortunately the only way at present that we can compute observables from such sites is to parameterize them in a simple way, and convolve the thermal output from them into a unified signal. The model for a nanoflare heated corona presented here adopts this approach, and makes some observational predictions. Future work on this model could include a parameterization of electron acceleration in current sheets leading to computation of hard X-ray and radio signatures. While it is imperative that calculations be continued to understand the basic physics of individual current sheets, resonance absorption region etc., such calculations may not lead to useful observables. We believe that the present approach hold out great hope in this regard.

Acknowledgements. This work was supported by ONR. I am grateful to many of the attendees of the 1993 Yosemite meeting for stimulating conversations and to Jack Zirker and Jim Klimchuk for reading a preliminary version of the manuscript.

REFERENCES

Amendt, P. and G. Benford, Solar coronal loop heating by cross-field wave transport, *Astrophys. J.*, *341*, 1082, 1989.

Antiochos, S. K., Radiative-dominated cooling of the flare corona and transition region, *Astrophys. J.*, *241*, 385, 1980.

components of the emission will then be approximately the ratio of the jet lifetime to the plasma cooling time, where the latter is of order 10^3 seconds (C94). The line-broadened component will then comprise a very small number of the total counts, and in particular would not be detectable by SMM.

Antiochos, S. K., The Topology of Force-free Magnetic Fields and its Implications for Coronal Activity, *Astrophys. J.*, *312*, 886, 1987.

Antiochos, S. K. and G. Noci, The structure of the static corona and transition region, *Astrophys. J.*, *301*, 440, 1986.

Antiochos, S. K. and P. A. Sturrock, Evaporative cooling of flare plasmas, *Astrophys. J*, *220*, 1137, 1978.

Athay, R. G., The relationship between R.M.S. doppler velocities and temperature in the solar transition region, *Solar Phys.*, *116*, 223, 1988.

Athay, R. G. and O. R. White, Chromospheric observations observed with OSO-8. IV. Power and phase spectra for CIV, *Astrophys. J.*, *229*, 1147, 1979.

Benford, G., Turbulent resistive heating of solar coronal arches, *Astrophys. J.*, *269*, 690, 1983.

Benz, A. O., Acceleration and energization by currents and electric fields, *Solar Phys.*, *111*, 1, 1987.

Browning, P. K., Mechanisms of solar coronal heating, *Plasma Phys. and Controlled Fusion*, *33*, 539, 1991.

Cargill, P. J., Some implications of the nanoflare concept, *Astrophys. J.*, , Feb. 10, 1994 issue (in press), 1994.

Cook, J. W., C.-C. Cheng, V. L. Jacobs and S. K. Antiochos, Effect of coronal elemental abundances on the radiative loss function, *Astrophys. J.*, *338*, 1176, 1989.

Davila, J., Heating of the solar corona by the resonant absorption of Alfven waves, *Astrophys, J.*, *317*, 514, 1987.

Dere, K. P., The XUV structure of solar active regions, *Solar Phys.*, *75*, 189, 1982.

Dere, K. P., H. E. Mason, K. G. Widing and A. K. Bhatia, XUV electron density diagnostics for solar flares, *Astrophys. J. (Supp)*, *40*, 341, 1979.

Doschek, G. A., U. Feldman and J. D. Bohlin, Doppler wavelength shifts of transition zone lines measured in Skylab solar spectra, *Astrophys. J. (Letters)*, *205*, L177, 1976.

Doschek, G. A., J. P. Boris, C.-C. Cheng, J. T. Mariska and E. S. Oran, A numerical simulation of cooling coronal plasma, *Astrophys. J.*, *258*, 373, 1982.

Emslie, A. G., Models of Flaring Loops, *Solar Phys.*, *121*, 105, 1989.

Forbes, T. G., Fast shock formation in line-ties reconnection models of solar flares, *Astrophys. J.*, *305*, 553, 1986.

Forbes, T. G. and E. R. Priest, A comparison of analtyical and numerical models for steadily driven magnetic reconnection, *Revs. Geophys.*, *25*, 1583, 1987.

Gabriel, A. H., A magnetic model of the solar transition region, *Phil. Trans. Royal Soc. (London)*, *A281*, 339, 1976.

Ganguli, G., Y. C. Lee, P. J. Palmadesso and S. L. Ossakow, Oscillations in a plasma with parallel currents and transverse velocity shears, in *Physics of Space Plasmas*, Scientific Publishers, Cambridge, MA, p. 231, 1988.

Gold, T. and F. Hoyle, On the origin of solar flares, *Mon. Notices of Royal Astron. Soc.*, *120*, 89, 1960.

Harrison, R. A., The coronal diagnostic spectrometer for SOHO, *Rutherford Appleton Lab report 91-0005*, 1991.

Heyvaerts, J. and E. R. Priest, Coronal heating by reconnection in DC current systems. A theory based on Taylor's hypothesis, *Astron. and Astrophys.*, *137*, 63, 1984.

Hinata, S., The role of turbulent heating in the solar atmosphere, *Astrophys. J.*, *232*, 915, 1979.

Hollweg, J. V., Resonance absorption of magnetohydrodynamic surface waves: physical discussion, *Astrophys. J.*, *312*, 880, 1987.

Hudson, H. S., Solar flares, microflares, nanoflares and coronal heating, *Solar Phys.*, *133*, 357, 1991.

Jordan, C., The structure and energy balance of solar active regions, *Phil. Trans. Royal Soc. (London)*, *A281*, 391, 1976.

Jordan, C., The energy balance of the solar transition region, *Astron. and Astrophys.*, *86*, 355, 1980.

Klimchuk, J. A., Large-scale structure and dynamics of solar active regions observed in the far ultraviolet, Ph.D. thesis, *NCAR/CT-96*, National Center for Atmospheric Research, Boulder, Colorado, 1986.

Lee, T. T., Petrosian, V. and J. M. McTiernan, The distribution of flare parameters and implications for coronal heating, *Astrophys. J.*, *412*, 401, 1993.

Lin, R. P., R. A. Schwartz, S. R. Kane, R. M. Pelling and K. Hurley, Solar hard X-ray microflares, *Astrophys. J.*, *283*, 421, 1984.

Offman, L., R. S. Steinolfson and J. Davila, Coronal Heating by the Resonant Absorption of Alfven Waves: The Effect of the Viscous Stress Tensor, *Astrophys. J.* (submitted), 1993.

Papadopoulos, K., A review of anomalous resistivity for the ionosphere, *Revs. Geophys. Space Phys.*, *15*, 113, 1977.

Parker, E. N., Topological dissipation and the small scale fields in turbulent gases, *Astrophys. J.*, *174*, 499, 1972.

Parker, E. N., Nanoflares and the solar X-ray corona, *Astrophys. J.*, *330*, 474, 1988.

Petschek, H. E., Magnetic field annihilation, in *The Physics of Solar Flares*, ed. W. Hess, NASA SP-50, p.425, 1964.

Porter, J. G., J. Toomre and K. B. Gebbie, Frequent ultraviolet brightenings observed in a solar active region with Solar Maximum Mission, *Astrophys. J.*, *283*, 879, 1984.

Priest, E. R., *Solar Magnetohydrodynamics*, Reidel, 1982.

Rosner, R., W. H. Tucker and G. S. Vaiana, Dynamics of the quiescent solar corona, *Astrophys. J.*, *220*, 643, 1978.

Roumeliotis, G., Joule heating as an explianation for the differential emission measure structure and systematic redshifts in the Sun's lower transition region, *Astrophys. J.*, *379*, 392, 1991.

Saba, J. L. R. and K. T. Strong, Coronal dynamics of a quiescent active region, *Astrophys. J.*, *375*, 789, 1991.

Scudder, J. D., On the cause of temperature change in inhomogeneous low-density astrophysical plasmas, *Astrophys. J.*, *398*, 299, 1992a.

Scudder, J. D., Why all stars should possess circumstellar temperature inversions, *Astrophys. J.*, *398*, 319, 1992b.

Serio, S., F. Reale, J. Jackimiec, B. Sylwester and J. Sylwester, Dynamics of flaring loops 1. Thermodynamic scaling laws, *Astron. and Astrophys.*, *241*, 197, 1991.

Spicer, D. S. and P. MacNiece, *Electromechanical coupling of the solar atmosphere*, AIP, 1992.

Steinolfson, R. S. and J. Davila, Coronal heating by the resonant absorption of Alfven waves: Importance of the global mode and scaling laws, *Astrophys. J.*, in press, 1993.

Sturrock, P. A., The role of eruptions in solar flares, *Solar Phys., 121*. 387, 1989.

Sturrock, P. A., W. W. Dixon, J. A. Klimchuk and S. K Antiochos, Episodic coronal heating, *Astrophys. L. (Letters), 356*, L31, 1990.

Taylor, J. B., Relaxation of torroidal plasma and generation of reverse magnetic fields, *Phys. Rev. Letters, 33*, 1139, 1974.

Ulmschneider, P., Rosner, R. and Priest, E. R., *Mechanisms of Chromospheric and Coronal Heating*, Springer-Verlag, 1991.

Underwood, J. H., S. K. Antiochos, U. Feldman and K. P. Dere, Evolution of the coronal and transition-zone plasma in a compact flare: the event of 1973 August 9, *Astrophys. J., 224*, 1017, 1978.

Van Ballegooijen, A. A., Electric current sheets in the corona and the existence of magnetostatic equilibrium, *Astrophys. J., 298*, 421, 1985.

Van Ballegooijen, A. A., Cascade of magnetic energy as a mechanism for coronal heating, *Astrophys. J., 311*, 1001.

Withbroe, G. L. and R. W. Noyes, Mass and energy flow in the solar chromosphere and corona, *Ann. Rev. Astron. Astrophys., 15*, 363, 1977.

Woods, D. T., T. E. Holzer and K. B. MacGregor, Lower solar chromosphere-corona transition region. 1. Theoretical models with small temperature gradients, *Astrophys. J., 355*, 298, 1990.

Zirker, J., Coronal Heating, *Solar Phys.*, (in press), 1993.

Zweibel, E. G. and H. Li, The formation of current sheets in the solar atmosphere, *Astrophys. J., 312*, 423, 1987.

Space-time Structure of Langmuir turbulence in the Lower Solar Corona

M.V. Goldman and D.L. Newman

Department of Astrophysical, Planetary and Atmospheric Sciences
Campus Box 391, University of Colorado, Boulder, Colorado, 80309

Langmuir turbulence driven by electron beams during Type III solar radio emissions above ≈100 MHz can arise at moderately-magnetized low altitudes in the solar corona. We have carried out numerical solutions of generalized magnetic "Zakharov" equations, which reveal that Langmuir wavepackets take the form of highly anisotropic real space "pancakes" when electron cyclotron and plasma frequencies are of the same order. Spatial collapse of the wavepackets is suppressed above a critical magnetic field. The anisotropy of Langmuir wavepackets may affect Type III emissions at high-frequenies.

1. INTRODUCTION

1.1 Langmuir waves in space plasmas

There is ample evidence that electron plasma (Langmuir) waves are commonly driven unstable by electron streams in a variety of space plasma environments. The electron streams may be associated with Type III emissions from the solar corona [Goldman, 1983; Newman, 1985; Goldman, 1989] and solar wind [Lin et al, 1981; Newman, 1985; Lin et al, 1986; Melrose and Goldman, 1987], created at planetary bow-shocks [Robinson and Newman, 1991, and references therein], or precipitating into Earth's auroral zone [Boehm, 1987; Ergun et al, 1991; Newman et al, 1993].

A challenging theoretical problem has been to determine how the Langmuir waves driven unstable by such streams saturate nonlinearly. This problem is of considerable importance to solar Type III emissions at the fundamental and 2nd harmonic, because such emission cannot be calculated without a knowledge of the turbulent Langmuir wave spectrum. It is of interest in the solar wind, planetary foreshocks and in Earth's auroral zone because the Langmuir waves have been measured directly, along with the electron streams, in these environments, so theoretical predictions can be directly compared with satellite and rocket data.

In this paper, we present a theoretical model for the spatial and temporal structure of Langmuir turbulence driven by electron beams which are presumed to be associated with Type III emission from the lower solar corona. We are particularly interested in the effects a typical active-region solar magnetic field in the range from 30-300 gauss, which, in our model, strongly influences the spatial structure of the turbulence. This, in turn may play a role in the paucity of observations of Type III bursts at frequencies much above 100 MhZ.

1.2 Parameters and degree of magnetization

Table 1 shows typical *measured* parameters associated with electron-beam-driven Langmuir turbulence in the solar wind, electron foreshock and lower auroral region, and the *assumed* parameters for the lower corona. The measured parameters characterize the (time-averaged) electron beam, and yield the mean Langmuir-wave energy for a given background magnetic field.

The ratio, r, of the electron cyclotron frequency, $\Omega_e \equiv eB/mc$, to the electron plasma frequency, $\omega_e \equiv (4\pi n e^2/m)^{1/2}$, is a critical parameter.

$$r \equiv \Omega_e/\omega_e \qquad (1)$$

TABLE 1. Parameters associated with Langmuir turbulence.

Where? ↓	Wave energy $|E_w|^2/8\pi n T_e$	e-beam n_b/n, v_b/v_e	Ω_e/ω_e
Solar Wind ($\approx$ 1 AU) $\lambda_e \approx$ 10 m, $f_p \approx$ 30 kHz	10^{-6} max	10^{-5}, 20	10^{-2}
Electron foreshock. $\lambda_e \approx$ 10 m, $f_p \approx$ 30 kHz	10^{-3} max	10^{-3}, 6	10^{-2}
Auroral region $\lambda_e \approx$ 3 cm, $f_p \approx$ 1 MHz	10^{-2} max	10^{-4}, 50	O(1)
Lower solar corona during Type III flare (present treatment) $n = 10^8\text{-}10^{10}$, B = 30-300g $\lambda_e \approx$ 0.2 cm, $f_p \geq$ 100 MHz		10^{-5}, 25	O(1)

In the solar wind and electron foreshock r<<1. At altitudes interrogated by rockets in the lower auroral zone [Boehm, 1987; Ergun et al, 1991; Newman et al, 1993], r can be of order 1. We will be interested in conditions during Type III burst in the lower solar corona for which r ≈1. We refer to such a regime as *moderately magnetized* for Langmuir waves (not to be confused with measures of magnetization for *particles,* such as the plasma β).

Figure 1 shows the magnetic fields and densities associated with certain space and astrophysical plasmas in relation to the *critical line* r = 1. Magnetic fields above the line are *strong* in relation to Langmuir waves, while magnetic fields below are *weak*. Both the lower solar corona and the lower auroral region of Earth's magnetosphere can serve as environments of moderate magnetization in terms of their effect on beam-driven Lanmguir turbulence.

From the top row in Table 1, we see that typical ratios of beam-to-background electron densities are 10^{-5} for Type III emission from the solar wind near 1 AU, and corresponding ratios of electron beam velocity, v_b to electon thermal velocity v_e are ≈ 20. The bottom row in Table 1 shows the parameters associated with our present model of Langmuir turbulence excited in the lower solar corona. Electron-beam parameter ratios are chosen to be of the same order as in the solar wind [Goldman, 1983]. The plasma density and solar magnetic field are chosen to assure

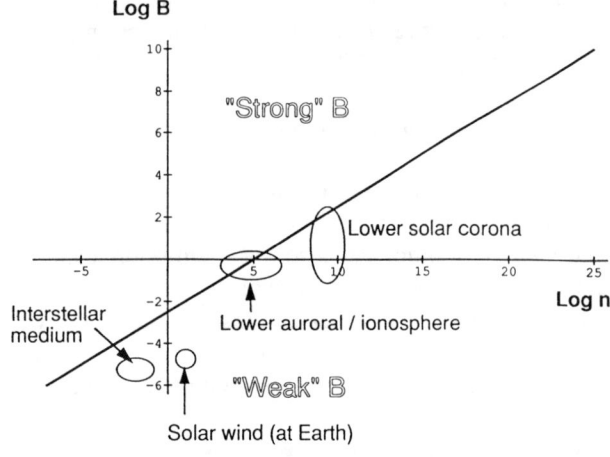

Fig. 1. Critical magnetic field versus density in space and astrophysical plasmas.

$r \approx 1$. The plasma frequency, $f_p = \omega_e/2\pi$ will typically be on the order of 100 MHz or more, and the Debye length, λ_e, a fraction of a centimeter.

2. THE THEORETICAL MODEL

Our model is a magnetic generalization of the Zakharov equations [Goldman, 1984]. Through nonlinear refraction and ponderomotive force, the Zakharov equations take into account the transfer of energy (wave-wave coupling) among plasma waves interacting with low-frequency density disturbances, such as ion-acoustic waves. The model neglects the reaction of the unstable waves back onto the driving electron beam, a process which is assumed to occur on a slower time scale [Goldman, 1984].

2.1 Zakharov equations

A heuristic derivation of the Zakharov equations is based on the linear frequency and growth/damping of the high-frequency Langmuir waves, $\omega(k) = \omega_k - iv_{ek}$. In regions of **k**-space which are not resonant with the electron beam, v_{ek}, is positive, and describes dissipation such as Landau damping. In *resonant* regions of **k**-space, v_{ek} is negative, and represents the excess of the growth rate of beam-unstable Langmuir waves over the damping rate.

We can express in **k**-space a *linear* wave eqn, for the *envelope*, E_k, of these longitudinal Langmuir waves, as:

$$i\mathbf{k}\cdot(i\partial_t - \omega_k + iv_{ek})\cdot\left(E_k e^{-i\omega_e t}\right) = 0$$

Assuming that the dominant *nonlinearity* is refraction by 2nd-order density deviations, $\delta n(\mathbf{r},t)$ [Goldman, 1984; Newman et al, 1993], it follows that the first Zakharov equation takes the following form in real space:

$$\nabla\cdot(i\partial_t - \delta\hat{\omega} + i\hat{v}_e - \omega_e\cdot\delta n/2n_0)\cdot E(\mathbf{r},t) = 0 \quad (2)$$

where n_0 is the uniform background density, and $\wedge$ indicates a nonlocal real-space operator transforming into k-space as, for example,

$$\delta\hat{\omega}E \rightarrow (\omega_k - \omega_e)\cdot E_k.$$

The $\delta\omega$ term therefore contains the description of wave dispersion.

The second Zakharov equation describes the response of the density deviation to the ponderomotive force of the Langmuir waves:

$$\left(\partial_t^2 + 2\cdot\hat{v}_i\partial_t - c_s^2\nabla^2\right)\cdot\delta n/2n_0 = c_s^2\nabla^2\frac{|E|^2}{32n_0\pi T_e} \quad (3)$$

The linear density response is governed by the ion-acoustic wave operator on the left side. Here, c_s is the ion-acoustic speed, v_{ik} is the ion-acoustic wave damping, and T_e is the background electron temperature.

2.2 Unmagnetized plasma limit.

In the unmagnetized limit, the $\delta\omega$-term in the first Zakharov equation describes thermal dispersion of the Lanmguir waves:

$$\delta\hat{\omega}E \rightarrow \frac{3\omega_e}{2}(k\lambda_e)^2\cdot E_k$$

Equations (2) and (3), with this ansatz, constitute the standard Zakharov equation model, which has been analyzed numerically in great detail in connection with the first two space plasma applications shown in Table 1.

An important physical process described by these *strong turbulence equations* is the spatial collapse of nonlinear Langmuir wavepackets [Goldman, 1984; Newman, 1985; Goldman, 1989]. Langmuir wave packets driven by the electron beam nucleate (Russell et al, 1988) in density cavities. Above a certain threshold, the ponderomotive force of the packet ejects a significant number of electrons from the cavity. Ions are also drawn out by the charge separation field. The cavity gets deeper, and nonlinearly refracts the Langmuir wave packet, which becomes spatially narrower and more intense.

This collapse process continues self-consistently until both the nonlinear wave packet, $E(\mathbf{r},t)$, and the density cavity, $\delta n(\mathbf{r},t)$, have self-focused down to small spatial dimensions on the order of tens of Debye lengths. At these scales there is efficient dissipation or "burn-out" of the Langmuir wave-packet, due to the Langmuir damping described by v_{ek}. This leaves the density cavity unsupported by ponderomotive force, allowing it to evolve hydrodynamically, with a lifetime determined by v_i.

The space-time history of beam-driven Langmuir turbulence in the solar wind and planetary foreshock regions is thought to exhibit such collapsing structures at the relatively smaller spatial scales [Goldman, 1983; Robinson and Newman, 1991]

2.3 Moderately magnetized turbulence model

Prompted by rocket data on Langmuir turbulence in the lower auroral zone [Boehm, 1987; Ergun, 1991], we have generalized the Zakharov equations to treat the moderately magnetized environment of that region. The reader is referred to [Newman et al, 1993] for further details specifically pertinent to the auroral zone measurements.

The final generalized equations are straightforward to describe:

The model again consists of equations (2) and (3), for the Langmuir wave envelope $\mathbf{E}(\mathbf{r},t)$ and density cavity, $\delta n(\mathbf{r},t)$. The new physics is contained entirely in the dispersion, $\omega_\mathbf{k}$, and damping, $\nu_{e\mathbf{k}}$ of the Langmuir wave in the magnetized plasma. Other high-frequency branches are neglected. Only the ion-acoustic low-frequency branch is retained, as it is believed to be the branch to which the Langmuir waves couple most strongly [Newman et al,

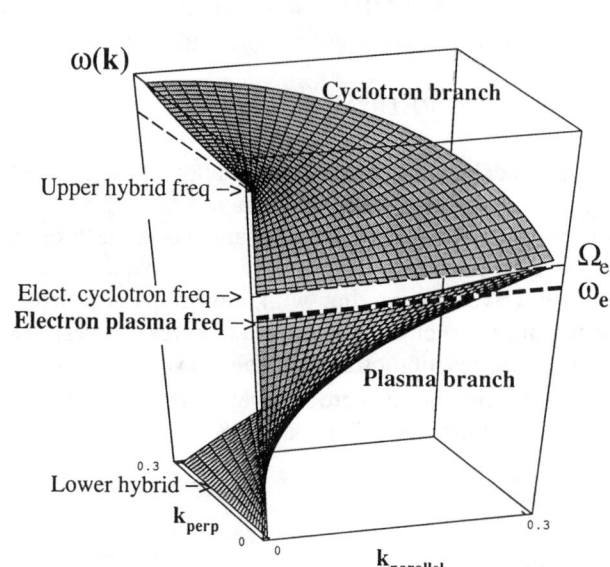

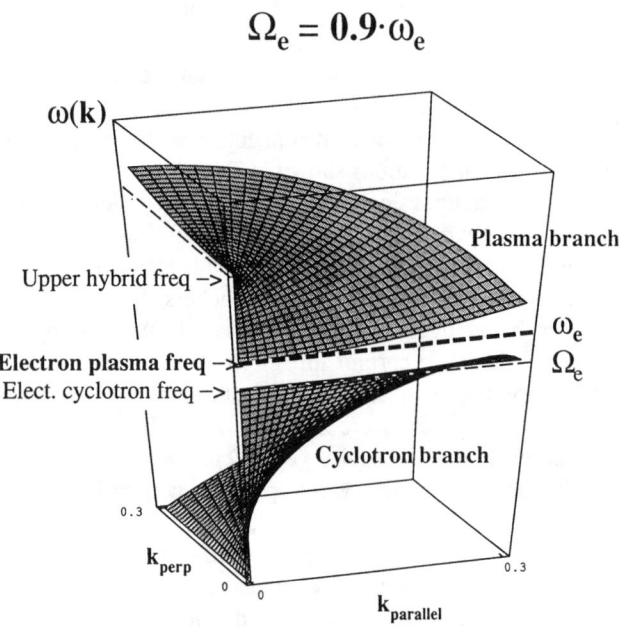

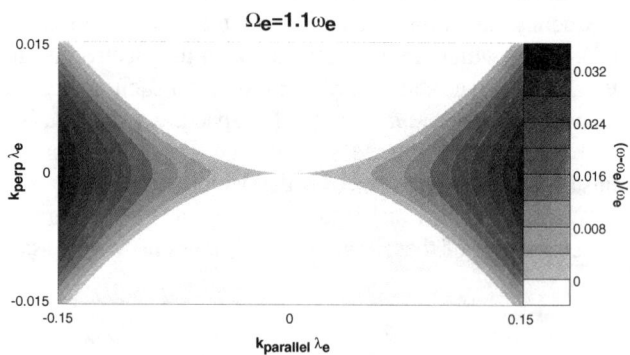

Fig. 3. Supercritical case: Langmuir wave dispersion, $\omega_\mathbf{k}$, for B above the critical value (r = 1.1). (a) Connectivity to other quasilongitudinal waves exhibited in perspective plot. (b) Contour map of contours of equal $\omega_\mathbf{k}$

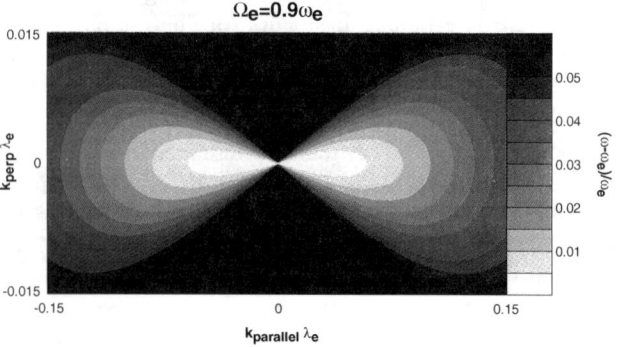

Fig. 2. Subcritical case: Langmuir wave dispersion, $\omega_\mathbf{k}$, for B below the critical value (r = 0.9). (a) Connectivity to other quasilongitudinal waves exhibited in perspective plot. (b) Contour map of contours of equal $\omega_\mathbf{k}$.

1993; Weatherall, 1980]. This branch is relatively unmodified by the magnetic field when $kc_s \gg \Omega_i$, except in a narrow angular region within $\sqrt{m_e/M_i}$ of perpendicular to **B**.

In Figs. 2 and 3 the dispersion, $\omega_\mathbf{k}$, is visualized as a surface over the $k_\parallel$-$k_\perp$ plane, where $k_\parallel$ is the component of **k** parallel to the magnetic field, and $k_\perp$ is the magnitude of the projection of **k** perpendicular to the magnetic field.

For the subcritical case, r = 0.9, (Fig. 2) the Langmuir branch connects to the upper-hybrid branch, and the contours of constant $\omega_\mathbf{k}$ are nested *closed* curves which shrink as $\omega_\mathbf{k}$ asymptotes down to ω_e. For the supercritical case, r = 1.1, (Fig. 3) the Langmuir branch connects to the low-frequency perpendicular modes, and the contours of constant ω_k are nested *open* curves which are spread to wider $k_\perp$ (for a given $k_\parallel$) as ω_k decreases to ω_e and below.

At $r = 1$, a topological change occurs in the dispersion relation, which will be significant for the numerical studies of turbulence in the lower solar corona.

The generalized Zakharov equations we have solved numerically for Langmuir turbulence in the lower auroral zone [Newman et al, 1993] and, in the present paper, for the lower solar corona, employ analytical expressions for the Langmuir dispersion, ω_k, in both the sub and supercrital limits [Newman et al, 1993]

2.4 Growth and damping rates

The expressions to be used for the growth/damping rates, ν_{ek}, will differ substantially for the lower corona, in comparison to expressions used for the lower auroral zone.

In the auroral zone, Langmuir waves out of resonance with the electron beam experience strong linear Landau damping due to the (measured) presence of scattered background electrons which accompany both auroral electrons and the (lower energy) beam which drives the Langmuir waves. These scattered background electrons have reflected upwards from lower altitudes (the D-region of the ionosphere).

In regions of negative ν_{ek}, the growth rate of beam-unstable waves was modified, in the case of the auroral zone, to simulate the nonlinear reaction of the resonant Langmuir waves back on the beam (leading to a saturation of the growth rate).

For the present treatment of the lower corona, we assume a gaussian-shaped beam-unstable region of k-space, centered about the beam-resonant wavenumber,

$$k_b = \omega_e/v_b,$$

where v_b is the beam velocity.

For the positive-damping regions of k-space we assume a power law of the following form:

$$\nu_{ek} = \nu_{ek}^+ \equiv 10 \cdot (k\lambda_e)^4 \omega_e \qquad (4a)$$

This damping can be regarded either as Landau damping from a power-law tail on the background electron distribution, or as a linear approximation to nonlinear transit time damping of collapsing Langmuir wavepackets. Such a model of collapsing wavepacket dissipation has been justified in the nonmagnetic strong Langmuir turbulence by comparing Zakharov simulations with 2-D particle-in-cell simulations [Newman et al, 1990].

In the negative-damping, beam-unstable region of k-space, ν_{ek} is the difference between positive damping, ν_{ek}^+, and the gaussian growth rate described earlier:

$$\nu_{ek} \equiv \nu_{ek}^+ - a \cdot \exp\left(-\frac{(k_\parallel - k_b)^2}{2\delta k_\parallel^2} + \frac{k_\perp^2}{2\delta k_\perp^2}\right). \qquad (4b)$$

3. LANGMUIR TURBULENCE IN THE MODERATELY-MAGNETIZED LOWER SOLAR CORONA.

3.1 Numerical parameters for lower corona

We chose lower solar corona parameters consistent with measured parameters associated with Type III emission in the solar wind (see Table 1) The beam is taken to have a velocity around 25 times thermal, $v_b \approx 25 \cdot v_e$, so that we take $k_b \lambda_e = 0.04$. The maximum growth rate of beam-resonant modes is set by fixing $a = 1.26 \cdot 10^{-4}$ in eqn (4b):

$$-\nu_{ek}|_{max} = 10^{-4} \cdot \omega_e.$$

The resonant region is characterized by the parameters

$$\delta k_\parallel = 0.31 \cdot k_b, \quad \delta k_\perp = 0.005 \cdot k_b.$$

Thus, the unstable region is confined to a narrow "tube" about k_b; there is not much spread in $k_\perp$ (see Fig. 8).

The electron dissipation is given by equation (4). The ion-acoustic dissipation is assumed $\nu_{ik} = 0.1 \cdot c_s k$, due Landau damping of ion-acoustic waves for a temperature ratio of $T_e/T_i = 4$.

Initial conditions consist of white noise in Langmuir waves and no initial density noise. Boundary conditions are periodic. The real-space grid consists of 256 points in x by 64 points in y, with maximum sizes of $x_{max} = 2,513 \cdot \lambda_e$ and $y_{max} = 40,212 \cdot \lambda_e$. The maximum wave numbers in k-space are $\lambda_e k_\parallel^{max} = 0.16$ and $\lambda_e k_\perp^{max} = 0.0025$.

Both sub and supercritical magnetic fields are assumed (**B** between 30 and 300 gauss in an active region). The parameter r is chosen to be 0.9 and 1.1, with appropriate analytic expressions for the dispersion, $\delta \omega$, in each case.

3.2 Numerical results in time and space.

Figure 4 shows the (asymptotic) time histories of Langmuir energy density, $|E|^2$, (solid lines) and *negative*

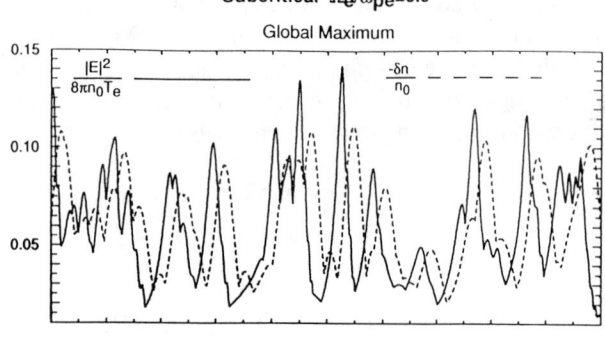

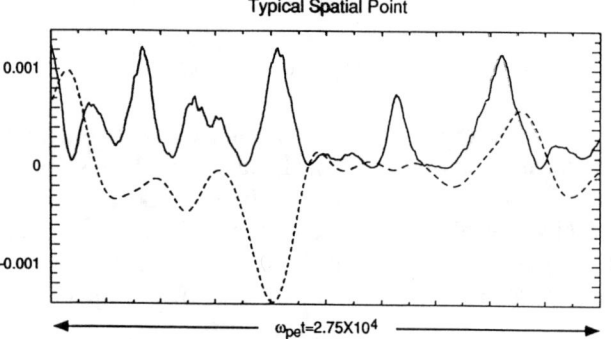

Fig. 4. Subcritical case: Time histories of Langmuir energy, $|E|^2$, (solid) and ion density deficit, $-\delta n$, (dashed) for $r = 0.9$, in the lower solar corona. (a) Global maximum. (b) At a random spatial location.

ion density, δn, (dashed lines) for the subcritical case $r = 0.9$. Both global maxima (within the entire spatial grid) and the time-dependence at an arbitrary fixed spatial point are shown. In Fig. 4a, we see the deeply-modulated time-dependence of $|E|^2_{max}$ which is indicative of the strong-turbulence phenomenon of spatial collapse. The peaks of the maximum density depression are seen to lag slightly behind the peaks of $|E|^2_{max}$. This is as expected, since after burnout of the Langmuir waves by electron dissipation, ion inertia continues to deepen the cavity, after which it relaxes on the ion time-scale. In Fig. 4b, the time-dependence at an arbitrary spatial point does not capture a collapse event (as evidenced by the absence of correlation between $|E|^2$ and $-\delta n$; peak values here are two orders of magnitude below global maximum values.

Figure 5, by contrast, shows the time-history of the turbulence for the supercritical case, $r = 1.1$. Here, values of the global maxima are lower and the temporal modulations of $|E|^2_{max}$ are shallower; there are no apparent correlations with $-\delta n_{max}$, suggesting that collapse has been suppressed. The behavior at a random local spatial point shows typical peak values of $|E|^2$ and δn not too different from the subcritical case, however, now typical local peak values are less than an order of magnitude below global maxima.

Time-averaged spatial snapshots confirm the interpretation that collapse is inhibited in the supercritical case. The left side of Fig. 6 shows a portion of the spatial structure of Langmuir turbulence, $|E|^2$, for the subcritical case, $r = 0.9$, which includes one collapsing wave packet. A more intense (global maximum) wavepacket sits elsewhere (outside the region selected, which is not the entire grid). It is evident that the wavepacket shown is highly elongated into a shape corresponding to an extremely wide and flat pancake in three-dimensions. Note the aspect ratio in Fig. 6 is not true — the perpendicular dimension has been compressed by a factor of 4.

The right side of Fig. 6 shows a corresponding spatial snapshot of $|E|^2$, for the supercritical case, $r = 1.1$. The pancake shape of localized packets is less extreme, but still highly anisotropic; the main distinguishing feature is the significantly larger volume occupied by wavepackets with intensity above that of the minimum contour; these wavepackets are *not* collapsing spatially. The strong

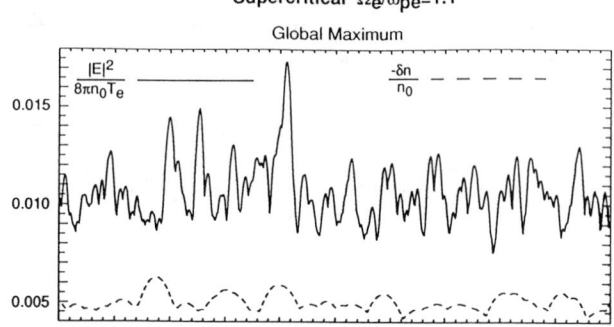

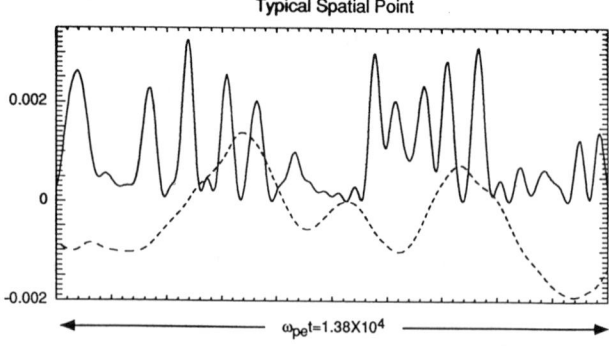

Fig. 5. Supercritical case: Time histories of Langmuir energy, $|E|^2$, (solid) and ion density deficit, $-\delta n$, (dashed) for $r = 1.1$, in the lower solar corona (a) Global maximum. (b) At a fixed random spatial location.

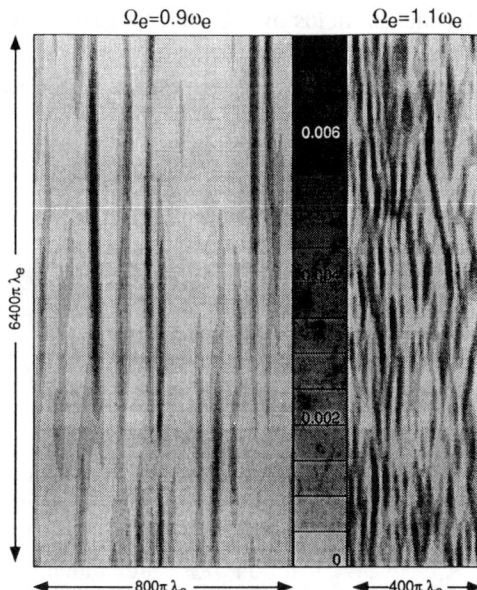

Fig. 6. Portions of spatial structure of Langmuir turbulence, $|E|^2$. Left: subcritical case, showing a collapsing nonlinear pancake-shaped wavepacket. The global maximum is not shown. Right: supercritical case in which collapse is suppressed. Note distorted aspect ratio.

turbulence phenomenon of collapse has been suppressed by the change in dispersive topology associated with the (slightly) stronger magnetic field.

3.3 Physical interpretation of k-space spectra

Time-averaged **k**-space spectra of $|E_\mathbf{k}|^2/|E_{\mathbf{k}\max}|^2$ illuminate other aspects of the sub and supercritical turbulence, and help explain why collapse is inhibited in the supercritical case.

Fig. 7 shows such numerically-determined **k**-space spectra. A noteworthy feature is the similarity of these *turbulent* **k**-space spectra to the *linear* dispersion exhibited in **k**-space in Figs. 2 and 3. In both Fig. 8 and Figs 2 and 3, the aspect ratio is distorted, so that these spectra are expanded in $k_\perp$, and, in true aspect ratio are almost 1-dimensional "tubes" oriented along the magnetic field direction.

The beam-resonant modes are indicated with arrows. The contours corresponding to the remaining, nonresonant, Langmuir waves have been filled in the turbulent spectra by 3-mode decay cascades which began with beam-resonant modes. A beam-resonant Langmuir wave $(\omega_1, \mathbf{k}_1)$ can decay into a slightly lower-frequency Langmuir wave $(\omega_2, \mathbf{k}_2)$ and an ion-acoustic wave, $(\omega_3, \mathbf{k}_3)$ provided that frequency and wavevector matching occurs:

$$\omega_1 = \omega_2 + \omega_3, \quad \mathbf{k}_1 = \mathbf{k}_2 + \mathbf{k}_3 \qquad (5)$$

For the *subcritical* case the lower-frequency ω_2 (daughter) Langmuir wave lies on a smaller closed curve than the one containing the beam-resonant wave. Note that all of the wavevectors on the daughter wave contour are of *smaller* magnitude than the wavenumber of the beam-resonant wave. This means that the damping at the daughter wavenumbers will be *less* than at the beam-resonant wavenumber, according to eqn. (4a) (assuming that the daughter wavevectors lies outside the resonant region of **k**-space defined by eqn. (4b)).

The cascade continues as the ω_2 mode decays into $\omega_4 + \omega_5$, etc. In this manner, the turbulent spectrum spreads to many nonresonant contours associated with lower Langmuir frequencies, lower wavenumbers, and hence less and less access to wave-dissipation by energy transfer to electrons (eqn. (4)).

The situation is similar for the *supercritical* case, with one important exception. Here, the ω_2 (daughter) Langmuir wavevectors lie on an open curve, so that some daughter wavevectors have a magnitude *larger* than the wavenumber of the beam-resonant wave. Such kinematically accessible daughter waves will be *more* heavily damped than the beam-resonant mode, according to

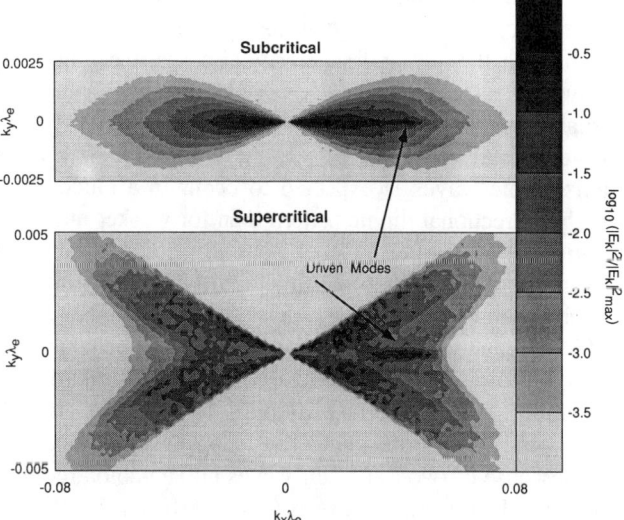

Fig. 7. Contour maps in k-space of numerically-determined beam-driven Langmuir energy density, $|E_\mathbf{k}|^2/|E_\mathbf{k}|^2_{\max}$, in the lower solar corona. (a) Subcritical, (b) Supercritical

eqn. (4a) (once again, assuming that they lie outside the beam-resonant region of **k**-space.)

The cascade again continues until the turbulent spectrum spreads down to nonresonant contours associated with lower Langmuir frequencies, but to larger as well as smaller wavenumbers, and hence to relatively more wave-dissipation by energy transfer to electrons than for the subcritical case. Access to stronger dissipation is one factor that accounts for the difference between the subcritical and supercritical cases. Collapse is suppressed in the supercritical case because the topologically different kinematics allows access to increased Langmuir wave damping, which inhibits the (inertial) collapse of nonlinear Langmuir wavepackets. By contrast, in the subcritical case, there is no efficient route to dissipation except collapse to shorter spatial scales.

A second factor which may be relevant to the suppression of collapse in the supercritical case is the topologically different disperion relation. This will be further discussed in future work by the authors.

4. CONCLUSION

We have numerically studied Langmuir turbulence excited by electron beams associated with Type III emission events in the lower solar corona, under conditions of moderate magnetization (active regions with B = 30-300 gauss, densities of 10^8-10^{10}, plasma frequencies above 100 MHz and comparable electron cyclotron frequencies).

Two major results were obtained concerning the space-time structure of the Langmuir turbulence:

In both the sub and supercritical cases, the Langmuir wavepackets have a wide, flat pancake shape, with the local solar magnetic field direction as axis of symmetry. Type III fundamental emission associated with such extremely anisotropic waves is expected to occur in a much more highly directional dipole pattern than for weaker magnetic regions. In the absence of spatially convoluted solar magnetic field structure, this could translate into an explanation for the relative paucity of Type III bursts above 100 MHz, although propagation effects must be taken into account to determine the effective angular distrubution of escaping fundamental emission. In addition, second harmonic emission could be suppressed, since the coalescense of two Langmuir waves into a photon requires non-parallel Langmuir wavevectors for efficient coupling.

A second result concerns differences in the space-time structure of the Langmuir waves in the subcritical case $\Omega_e<\omega_e$, compared to the supercritical case, $\Omega_e>\omega_e$. In the subcritical case, spatial collapse of nonlinear Langmuir wavepackets can occur, leading to isolated examples of extremely intense fields over small space and time scales. In the supercritical case the wavepackets are less coherent, and much more homogeneous in space and time. The suppression of collapse is thought to be associated with the topologically different dispersion and increased access to wave damping by electrons in the supercritical case.

Acknowledgement. This work has been supported by NSF under grant ATM-9020283 and by Ecole Polytechnique, Paris, France. Computing was carried out at the National Center for Atmospheric Research.

REFERENCES

Boehm, M.H., Waves and static electric fields in the auroral acceleration region, PhD. thesis, Department of Physics, University of Califonia at Berkeley, 1987

Ergun, R.C., C.W. Carlson, J.P. McFadden, and J.H. Clemmons, Langmuir wave groth and electron bunching: results from a wave-particle correlator, *J. Geophys. Res.*, **96**, 225, 1991

Goldman, M.V., Progress and problems in the theory of Type III solar radio emission, *Solar Physics* **89**, 403-442, 1983.

Goldman, M.V., Strong turbulence of plasma waves, *Rev. Mod. Phys.* **66**, 709-735, 1984.

Goldman, M.V., Electron Beams and Instabilities during Solar Radio Emission, *Solar System Plasma Physics, Geophysical Monograph* **54**, 229-236, 1989

Lin, R.P., D.W. Potter, D.A. Gurnett, & F.L. Scarf, Energetic Electrons and Plasma Waves Associated With a Solar Type III Radio Burst, *Ap. J*, **251**, 364 1981.

Lin, R.P., W.K. Levedahl, W. Lotko, D.A. Gurnett, and F.L. Scarf, Evidence for Parametric Decay of Langmuir Waves in Solar Type III Radio Bursts, *Ap. J*, **308**, 954 (1986)

Melrose, D.B., & M.V. Goldman, Microstructures in Type III Events in the Solar Wind, *Solar Physics,* **107**, 329-350, 1987

Newman, D.L., Emission of Electromagnetic Radiation from Beam-Driven Plasmas, PhD thesis, Department of Astrophysical, Planetary and Atmospheric Sciences, University of Colorado at Boulder, 1985

Newman, D.L., R.M. Winglee, P.A. Robinson, J. Glanz, and M.V. Goldman, Simulation of the collapse and dissipation of Langmuir wave packets, *Phys. Fluids*, **B2**, 2600, 1990

Newman, D.L., M.F. Goldman, R.E. Ergun and M.H. Boehm, Simulation and Theory of Langmuir Turbulence in the Auroral Ionosphere, submitted to *J. Geophs. Res.*, 1993

Robinson, P.A. and D.L. Newman, Strong Plasma Turbulence in the Earth's Electron foreshock, *J. Geophs. Res.* **96**, 17,733-17,749, 1991

Russell, D., D.F. DuBois, and H.A. Rose, Nucleation in Two-Dimensional Langmuir Turbulence, *Phys. Rev. Lett.* **60**, 581, 1988

Weatherall, J.C., Nonlinear Langmuir waves in weak magnetic fields, Ph.D. thesis, Department of Astrogeophysics, University of Colorado, Boulder, 1980

Development of Fractal Structure in the Solar Wind and Distribution of Magnetic Field in the Photosphere

A.V. Milovanov and L.M. Zelenyi

Space Research Institute, Russian Academy of Sciences, Moscow, Russia

A fractal model for the solar wind is presented. Basic physical relations between fractal properties of the interplanetary magnetic field turbulence, and the state of large-scale magnetic configurations on the solar surface are discussed. Development of fractal structure of the solar wind fluctuations is considered from the viewpoint of nonlinear dynamical processes in the solar convection zone and photosphere. Direct spacecraft measurements of the interplanetary magnetic field and bulk velocity profiles at different heliocentric distances are analyzed. It is proved that fractal distribution of magnetic flux tubes across the solar surface covering the range of scales from ~400 km to ~40,000 km [*Tarbell et al.*, 1990], is responsible for the multifractal properties of the solar wind fluctuations between ~0.2 AU and ~8 AU. The monofractal structure of the interplanetary magnetic field turbulence observed by Voyager 2 spacecraft near 8.5 AU over more than 4 decades in frequency (from $3 \cdot 10^{-6}$ Hz to $5 \cdot 10^{-2}$ Hz) is shown to agree well with basic model prediction. Possible existence of gigantic convection cells in the photosphere as well as the properties of the solar wind fluctuations at large heliocentric distances ($R \sim 60$ AU) are discussed.

1. INTRODUCTION

Many of the recent advances and outstanding problems in the solar wind research deal with the statistical properties of the interplanetary magnetic field (IMF) fluctuations with the goal of understanding the fluctuation spectra of the magnetic and velocity fields and their relationship to macroscopic structure. The basic theoretical model which we use to organize this investigation is to consider the solar wind as a stream of magnetic clusters with the considerable fine structure on different spatial scales. Moreover, the solar wind plasma certainly exhibits dynamical effects, and many of the phenomena in the wind can be understood in terms of a dynamical fractal model. There are theoretical arguments suggesting that the nonlinear turbulence processes significantly originate from the state of the large-scale fractal configurations on the solar surface [*Milovanov and Zelenyi*, 1992; *Milovanov and Zelenyi*, 1993]. It is entirely possible that the large-scale structure of a nonlinearly active solar wind can be understood only by including such features of turbulence. Below we assume that the IMF is "frozen in" the solar wind plasma, i.e., basic properties of the turbulence derived either from magnetic field or bulk velocity measurements, are equivalent.

As spacecraft observations have become available, it has been recognized that the IMF is variable on many time scales, and spectral analysis of the time series for both the magnitude and the components of the field shows a power law spectrum $f^{-\alpha}$ over at least one decade in frequency, with a slope depending on the frequency range and the time interval that is analyzed (see for example, the papers by *Burlaga and Klein*, 1986; *Burlaga*, 1991a). Indeed, large-scale fluctuations (with spacecraft frame periods between several hours and approximately the solar rotation period) in the magnetic field strength $B(t)$ and bulk velocity $V(t)$ observed at 1 AU were analyzed using data from IMP 8 and ISEE 3 [*Burlaga et al.*, 1989]. In the intermediate frequency range ($3 \cdot 10^{-6}$ to $\sim 3 \cdot 10^{-5}$ Hz), the spectra of $B(t)$ and $V(t)$ computed for successive ~6-month intervals from 1978 to 1982 had the form $f^{-\alpha}$, where $\langle \alpha_B \rangle = 1.92 \pm 0.06$ for the magnetic field spectra, and $\langle \alpha_V \rangle = 1.92 \pm 0.10$ for

Solar System Plasmas in Space and Time
Geophysical Monograph 84
Copyright 1994 by the American Geophysical Union.

the bulk velocity spectra. Meanwhile, at the frequencies greater than approximately $3 \cdot 10^{-5}$ Hz, the spectra of the large-scale fluctuations in $B(t)$ tend towards the form $f^{-5/3}$ corresponding to the Kolmogorov spectrum for homogeneous, isotropic, stationary turbulence in an incompressible fluid [*Landau and Lifshitz*, 1988].

The function $B(t)$ representing direct measurements of the magnitude of the IMF or any of its components, can be viewed geometrically as a curve which has structure on every scale and which is "statistically self-affine" [*Burlaga and Klein*, 1986]. Statistically self-affine means that each part of the curve can be considered as a reduced scale image of the whole, that is $h^{-2H}B(ht)$ is statistically identical to $B(t)$. A statistically self-affine curve could be treated as a fractal set with the fractal dimension $\delta = 2 - H$, where H is a constant between 0 and 1 [*Mandelbrot*, 1993]. For smooth rectifiable curves, δ is equal to the topological dimension: $\delta = 1$. However, for statistically self-affine curves in a plane, δ is a fraction $1 < \delta < 2$. Roughly speaking, the fractional values of δ describe highly ragged curves with the considerable fine structure on many spatial scales. The quantity δ (which should be called "temporal fractal dimension" hereafter) determines the exponent of the power spectrum α:

$$\delta = (5 - \alpha)/2. \qquad (1)$$

A derivation of this relation was given by *Berry* [1979].

From equation (1) we immediately conclude that the Kolmogorov spectrum, $f^{-5/3}$, which describes inertial range turbulence in an incompressible fluid, corresponds to the fractal dimension $\delta = 5/3$. This dimension must be dominant in the high-frequency range ($f > 3 \cdot 10^{-5}$ Hz) when considering the structure of large-scale fluctuations in the solar wind at $R \sim 1$ AU. Moreover, measurements of the IMF fluctuations made by Voyager 2 spacecraft near 8.5 AU over more than 4 decades in frequency, from $f = 5 \cdot 10^{-2}$ Hz to $f = 3 \cdot 10^{-6}$ Hz, also exhibit fractal behaviour with the temporal fractal dimension $\delta = 5/3$ [*Burlaga and Klein*, 1986]. Recently *Tu et al.* [1989] gave some arguments suggesting that an isotropizing cascade process nearly equalizes the component spectral densities and also produces a spectral slope close to $\alpha = 5/3$. Meanwhile, the slope $\alpha = 1.92$ which arises in the intermediate frequency range ($3 \cdot 10^{-6}$ to $3 \cdot 10^{-5}$ Hz) at $R \sim 1$ AU, corresponds to the temporal fractal dimension $\delta = 1.54$. This dimension could be related to Burger's turbulence resulting from an infinite number of "jumps", shocks, and discontinuities in the magnetic field strength and bulk velocity profiles [*Burlaga et al.*, 1989]. The fractal dimension $\delta = 2$ corresponds to an f^{-1} spectrum, but this case must be treated with care because it is valid only as a limit [*Berry*, 1979].

In the next section, we will consider development of fractal structure of the solar wind turbulence from the viewpoint of nonlinear dynamic processes in the solar convection zone and photosphere. Our goal is to find basic physical relations between fractal properties of the IMF and bulk velocity fluctuations, and the characteristics of large-scale magnetic configurations on the solar surface. This idea is schematically illustrated by Figure 1. We will also discuss spacecraft observations of the IMF turbulence in the light of a dynamical fractal model for the solar wind.

2. HIERARCHY OF CONVECTION CELLS IN THE PHOTOSPHERE

The basic theoretical point is that turbulent flows in the convection zone and photosphere transport solar magnetic field (which is concentrated in the intense magnetic flux tubes) and distribute it across the solar surface in very intermittent, highly structured patterns. As is evident from direct observations, these patterns can be considered as fractal sets over the range of spatial scales from 400 to 40,000 km with the fractal dimension D of 1.4–1.7 [*Tarbell et al.*, 1990]. A two-parameter, least squares search over D yields the value $D = 1.56 \pm 0.08$ [*Lawrence*, 1991]. Fractal distribution of magnetic flux tubes implies a remarkable property of self-similarity. Actually the spatial distribution of the magnetic tubes at the scales of ~ 400 km is similar to the spatial distribution of aggregates of the tubes at the scales of 40,000 km. The quantity D (which should be called "spatial fractal dimension" in contrast to the temporal fractal dimension δ) determines the number of the tubes inside a circle of given radius r [*Mandelbrot*, 1983]:

$$N(r) \propto r^D, \qquad 1 < D < 2. \qquad (2)$$

It seems reasonable that the fractal distribution of magnetic flux tubes across the solar surface originates from the fractal distribution of the turbulent flows (convection cells) in the photosphere (cf. *Milovanov and Zelenyi*, 1993]. Following *Richardson and Schwarzschild* [1950], let us consider the convection cells as turbulent eddies with the size, l_n; speed, v_n (the difference in speed across an eddy); lifetime, $\tau_n \sim l_n / v_n$; mass, $m_n \sim \rho l_n^2$; and energy, $e_n \sim m_n v_n^2$. It is generally assumed that a large eddy gives birth to several smaller eddies and then dies, with a resultant energy transfer from the scale of the eddy to a smaller scale at a rate e_n/τ_n. Under this assumption (which leads to a hierarchical structure like a family tree) the daughter eddies, l_{n+1}, form a fractal set inside the mother eddy, l_n. Using Eq. (2), we obtain the total number of the daughter eddies, $N = (l_n/l_{n+1})^{D_n}$, where D_n is the spatial

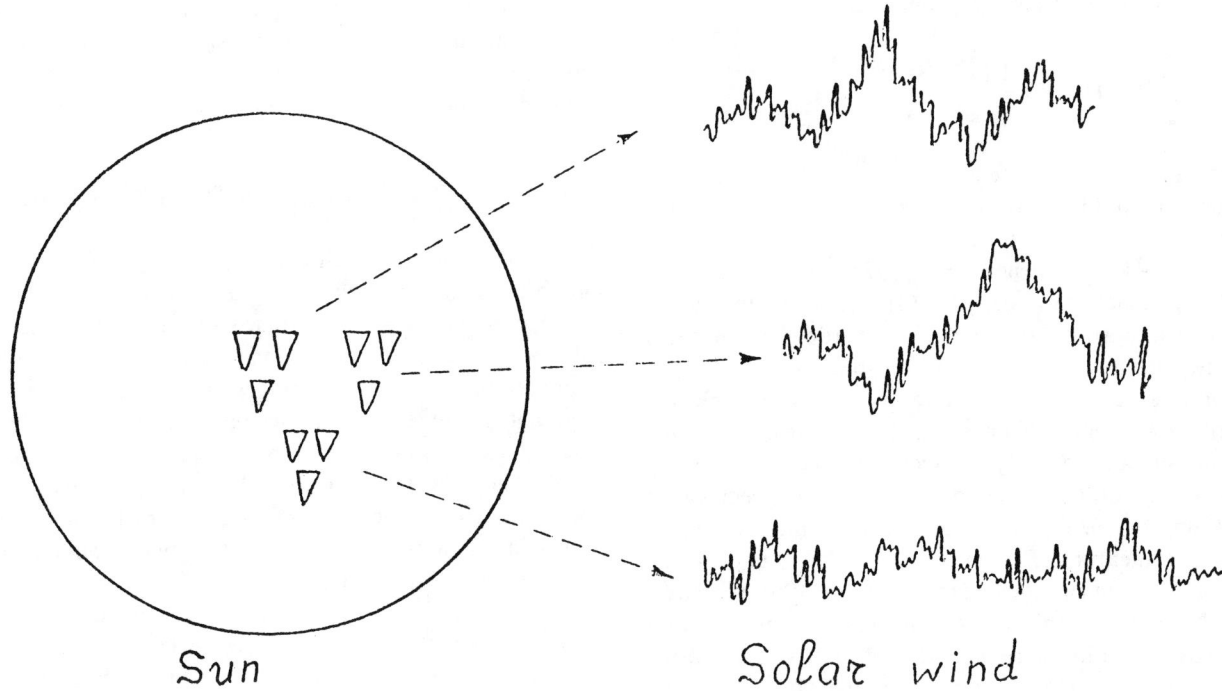

Fig. 1. Schematic of the relations between fractal properties of the IMF turbulence and the distribution of magnetic field in the photosphere.

fractal dimension of the set. Let β_{n+1} be the ratio of the volume occupied by the offspring to that of the mother eddy: $\beta_{n+1} = V_{n+1} / l_n^2 = (l_n / l_{n+1})^{D_n - 2}$. Hence the required fractal dimension is given by [cf. *Burlaga*, 1991*b*]:

$$D_n = 2 + \frac{\ln \beta_{n+1}}{\ln(l_n / l_{n+1})}. \qquad (3)$$

Exact self-similarity implies that $l_n / l_{n+1} = \beta_{n+1}^\mu$ where μ is a constant. Finally the fractal dimension of the set occupied by the eddies is $D_n = D = 2 + \mu^{-1}$. To estimate the value of D, let us note that the average diameter of the large-scale turbulent eddies in the photosphere is of the order of that of supergranules, $l_0 \approx 3.2 \cdot 10^4$ km [*Priest*, 1982]. When a supergranule dies, it gives birth to several mesogranules with the average diameters close to $l_1 \approx 6 \cdot 10^3$ km. The mesogranules, in turn, give birth to ordinary granules with typical diameters of the order of $l_2 \sim 10^3$ km. Thus, one can view the evolution of solar granules as a fragmentation process. Let us assume the average distance between the eddies is half again their average size: $a_{n+1}/l_{n+1} \approx 1.5$. (e.g., for ordinary granules, this ratio is close to 1.8 [*Priest*, 1982]). This yields the parameter $\beta_{n+1} = (l_{n+1}/a_{n+1})^2 = 4/9$. Taking into account that the ratio $l_n/l_{n+1} \approx 6$ independently of n, from equation (3) one finds $D_n = D \approx 1.54$ which is consistent with the observational values of D reported by *Tarbell et al.* [1990] and *Lawrence* [1991].

3. PARAMETERS OF THE PLASMA DISTRIBUTION FUNCTION

From the previous consideration it follows that the minimum spatial scale of the highly structured patterns with fractal distribution of the intense magnetic flux tubes is determined by the size of ordinary granules: $L_{min} \approx l_2 \sim 10^3$ km. This conclusion is consistent with the value $L_{min} \approx 400$ km obtained by *Tarbell et al.* [1990]. Meanwhile, the maximum scale of these patterns is correlated with typical diameters of supergranules, $l_0 \sim 3 \cdot 10^4$ km, and actually is close to $L_{max} \approx 40,000$ km [*Tarbell et al.*, 1990]. Below we will assume the value of the spatial fractal dimension D for the distribution of the flux tubes within these patterns is equal to that derived from the hierarchy of the convection cells in the photosphere: $D = 1.54$.

It is theoretically important to note that the minimum fractal patterns on the solar surface corresponding to the minimum scale $L_{min} \sim 10^3$ km, could be considered as fractal aggregates of the intense flux tubes with the fractal dimension D depending on the parameters of the plasma distribution function [*Milovanov and Zelenyi*, 1992; *Milovanov and Zelenyi*, 1993]. Indeed, the fractal dimension D of these aggregates is given by the relation

$$D = \frac{4}{3} + \frac{1}{3} \frac{\ln(3/8\pi\Delta^2 \ln 2)}{\ln \Lambda} \quad (4)$$

where Λ is the size of the aggregates in units of the tube's radius. It is well known [*Priest*, 1982] that the radius of the intense magnetic flux tubes in the photosphere is of the order of $a \sim 10^2$ km, hence $\Lambda = L_{min}/2a \approx 5$.

The parameter Δ in equation (4) depends on the particular plasma distribution function for particles of species j (where j denotes electron or ion component of the plasma). Without regard for dissipative processes, the plasma distribution function should depend on constants of the motion. For cylindrically symmetric magnetic flux tubes, the constants of the motion for the particles of species j are the Hamiltonian, W_j; the z-component of the generalized linear momentum, P_j; and the z-component of the generalized angular momentum, M_j [*Marx*, 1968; *Channell*, 1976]. In the simplest case, the distribution function depends on the linear combination of the constants of the motion with constant coefficients u_j and Ω_j [*Zelenyi and Milovanov*, 1992a],

$$f_j \sim \left[\frac{1}{T_j}(-W_j + P_j u_j + M_j \Omega_j)\right]^{p-1/2} \quad (5)$$

The finite power exponent p (which is assumed to be more than 1/2) characterizes the degree of deviation from the Maxwellian distribution; when $p \to \infty$, we obviously obtain the displaced Maxwellian plasma distribution function.

For the power law plasma distribution function of the form (5), the parameter Δ is given by [*Zelenyi and Milovanov*, 1991]

$$\Delta \sim \frac{3p+2}{(p+1)(p+2)} \left[\frac{2(p+1)(p+2)}{p^2}\right]^{(p+1)/p} \quad (6)$$

Equations (4)–(6) establish the relation between the photospheric value of the power exponent p in equation (5), and the spatial dimension of the fractal aggregates of the intense flux tubes, related to the minimum scale $L_{min} \sim 10^3$ km. Setting $D = 1.54$ in equation (4), one finds $\Delta \approx 0.25$. Further, it is easy to prove that in the limiting case $p \to \infty$, equation (6) becomes: $\Delta \approx 6/p$, hence the photospheric value of the parameter p is close to $p \approx 24 \gg 1$. This result is consistent with the value $p \approx 21$ derived from the dynamical model for sunspot evolution [*Zelenyi and Milovanov*, 1992b; *Milovanov and Zelenyi*, 1992]. The inequality $p \gg 1$ shows that the photospheric plasma distribution function which determines the fine scale structure of the fractal aggregates of intense magnetic flux tubes, could be reasonably considered as the Maxwellian one.

4. A FRACTAL MODEL FOR THE SOLAR WIND

In this section, we will discuss formation of the structure in the solar wind induced by the fractal distribution of convection cells across the solar surface. First of all, the solar wind, i.e. an extension and expansion of the solar atmosphere into interplanetary space, is generated preferably at the regions with the open magnetic field lines, where the solar corona is not in equilibrium [*Hundhausen*, 1972]. The solar magnetic field lines concentrated in the open intense flux tubes [*Priest*, 1982], are picked up by the solar wind streams and then are carried away from the Sun to the interplanetary medium.

When rising to the upper solar atmosphere, the open intense flux tubes undergo radial expansion due to nonzero variance of radial velocities, $\bar{v}_r^2$. Assuming isotropic temperature, we may write $\bar{v}_r^2 = \bar{v}^2/3 = 2kT/3m_p$ where m_p is the proton mass. Hence the typical angle ϑ characterizing the flux tube's expansion, is determined from the relation

$$\frac{\vartheta}{2} = \frac{\sqrt{2kT/3m_p}}{v_R} \quad (7)$$

where $v_R \approx 400$ km·s^{-1} is the average solar wind speed [*Priest*, 1982], and T is the plasma temperature at the foots of the tubes. Setting $T \sim 10^4$ K in equation (7), one finds $\vartheta \sim 4 \cdot 10^{-2}$. A more accurate consideration of the radial expansion was given by *Tamano* [1991]. Let us now remind that the intense magnetic flux tubes are concentrated in fractal aggregates corresponding to the minimum spatial scale of very intermittent, highly structured patterns on the solar surface $L_{min} \sim 10^3$ km. Thus we conclude that these aggregates freely expand with some radial velocity $v_r \sim \vartheta v_R/2$ as they are carried heliocentrically away from the Sun by the solar wind streams. It is easy to see that $v_r/v_R \sim \vartheta/2 \sim 2 \cdot 10^{-2}$ which is within a factor of two compared with the corresponding observational value [*Tamano*, 1991]

Radial expansion of the fractal aggregates implies existence of some heliocentric distance R_* where the neighbouring aggregates start to overlap with each other. Indeed, let β be the angular separation between these aggregates on the solar surface: $\beta \sim L_{min}/R_s$, where R_s is the Sun's radius. Hence the time it takes for the neighbouring aggregates to overlap, is of the order of βT_s

$v_R/2v_r$, where T_s is the solar rotation period. Therefore, the distance R_* is easily seen to be given by

$$R_* = \beta T_S v_R^2 / 2v_r \qquad (8)$$

or, equivalently,

$$R_* = \beta T_S v_R / \vartheta. \qquad (9)$$

Substituting $\vartheta \sim 4 \cdot 10^{-2}$, $\beta \sim 1/700$, $T_S \sim 26$ days, and $v_R \sim 400$ km·s^{-1} into equation (9), one finds $R_* \approx 0.2$ AU.

Thus, at the heliocentric distances exceeding approximately $R_* \approx 0.2$ AU, the overlapping fractal aggregates related to the minimum spatial scale of the highly structured patterns $L_{min} \sim 10^3$ km, give rise to formation of small-scale fractal clusters in the solar wind with at least the doubled dimensionless size compared with that of the original aggregates: $\Lambda(R_*) \approx 2\Lambda_0 \sim 10$. (The subscript zero shows hereafter that the corresponding value of the physical quantity is taken at the photosphere level.) The spatial fractal dimension of the small-scale clusters is determined from equation (4):

$$D(R_*) = \frac{4}{3} + \frac{1}{3} \frac{\ln(3/8\pi\Delta_0^2 \ln 2)}{\ln \Lambda(R_*)} \qquad (10)$$

where we have assumed the plasma distribution function is formed in the lower solar atmosphere with relatively high densities and significant Coulomb collisions and then is carried by the solar wind to the interplanetary space. Substituting $\Lambda(R_*) \sim 10$ and $\Delta_0 \approx 0.25$ (see the very end of the previous section), we get $D(R_*) \approx 1.48$. The result obtained shows that the small-scale clusters of the flux tubes arising at the heliocentric distances $R > 0.2$ AU, could be considered as fractal sets with the spatial fractal dimension $D \approx 1.48$. On the other hand, spatial distribution of these clusters in the solar wind originates from the hierarchy of the convection cells in the photosphere corresponding to the spatial fractal dimension $D \approx 1.54$. Thus, exact description of the structure of the IMF requires more than one fractal dimension at $R > 0.2$ AU and therefore, becomes multifractal [*Feder*, 1988]. Multifractal properties of the IMF turbulence are of great theoretical and conceptual interest in the solar wind research [*Burlaga*, 1991a,b].

It should be emphasized, however, that the multifractal structure of the IMF turbulence (characterized by at least the two fractal dimensions $D \approx 1.48$ and $D \approx 1.54$) exists until all fractal clusters originating from the common large-scale pattern on the solar surface, overlap with each other.

Since the characteristic size of these patterns is close to $L_{max} \sim 4 \cdot 10^4$ km [*Tarbell et al.*, 1990], from equations (8)–(9) it follows that the small-scale clusters overlap at the heliocentric distances $R'_* \sim (L_{max}/L_{min}) R_* \sim 40 R_* \approx 8$ AU. It is important to note that the small-scale clusters overlapping at $R \sim 8$ AU, give rise to formation of large-scale clusters of the flux tubes, which could be called hereafter "superclusters". The spatial fractal dimension of superclusters can be easily estimated from equation (10). Indeed, setting the dimensionless size of the superclusters $\Lambda(R'_*) \sim (L_{max}/L_{min}) \Lambda (R_*) \sim 200$, one finds $D(R'_*) \approx 1.39$. Thus, at the heliocentric distances exceeding $R'_* \approx 8$ AU, statistical properties of the solar wind fluctuations become monofractal in a wide range of frequencies related to the fine structure of the superclusters. This conclusion is in good agreement with direct spacecraft measurements of the IMF made by Voyager 2 near 8.5 AU [*Burlaga and Klein*, 1986]. Spacecraft observations of the solar wind turbulence will be discussed more accurately in the next section. Let us, however, note that direct spacecraft measurements of the solar wind magnetic field and bulk velocity deal with the temporal fractal dimension δ, but not with the spatial dimension D. Now we turn to derivation of the relation between these two dimensions.

Let us introduce the temporal "scale" τ which is determined by the averaging interval that we choose when measuring e.g., the IMF strength or any of its components. Then the function $B(t)$ representing direct measurements of the IMF, may be approximated by a histogram, where the width of each bar is τ and the height of each bar is the average value of $B(t)$ between $t = t_k$ and $t_k + \tau$, which we denote by $\overline{B}(t_k)$. The "length" of the curve defined by the histogram over some interval $0 \leq t \leq T_0$ (where $T_0 = N\tau$ and N is integer) is

$$L(\tau) = \sum_{k=1}^{N} \left|\overline{B}(t_k + \tau) - \overline{B}(t_k)\right|. \qquad (11)$$

The "length" $L(\tau)$ is a function of τ, and for statistically self-affine curves [*Burlaga and Klein*, 1986],

$$L(\tau) = L_0 \tau^{-S} \qquad (12)$$

where L_0 and S are constants. The temporal fractal dimension δ is related to S by the following equation:

$$\delta = S + 1. \qquad (13)$$

In the solar wind frame of reference, different time moments t_k correspond to different radial coordinates $x_k = $

$v_R t_k$, hence one may easily rewrite equation (11) in the "spatial" form:

$$L(\chi) = \sum_{k=1}^{N} |\overline{B}(x_k + \chi) - \overline{B}(x_k)| \quad (14)$$

From equation (2) it is evident that the number of the magnetic flux tubes inside a fractal aggregate of radius x is proportional to x^D where D is the spatial fractal dimension. Since the magnetic clusters in the solar wind are considered as fractal sets of the magnetic flux tubes, the average magnetic field $\overline{B}(x)$ is obviously proportional to x^{D-2}, then from equation (14) it follows that

$$L(\chi) = L(1)\chi^{D-2} \quad (15)$$

or

$$L(\tau) = L_0 \tau^{D-2} \quad (16)$$

Equations (12),(13) and (16) yield the required relation between spatial and temporal fractal dimensions:

$$D = 3 - \delta. \quad (17)$$

Taking into account equation (1), we obtain the relation between the spatial fractal dimension D and the exponent of the power spectrum α:

$$D = (\alpha + 1)/2. \quad (18)$$

Equation (18) shows that the Kolmogorov spectrum for homogeneous, isotropic, stationary turbulence in an incompressible fluid ($\alpha = 5/3$) corresponds to the spatial fractal dimension $D = 4/3$. Meanwhile, the Burger's turbulence which originates from an infinite number of discontinuities (planes) whose positions, orientations, and intensities are given by three infinite sequences of mutually independent random variables [*Mandelbrot*, 1983], is related to the spatial fractal dimension $D = 3/2$.

5. COMPARISON WITH OBSERVATIONS AND DISCUSSION

First of all, theoretical results obtained in the previous section, indicate that temporal variations of the IMF and bulk velocity at the heliocentric distances $R \leq 0.2$ AU, could be characterized by the (temporal) fractal dimension $\delta = 3 - 1.54 = 1.46$. This dimension results from the fractal distribution of convection cells across the solar surface, corresponding to the spatial dimension $D = 1.54$ and the exponent of the power spectrum $\alpha = 2.08$ (see equation (8)). Therefore, we conclude that the early life of the solar wind fluctuations is the Burger's turbulence at $R_S \leq R \leq R_*$, related to the hierarchy of the convection cells in the photosphere.

In the range of heliocentric distances 0.2 AU $\leq R \leq 8$ AU, statistical properties of the IMF and bulk velocity fluctuations exhibit strongly pronounced multifractal structure. Indeed, the fractal dimension $D = 1.54$, which describes spatial distribution of the small-scale clusters in the solar wind, again corresponds to the temporal dimension $\delta = 1.46$. Meanwhile, fine structure of these clusters considered as fractal sets of magnetic flux tubes, is related to the temporal fractal dimension $\delta = 3 - D(R_*) = 1.52$, which exceeds the value $\delta = 1.46$. The frequency f_S separating these two dimensions, is determined by the spatial size of the small-scale clusters in the solar wind. Since these clusters radially expand with some constant velocity $v_r \sim \vartheta\, v_R/2$ as they move heliocentrically away from the Sun, the frequency f_S varies as $1/R$ with the heliocentric distance R. It is easy to estimate the frequency f_S in the vicinity of the Earth's orbit ($R \sim 1$ AU). In fact, the typical size of the small-scale clusters at $R \sim 1$ AU is of the order of $r_{(1)} \sim (v_r/v_R) \cdot 1$ AU $\sim 3 \cdot 10^6$ km. Hence $f_S \sim v_R/2r_{(1)} \sim 6 \cdot 10^{-5}$ Hz which is in good agreement with the observational value $f_S \sim 3 \cdot 10^{-5}$ Hz [*Burlaga et al.*, 1989].

Note, by the way, that the frequency f_S determines only the lower frequency limit for the spatial fractal dimension $D(R_*) = 1.48$ and the temporal dimension $\delta = 1.52$ at given heliocentric distance 0.2 AU $\leq R \leq 8$ AU. The upper frequency limit, $\tilde{f}_S$, is related to the minimum magnetic elements inside the clusters of the flux tubes. These minimum elements are simply the domains where the neighbouring flux tubes overlap. As shown in the Appendix, the typical size of these domains is approximately εa, where a is the tube's radius, and ε is a small parameter depending on the particular plasma distribution function. For the distribution function of the form (5) with the power exponent $p \approx 24$, this parameter equals $\varepsilon \approx 1/2 \cdot 10^{-3}$. Hence $\tilde{f}_S \sim v_R/\varepsilon a$, varying as $1/R$ with the heliocentric distance R.

Now let us estimate the frequency $\tilde{f}_S$ at $R \sim 1$ AU. First of all, the number of magnetic flux tubes inside the small-scale fractal clusters can be easily derived from equation (2): $N = (2r_{(1)}/a_{(1)})^{1.48}$ where $a_{(1)}$ is the tube's radius at $R \sim 1$ AU, and 1.48 is the spatial fractal dimension of the cluster. On the other hand, N is evidently equal to the number of the intense flux tubes concentrated within the minimum fractal patterns on the solar surface. (Remember the spatial scale of these patterns is of the order of $L_{\min} \sim 10^3$ km). Considering equation (2) once again, we write $N = (L_{\min}/a_{(0)})^{1.54}$. Here $a_{(0)} \sim 10^2$ km is the typical radius of the intense flux tubes in the photosphere, and the fractal dimension 1.54 is related to the hierarchy of the convection cells. Thus we obtain $N \approx 32$, and $a_{(1)} \approx 6 \cdot 10^5$ km. Finally at $R \sim 1$ AU, we have $\tilde{f}_S \sim v_R \varepsilon a_{(1)} \sim 0.56$ Hz.

The characteristic strength of the solar wind magnetic field can be estimated from the conservation of the magnetic flux in the expanding magnetic flux tubes. Indeed, at $R \sim$ AU, one finds $B_{(1)} \sim B_{(0)} a_{(0)}^2 / a_{(1)}^2$. Since the typical magnetic field of the intense flux tubes is $B_{(0)} \sim 2 \cdot 10^3$ G [Priest, 1982], we get $B_{(1)} \sim 6 \cdot 10^{-5}$ G in close agreement with direct observations [Burlaga et a., 1981].

At the heliocentric distances $R'_* \sim 8$ AU, all small-scale fractal clusters originating from the common large-scale pattern on the solar surface, start to overlap with each other, which results in formation of superclusters of magnetic flux tubes. The formation of superclusters leads to considerable extension of the frequency f_S to the low frequency range ($f \leq 3 \cdot 10^{-6}$ Hz). At $R \geq 8$ AU, this frequency determines the lower limit for the spatial fractal dimension of superclusters, $D(R'_*) \approx 1.39$. From equation (10) it follows that the spatial fractal dimension of magnetic clusters in the solar wind tends towards the universal value 4/3 as the dimensionless size of these clusters increases. Let us remind that this value corresponds to the Kolmogorov spectrum, $f^{-5/3}$, for homogeneous, isotropic, stationary turbulence in an incompressible fluid. Note, that the fractal dimension of superclusters, $D(R'_*) \approx 1.39$, is rather close to this universal value. In this regard, the frequency f_S yields the lower frequency limit for the Kolmogorov spectrum of the solar wind turbulence at the heliocentric distances $R \geq 8$ AU. In the range of frequencies $f < f_S$, the spectrum of the turbulence is related to the original distribution of the large-scale fractal patterns over the solar surface. Measurements of the IMF and bulk velocity profiles in this range of frequencies could provide important results concerning large-scale dynamics of the solar magnetic field.

To estimate the frequency f_S at $R \sim 8$ AU, let us note that the corresponding (dimensional) size of the superclusters is of the order of $\Lambda(R'_*)a_{(8)} \sim 8\Lambda(R'_*)a_{(1)} \sim 10^9$ km (see previous section). Hence $f_S \sim v_R/\Lambda(R'_*)a_{(8)} \sim 4 \cdot 10^{-7}$ Hz. Meanwhile, the upper frequency limit, $\tilde{f}_S$, is close to $\tilde{f}_S \sim v_R / \varepsilon a_{(8)} \sim 7 \cdot 10^{-2}$ Hz. Direct spacecraft measurements of the IMF made by Voyager 2 near 8.5 AU, did proved the existence of the Kolmogorov $f^{-5/3}$ spectrum over more than 4 decades in frequency, from $f = 5 \cdot 10^{-2}$ Hz to $f = 3 \cdot 10^{-6}$ Hz [Burlaga and Klein, 1986].

So far we neglected possible existence of the gigantic convection cells with the typical size of the order of the depth of the solar convection zone, $L_g \sim 3 \cdot 10^5$ km [Priest, 1982]. The gigantic cells could be responsible for the distribution of the large-scale patterns across the solar surface. However, the existence of these cells hasn't been clearly established. Nevertheless, fractal properties of the solar wind turbulence must be sensitive to dynamical phenomena in the solar wind induced by the gigantic cells. In fact, the gigantic cells involve overlapping of the magnetic superclusters at the heliocentric distances $R''_* \sim (L_g/L_{max})R'_* \sim 60$ AU which probably leads to formation of gigantic clusters (super-superclusters) in the solar wind. This implies a change in the slope of the power spectrum of the IMF turbulence, as well as in the temporal fractal dimension of the time series δ, representing direct measurements of the solar wind magnetic field in the distant heliosphere. Actually the dimensionless size of the gigantic clusters is of the order of $(L_g/L_{max})\Lambda(R'_*) \sim 2 \cdot 10^3$, hence their spatial fractal dimension is $D(R''_*) \approx 1.37$ (see equation (10)). From equations (17) and (18) we find the temporal dimension $\delta(R''_*) \approx 1.63$, and the slope of the power spectrum $\alpha \approx 1.74$. The numerical values for the fractal dimensions $D(R''_*)$ and $\delta(R''_*)$ only slightly differ from that obtained at $R \sim R'_*$ due to logarithmic dependence of these parameters on the linear size of the clusters. However, the formation of the gigantic clusters produces considerable impact on the frequency range related to the Kolmogorov spectrum, $f^{-5/3}$, of the turbulence. Indeed, the frequency f_S which determines the lower frequency limit for the Kolmogorov turbulence in the solar wind, should extend to $f_S \sim v_R / (L_g/L_{max})^2\Lambda(R'_*)a_{(8)} \sim 10^{-8}$ Hz at $R \geq R''_* \sim 60$ AU, but this case must be treated accurately because of the yearly variations of the solar magnetic field (for details, see [Priest, 1982]).

Basic dynamical phenomena in the solar wind fluctuations are shown at the diagram $f(R)$ (see Figure 2) representing the relationship between the frequencies of the solar wind fluctuations and fractal dimensions of the time series at different heliocentric distances R. This diagram summarizes theoretical results obtained in the previous

6. CONCLUSIONS

Theoretical results obtained in the previous sections, indicate that the Burger's turbulence and the f^{-2} spectrum appear to be dominant in the lower frequency range which is formally determined by the spatial size of magnetic clusters (superclusters) at 0.2 AU $\leq R \leq 8$ AU (accordingly, at $R \geq 8$ AU). The Burger's f^{-2} turbulence originates from the hierarchy of the convection cells in the photosphere and, therefore, describes spatial distribution of the magnetic clusters and superclusters in the solar wind. On the other hand, the Kolmogorov spectrum arises in the

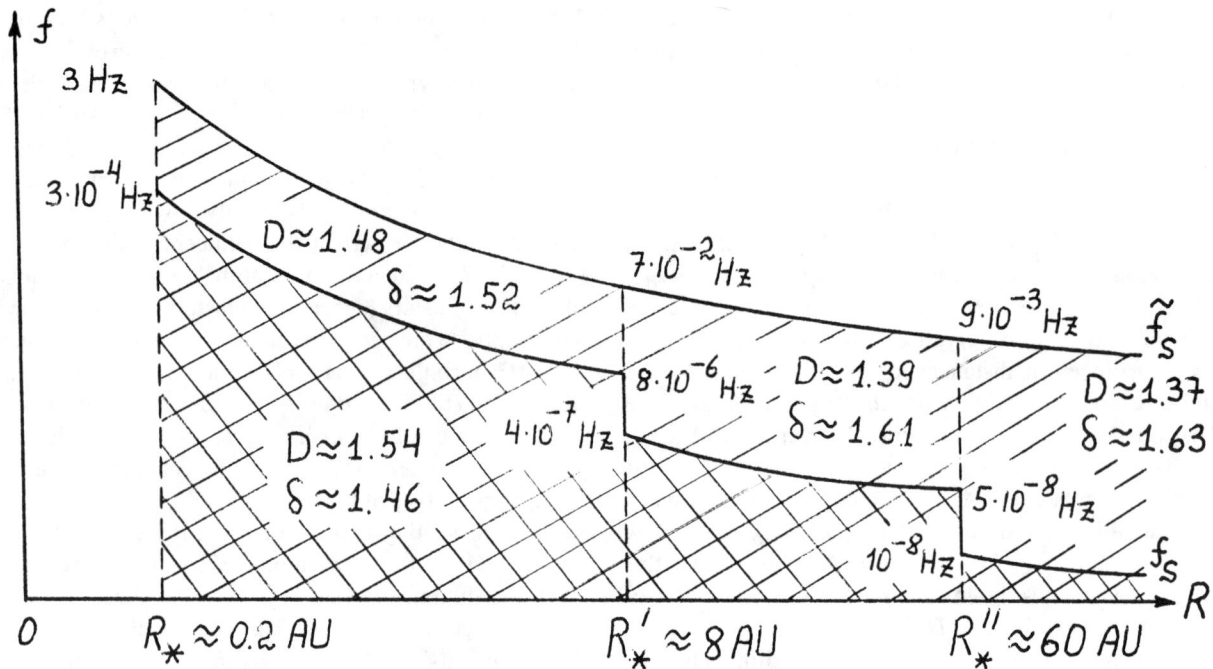

Fig. 2. Evolution of the IMF turbulence with the heliocentric distance. The area corresponding to the Kolmogorov turbulence, is dashed; meanwhile the area corresponding to the Burger's turbulence, is dashed twice.

higher frequency range which corresponds to the internal structure of the magnetic clusters (superclusters) considered as fractal sets of magnetic flux tubes. However, the spatial fractal dimension D which directly quantifies fine structure of these sets, depends on the parameters of the solar wind plasma distribution function and on the spatial size Λ of the sets. This result implies that the exponent of the power spectrum α at the higher frequency range varies slightly with the heliocentric distance R and tends towards the universal value $\alpha = 5/3$ at relatively large heliocentric distances (actually at $R \sim 60$ AU). The evolution of the IMF turbulence with the heliocentric distance R is illustrated by Figure 2. Since the magnetic flux tubes are assumed to expand radially with some constant velocity v_r as they move heliocentrically away from the Sun, the frequency f_S separating the Burger's turbulence and the Kolmogorov one, decreases as $1/R$ with the heliocentric distance R. This, in turn, indicates that the development of magnetic clusters and superclusters in the solar wind makes finally statistical properties of the IMF fluctuations independent of the boundary conditions setted at the photosphere level by the hierarchical structure of the convection cells. One could say that the solar wind "loses its memory" about the properties of the solar convection as it moves heliocentrically away from the Sun. This conclusion is not in contradiction with the previously reported results [e.g., see *Burlaga* 1987].

When considering the fractal model for the solar wind, we have presupposed that the quasi-stationary fractal distribution of the photospheric magnetic field is mapped directly into a quasi-stationary solar wind structure. It should be emphasized, however, that a large fraction of the photospheric magnetic field consists of closed magnetic flux tubes that are not mapped into the solar wind. Moreover, it seems reasonable to assume that fractal properties of the regions with the closed magnetic flux tubes could differ considerably from that with the open flux tubes, which, therefore, could involve a change in the statistical properties of the wind originating from the regions on the solar surface with different topology of the magnetic field. This could be an explanation for certain difference in statistical behavior of fast and slow solar wind streams at different heliocentric distances (e.g., see the reviews by *Hundhausen* [1972], *Gosling* [1981], *Burlaga et al.* [1983], and *Fainshtein* [1991]).

In the previous consideration we have neglected the solar cycle variations of the multifractal structure of the solar wind, which could induce considerable temporal modulation of the IMF turbulence. Recently *Burlaga et al.* [1993] obtained promising observational results indicating that the large-scale magnetic field strength fluctuations have multifractal structure with different parameters during the time intervals when the solar activity is declining; when the solar activity is low; and when the solar activity is increasing. These attractive observations support the basic theoretical idea claiming the existence of physical relations between the fractal properties of the IMF and bulk velocity fluctuations at different heliocentric distances, and the characteristics of large-scale magnetic configurations in the photosphere. A more detailed study of the dynamical phenomena governing yearly variations of the solar wind turbulence, we will address to a forthcoming paper.

The important point is that in the framework of the present study, the evolution of the turbulence in the solar wind is considered from the viewpoint of the self-organizing dynamical processes. A more comprehensive analysis of such processes (which is, however, beyond the model discussed) shouldn't ignore e.g., the presence of transient and corotating streams and shocks and their interactions, which are known to strongly influence the structure of the solar wind [*Burlaga and Goldstein, 1984*] as well as the generation of turbulence in the wind by velocity shears [*Coleman, 1968*]. An adequate consideration of these phenomena in addition to the fractal approach could offer new and attractive possibilities in the theoretical studies of the solar wind.

In conclusion let us stress that the fractal model for the solar wind discussed in the previous sections, is based considerably on the hierarchy of the convection cells. This fact entails an important ramification concerning statistical properties of the stellar wind. In fact, the stars of spectral classes *F* and *G* having the outer convection zone (e.g., the Sun), must create the wind with spatial fractal structure of the same origin as that in the solar wind. Meanwhile, the hot stars of spectral classes *O* and *B* actually have the convection core, but not the outer convection zone. It is entirely possible that these stars create the wind with somewhat different statistical properties depending on the particular mechanism of the magnetic field generation. Fractal configurations in the stellar wind (if any) with the considerable fine structure in a wide range of spatial scales might contain an important information about physical conditions at the inner layers of the stars.

7. APPENDIX

Calculation of the small parameter ε

Let us introduce the small parameter $\varepsilon \ll 1$, so that the mean separation between the neighbouring magnetic flux tubes inside the cluster is equal to $b = 2a(1 - \varepsilon)$, where a is the tube's radius. The area of the domain U where the (cylindrically-symmetric) flux tubes overlap, is approximately $S_U \approx 8\sqrt{2}\, \varepsilon^{3/2} a^2/3$ provided $\varepsilon \ll 1$. Then, the total magnetic energy concentrated in the domain U is

$$E_m = \frac{(2B)^2}{8\pi} S_U \quad (A1)$$

where B is the magnetic field of the flux tube. The excess of magnetic energy in the domain U stops attraction of the tubes to each other caused by the longitudinal currents [*Zelenyi and Milovanov, 1992a*]. The reciprocal energy of these currents is given by [*Landau and Lifshitz, 1975*]:

$$E_r = \frac{I^2}{2c^2}\left(1 + 4\ln\frac{b}{a}\right) \quad (A2)$$

where I is the total electric current along the tube. To find the equilibrium value of the parameter ε, one should minimize the sum of the magnetic and reciprocal energy to give

$$\varepsilon \approx \frac{\pi^2}{2}\left(\frac{1}{Bac}\right)^4 \quad (A3)$$

For the plasma distribution function of the form (5), the ratio I/Bac equals exactly to $\Delta/2$ [*Zelenyi and Milovanov, 1991*] where Δ is defined by equation (6). Hence equation (A3) becomes

$$\varepsilon \approx \frac{\pi^2}{32}\Delta^4(p). \quad (A4)$$

Substituting $\Delta(p) \approx 0.25$, one finds $\varepsilon \approx 1.2 \cdot 10^{-3}$.

Acknowledgments. We would like to thank Prof. I. Veselovskyi for helpful discussions of various aspects of this work.

REFERENCES

Berry, M.V., Diffractals, *J. Phys. A Math. Gen, 12*, 781, 1979.

Burlaga, L.F., Interaction regions in the distant heliosphere, *Solar wind six*, edited by V.J. Pizzo, T.E. Holzer, and D.G. Sime, NCAR Technical Note NCAR/TN-306 + Proc, 547, 1987.

Burlaga, L.F., Multifractal structure of speed fluctuations in recurrent streams at 1 AU and near 6 AU, *Geophys. Res. Lett., 18*, 1651, 1991a.

Burlaga, L.F., Intermittent turbulence in the solar wind, *J. Geophys. Res., 96*, 5847, 1991b.

Burlaga, L.F. and M.L. Goldstein, Radial variations of large-scale magnetohydrodynamic fluctuations in the solar wind, *J. Geophys. Res., 89*, 6813, 1984.

Burlaga, L.F. and L.W. Klein, Fractal structure of the interplanetary magnetic field, *J. Geophys. Res., 91*, 347, 1986.

Burlaga, L.F., E. Sittler, F. Mariani and R. Schwenn, Magnetic loop behind an interplanetary shock: Voyager, Helios, and IMP-8 observations, *J. Geophys. Res., 86*, 6673, 1981.

Burlaga, L.F., R. Schwenn, and H. Rosenbauer, Dynamical evolution of interplanetary magnetic fields and flows between 0.3 AU and 8.5 AU: Entrainment, *Geophys. Res. Lett., 10*, 413, 1983.

Burlaga, L.F., W.H. Mish and D.A. Roberts, Large-scale fluctuations in the solar wind at 1 AU: 1978-1982, *J. Geophys. Res., 94*, 177, 1989.

Burlaga, L.F., J. Perko, and J. Pirraglia, Cosmic-ray modulation, merged interaction regions, and multifractals, *Ap. J., 407*, 347, 1993.

Channell P.J., Vlasov-Maxwell equilibria with sheared magnetic fields, *Phys. Fluids, 19*, 1541, 1976.

Coleman, P.J., Turbulence, viscosity and dissipation in the solar wind plasma, *Ap. J., 153*, 371, 1968.

Fainshtein, V.G., The interaction effect of fast and slow solar wind streams in interplanetary space on wind characteristics at the earth's orbit, *Solar Phys., 136*, 169, 1991.

Feder J., *Fractals*, Plenum Press, New York, 1988.

Gosling, J.T., Solar wind stream evolution, in *Solar Wind Four*, edited by H. Rosenbauer, Max-Planck-Institut für Aeronomie, Garching, Federal Republic of Germany, 107, 1981.

Hundhausen, A.J., *Coronal expansion and solar wind*, Springer-Verlag, New York, 1972.

Landau, L.D. and E.M. Lifshitz, *Electrodynamics of continuous media*, Pergamon, New York, 1975.

Landau, L.D. and E.M. Lifshitz, *Hydrodynamics*, Nauka, Moscow, 1988. (in Russian).

Lawrence, J.K., Diffusion of magnetic flux elements on a fractal geometry, *Solar Phys., 135*, 249, 1991.

Mandelbrot, B.B., *The fractal geometry of nature*, W.H. Freeman, New York, 1983.

Marx, K.D., Equilibria and stability of collisionless plasmas in cylindrical geometry, *Phys. Fluids, 11*, 357, 1968.

Milovanov, A.V. and L.M. Zelenyi, Fractal model for sunspot evolution, *Geophys. Res. Lett., 19*, 1419, 1992.

Milovanov, A.V. and L.M. Zelenyi, Applications of fractal geometry to dynamical evolution of sunspots, *Phys. Fluids B, 5(7)*, 2609, 1993.

Priest, E.R., *Solar Magnetohydrodynamics*, Reidel, Dordrecht, 1982.

Richardson, R.S. and M. Schwarzschild, On the turbulent velocities of solar granules, *Ap. J., 111*, 351, 1950.

Tamano, T., A Plasmoid model for the solar wind, *Solar Phys., 134*, 187, 1991.

Tarbell, T., L. Acton, K. Topka, A. Title, W. Schmidt and G. Scharmer, Intermittency of fine scale solar magnetic fields in the photosphere, *Bull. Am. Astron. Soc., 22*, 878, 1990.

Tu, C.-Y., E. March, and K.M. Thieme, Basic properties of solar wind MHD turbulence near 0.3 AU Analyzed by Means of Elsasser Variables, *J. Geophys. Res., 94*, 11739, 1989.

Zelenyi, L.M. and A.V. Milovanov, Fractal properties of sunspots, *Sov. Astron. Lett., 17(6)*, 425, 1991.

Zelenyi, L.M. and A.V. Milovanov, Applications of lie groups to the equilibrium theory of cylindrically symmetric magnetic flux Tubes, *Sov. Astron., 36(1)*, 74, 1992a.

Zelenyi, L.M. and A.V. Milovanov, Evolution of sunspots: the cluster model, *Sov. Astron. Lett., 18(4)*, 249, 1992b.

A.V. Milovanov and L.M. Zelenyi, Space Research Institute, Russian Academy of Sciences, Profsoyuznaya 84/32, 117810 Moscow, GSP-7, Russia

Evolution of the Interplanetary Magnetic Field

DAVID J. MCCOMAS

Space and Atmospheric Sciences Group, Los Alamos National Laboratory, Los Alamos, New Mexico

Remote observations of magnetic field topologies in the solar corona and in situ observations of the solar wind and interplanetary magnetic field (IMF) are used to examine the temporal evolution of open and closed field regions linking the solar corona out into interplanetary space. The simple "open" configuration of inward and outward pointing IMF sectors is periodically disrupted by magnetically distinct coronal mass ejections (CMEs) which erupt from previously magnetically closed coronal field regions. At 1 AU, CMEs contain counterstreaming halo electrons which indicate their distinct magnetic topology. This topology is generally thought to be either: (1) plasmoids that are completely disconnected from the Sun; (2) magnetic "bottles," still tied to the corona at both ends; or (3) flux ropes which are only partially disconnected. Fully disconnected plasmoids would have no long term effect on the amount of open flux in interplanetary space. However, both *in situ* halo electron and remote coronagraph observations indicate that CMEs probably retain at least partial magnetic attachment to the Sun. Both the magnetic-bottle and flux rope geometries require some mitigating process to close off previously open fields in order to avoid a "flux catastrophe." In addition, the average amount of magnetic flux observed in interplanetary space varies over the solar cycle by a factor of ~50%, also indicating that there must be ways of opening new flux and closing off previously open flux. The most likely scenario for closing off open magnetic fields is for reconnection to occur above helmet streamers, where oppositely directed field regions are juxtaposed in the corona. Such reconnection would serve to return closed field arches to the Sun and release open, U-shaped structures into the solar wind; disconnection events of this nature have been remotely observed with the Solar Maximum Mission coronagraph. In interplanetary space, disconnection events may manifest themselves as heat flux dropouts (HFDs), in which the solar wind electron heat flux is absent or greatly reduced. This paper reviews and examines these issues and shows how the synthesis of remote global, and local, in situ, observations have been used to develop an understanding of the temporal evolution of open and closed magnetic field regions linking the corona into interplanetary space.

1. INTRODUCITON

Over the past three and a half decades of space flight *in situ* observations of the local space environs and remote imaging of distant regions have been used to examine a wide variety space physics phenomena. Regions studied span the gamut from the solar photosphere through the interplanetary medium to planetary ionospheres and magnetospheres.

Recently, considerable progress has been made on understanding the ways in which the open and closed field regions of the solar corona evolve and the effects of this evolution on the magnetic structure of the interplanetary medium. This progress has been driven by a combination of information from single point, *in situ*, observations in interplanetary space and remote coronal imaging; together they provide an example of how a global process in space and time can be at least partially resolved by the fusion of local detailed and remote synoptic observations.

Coronagraph and soft X-ray observations over the past several decades have shown that the solar corona is a highly dynamic and variable region. Short- and long-lasting coronal holes are observed in coronagraph images as dark regions; in these holes the solar magnetic field is "open" to interplanetary space and the solar wind plasma flows out with low density and high speed. In contrast, bright looplike structures are indicative of closed field regions on the Sun which contain high density plasma. Helmet streamers (closed field loops overlaid by radial high density streamer structures) map out to the heliospheric current sheet in interplanetary space [e.g., *Hundhausen*, 1977; *Gosling et al.*, 1981]. These various magnetically open and closed regions evolve over timescales as short as several hours and as long as the 22 year solar cycle.

The amount of magnetic flux present in interplanetary space can vary substantially. The top panel of Figure 1

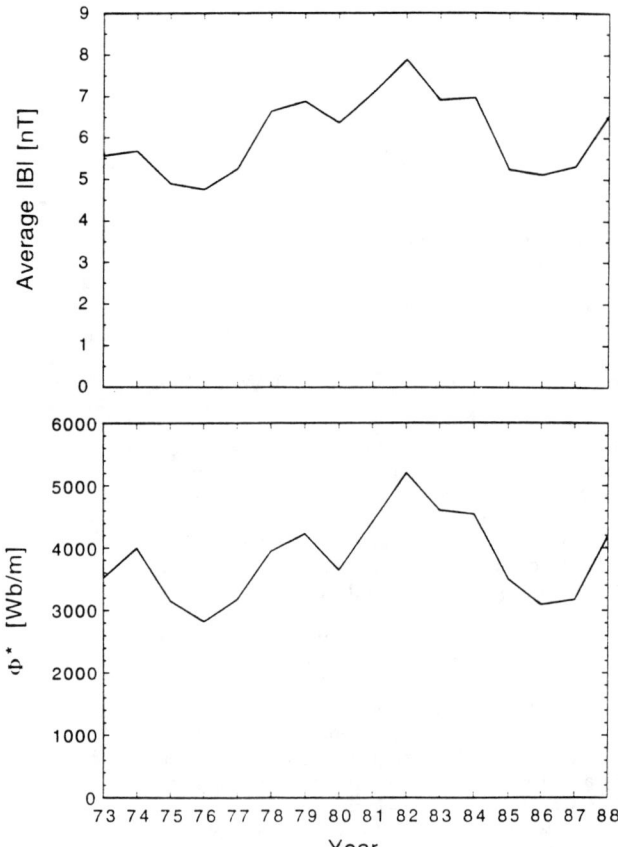

Fig. 1. Annual averages of |B| (top) and the 2-dimensional magnetic flux, Φ, (bottom) in the ecliptic plane over solar cycle 21.

displays the variation in annual averages of the magnitude of the magnetic field, |B|, from 1973 through 1988. This trace displays ~50% variation over solar cycle 21 [*Slavin et al.*, 1984; 1986], however, the variation over solar cycle 20 was much smaller [*King*, 1979]. The lower panel displays the variation in the 2-dimensional "total flux integral," Φ, which removes the effects of variations in the solar wind speed to provide a more reliable measure of the amount of magnetic flux in the ecliptic plane at 1 AU [*McComas et al.*, 1992a]. This integral similarly displays ~60% variation over the solar cycle.

Both |B| and Φ show basically the same variation over solar cycle 21; the maximum amount of magnetic flux observed in the ecliptic plane at 1 AU occurs in 1982, two years after solar maximum while the minima occur in 1976 and 1986-1987, shortly after solar minima. While not all of the flux measured at 1 AU is necessarily open, the variations observed in Figure 1 are so large that clearly the amount of magnetic field open from the Sun into interplanetary space can and does vary substantially. In order for this to be true, there must be processes for both 1) opening new, previously closed field regions in the corona and 2) closing off previously open field regions and returning closed loops to the Sun. Since the solar wind will carry away from the Sun any field structures formed past the Alfvenic point (≤ solar radii), closing off of previously open fields must occur inside this distance.

This paper examines the processes by which magnetic fields are opened and closed from the Sun into interplanetary space and reviews and synthesizes the observations of four phenomena which may be manifestations of these processes. Sections 2 and 3 discuss coronal mass ejections (CMEs) as observed remotely with coronagraphs and counterstreaming electron events observed with plasma data in the solar wind at 1 AU. This latter signature provides the best indicator of CMEs in the interplanetary medium; here, both data sets are used to examine the topologies of mass ejections. Section 4 examines the possibility that heat flux dropouts (HFDs) in the solar wind at 1 AU could be the interplanetary manifestation of the disconnection of previously open magnetic fields. Section 5 discusses coronal disconnection events observed, again, remotely with coronagraphs. We argue that at least some fraction of the HFDs observed at 1 AU are the interplanetary manifestations of coronal disconnections, and examine the implications of these topologically detached structures. Finally, Section 6 examines these phenomena in terms of the opening and closing magnetic fields from the sun and summarizes this papers results.

2. CORONAL MASS EJECTIONS AND THE OPENING OF NEW MAGNETIC FIELDS

Coronal mass ejections were first discovered in OSO-7 coronagraph data [*Tousey*, 1973] and were extensively examined using Skylab observations [*Gosling et al.*, 1974; 1976]. CMEs originate from closed field regions of the corona that were not previously participating in solar wind expansion [*Gosling et al.*, 1976]. As such, magnetic field loops from a closed region erupt and expand outward into interplanetary space.

Figure 2, adapted from Gosling *et al.* [1974], shows two images from the CME eruption of August 10, 1973. The bright loops on the west limb of the Sun indicate higher plasma densities than in the surrounding corona. Due to the very high electrical conductivity of the coronal plasma, such loops have generally been interpreted as magnetic flux tubes. While this interpretation is not critical for the purposes of this paper, it is consistent with the notion that CMEs arise from previously closed field regions on the Sun. The left and right panels in Figure 2 were taken 24 minutes apart, indicating that over this interval the CME expanded outward with an apparent radial speed of ~400 km s^{-1}.

At later times in this event the CME continues to expand outward with the leading edge quickly passing beyond the Skylab field-of-view [*Gosling et al.*, 1974]. After the passage of the CME, long lasting, bright radial structures are observed to persist for at least another half day. These structures, called the "legs" of the CME, are typically found along

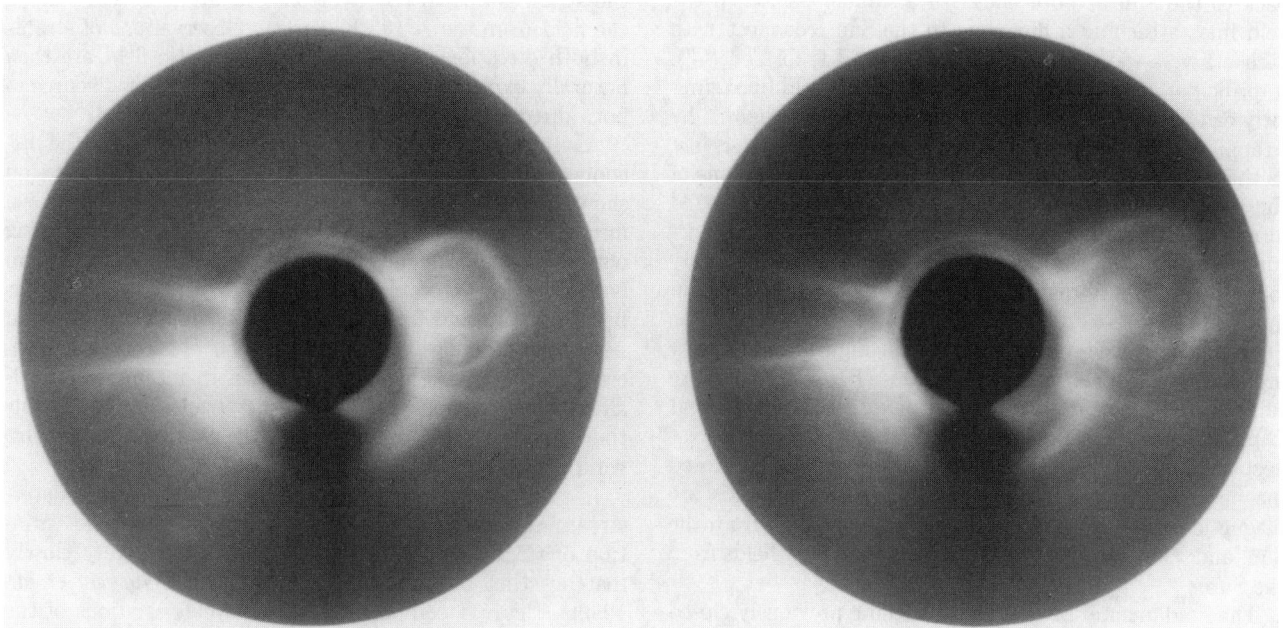

Fig. 2. Coronagraph images showing the release of a coronal mass ejection on 10 August, 1973; the two images are separated by only 24 minutes. These images indicate the rise of closed magnetic field loops which appear to maintain magnetic attachment to the Sun. Adapted from Gosling et al. [1974].

the edges of the region of the CME eruption. In some cases, legs have been observed to persist for several days after a CME. While not a unique interpretation [e.g., *Kahler and Hundhausen*, 1992], such structures have generally been interpreted as evidence that CMEs generally maintain some magnetic connection to the Sun.

Because the electrical conductivity of the solar wind is extremely high, magnetic field lines provide a useful paradigm for understanding the magnetic field configuration in interplanetary space. The only four fundamentally distinct magnetic topologies in the heliosphere are displayed schematically in panels A-D in a "flat heliosphere" representation in Figure 3. Panel A shows normal "open" field lines; open field lines extend outward from the Sun and close somewhere far away in the outer heliosphere. Panel B displays a magnetic tongue or bottle geometry in which the magnetic field closes to the Sun at both ends [*Gold*, 1959, *Hundhausen*, 1972]. Panel D shows a disconnected region in which the magnetic field lines connect to the outer heliosphere at both ends [*McComas et al.*, 1989]. Panel C displays a fully disconnected plasmoid which closes on itself, that is it is not magnetically connected to either the Sun or the outer heliosphere. Both the bottle (B) and the plasmoid (C) topologies are typically called "closed" because the fields in them do not extend into the outer heliosphere. Beyond the Alfvenic point (dashed line), the solar wind flows everywhere outward from the Sun (bottom to top in Figure 3) so that over time the field lines in panel B naturally evolve into open field lines (panel A) while field lines in panels C and D simply propagate out (upward) through the system.

Panel B' displays a magnetic flux rope that is in some ways intermediate between the bottle (B) and the plasmoid (C) geometries [*Gosling*, 1990]. In the flux rope geometry the field loops around or possibly even through itself instead of being simply connected. Fundamentally, however, this topology is equivalent to panel B in that it connects

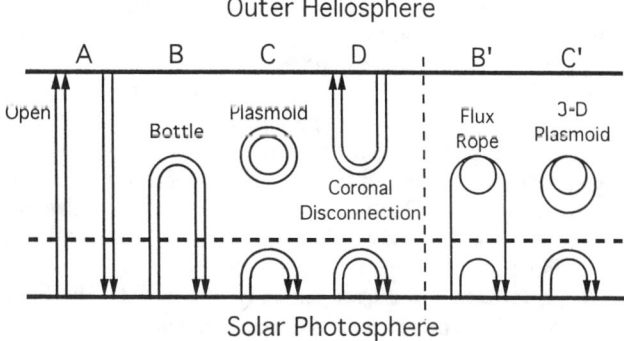

Fig. 3. Schematic diagram showing normal open magnetic fields (A), a magnetic bottle closed to the Sun at both ends (B), a plasmoid magnetically closed on itself (C), and a coronal disconnection in which the field is open to the outer heliosphere at both ends (D). The flux rope (B') and 3-D plasmoid (C') are simply more complicated versions of the topologies shown in B and C, respectively.

back to the Sun at both ends. In a similar fashion, if the field lines attaching a flux rope to the Sun reconnect with each other, a more general (3-D) plasmoid is formed (C'). In principal, open (A) and disconnected (D) field lines similarly can all be more complicated than drawn in Figure 3 by wrapping around and/or through themselves. However, just as the flux rope and 3-D plasmoids are simply extensions of the bottle and 2-D plasmoid topologies, such complicated forms would simply be complications of these other structures. The structures shown in panels A-D are the only four fundamentally distinct field line topologies possible in interplanetary space.

Four of the six geometries shown in Figure 3 have been proposed for CMEs: magnetic bottles (B) and their more complicated forms, flux ropes (B'), and plasmoids (C and C'). Bottle and flux rope geometries maintain magnetic connection to the Sun and, consequently, open previously closed magnetic fields into interplanetary space. Plasmoids are magnetic structures which become fully detached from the Sun, and, as a consequence, do not open any new fields from the Sun.

The fundamental problem with opening previously closed magnetic field regions on the Sun to interplanetary space is that in the absence of other mitigating processes, the amount of open magnetic field in interplanetary space, and hence IMF magnitude, would grow without bound. This "flux catastrophe" was first pointed out by Gosling [1975]. MacQueen [1980] used Skylab CME observations to very roughly estimate that the IMF magnitude should approximately double over ~ 100 days if all of the magnetic loops in the observed CMEs remained rooted in the Sun (B as opposed to B' in Figure 3).

3. COUNTERSTREAMING ELECTRON EVENTS: THE INTERPLANETARY SIGNATURE OF CMES

A number of distinct signatures in interplanetary space have been previously associated with CMEs. These various signatures, which were recently summarized by Gosling [1990], include 1) helium abundance enhancements, 2) ion and electron temperature depressions, 3) unusual ionization states, 4) strong magnetic fields, 5) low magnetic field variances, 6) low plasma beta, 7) anomalous field rotations, 8) counterstreaming energetic protons, and 9) counterstreaming suprathermal electrons. Of all of these signatures, counterstreaming suprathermal (halo) electrons [e.g., *Gosling et al.*, 1987] is the most robust signature of CMEs in the interplanetary medium [*Gosling*, 1990].

If CMEs remain at least partially attached to the Sun, as indicated in the previous section, then counterstreaming suprathermal electrons would be expected. In the normal solar wind, the electron heat flux is carried by halo electrons streaming outward along the local magnetic field. This beam of energetic electrons, or "strahl" [*Rosenbauer et al.*, 1977], is observed nearly continuously in the solar wind. Since the normal IMF is effectively connected to the Sun at one end and the outer heliosphere at the other (A in Figure 3), the strahl is due to hot electrons streaming outward along the field from the $\sim 10^6$ K corona. Observations of strahls in both directions along the local magnetic field are then naturally explained as connection to a hot coronal source in both directions back along the local magnetic field.

A similar explanation has also been proposed for the disconnected, plasmoid topology (either the simple 2-D version shown in panel C or the more generic 3-D version shown in panel C'); in either case hot electrons initially streaming outward along both ends of the loops continue to circulate in both directions around the plasmoid even after the field has been disconnected from the Sun. One important question with this explanation is whether such electrons are sufficiently non-interacting that they can maintain their field-aligned distributions over the 2-4 day transit to 1 AU while they are convected outward from the Sun at normal solar wind speeds.

In addition to the topological explanation of counterstreaming events as CMEs, the overall solar cycle variation of CMEs as observed with coronagraphs very closely matches that of counterstreaming events [*Gosling et al.*, 1992]. Figure 4 compares the solar cycle variations of the percentage of the time that counterstreaming electrons are observed at 1 AU (top panel) and an estimate of the annual whole-Sun occurrence rates of CMEs (middle panel). The smoothed sunspot number (bottom panel) indicates the phases of the solar cycle. Clearly the solar cycle variations in CMEs and counterstreaming electron events closely track each other. A quantitative comparison of the rates of these two phenomena also indicates reasonable agreement [*Gosling et al.*, 1992].

In order to assess whether counterstreaming halo electron events represent magnetic structures which are still connected to or disconnected from the Sun, details of the electron distributions in counterstreaming events at 1 AU were examined [*Phillips et al.*, 1992]. In particular, these authors examined asymmetries between the counterstreaming beams and variations in the field orientation during transit through CMEs. They concluded that this evidence suggests that CMEs maintain at least some magnetic connection to the corona.

In addition to statistical evidence that favors a bottle or flux rope geometry over plasmoids, Phillips *et al.*[1992] also showed the first examples of a new type of distribution that they refer to as "strahl-on-strahl" (SOS). Figure 5 displays a SOS distribution in which halo electrons are observed in two broad counterstreaming beams with an additional narrow strahl superposed in only the direction running most nearly outward from the Sun along the local magnetic field. These authors interpret SOS distributions as additional evidence for magnetic bottles or flux ropes in which counterstreaming halo electrons mirror repeatedly to form the broad counterstreaming beams while new suprathermal electrons making their first outward transits constitute the narrow, pristine strahl.

The monotonically increasing trace in Figure 6 shows the rate at which 2-D magnetic flux was opened from the Sun

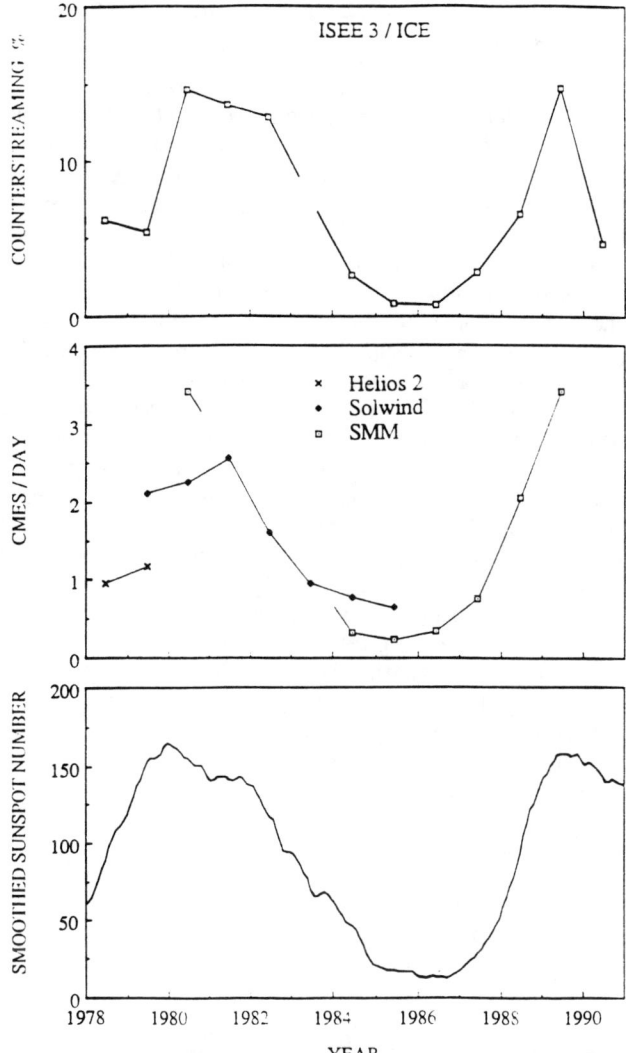

Fig. 4. Greyscale coded electron counts in the energy ranges from 137-362 eV and 500-958 eV from March 22 and 23, 1979. Prior to ~14:00 UT a beam or strahl carrying the bulk of the solar wind electron heat flux is observed. After this time the spectrogram displays an interval of counterstreaming halo electrons. From Gosling et al. [1986].

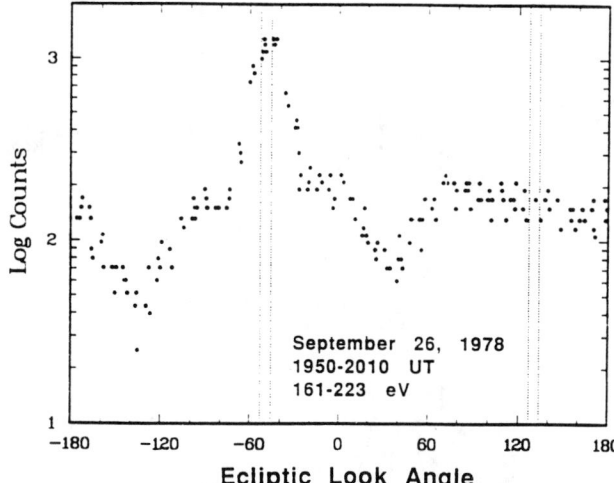

Fig. 5. "Strahl-on-strahl" distribution showing counts in the 161-223 eV range of halo electrons versus in-ecliptic angle. Dotted lines indicate the range of magnetic field orientations during the measurement interval. Taken from Phillips et al. [1992].

and added to the IMF, in the 1978-1980 timeframe, assuming that all counterstreaming electron events represent simply connected magnetic bottles (B in Figure 3). The dashed line indicates the 2-D magnetic flux (described above and defined by McComas et al. [1992a]) present, on average, in the ecliptic plane at 1 AU. This figure shows that for simple bottle geometry CMEs (assuming solar maximum CME rates) the amount of flux would double over ~9 months in the absence of any mitigating processes.

A twist on the flux catastrophe is that if CMEs remain magnetically attached, but are flux ropes (B' in Figure 2) [Gosling et al., 1990], that rate of build-up would be reduced. If, for example, a field line in a typical CME came out from the Sun, looped around once, and then returned to the Sun, the rate of flux build-up would be only one half as fast as that shown for simple bottles (Figure 6). The above estimate for magnetic flux doubling would then be increased from ~9 months to one and one half years. For very complicated geometries with many many loops, the rate of build-up would be only a tiny fraction of that calculated for simple bottles. However, it must be stressed that if CMEs retain

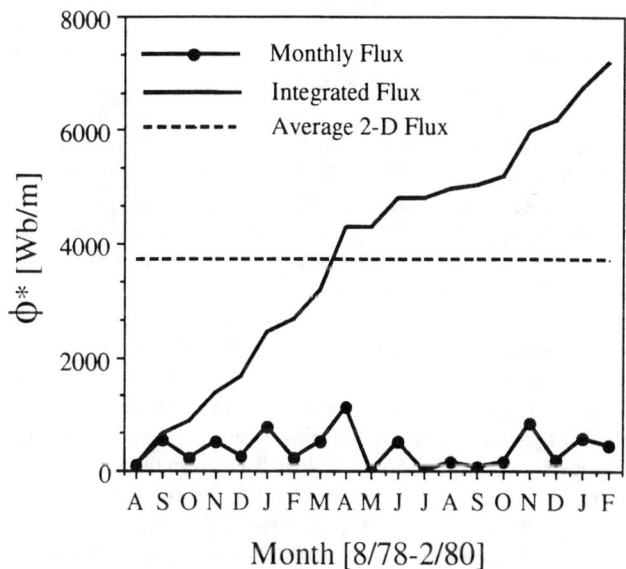

Fig. 6. Monthly and integrated 2-D magnetic flux (Φ) observed in CMEs (bidirectional electron events) for the interval from August 1978 through February 1980. The dashed line indicates the average amount of 2-D flux present in the ecliptic plane on average. From McComas et al. [1992b].

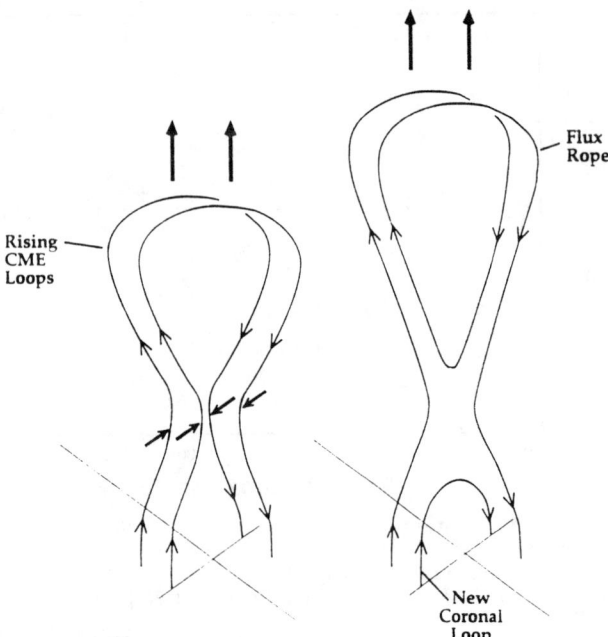

Fig. 7. Schematic diagram of the formation of a magnetic flux rope. Closed field arches are sheared by motions of their footprints down in the photosphere. As the arches begin to expand upward, compression of the field out of the plane of the field lines will cause them to become non-planar and to reconnect with adjacent lines. This process leads to the formation of a flux rope and underlying closed field arches (bottom panel). Taken from Gosling [1993].

any attachment to the Sun whatsoever, the flux catastrophe will ultimately occur in the absence of some other process to close off previously open fields.

A schematic representation of how CMEs could form as flux ropes is displayed in Figure 7. The process begins with sheared coronal magnetic field loops which expand upward. The shear is created by motions of the magnetic footprints down in the photosphere. The same sort of footprint motions can also bring the feet closer together increasing the probability of reconnection. When the loops are pushed together by some force which is out of the plane of the individual field lines, the lines become non-planar, and reconnection can occur between adjacent field lines (top panel). This process leads to both a rising flux rope topology CME and the creation of underlying magnetic loops (bottom panel). Since the ends of the newly formed flux rope, which still connect back to the Sun, are not located physically near to each other, it is hard to imagine how these fields could reconnect to form a fully detached 3-D plasmoid. While the simple example shown in Figure 7 displays formation of only a single loop flux rope, this process could progress along a series of sheared arches to produce a long (and potentially magnetically complicated) flux rope and an arcade of underlying coronal loops. Observations of the corona and in interplanetary space indicate that flux ropes may account for ~1/3 of all CMEs [Gosling, 1990; 1993].

4. HEAT FLUX DROPOUTS: A SIGNATURE OF RECONNECTION ON OPEN FIELD LINES?

The interpretation of 1) the normal mono-directional heat flux as evidence for simple magnetic connection to the Sun along open field lines (Figure 3-A) and 2) counterstreaming halo electron events as evidence for closed magnetic structures (Figure 3-B/B', C/C') led naturally to the question of whether there were also intervals devoid of electron heat flux in the solar wind. If so, these intervals might provide evidence for magnetic structures that are disconnected from the Sun and open to the outer heliosphere at both ends [McComas et al., 1989]. Panel D of Figure 3 schematically displays this sort of magnetic topology.

Heat flux dropouts (HFDs), where outward streaming halo electrons are not evident in the solar wind, have been found using data from the ISEE-3 solar wind electron experiment [McComas et al., 1989]. Figure 8 displays cuts through the electron distribution function, f(v), for an interval of normal solar wind (circles) and during an HFD (squares). The vast majority of the electrons lie within the core population (seen in Figure 8 as the tall central peak with |electron speeds| < ~5000 km s^{-1}). The superposed halo population is broader and hotter (|electron speeds| > ~5000 km s^{-1}). The substantially enhanced component in the negative velocity direction along the field in the normal solar wind is absent during the HFD.

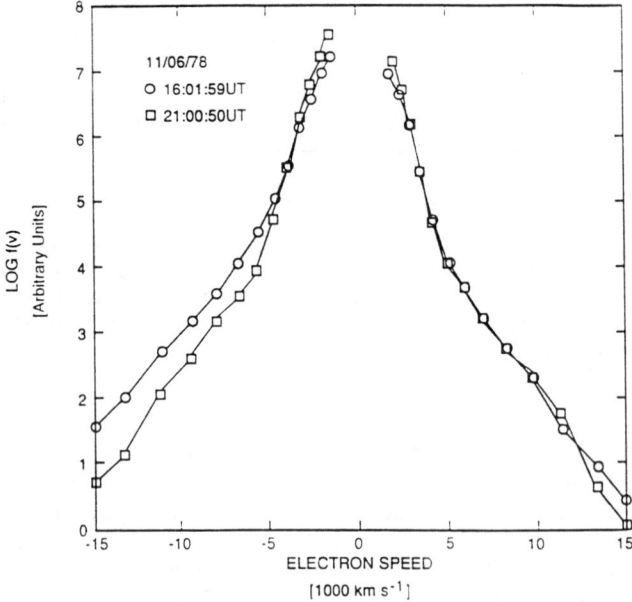

Fig. 8. Cuts through the electron distribution function, f(v), along the local magnetic field direction prior to (16:01:59 UT) and during (21:00:50 UT) a heat flux dropout. The difference between the two curves at large negative speeds (left third of figure) represents the missing strahl electrons which carry the majority of the heat flux out from the Sun. Taken from McComas et al. [1989].

Heat flux dropouts were identified purely on the basis of the electron distributions [*McComas et al.*, 1989; 1992b]. However, when these HFD intervals were compared to the interplanetary magnetic field, a surprisingly good correlation was discovered. Figure 9 shows the correlation of HFDs (black squares) with the in-ecliptic spiral angle. Over the ~18 months covered by this plot, the sector structure is evident as the quasi-regular transitions back and forth between the normal spiral angles at 1 AU (135° and -45°). While not all sector boundary crossings have associated HFDs, when HFDs do occur, they are generally associated with large rotations of the IMF in general, and sector boundary crossings in particular.

The obvious explanation for HFDs is that they represent structures magnetically disconnected from the hot solar corona. Figure 10 displays a time sequence of diagrams which shows schematically how reconnection between oppositely directed open magnetic fields can lead to the formation of disconnected U-shaped magnetic structures. After disconnection, such structures would be carried outward with the normal solar wind, passing 1 AU in typically 2-4 days. The hot halo electrons run out along the field and pass 1 AU in only a few hours. Consequently, there is a very large probability that once such a structure is formed, it would be observed as an HFD in interplanetary space [*McComas et al.*, 1989; 1992a; 1992b]. The association of HFDs with the heliospheric current sheet is naturally explained by this process since oppositely directed fields are juxtaposed above helmet streamers, at the base of the heliospheric current sheet.

In an independent study of HFDs, Lin and Kahler [1992] examined the electron distributions at higher energies (2-8.5 keV solar electrons). Due to their higher energies, such electrons are superior tracers of magnetic topologies compared

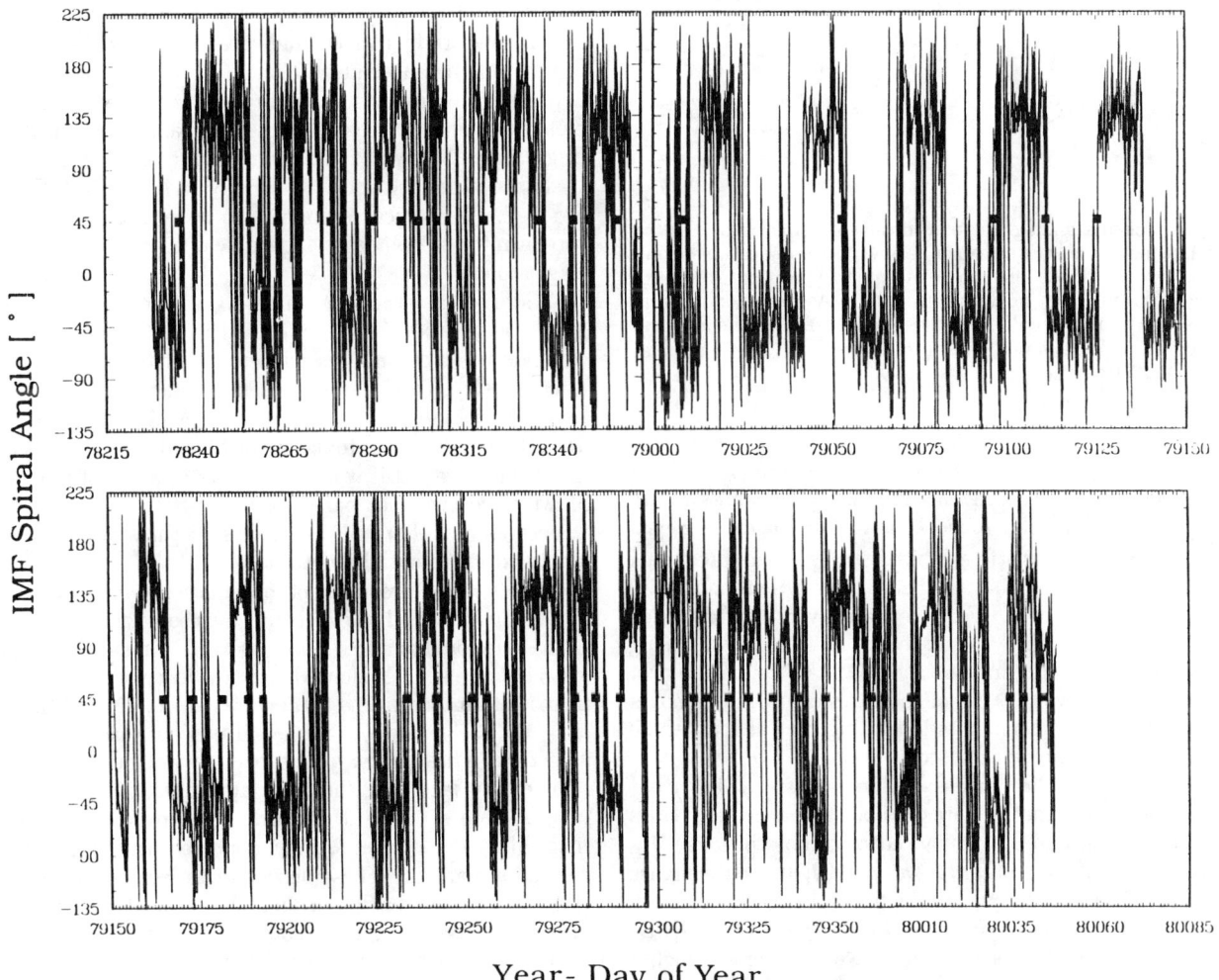

Fig. 9. Hourly averaged IMF spiral angle plotted as a function time for the 18 month interval from 8/78-2/80. The squares indicate times of 83 HFD events. Note the excellent correlation between HFDs and large rotations of the field, especially sector boundary crossings. From McComas *et al.* [1992d].

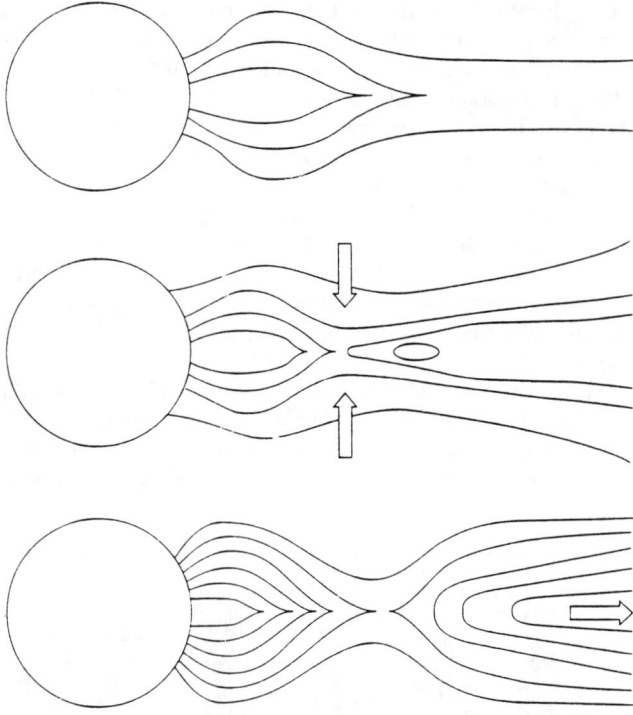

Fig. 10. Schematic diagrams of a coronal disconnection. The top panel displays a helmet streamer geometry. Reconnection begins, in this case on a closed field loop, but quickly propagates through to the adjacent open field lines (middle). As reconnection continues (bottom), closed field arches are returned to the Sun and a disconnected, U-shaped structure is released. From McComas et al. [1992c].

to 0.1-1 keV halo electrons, whenever they are present. Of the 25 events on the original HFD list [McComas et al., 1989], Lin and Kahler found that in nine of these events the >2 keV electrons still exhibited strong mono-directional streaming indicating that the field lines were probably still connected to the Sun. For the majority of the identified HFDs (13 of 25), only a weak mono- directional streaming was observed, leaving in doubt the ultimate connectivity of these events. However, two of the cases exhibited dropouts even at these higher energies indicating that at least these events were probably real disconnections.

In addition to the obvious explanation for HFDs noted earlier, other explanations are possible. In particular, HFDs may simply be caused by enhanced coulomb collisions [McComas et al., 1989; Lin and Kahler, 1992] due to the higher densities often associated with the heliospheric current sheet in interplanetary space [Borrini et al., 1981]. In fact, HFD-associated current sheet crossings have, on average, higher plasma density than those not associated with HFDs [McComas et al., 1989]. While scattering may account for some HFDs, Lin and Kahler [1992] showed that it cannot account for all of the HFD observations. Perhaps both disconnected U-shaped field structures and enhanced collisions produce HFDs.

5. CORONAL DISCONNECTIONS: DIRECT EVIDENCE FOR CLOSING OF OPEN IMF

As can be seen in Figure 10, the topological effect of reconnecting oppositely directed, previously open fields near to the Sun is to 1) return closed field loops to the Sun and 2) release disconnected U-shaped field structures into interplanetary space. It was, in fact, the interpretation of HFDs as disconnected magnetic structures that motivated an examination of coronagraph observations for direct evidence of such disconnections in the solar corona [McComas et al., 1991].

Figure 11 displays a sequence of four SMM coronagraph images from 1 June 1989. In this sequence a large coronal streamer (straight up in the 6:32 UT image) appears to disconnect. By the 11:10 UT image, the formation of a dark band across the structure at a heliocentric distance of ∼2 solar radii (arrow) indicates that reconnection has begun. In the subsequent image (14:16 UT), reconnection appears to continue as the separation, indicated by the dark band, continues to grow. The resulting appearance is that of bright loops below the dark band and a large, disconnected U-shaped structure above it. Finally, in the 17:05 UT image the U-shaped structure appears to have lifted-off from the corona into interplanetary space, leaving only an arcade of bright loops.

A statistical study of three months of SMM coronagraph observations was performed in order to assess how common these coronal disconnection events are [McComas et al., 1992c]. The interval chosen extended from 1 January through 26 March 1988. The initial SMM survey [St. Cyr and Burkepile, 1990] found no obvious disconnections during this interval. It is worth noting, however, that the initial SMM survey was not particularly sensitive to disconnected structures and that only ∼3% of structures observed were identified as potentially detached.

Over the ∼3 months of observations, St. Cyr and Burkepile identified 53 events in which some outward motion of material or structures were observed. On careful reanalysis, 6 of the 53 (11%) showed some evidence of disconnection in more than one frame [McComas et al., 1992c]. In addition, 13 (23%) showed 1 frame with an outward "U" or "V" structure. These latter structures may not be significant in light of the fact that there are ambiguities in interpreting images from optically thin media. In any case, of the six more likely detachments, most appeared as detachments of coronal streamers. These observations led McComas et al. [1992c] to conclude that magnetic disconnection events on previously open field lines (above helmet streamers) may be far more common than previously appreciated.

There is an documented example of a possible coronal disconnection as far back as the 16 April 1893 solar eclipse. Figure 12, taken from Cliver et al. [1989], shows both the locations of three observing sights (Chile, Brazil, and Senegal)

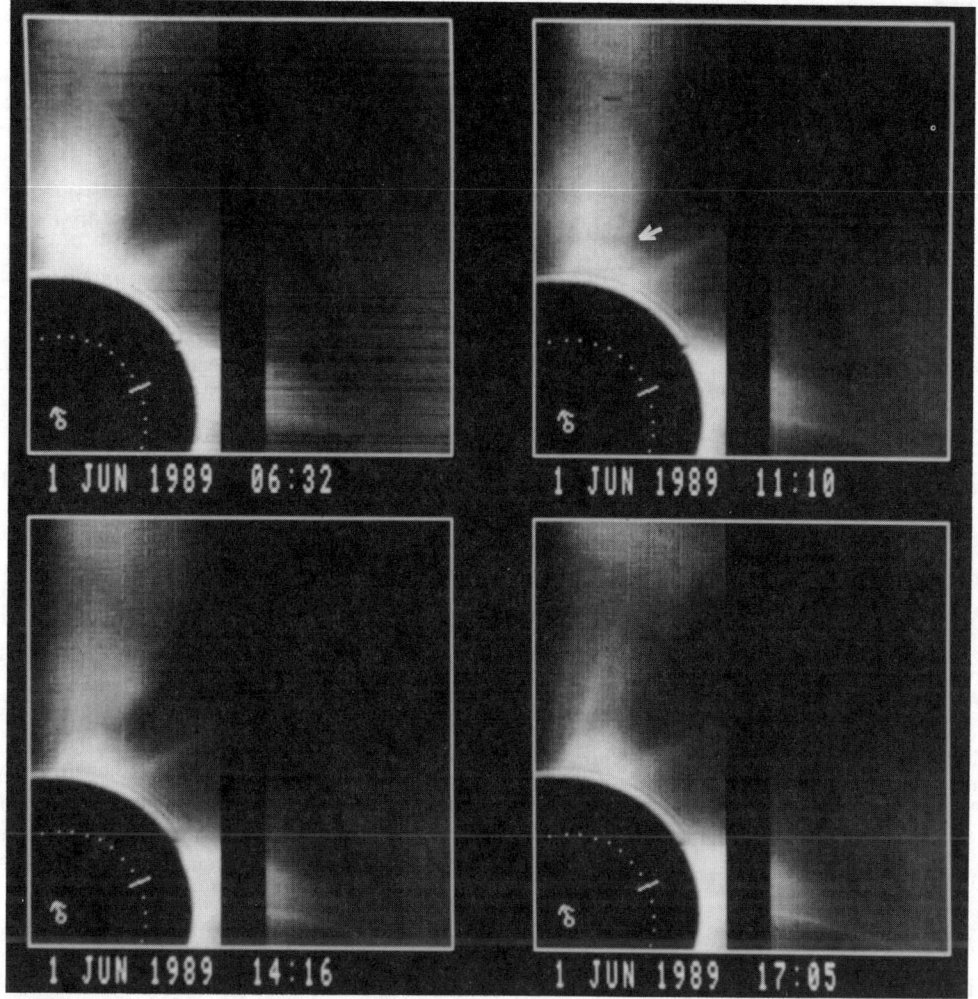

Fig. 11. Sequence of four coronagraph images from SMM displaying a coronal disconnection event on 1 June 1989. The arrow in the 11:10 UT image indicates where the reconnection line is presumably forming. By 17:05 UT the entire structure has lifted off leaving only small underlying arches. From McComas et al. [1992c].

and sketches made at each of these sights, indicating the outward motion of a large U-shaped structure. This structure was originally interpreted as a comet, but Cliver et al. [1989] reinterpreted it as evidence of a fully detached CME (e.g., a plasmoid). While this latter interpretation may be correct, the sketches do not display any leading edge of a CME. Rather, the size and shape of this event are highly reminiscent of coronal disconnection events found in the SMM data (i.e., the 27 June 1988 event [McComas et al., 1991]).

Another independent line of evidence for magnetic reconnection above helmet streamers is provided by "high coronal flares" [Cliver and Kahler, 1991]. These impulsive low energy (<10 keV) solar electron events have been inferred to arise from acceleration in the corona at heights above the solar surface >0.5 solar radii [Potter et al. 1980]. Reconnection across the base of the heliospheric current sheet, above helmet streamers, is the most likely location for such high altitude reconnection [Cliver and Kahler, 1991]. Lin [1985] found 135 such events over a 15 month period, corresponding to one event roughly every three days; such a high rate suggests that the formation of detached coronal magnetic structures, may be a common feature of the solar corona.

6. SUMMARY AND DISCUSSION

Coronal mass ejections have been inferred to be magnetic 1) bottles, 2) flux ropes, or 3) plasmoids. A number of lines of evidence indicate that all CMEs cannot be fully detached plasmoids (2-D or 3-D). First, if this were the case, some other process would need to account for the long-term increase in $|B|$ and the 2-D flux in the ecliptic plane, Φ, observed over the rising portion of solar cycle 21. In addition, the persistence of long lasting "legs" along the flanks of

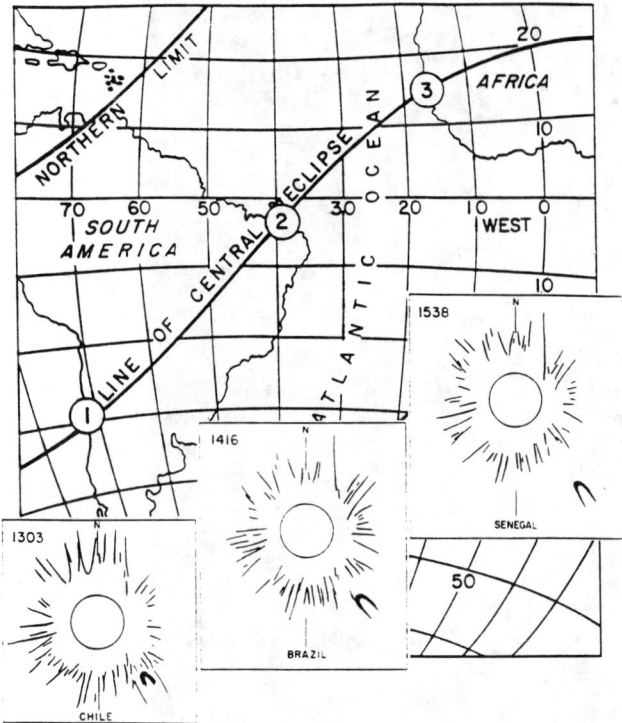

Fig. 12. Possible coronal disconnection observed from Chile, Brazil, and Senegal during the 16 April 1893 solar eclipse. The sketches indicate the time sequence of coronal structure observed from the three independent locations. Taken from Cliver et al. [1989].

the regions that CMEs rise out of have generally been interpreted as continued magnetic connection to the Sun. Details of the particle distributions inside the interplanetary manifestations of CMEs (counterstreaming halo electron events) also indicate that CMEs remain magnetically attached to the Sun. Finally, although it is possible to form completely closed magnetic structures at the Sun, it is hard for me to imagine how all of the magnetic field lines in every CME could somehow find precisely the right partners to reconnect with to form such truly disconnected structures.

If CMEs are either magnetic bottles or flux ropes (or both), then they serve to open previously closed coronal magnetic fields to interplanetary space. In the absence of some method for closing off previously open flux, the magnetic field in interplanetary space would build-up without bound leading to the so-called flux catastrophe. Both in order to avoid the flux catastrophe, and to explain the reduction in |B| and Φ over the declining phase of solar cycle 21, some other process must be closing off previously open field regions from the Sun. The most likely method for removing open field regions and returning closed field arches to the Sun is by reconnecting oppositely directed fields at the base of the heliospheric current sheet above helmet streamers. This process, combined with CMEs opening new flux from the Sun provides a ready method for moderating the amount of coronal magnetic flux open to interplanetary space.

There is also independent evidence for reconnection above helmet streamers in the corona. Coronal disconnection events remotely observed in sequences of coronagraph (and eclipse) images appear to provide direct evidence for such reconnection. In addition, at least some heat flux dropouts appear to represent the interplanetary manifestation of magnetic field structures which are disconnected from the Sun and connected to the outer heliosphere at both ends. Finally, impulsive electron events with energies <15 keV appear to be best explained in terms of acceleration in reconnection regions above helmet streamer at heights >0.5 R_S above the solar surface.

While there are long-term, solar cycle variations in the amount of magnetic flux in interplanetary space, the fact that this flux neither grows nor shrinks without bound suggests that there must be some feedback process between the opening and closing of magnetic field regions from the Sun. The transverse magnetic pressure in the solar corona has been proposed as one possible mediator of the amount of open flux [McComas et al., 1989; 1991]. In this model, the expansion of newly opened field regions would enhance the transverse pressure around the Sun. This transverse pressure could build-up until it was just large enough that oppositely directed magnetic fields were marginally stable against reconnection. Thereafter, the opening of more regions would increase the pressure to the point that the oppositely directed fields would be unstable, and reconnection would commence until sufficient flux was removed to again attain a marginally stable configuration.

As an example of this feedback process, the coronal disconnection event on 27 June 1988 was preceded by a substantial disruption of the streamers in the corona in which these structures were observed to fold around the Sun toward the region of the disconnection [McComas et al., 1991]. These observations provide direct evidence that the coronal pressure was enhanced in the region of the disconnection just prior to the detachment.

The suggestion that opening and closing magnetic fields maintain some sort of equilibrium has also been examined numerically [Linker et al., 1992]. These authors used a 2-D MHD simulation of the corona to examine the effect of increasing the overall magnetic pressure around an equilibrium helmet streamer. Figure 13 displays a sequence of magnetic field traces at different times, t, following the increase in pressure. Once the pressure is enhanced, the streamer configuration reconnects until sufficient flux has been removed to decrease the pressure to the point where an equilibrium can again be maintained.

In addition to accounting for 1) the variation in the amount of magnetic flux observed in interplanetary space over the solar cycle and 2) all of the observed features of CMEs and coronal disconnection events, the opening and closing processes described here could also account for how coronal holes and closed loop structures can quickly move

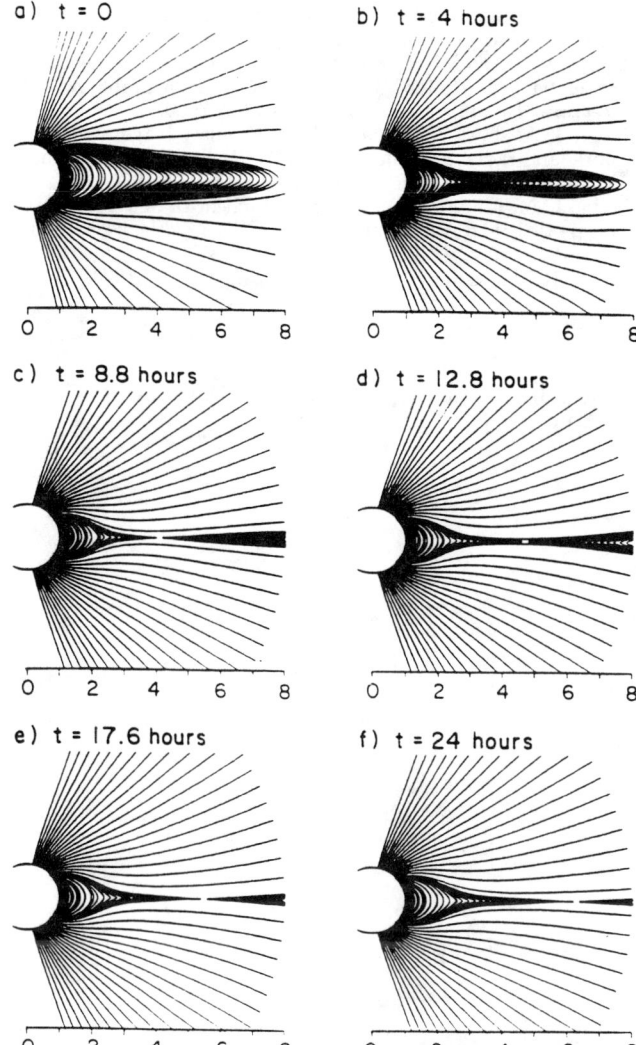

Fig. 13. 2-D MHD simulation of the effect of increasing the magnetic pressure around an equilibrium helmet streamer. The panels display magnetic field traces at times, t, following the increase. Reconnection begins shortly after the increase and proceeds until sufficient magnetic pressure has been removed to again achieve an equilibrium. Taken from Linker et al. [1992].

around between various widely spaced locations on the Sun. Large modifications in the overall magnetic configuration of the corona can occur on relatively short timescales; it seems more reasonable that such large changes be accomplished primarily by the opening and closing processes described here than the large scale motion of magnetic footprints across the Sun.

Acknowledgments. Many valuable discussions with and comments on this manuscript by J.T. Gosling are gratefully acknowledged. This work was carried out under the auspices of the United States Department of Energy.

REFERENCES

Borrini, G., J. T. Gosling, S. J. Bame, W. C. Feldman, and J. M. Wilcox, Solar wind helium and hydrogen structure near the heliospheric current sheet: A signal of coronal streamers at 1 AU, *J. Geophys. Res.*, 86, 4565, 1981.

Cliver, E. W., Was the eclipse comet of 1893 a disconnected coronal mass ejection?, *Sol. Phys.*, 122, 319, 1989.

Cliver, E. and S. Kahler, High coronal flares and impulsive acceleration of solar energetic particles, *Astrophys. J.*, 366, L91-L94, 1991.

Gold, T., Plasma and magnetic fields in the solar system, *J. Geophys. Res.*, 64, 1665, 1959.

Gosling, J. T., Large-scale inhomogeneities in the solar wind of solar origin, *Rev. of Geophys. and Space Phys.*, 13, 1053, 1975.

Gosling, J. T., Coronal mass ejections and magnetic flux ropes in interplanetary space, Physics of Magnetic Flux Ropes, ed. C. T. Russell, E. R. Priest, and L. C. Lee, *Geophys. Mono.* 58, AGU, 343, 1990.

Gosling, J. T., The solar flare myth, submitted to *J. Geophys. Res.*, 1993.

Gosling, J. T., E. Hildner, R. M. MacQueen, R. H. Munro, A. I. Poland, and C. L. Ross, Mass ejections from the Sun: a view from Skylab, *J. Geophys. Res.*, 79, 4581, 1974.

Gosling, J. T., E. Hildner, R. M. MacQueen, R. H. Munro, A. I. Poland, and C. L. Ross, The speeds of coronal mass ejection events, *Solar Phys.*, 48, 389, 1976.

Gosling, J. T., G. Borrini, J. R. Asbridge, S. J. Bame, W. C. Feldman, and R. T. Hansen, Coronal streamers in the solar wind at 1 AU, *J. Geophys. Res.*, 86, 5438, 1981.

Gosling, J. T., D. N. Baker, S. J. Bame, W. C. Feldman, and R. D. Zwickl, Bidirectional solar wind electron heat flux and hemispherically symmetric polar rain, *J. Geophys. Res.*, 91, 11352–11358, 1986.

Gosling, J. T., D. N. Baker, S. J. Bame, W. C. Feldman, R. D. Zwickl, and E. J. Smith, Bidirectional solar wind electron heat flux events, *J. Geophys. Res.*, 92, 8519, 1987.

Hundhausen, A. J., Coronal expansion and solar wind, Springer-Verlag, New York, Heidelberg, Berlin, 1972.

Hundhausen, A. J., An interplanetary view of coronal holes, Coronal Holes and High Speed Wind Streams, ed. J. Zirker, Colorado Assoc. Univ. Press, Boulder, 1977.

Kahler, S. W. and A. J. Hundhausen, The magnetic topology of solar coronal structures following mass ejections, *J. Geophys. Res.*, 97, 1619–1631, 1992.

King, J. H., Solar cycle variations in IMF intensity, *J. Geophys. Res.*, 84, 5938, 1979.

Lin, R. P., Energetic solar electrons in the interplanetary medium, *Solar Phys.*, 100, 537, 1985.

Lin, R. P. and S. W. Kahler, Interplanetary magnetic field connection to the Sun during electron heat flux dropouts in the solar wind, *J. Geophys. Res.*, 97, 8203–8209, 1992.

MacQueen, R. M., Coronal transients: a summary, *Phil. Trans. R. Soc. Lond. A*, 297, 605, 1980.

McComas D. J., J. T. Gosling, J. L. Phillips, S. J. Bame, J. G. Luhmann, and E. J. Smith, Electron heat flux dropouts in the solar wind: evidence for interplanetary magnetic field reconnection?, *J. Geophys. Res.*, 94, 6907, 1989.

McComas, D. J., J. L. Phillips, A. J. Hundhausen, and J. T. Burkepile, Observations of disconnection of open coronal magnetic structures, *Geophys. Res. Lett.*, 18, 73, 1991.

McComas, D. J., J. T. Gosling, and J. L. Phillips, Interplanetary magnetic flux: Measurement and balance, *J. Geophys. Res.*, 97, 171, 1992a.

McComas, D. J., J. T. Gosling, and J. L. Phillips, Regulation of the Interplanetary Magnetic Field, Solar Wind 7, ed. E. Marsch and R. Schwenn, Pergamon Press, Oxford, 643–646, 1992b.

McComas, D. J., J. L. Phillips, A. J. Hundhausen, and

J. T. Burkepile, Disconnection of Open Coronal Magnetic Structures, Solar Wind 7, ed. E. Marsch and R. Schwenn, Pergamon Press, Oxford, 225–228, 1992c.

McComas, D. J. and J. L. Phillips, The Extension of Solar Magnetic Fields into Interplanetary Space, *Proc. of the first SOLTIP Symposium, Vol. 1*, edited by S. Fischer and M. Vandas, Czechoslovak Academy of Sciences, Prague, 180–191, 1992d.

Phillips, J. L., J. T. Gosling, D. J. McComas, S. J. Bame, and W. C. Feldman, Quantitative Analysis of Bidirectional Electron Fluxes Within Coronal Mass Ejections at 1 AU, Solar Wind 7, ed. E. Marsch and R. Schwenn, Pergamon Press, Oxford, 651–656, 1992.

Potter, D. W., R. P. Lin, and K. A. Anderson, Impulsive 2-10 keV solar electron events not associated with flares, *Astrophys. J., 236*, L97, 1980.

Rosenbauer, H., R. Schwenn, E. Marsch, B. Meyer, H. Miggenrieded, M. D. Montgomery, K.-H. Muhlhauser, W. Pilipp, W. Voges, and S. M. Zink, A survey on initial results of the Helios plasma experiment, *J. Geophys., 42*, 561, 1977.

St. Cyr, O. C. and J. T. Burkepile, A catalog of mass ejections observed by the Solar Maximum Mission coronagraph, *NCAR Technical Note, NCAR/TN-352-STR*, 1990.

Slavin, J. A., E. J. Smith, and B. T. Thomas, Large scale temporal and radial gradients in the IMF: Helios 1,2, ISEE-3, and Pioneer 10, 11, *Geophys. Res. Lett., 11*, 279, 1984.

Slavin, J. A., G. Jungman, and E. J. Smith, The interplanetary magnetic field during solar cycle 21: ISEE-3/ICE Observations, *Geophys. Res. Lett., 13*, 513, 1986.

Tousey, R., The solar corona, *Space Res., 13*, 713, 1973.

David J. McComas, Space and Atmospheric Sciences Group, Los Alamos National Laboratory, Los Alamos, NM 87545.

The Solar Flare Myth in Solar-Terrestrial Physics

J. T. Gosling

Los Alamos National Laboratory, Los Alamos, New Mexico

Early observations found an association between solar flares and large non-recurrent geomagnetic storms, large 'solar' energetic particle events, and transient shock wave disturbances in the solar wind. These results led to the development of a paradigm of cause and effect that gave flares a central position in the chain of events leading from solar activity to major transient disturbances in the near-Earth space environment. However, research in the last two decades shows that this emphasis on flares is misplaced. In this paper I outline briefly the rationale for a different paradigm of cause and effect in solar-terrestrial physics that removes solar flares from their central position as the "cause" of major disturbances in the near-Earth space environment. Instead, this central role of "cause" is played by events now known as coronal mass ejections, or CMEs.

1. INTRODUCTION

In 1859, R. Carrington observed an intense, short-lived brightening of the surface of the Sun in the vicinity of a sunspot [*Carrington*, 1860]. Such brightenings on the surface of the Sun are now known as solar flares and have been the objects of extensive research during the present century. Major transient shock wave disturbances in the solar wind, energetic particle events in interplanetary space, and large non-recurrent geomagnetic storms often occur in close association with large solar flares such as the one Carrington observed. Over the years such observed associations led to a paradigm of cause and effect - that large solar flares are the fundamental cause of these major events in the near-Earth space environment [e.g., *Hale*, 1931; *Chapman*, 1950; *Parker*, 1963]. This paradigm, which I call "The Solar Flare Myth", not only dominates the popular perception of the relationship between solar activity and interplanetary and geomagnetic events, but also provides much of the pragmatic rationale for the study of the solar flare phenomenon [e.g., *Haisch et al.*, 1991].

2. A HISTORICAL PARADIGM OF CAUSE AND EFFECT IN SOLAR-TERRESTRIAL PHYSICS

Figure 1 provides an outline of this paradigm of cause and effect whose major elements have all been firmly in place since at least the early 1960's. A simple elaboration of the paradigm follows. Solar activity is associated with the evolution of the solar magnetic field. Large solar flares occur in magnetically complex regions where the field is often strongly sheared. The actual energy release mechanism associated with flaring activity is uncertain but is usually thought to include some form of magnetic reconnection. During the flare process some fraction of the charged particles present in the vicinity of the flare site are accelerated to high energy (right hand branch in the figure). Some of these accelerated particles escape quickly into space along the interplanetary magnetic field; others are trapped in closed field regions at the Sun, diffuse slowly across field lines in the solar atmosphere, and leak out into interplanetary space over a period of several days. When the energetic particles arrive at 1 AU (or at a spacecraft) they cause a solar energetic particle event; when they impinge upon the upper atmosphere in the polar regions of the Earth they cause a polar cap absorption event. The flare process also substantially heats the chromosphere and the corona in the region immediately surrounding the flare site (left hand branch in the figure). This heating, in possible conjunction with magnetic forces, produces a rapid expansion of the chromosphere and corona around the flare site. When the speed of the rapidly expanding corona and/or chromosphere material is sufficiently high, a shock disturbance is produced in interplanetary space. A large geomagnetic storm and auroral disturbance results when this interplanetary disturbance impinges upon the Earth's magnetosphere.

3. EVIDENCE OF THE NEED FOR A NEW PARADIGM OF CAUSE AND EFFECT

The foregoing paradigm appears reasonable in the light of observational knowledge available by the early 1960's, but

Solar System Plasmas in Space and Time
Geophysical Monograph 84
Copyright 1994 by the American Geophysical Union.

A Paradigm of Cause and Effect

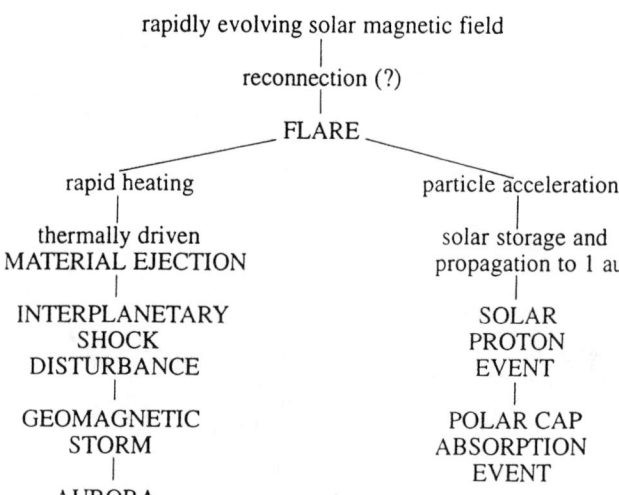

Fig. 1. The solar flare myth - a paradigm of cause and effect illustrating the supposed central position of solar flares as a "cause" of major disturbances in the near-Earth space environment. Capitol letters indicate observational phenomena and lower case letters indicate physical processes or descriptive characteristics.

today there is good evidence that this paradigm is wrong and that solar flares do not generally play a fundamental role in producing major transient disturbances in the near-Earth space environment [e.g., *Kahler*, 1992]. The evidence, which is elaborated upon more fully elsewhere [*Gosling*, 1993], consists primarily of the following:

1. No one-to-one relationship exists between solar flares and either transient shock disturbances in the solar wind at 1 AU or major solar energetic particle events or large non-recurrent geomagnetic storms. Indeed, all of these phenomena can and do occur in the absence of any substantial flaring activity [e.g., *Newton*, 1943; *Joselyn and McIntosh*, 1981; *Hundhausen*, 1972; *Gosling et al.*, 1980; *Domingo et al.*, 1979; *Cliver et al.*, 1983].

2. Although the Sun does commonly and often somewhat impulsively eject large quantities of material (10^{+15} - 10^{+16} g) into interplanetary space in events now known as coronal mass ejections, CMEs, [e.g., *Tousey*, 1973; *Gosling et al.*, 1974], solar flares are not fundamentally responsible for these ejections. Many CMEs, including some of the more spectacular ones, occur in the absence of any substantial flaring activity [e.g., *Munro et al.*, 1979; *Webb*, 1992]. Even when they occur in conjunction with flares the CMEs generally lift off from the Sun before any substantial flaring activity has occurred. Moreover, any associated flaring that does occur often lies to one side of the much broader (typically many 10s of degrees) CME span [e.g., *Harrison*, 1986; *Hundhausen*, 1988; *Harrison et al.*, 1990]. All of which indicates that CMEs are not generally the result of solar flares even though these different aspects of solar activity can occur together. It seems likely that both CMEs and solar flares arise from instabilities connected with the temporal and spatial evolution of the magnetic field in the solar atmosphere, with CMEs resulting more from changes in the large scale magnetic field that permeates the solar corona [e.g., *Low*, 1993] and flares resulting more from changes in the stronger, but smaller scale, fields associated with solar active regions.

3. Virtually all transient shock wave disturbances in the solar wind at 1 AU are driven by fast CMEs [e.g., *Sheeley et al.*, 1985; *Cane et al.*, 1987]; however, as noted above, solar flares play no fundamental role in producing CMEs.

4. Large, non-recurrent geomagnetic storms are produced almost exclusively by Earth-passage of interplanetary disturbances driven by fast CMEs [*Gosling et al.*, 1990; 1991]; however, again as noted above, solar flares play no fundamental role in producing CMEs.

5. There are at least two fundamentally different types of solar energetic particle events: impulsive events and gradual events [e.g., *Cane et al.*, 1986; *Lin*, 1987; *Mason et al.*, 1989; *Reames* 1992a,b]. Impulsive events are common (~1000 events yr-1 near solar activity maximum), are usually weak events with durations of several hours, are rich in electrons, ^{3}He, and Fe, have ionization states characteristic of flare temperatures, are commonly observed in association with impulsive optical and X-ray flares, and are detected almost exclusively in association with solar events that are relatively well connected to the observer along the interplanetary magnetic field spiral. It is clear that such events are a direct product of the same process that produces flares and that the energetic particles in these events are accelerated near the flare site. But these events are usually not major events in terms of particle intensity or event duration [e.g., *Reames*, 1993]. By way of contrast, gradual events occur infrequently (~10 events yr-1 near solar activity maximum), are often intense events that last for several or more days, are rich in protons and have elemental abundances and ionization states characteristic of the solar corona and solar wind [e.g., *Mason et al.*, 1984], have spectra that emerge smoothly from the solar wind thermal population [e.g., *Gosling et al.*, 1981], are strongly associated with CMEs that drive interplanetary shock disturbances [e.g., *Kahler et al.*, 1984], can arise from disturbances that originate anywhere on the visible disk of the Sun [e.g., *Reames*, 1992a], and are observed primarily on field lines that connect the observer to an interplanetary shock [e.g., *Cane et al.*, 1988]. All presently available evidence indicates that gradual events are the product of the shock acceleration of coronal and solar wind particles in interplanetary space. That is, gradual events are not related in a fundamental way to solar flares. Most of the major (that is, intense and long-lasting) solar energetic particle events observed in interplanetary space are gradual events or composites of gradual and impulsive events.

4. A MODERN PARADIGM OF CAUSE AND EFFECT IN SOLAR-TERRESTRIAL PHYSICS

The foregoing evidence indicates that the paradigm of cause and effect in solar-terrestrial physics outlined in Figure 1 is incorrect, primarily with regard to the central importance given to solar flares. Figure 2 outlines a more modern paradigm that is, I believe, far more consistent with present knowledge. The underlying cause of solar activity appears to be the evolution of the solar magnetic field. Solar flares occur in magnetically complex regions, perhaps as a result of magnetic reconnection. Energetic particles are often produced during the impulsive phase of solar flares; these particles escape from the Sun along field lines originating close to the flare sites to produce impulsive energetic particle events in interplanetary space. Impulsive events are observed near Earth only for flares in the western solar hemisphere [e.g., *Reames*, 1992a,b], indicating that there is little diffusion of the energetic particles in these events across the spiral interplanetary magnetic field. These events have characteristic durations of a few hours at 1 AU and, with a few exceptions, typically are weak events.

Coronal mass ejections also appear to be a result of the spatial and temporal evolution of the solar magnetic field, although the processes that trigger the release of CMEs and the factors that determine the timing, the size, and the speed of the ejections are still not well understood (see, for example, the review by *Low* [1993]). It does seem clear, however, that flares do not play a fundamental role in producing CMEs. CMEs may result from global instabilities in the coronal magnetic field [e.g., *Priest*, 1988], and buoyancy may be important in accelerating the plasma outward into interplanetary space, but this is uncertain. Solar prominence material or material ejected from a flaring region is often embedded within CMEs; however, most of the material within CMEs usually originates from the corona rather than from prominences or the chromosphere [e.g., *Hildner et al.*, 1975]. Further, there is no observational evidence to suggest that prominences or chromospheric material drive the CMEs outward from the Sun. CMEs exhibit a wide range of outward speeds [e.g., *Gosling et al.*, 1976]; those that move at the same speed as or slower than the ambient solar wind ahead do not produce significant disturbances in the solar wind. The fastest CMEs, on the other hand, often produce very large interplanetary disturbances, characterized by high solar wind speeds and strong magnetic fields, often with strong southward components. The strong fields in these disturbances are primarily a result of compression in interplanetary space. An interplanetary shock usually, but not always, is an integral part of such disturbances, depending primarily on the relative speed between the CME and the ambient solar wind ahead. When these major interplanetary disturbances are directed earthward, large geomagnetic storms and auroral disturbances usually result, the most crucial element being the presence of a strong southward directed field somewhere within the interplanetary disturbance [e.g., *Gonzalez and Tsurutani*, 1987; *Tsurutani et al.*, 1988; *Gosling et al.*, 1990].

The strong shocks driven by the fastest CMEs are also effective in accelerating a small fraction of the particles they intercept to very high energies [e.g., *Lee and Ryan*, 1986]. Only a small fraction of the solar wind particles encountering these shocks are accelerated to high energies, but the flux of these particles relative to the cosmic ray background is quite high, and the accelerated particles are found on all field lines intersecting the shocks. The largest number of accelerated particles probably are produced near the Sun where the CME-driven shocks are strongest and the ambient density is highest, but acceleration takes place over a prolonged period of time as the shocks propagate outward through the solar wind to the Earth and beyond. Throughout the outward journey of the disturbance, accelerated particles continually leak away from the acceleration region near the shock along the interplanetary magnetic field. CMEs typically are large structures with broad latitudinal and longitudinal extents and the shocks they drive often spread over more than 90° in solar latitude and longitude. The gradual, but intense, energetic particle events produced by CME-driven shocks typically last for several days or longer and are found in association with disturbances originating from virtually anywhere on the visible solar disk. The detailed temporal intensity profiles that are observed depend sensitively on the longitude where

CAUSE AND EFFECT IN SOLAR-TERRESTRIAL PHYSICS

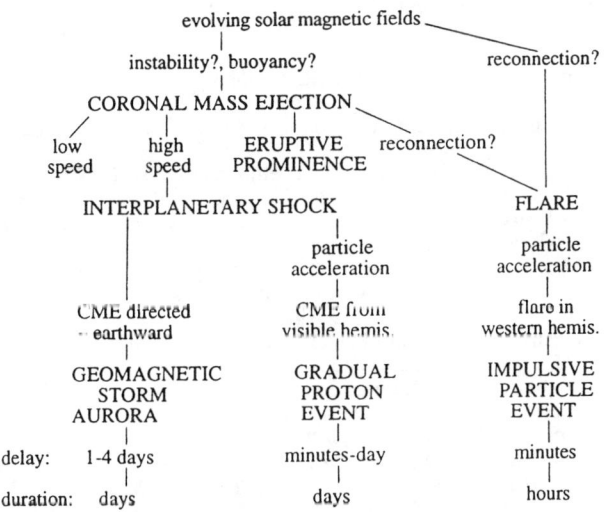

Fig. 2. A modern paradigm of cause and effect in solar-terrestrial physics emphasizing the central importance of coronal mass ejections, CMEs, in producing major events in the near-Earth space environment and deemphasizing the importance of solar flares in this respect. Capitol letters indicate observational phenomena and lower case letters denote processes or descriptive characteristics. This new paradigm appears to be consistent with a wide variety of observations.

the CMEs originate relative to the observer [e.g., *Cane et al.*, 1988]. According to *Reames* [1992a,b], most major solar energetic particle events observed in the vicinity of the Earth are gradual events associated with fast CMEs, although some fraction of major particle events are composites of the gradual and impulsive types because of the overall association between CMEs and flaring activity.

5. CONCLUSION

Over the last twenty years a major change in paradigm has occurred concerning the solar "cause" of major transient disturbances in the near-Earth space environment. This change appears to be based firmly upon present observational knowledge from a wide variety of sources. Yet it is clear to me that this new paradigm has not yet caught the attention of many members of the solar-terrestrial physics community and that "the solar flare myth" continues to propagate widely in scientific articles and books, in presentations at scientific meetings and colloquia, in posters and other material released for educational purposes, and in the popular press. It is my belief that the time has come to lay the myth to rest, and it is my hope that the present paper will play some small role in achieving its burial.

Acknowledgments. This paper is based upon a considerably more detailed paper submitted to the *Journal of Geophysical Research*. A large American Geophysical Union exhibit, prepared for the Smithsonian Air and Space Museum and titled "Electric Space: Our Sun-Earth Environment" provided the original impetus for the paper. In its original form this exhibit, which highlights connections between solar events and events in the near-Earth space environment, made no mention of CMEs. Further inspiration for the paper has come from statements made and not made in a variety of scientific papers and books, in scientific presentations, seminars, and colloquia, in casual conversations, in educational materials released for popular consumption, and in the popular press. The author has profited from discussions on this topic with a number of individuals, including E. Cliver, N. Crooker, D. Hamilton, A. Hundhausen, S. Kahler, G. Mason, D. Reames, and D. Webb among others. He thanks N. Crooker for help in the preparation of Figure 2 and M. Moldwin for his comments on the manuscript. This work was performed under the auspices of the U.S. Department of Energy with partial support from NASA.

REFERENCES

Cane, H. V., R. E. McGuire, and T. T. von Rosenvinge, Two classes of solar energetic particle events associated with impulsive and long-duration soft x-ray flares, *Astrophys. J., 301*, 448, 1986.

Cane, H. V., N. R. Sheeley, and R. A. Howard, Energetic interplanetary shocks, radio emission, and coronal mass ejections, *J. Geophys. Res., 92*, 9869, 1987.

Cane, H. V., D. V. Reames, and T. T. von Rosenvinge, The role of interplanetary shocks in the longitude distribution of solar energetic particles, *J. Geophys. Res., 93*, 9555, 1988.

Carrington, R. C., Description of a singular appearance seen on the Sun on September 1, 1859, *Mon. Not. R. Astron. Soc., 20*, 13, 1860.

Chapman, S., Corpuscular influences upon the upper atmosphere, *J. Geophys. Res., 55*, 361, 1950.

Cliver, E. W., S. W. Kahler, and P. S. McIntosh, Solar proton flares with weak impulsive phases, *Astrophys. J., 264*, 699, 1983.

Domingo, V., R. J. Hynds, and G. Stevens, A solar proton event of possible non-flare origin, *Proc. Int. Conf. Cosmic Rays 16th, 5*, 192, 1979.

Gonzalez, W. D., and B. T. Tsurutani, Criteria of interplanetary parameters causing intense magnetic storms (Dst < -100 nT), *Planet. Space Sci., 35*, 1101, 1987.

Gosling, J. T., The solar flare myth, *J. Geophys. Res.*, in press, 1993.

Gosling, J. T., E. Hildner, R. M. MacQueen, R. H. Munro, A. I. Poland, and C. L. Ross, Mass ejections from the Sun: A view from Skylab, *J. Geophys. Res., 79*, 4581, 1974.

Gosling, J. T., E. Hildner, R. M. MacQueen, R. H. Munro, A. I. Poland, and C. L. Ross, The speeds of coronal mass ejection events, *Sol. Phys., 48*, 389, 1976.

Gosling, J. T., J. R. Asbridge, S. J. Bame, W. C. Feldman, and R. D. Zwickl, Observations of large fluxes of He$^+$ in the solar wind following an interplanetary shock, *J. Geophys. Res., 85*, 3431, 1980.

Gosling, J. T., J. R. Asbridge, S. J. Bame, W. C. Feldman, R. D. Zwickl, G. Paschmann, N. Sckopke, and R. J. Hynds, Interplanetary ions during an energetic storm particle event: The distribution function from solar wind thermal energies to 1.6 MeV, *J. Geophys Res., 86*, 547, 1981.

Gosling, J. T., S. J. Bame, D. J. McComas, and J. L. Phillips, Coronal mass ejections and large geomagnetic storms, *Geophys. Res. Lett., 17*, 901, 1990.

Gosling, J. T., D. J. McComas, J. L. Phillips, and S. J. Bame, Geomagnetic activity associated with Earth passage of interplanetary shock disturbances and coronal mass ejections, *J. Geophys. Res., 96*, 7831, 1991.

Hale, G. E., The spectrohelioscope and its work, Part III. Solar eruptions and their apparent terrestrial effects, *Astrophys. J., 73*, 379, 1931.

Haisch, B., K. T. Strong, and M. Rodono, Flares on the Sun and other stars, *Ann. Rev. Astron. Astrophys., 29*, 275, 1991.

Harrison, R. A., Solar coronal mass ejections and flares, *Astron. Astrophys., 162*, 283, 1986.

Harrison, R. A., E. Hildner, A. J. Hundhausen, D. G. Sime, and G. M. Simnett, The launch of solar coronal mass ejections: Results from the coronal mass ejection onset program, *J. Geophys. Res., 95*, 917, 1990.

Hildner, E., J. T. Gosling, R. T. Hansen, and J. D. Bohlin, The sources of material comprising a mass ejection coronal transient, *Sol. Phys., 45*, 363, 1975.

Hundhausen, A. J., Interplanetary shock waves and the structure of solar wind disturbances, in *Solar Wind*, edited by C. P. Sonett, P. J. Coleman, and J. M. Wilcox, NASA Spec. Publ., SP-308, pp. 393-417, 1972.

Hundhausen, A. J., The origin and propagation of coronal mass ejections, in *Proceedings of the Sixth International Solar Wind Conference*, TN 306+Proc, edited by V. Pizzo, T. E. Holzer, and D. G. Sime, National Center for Atmospheric Research, Boulder, pp. 181-214, 1988.

Joselyn, J. A., and P. S. McIntosh, Disappearing solar filaments: A useful predictor of geomagnetic activity, *J. Geophys. Res., 86*, 4555, 1981.

Kahler, S. W., Solar flares and coronal mass ejections, *Ann. Rev. Astron. Astrophys, 30*, 113, 1992.

Kahler, S. W., N. R. Sheeley, R. A. Howard, M. J. Koomen, D. J. Michels, R. E. McGuire, T. T. von Rosenvinge, and D.

V. Reames, Associations between coronal mass ejections and solar energetic proton events, *J. Geophys. Res.*, *89*, 9683, 1984.

Lee, M. A., and J. M. Ryan, Time-dependent coronal shock acceleration of energetic solar flare particles, *Astrophys. J.*, *303*, 829, 1986.

Lin, R. P., Solar particle acceleration and propagation, *Revs. Geophys.*, *25*, 676, 1987.

Low, B. C., Mass acceleration processes: The case of the coronal mass ejection, *Adv. Space Res.*, in press, 1993.

Mason, G. M., G. Gloeckler, and D. Hovestadt, Temporal variations of nucleonic abundances in solar flare energetic particle events. II. Evidence for large-scale shock acceleration, *Astrophys. J.*, *280*, 902, 1984.

Mason, G.M., C. K. Ng, B. Klecker, and G. Green, Ion acceleration and scatter-free transport of ~1 Mev per nucleon ions in ^{3}He-rich solar particle events, *Astrophys. J.*, *339*, 529, 1989.

Munro, R. H., J. T. Gosling, E. Hildner, R. M. MacQueen, A. I. Poland, and C. L. Ross, The association of coronal mass ejection transients with other forms of solar activity, *Sol. Phys.*, *61*, 201, 1979.

Newton, H. W., Solar flares and magnetic storms, *Mon. Not. Roy. Astron. Soc.*, *103*, 244, 1943.

Parker, E. N., *Interplanetary Dynamical Processes*, John Wiley and Sons, New York, 1963.

Priest, E. R., The initiation of solar coronal mass ejections by magnetic nonequilibrium, *Astrophys. J.*, *328*, 848, 1988.

Reames, D. V., Trapping and escape of the high energy particles responsible for major proton events, in *Eruptive Solar Flares, Lecture Notes in Physics 399*, edited by Z. Svestka, B. V. Jackson, and M. E. Machado, Springer-Verlag, Berlin, pp. 180-185, 1992a.

Reames, D. V., Particle acceleration in solar flares: Observations, in *Particle Acceleration in Cosmic Plasmas*, AIP Conf. Proc. 264, pp. 213-222, 1992b.

Reames, D. V., Recent observations and the modeling of solar proton events, in *Solar-Terrestrial Predictions Workshop: IV*, edited by J. Hruska et al., NOAA, Boulder, in press, 1993.

Sheeley, N. R., R. A. Howard, M. J. Koomen, D. J. Michels, R. Schwenn, K.-H. Muhlhauser, and H. Rosenbauer, Coronal mass ejections and interplanetary shocks, *J. Geophys. Res.*, *90*, 163, 1985.

Tousey, R., The solar corona, *Adv. Space Res.*, *13*, 713, 1973.

Tsurutani, B. T., W. D. Gonzalez, F. Tang, S.-I Akasofu, and E. J. Smith, Origin of interplanetary southward magnetic fields responsible for major magnetic storms near solar maximum (1978-1979), *J. Geophys. Res.*, *93*, 8519, 1988.

Webb, D. F., The solar sources of coronal mass ejections, in *Eruptive Solar Flares, Lecture Notes in Physics 399*, edited by Z. Svestka, B. V. Jackson, and M. E. Machado, Springer-Verlag, Berlin Heidelberg, pp. 234-247, 1992.

J. T. Gosling, MS D466, Los Alamos National Laboratory, Los Alamos, NM 87545.

Generation and Nonlinear Evolution of Oblique Magnetosonic Waves: Application to Foreshock and Comets

N. Omidi[1], H. Karimabadi,
D. Krauss-Varban, and K. Killen

Department of Electrical and Computer Engineering, University of California, San Diego, California

In this study, a combination of linear theory and hybrid (fluid electrons, particle ions) simulations are used to address two outstanding issues regarding the formation and stability of steepened magnetosonic waves (shocklets) observed upstream of planetary bow shocks and at Comet Giacobini-Zinner (GZ). In regards to the former issue, previous studies based on linear theory had predicted the dominance of parallel propagating waves which do not steepen. However, our new study using linear theory shows that ions with a beam-ring velocity distribution function can excite magnetosonic waves via two separate instabilities driven by the beam and the ring parts of the distribution function, respectively. The beam driven instability has maximum growth along the magnetic field, while the ring (anisotropy) driven one has maximum growth in the oblique direction. By performing 2-dimensional simulations the nonlinear evolution of these two instabilities and the conditions under which each becomes dominant is investigated. In the case of field aligned beams, parallel propagating waves are observed to grow. The nonlinear evolution of these waves does not include steepening. On the other hand, beam-ring distribution functions can drive oblique waves unstable which steepen and form shocklets similar to those seen in the planetary foreshocks and Comet GZ. This suggests that the observed shocklets are not excited by the field aligned beams in the ion foreshock and that another beam population, e.g. gyrating ions, may be responsible for their presence. In regards to the stability of shocklets, it is shown that they are inherently unstable and evolve into less coherent structures with complex waveforms. This evolution can be due to a number of mechanisms including noncoplanarity of the magnetic field, dispersive effects, and shocklet-shocklet collisions. The late time evolution of shocklets is characterized by randomization of the phase between the two transverse components of the magnetic field and a substantial drop in the correlation between the density and the magnetic field oscillations. As a result of this process the helicity of the waves can change sign and eventually becomes ill defined.

1. INTRODUCTION

Spacecraft observations upstream of planetary bow shocks and at Comet Giacobini-Zinner (GZ) have shown the presence of steepened magnetosonic waves referred to as shocklets [e.g. *Hoppe et al.*, 1981; *Hoppe and Russell*, 1983; *Russell and Hoppe*, 1983; *Tsurutani and Smith*, 1986; *Tsurutani et al.*, 1987]. In the case of the Earth's bow shock, the waves are excited by the interaction between the solar wind and the ion beams moving back upstream. Observations have shown that these ions have a variety of distribution functions ranging from a cool field aligned beam to a much hotter and diffuse population [e.g. *Gosling et al.*, 1978; *Paschmann et al.*, 1979; *Gurgiolo et al.*, 1981; *Fuselier et al.*, 1986]. A recent review of these ion populations is given by *Fuselier* [1993]. According to *Gary et al.* [1981], the field aligned beams can excite magnetosonic waves via a cyclotron resonance instability referred to as the right-hand res-

---
[1] Also at California Space Institute, University of California, San Diego, California.

Solar System Plasmas in Space and Time
Geophysical Monograph 84
Copyright 1994 by the American Geophysical Union.

onant ion/ion instability. This instability can excite waves in a relatively large range of wave normal angles, although the maximum growth occurs at parallel propagation. Simulation studies of this instability by *Winske and Leroy* [1984] have shown that during their nonlinear phase the waves scatter the cool ion beam, and form a more diffuse population. In the study of *Hoppe et al.* [1981] the association between the various ion beam populations and hydromagnetic waves was investigated. It was found that the field aligned beams are not associated with magnetosonic wave activity, while the intermediate population is associated with magnetosonic waves with sinusoidal waveform. The diffuse ion distributions are observed when the waves have steepened to form shocklets. Based on theory and observations a model (or a paradigm) has emerged in which the field aligned beams generate magnetosonic waves whose growth and nonlinear evolution results in the formation of intermediate and diffuse ions. According to this model, the waves are too small to be observed during their linear phase, thereby explaining their absence while detecting cool field aligned beams. On the other hand, by the time the waves reach large enough amplitudes to be observed they would have partially scattered the ions into an intermediate form. Finally, shocklets which represent the fully nonlinear phase of the waves would naturally be associated with the diffuse ion population.

Formation of shocklets at comets (specifically Comet GZ) has been thought to be through a similar instability (i.e. right-hand resonant ion/ion), although in this case the ion beam is formed by the ionization of the cometary neutral gasses. As a result, their distribution function can range from a field aligned beam to a pure ring as the angle between the interplanetary magnetic field and the flow velocity varies from zero to 90°, respectively. Also, because of their cometary origin some of the ions are more massive (water group) as compared to the proton beams present in the planetary foreshocks. As a consequence, the excited waves have a lower frequency and a longer wavelength. The linear properties of the instabilities occurring at comets has been investigated in considerable detail [e.g. *Wu and Davidson*, 1972; *Winske et al.*, 1985; *Winske and Gary*, 1986; *Gary and Madland*, 1988; *Brinca and Tsurutani*, 1988; *Goldstein et al.*, 1990]. It was generally assumed that parallel propagating magnetosonic waves have the largest growth rate, as is the case with field aligned beams. However, the growth of oblique waves was investigated by *Brinca and Tsurutani*, [1988] and *Goldstein et al.*, [1990] who showed maximum growth can occur at finite wave normal angles. Since parallel propagating waves are linearly noncompressional their nonlinear evolution is different [see e.g. *Akimoto et al.*, 1991] from oblique waves which steepen to form shocklets. Assuming the dominance of parallel propagating waves in the linear regime, the question of formation of shocklets in the foreshock as well as the cometary environment has remained unresolved and various other effects such as refraction [*Hada et al.*, 1987] have been invoked to explain the observed obliquity of the shocklets. The refraction model was tested by *Le* [1991] during radial IMF conditions upstream of the bow shock. The results showed that refraction does not play a role in the formation of shocklets. In this paper we have performed a linear analysis of the instabilities associated with the beam-ring distribution function which show two separate instabilities exciting magnetosonic waves, one of which has maximum growth at oblique angles. By performing 2-D electromagnetic hybrid (fluid electrons, particle ions) simulations we show the dominance of oblique magnetosonic waves which then steepen to form shocklets.

Another very interesting and important issue regarding shocklets is whether they constitute the last phase in the nonlinear evolution of the magnetosonic waves or if they further evolve into other structures. Such a question is motivated not only from the plasma physics point of view but also to address observations near the quasi-parallel portion of the bow shock as well as the cometosheath region. Both observations and simulations of high Mach number quasi-parallel shocks have shown that they are not steady and undergo cyclic reformation [e.g. *Gosling et al.*, 1989; *Burgess*, 1989; *Thomsen et al.*, 1990a,b; *Thomas et al.*, 1990; *Scholer and Terasawa*, 1990; *Winske et al.*, 1990]. Magnetic field measurements near the shock have shown the presence of pulsations [e.g. *Greenstadt et al.*, 1970; *Thomsen et al.*, 1990a] which constitute the overall turbulent shock structure. As shown by *Thomsen et al.* [1990a], some of these pulsations have a convective signature, while others are due to the in-and-out motion of the shock. In addition to the magnetic pulsations, *Schwartz* [1990] and *Schwartz et al.* [1992] have identified more coherent isolated regions of enhanced magnetic field which they refer to as 'Short Large Amplitude Magnetic Structures' or SLAMS. SLAMS are characterized by a shorter period as compared to magnetic pulsations and they always exhibit a convective signature. Of interest is the fact that SLAMS are at times associated with a whistler wave train as are shocklets. This, plus their convective signature, suggests that SLAMS may be remnants of shocklets which have undergone further evolution. Such an interpretation seems reasonable *in lieu* of the fact that shocklets are convected by the solar wind towards the quasi-parallel portion of the bow shock. In the case of Comet GZ, a region (cometosheath) was identified in which the solar wind was heated (or shocked) and the magnetic field oscillations exhibited a turbulent behavior. Although not well understood, it is plausible that the turbulent field fluctuations were due to the evolution of shocklets observed further upstream.

To address the above mentioned issues, we have conducted linear theory as well as one- and two-dimensional electromagnetic hybrid simulations [e.g. *Winske and Omidi*, 1992]. The 1-D simulations are used to investigate further evolution of the shocklets while the 2-D simulations are utilized to see how obliquely propagating waves which steepen to form shocklets are excited. This paper presents the preliminary results from this investigation. A more complete description of linear theory and 2-D simulations is given by Killen et al. (paper submitted to *J. Geophys. Res.*, 1994).

In section 2 the results of linear theory for a beam-ring distribution function are presented where it is shown that two distinct instabilities can excite magnetosonic waves along and oblique to the ambient magnetic field. The results of two-dimensional hybrid simulations in which these two instabilities are allowed to compete are given in section 3. The results of our study of further evolution of shocklets is shown in section 4. Finally, section 5 provides a summary.

2. LINEAR THEORY

Although it is generally believed that the shocklets observed upstream of planetary bow shocks and at comets are formed via the same linear and nonlinear processes, the nature of the ion beam distribution function is not necessarily the same in the two environments. In the planetary case, the bow shock populates the upstream region with ion beams which, as discussed earlier, have a variety of velocity distribution functions. In contrast, cometary ions are created over a large region of space with their production rate varying with distance from the nucleus. While the velocity distribution function of the fresh ions (i.e. those created locally) is a cold beam-ring, the total distribution function is generally a scattered beam-ring due to wave-particle interactions and the convection of ions from upstream. In this section we show the results of linear theory for a distribution function of the form:

$$f_b(v_\parallel, v_\perp) \propto exp(-(v_\parallel - v_{\parallel b})^2/v^2{}_{th})\delta(v_\perp - v_{\perp b}) \quad (1)$$

which represents a beam-ring with a thermal spread in the parallel direction. Here $v_\parallel$ and $v_\perp$ are the velocities parallel and perpendicular to the magnetic field, respectively, v_{th} is the thermal velocity in the parallel direction and $\vec{v}_b$ is the beam velocity which has components parallel and perpendicular to the magnetic field. Although not shown in this paper, the stability properties of this distribution function are similar to a drifting bi-Maxwellian with temperature anisotropy $T_\perp > T_\parallel$. As will be discussed later, this may be important in explaining the generation of shocklets upstream of planetary bow shocks. The results of linear theory and simulations shown here are for a beam of singly ionized oxygen O^+, with velocity $v_b = 4V_A$ (where V_A is the Alfvén speed) which unless stated otherwise makes an angle of $30°$ with the ambient magnetic field. The background (solar wind) electron and ion beta (ratio of plasma to magnetic pressure) are both taken to be unity, while $v_{th} = v_{\perp b}/7$. Finally, beam densities are taken to be in the range of $0.05n_o \geq n_b \geq 0.005n_o$ where n_o is the background density.

Figure 1 shows the real (ω) and the imaginary (γ) parts of the frequency as a function of wave number k for $n_b = 0.05n_o$ and wave normal angle $\theta_{bk} = 30°$. Here the frequencies are normalized to the proton gy-

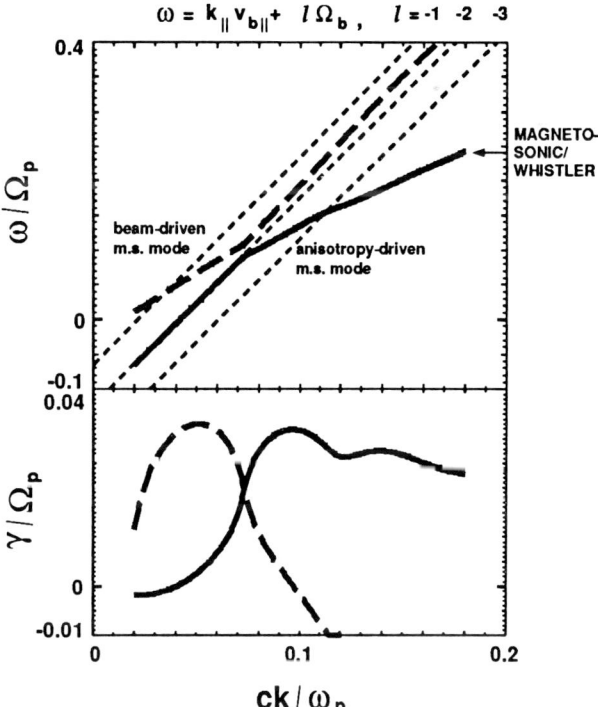

Fig. 1. Plots of the real (ω) and the imaginary (γ) parts of the frequency as a function of wave number K. Two unstable solutions are shown with solid and dashed lines. The dotted lines are the cyclotron resonances at the first, second, and the third harmonics.

rofrequency (Ω_p) and the wave number is normalized to the proton inertial length (c/ω_p). The figure shows two unstable solutions represented by the dashed and solid lines. Also shown in the top panel (dotted lines) are the cyclotron resonance conditions for the beam at the first, second, and the third harmonics. One characteristic of both solutions is the presence of a "knee" at $k \sim 0.07$ which comes about as a result of coupling between two modes. One of these modes is a combination of the solid line below $k \sim 0.07$ and the dashed line above it, while the other consists of the dashed line below $k \sim 0.07$ and the solid line beyond that point. The former corresponds to a beam mode (i.e. a mode which exists only when a beam is present in the plasma), while the latter is the magnetosonic whistler branch. It can be seen from the bottom panel that the regions of large growth for both of the unstable solutions occur at frequencies and wave numbers which fall on the magnetosonic whistler branch. In other words, the net result of the two unstable solutions is to excite fast magnetosonic waves over a relatively large range of frequencies. This unstable band is sufficiently broad that cyclotron resonances at the fundamental as well as the second and third harmonics (in this case) are important. Similar plots at smaller wave normal angles ($\theta_{bk} < 27°$) confirm this interpretation (see Killen et al., paper submitted to *J. Geophys. Res.*, 1994). It is found that at these smaller angles the magnetosonic branch and the beam mode remain completely separated with the former having larger growth rates which exhibits a double hump. We should also note that in addition to the excitation of magnetosonic waves the beam-ring distribution function can exhibit other instabilities [e.g. the nonresonant, *Winske and Gary*, 1986] depending on the plasma parameters. For the parameters chosen here the nonresonant mode is also unstable and has the largest growth rate (for more detail see paper by Killen et al., submitted to *J. Geophys. Res.*, 1994). To identify the source(s) of free energy resulting in the two unstable solutions, we have investigated the effects of the beam and the ring parts of the distribution function on the growth rates. The results show that the dashed solution is driven unstable mainly by the beam part of the distribution function, while the solid branch grows due to the the ring part although the beam component is still required to satisfy the resonance condition. In other words, the beam and the ring parts of the distribution function drive the magnetosonic mode unstable at lower and higher frequencies, respectively. Accordingly, we have labelled the low frequency portion of the dashed solution beam-driven magnetosonic mode, while the high frequency portion of the solid solution is called anisotropy-driven magnetosonic mode.

As shown in the bottom panel of Figure 1 the growth rates of these two instabilities are comparable at $\theta_{bk} = 30°$. To get a better understanding of how the growth rates compare in general, Figure 2 shows the growth rates of the beam and the anisotropy driven modes maximized over k as a function of θ_{bk}. As can be seen in the top panel, the beam-driven magnetosonic mode has maximum growth at parallel propagation, as is the case for a field-aligned beam. On the other hand, the anisotropy-driven magnetosonic mode has maximum growth at $\theta_{bk} \sim 25°$. A parameter study (not shown here) of the anisotropy-driven instability has shown that the maximum growth can occur over a relatively large range of propagation directions. Figure 2 shows that γ_{max} of the beam-driven mode is about $0.044 \Omega_p$ while that of the anisotropy-driven mode is $\sim 0.037 \Omega_p$, implying that the two instabilities have comparable growth rates. To address the question of the relative importance of the two instabilities and their nonlinear evolution, we

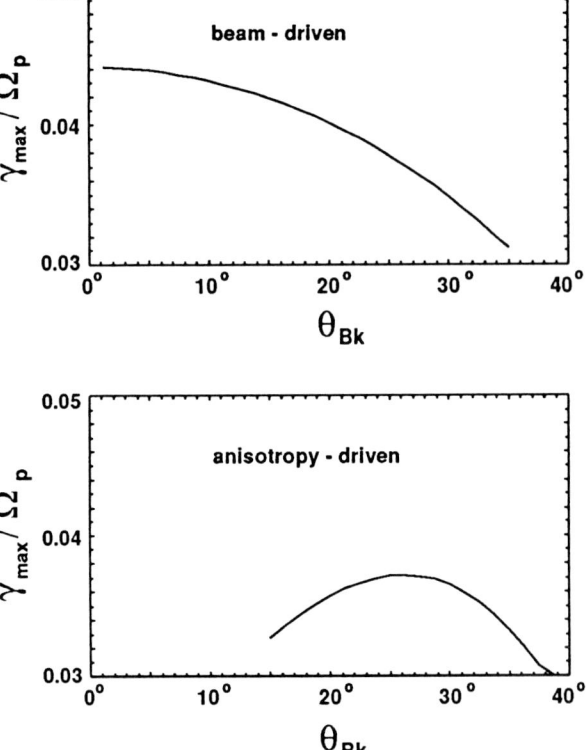

Fig. 2. Plots of the growth rates of the beam-driven (top panel) and the anisotropy-driven instabilities maximized over K (γ_{max}) as a function of wave normal angle. The beam-driven instability has maximum growth along the field while the anisotropy-driven mode has maximum growth at oblique angles.

have performed 2-D hybrid simulations which we now discuss.

3. FORMATION OF SHOCKLETS IN 2-D

The simulations presented in this section are doubly periodic in the X-Y plane with a system size of 400×85 c/ω_p and cell size of 1 c/ω_p in each direction. The solar wind ions are represented by 20 particles per cell. The background magnetic field lies in the X-Y plane and makes an angle of 30° with the X-axis. The simulations are either initial value or driven, with the former case corresponding to the beam ions being present at $t = 0$ and the latter case corresponding to the beam ions being injected uniformly in space and time throughout the run. This type of injection models a relatively small part of cometary environment in which the ion production rate is constant in space. In the initial value simulations the ion beam has a density $n_b = 0.05 n_o$ (represented by 20 particles per cell) while the driven simulations have an ion production rate which results in $n_b = 0.05 n_o$ at $\Omega_p t = 200$. The beam and plasma parameters used in the simulations are similar to those used in the linear theory presented in the last section.

As discussed earlier, linear analysis of the instabilities driven by a field aligned beam (i.e. the resonant and the nonresonant instabilities) has shown that maximum growth occurs along the magnetic field. As a result, one would expect the parallel propagating waves to grow the fastest and dominate the wave spectrum. To examine this issue, we have conducted a 2-D initial value simulation in which a cool field aligned beam is present in the simulation box at $t = 0$. Plate 1 shows one transverse component of the magnetic field (B_y) as a function of X and Y at $\Omega_p t = 112.5$, which is just before the saturation of the instability at $\Omega_p t \sim 125$. Although the waves have reached large amplitudes ($\delta B/B_o \sim 1$), their form is mainly sinusoidal and there is no evidence of steepening. To explain the lack of steepening in this case, we show in Figure 3 the power in B_y as a function of the propagation angle θ_{kx} (with respect to the X-axis). It is evident from this figure that maximum power occurs at $\theta_{kx} = 30°$, which implies the dominance of waves propagating parallel to the ambient magnetic field (recall that the background magnetic field makes an angle of 30° with the X-axis). This is in agreement with linear theory which predicts maximum growth rates along the magnetic field for a field aligned beam. As noted earlier, parallel propagating waves are linearly noncompressional and are not expected to steepen, which explains why the waves in Plate 1 have not steepened. Inspection of power vs. propagation angle (i.e. plots similar to Figure 3) at later times shows the excitation of obliquely propagating waves, which may be due to a linear process or caused by nonlinear wave-wave interactions. For example, two-dimensional studies of parametric instabilities [see e.g. *Vinas and Goldstein*, 1991] have shown that a parallel propagating (Alfvén or magnetosonic) pump wave can excite oblique daughter waves. A more detailed analysis of the simulation results is currently underway to establish the exact nature of this process. For our present purposes we note that steepening does not seem to be the dominant process in the nonlinear evolution of the excited waves.

To understand the nonlinear behavior of the instabilities associated with a beam-ring distribution function we first consider the results of an initial value simulation. The top panel in Figure 4 shows the time history of the fluctuating magnetic field (δB) as a function of time. It can be seen that the instability develops in two stages. In the first stage the waves grow and saturate at $\Omega_p t \sim 130$, which is then followed by a second growth phase saturating at $\Omega_p t \sim 185$. The bottom left and right-hand panels in Figure 4 show the power as a function of propagation direction (θ_{kx}) at the time of the first and second field saturation, respectively. It is evident in the left panel that while, at early times, the angular distribution of power is relatively broad, the peak occurs at $\theta_{kx} \sim 40°$ which corresponds to nearly parallel propagating waves. Further analysis of the waves (by Killen et al., paper submitted to *J. Geophys. Res.*, 1994) has shown the excitation of both the

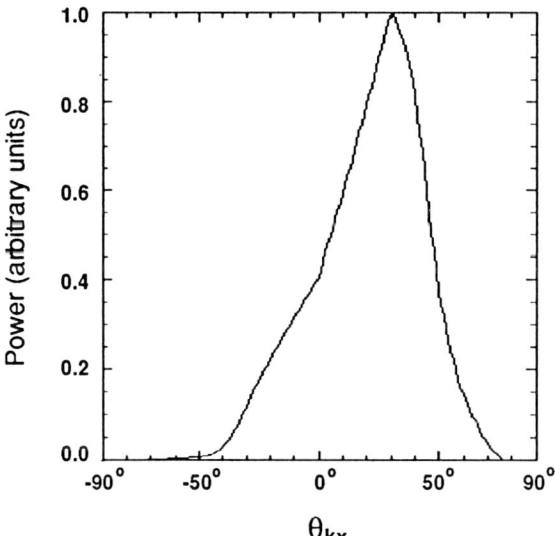

Fig. 3. Power as a function of θ_{kx}, the angle between $\vec{K}$ and the X-axis, for an initial value field aligned beam. Maximum power occurs at $\theta_{kx} \sim 30°$ which corresponds to parallel propagation.

nonresonant and the beam-driven magnetosonic modes, with the former having a larger amplitude. This is consistent with linear theory which shows the nonresonant instability having the largest growth rate. At later times, however, the obliquely propagating waves become dominant as illustrated in the bottom right panel. As can be seen, the peak power occurs at $\theta_{kx} \sim 0°$ implying a $\theta_{bk} \sim 30°$ which nearly corresponds to the angle at which the anisotropy-driven mode has maximum growth. Analysis of the waves shows power on both the beam-driven and the anisotropy-driven magnetosonic modes. This suggests that the first growth phase is mostly due to the nonresonant instability while the second growth phase is due to the beam and the anisotropy-driven instabilities.

To further examine this run, Figure 5 shows the distribution of the beam-ring ions in $(v_\parallel, v_\perp)$ phase space at three separate times. The top panel corresponds to

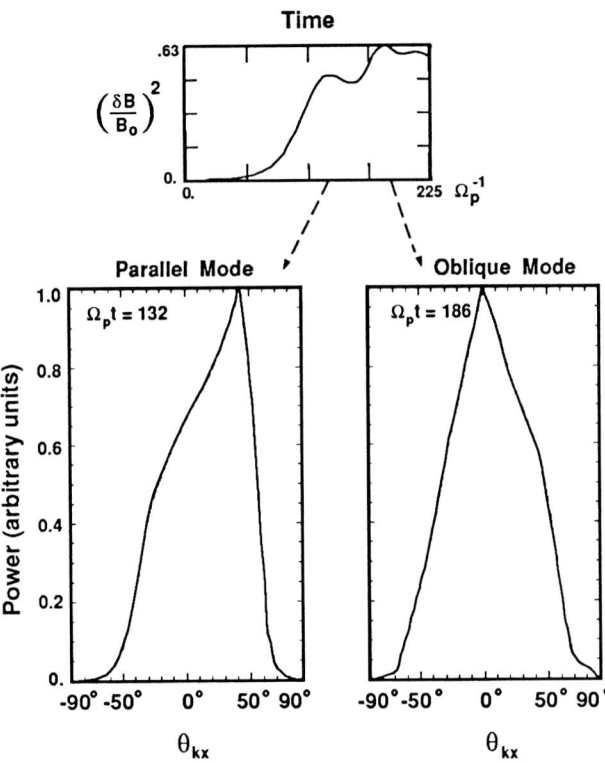

Fig. 4. The top panel shows the time history of the fluctuating magnetic field energy for an initial value beam-ring. Two peaks can be seen at times $\Omega_p t \sim 132$ and 186. Plots of power as a function of θ_{kx} at these two times (bottom panels) show the initial dominance of parallel propagating waves (beam-driven instability) followed by the dominance of oblique waves (anisotropy-driven instability).

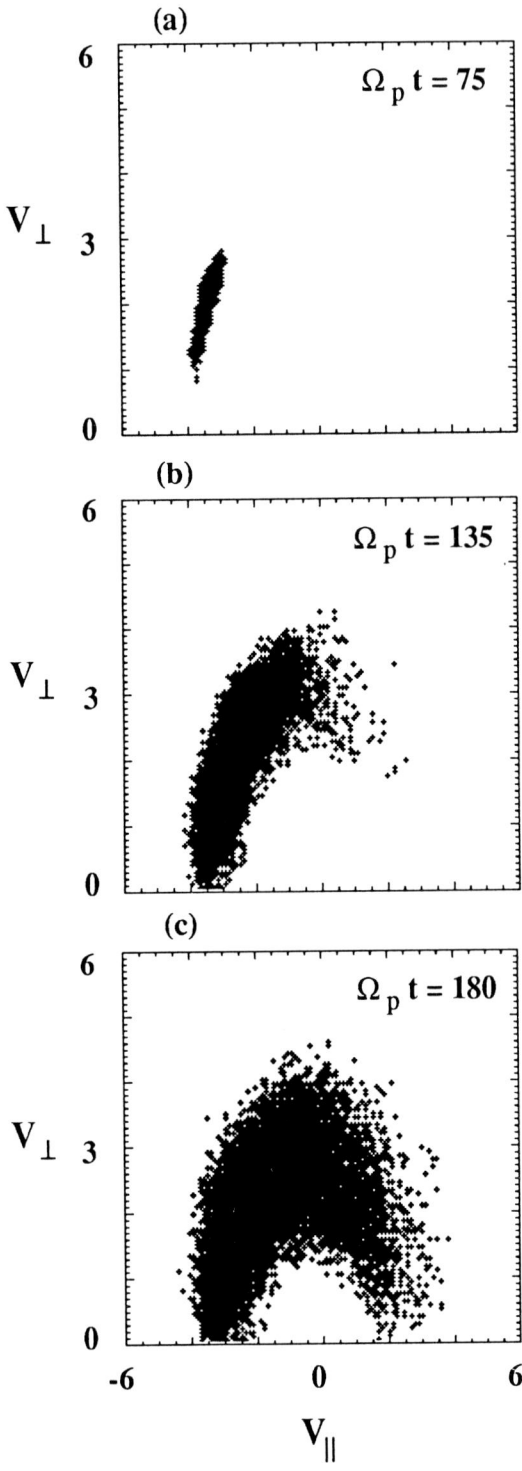

Fig. 5. Phase space density plots of the oxygen at the three indicated times. At time $\Omega_p t = 135$ the ions have undergone sufficient pitch angle scattering as well as velocity diffusion for the beam-driven instability to have saturated. But the anisotropy-driven instability is still operative and becomes saturated when the ions have formed a shell distribution ($\Omega_p t \sim 180$).

a relatively early time when the waves are growing and the amplitudes are small. It can be seen that while the original beam-ring has been scattered, the average parallel velocity of these ions is comparable to their initial speed. Similarly, the scattering is mostly in pitch angle with no significant velocity diffusion. The middle panel corresponds to a time after the first field saturation. By this time, the ions have undergone further pitch angle scattering and energy diffusion forming a half shell, resulting in the saturation of the nonresonant instability. On the other hand, it is evident that the distribution function still possesses a large anisotropy which can result in the growth of the anisotropy-driven mode. Apparently, this distribution function is still capable of exciting the beam-driven magnetosonic mode as well, however, the maximum power goes into the oblique direction. Finally, the bottom panel shows the phase space after the final wave saturation. By this time, pitch angle scattering and velocity diffusion have resulted in the formation of a thick shell which is no longer unstable. This result suggests the following scenario for the evolution of the instabilities associated with the beam-ring distribution function. Initially, the nonresonant waves grow the fastest with maximum power close to the field direction. Excitation of these waves results in the pitch angle scattering of the beam-ring ions until they form a half shell and the nonresonant mode becomes stable. At this time, the ions still possess enough free energy for the beam and the anisotropy-driven modes to con-

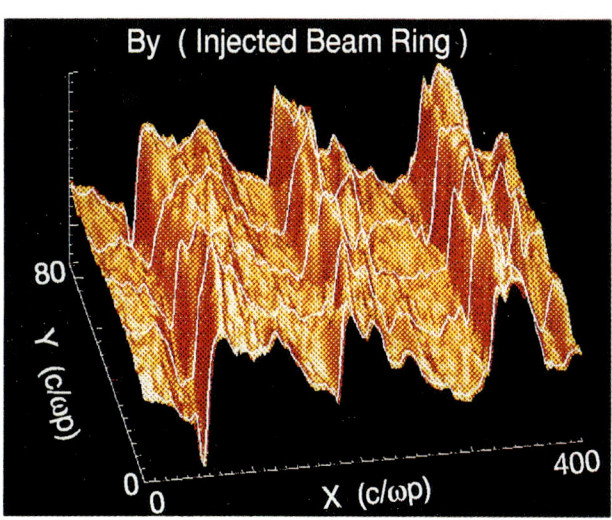

Plate 2. Same as Plate 1 except for a driven beam-ring distribution function. The 5 white lines are cuts in Y showing the waveform as a function of X at constant values of Y. The waves have steepened to form three shocklets.

tinue to grow resulting in a second growth phase. As the second instability develops, the maximum power shifts to oblique angles and the saturation occurs when the ions form a thick shell in velocity space. Thus, although the nonresonant instability has a larger growth rate, it also saturates faster resulting in the dominance of the oblique magnetosonic waves.

The above results are for an initial value system and may not be directly applicable to the settings such as comets or upstream of planetary shocks. Next we consider the results from a more realistic simulation in which the cometary ions are created continuously with time. The parameters are the same as in the previous run except that the cometary ions are now injected with a velocity along the X-axis. Given that the background magnetic field makes an angle of 30° with the X-axis, these ions have velocities of $4 \times \cos 30\ V_A$ and $4 \times \sin 30\ V_A$ parallel and perpendicular to the magnetic field, respectively. The production rate of the cometary ions is such that their density at the end of the run (i.e. $\Omega_p t = 200$) is 0.05. Thus on the average the beam density is smaller than the previous run. Figure 6 shows the maximum power as a function of θ_{kx} at $\Omega_p t = 200$, illustrating that the dominant waves have a wave normal angle $\theta_{bk} \sim 30°$. This may seem similar to the previous run where the oblique waves excited by the anisotropy-driven instability dominated at late times in the simulation. An interesting contrast between the initial value and the driven simulations, however, is the

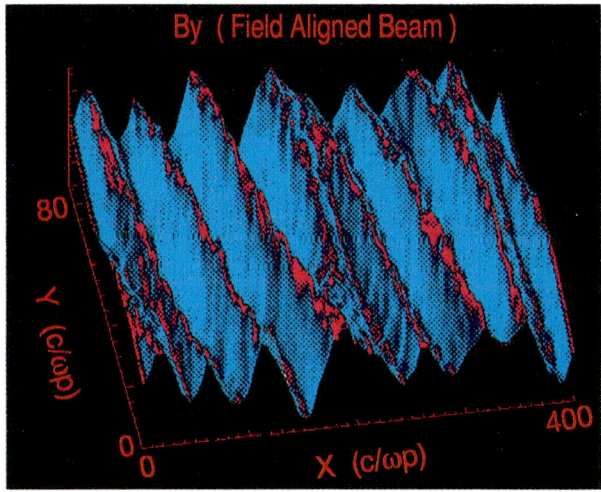

Plate 1. Surface plot of a transverse component of the magnetic field B_y as a function of X and Y for an initial value field aligned beam. Despite their large amplitude the waves have a sinusoidal form and have not steepened.

fact that in the latter case the oblique waves become dominant from a much earlier time. The fact that the parallel propagating nonresonant waves are not excited can be explained by noting that in the driven simulation the time averaged beam density is smaller. It has been shown by *Winske and Gary* [1986] that at smaller beam densities the nonresonant mode is either stable or has a growth rate which is smaller than that of the magnetosonic modes. The reason for the absence of parallel propagating magnetosonic waves is not as clear. One plausible explanation is the fact that at any given time, in a driven simulation, the injected ions have experienced a different degree of pitch angle scattering. Those injected earlier would have scattered more, while those injected later would have suffered less scattering. As a result, the beam forms a half shell (similar to the middle panel of Figure 5) much quicker. However, as shown already, a partial or a half shell distribution favors the excitation of oblique magnetosonic waves. A more detailed investigation of this point is currently underway by Killen et al. (paper submitted to *J. Geophys. Res.*, 1994).

To see the result of the nonlinear evolution of the oblique waves excited in the driven simulation, Plate 2 shows one transverse component of the magnetic field (B_y) as a function of X and Y at $\Omega_p t = 200$. Also present are five white lines which show the variation of B_y as a function of X at fixed values of Y (i.e. cuts along the X-axis). It is clearly evident from this figure that the waves have steepened to form three shocklets. This nonlinear behavior of the waves is consistent with their oblique propagation direction. Although variations with Y can be seen for each shocklet, the overall structure is quite coherent in that direction. The wavelength of each shocklet is about $133 c/\omega_p$ which is larger than the size of the simulation box in the Y-direction ($85 c/\omega_p$). Analysis of the shocklets upstream of the Earth's bow shock by *Le* [1991] has shown that their coherence length (perpendicular to the direction of the solar wind velocity) is of the order of an Earth radii, which is comparable to a wavelength. The results shown here demonstrate that the coherence length is at least 65% of the wavelength, which is consistent with the observational results. However, simulations with a larger size in the Y-direction are needed in order to make a more meaningful comparison with the observations.

Given the similarity between the beam injection process in our driven simulation and the actual cometary environment, it seems reasonable to assume that the observed shocklets at Comet GZ were excited in a similar manner. On the other hand, the beam injection model in our simulations is not directly applicable to the foreshock and therefore, we cannot draw any firm conclusions as to how the shocklets are formed in that environment. Nevertheless, our findings do provide us with further insights into the possible role of various beam populations in generating the shocklets. For example, it is unlikely that the field aligned beams are responsible for the excitation of oblique magnetosonic waves which steepen up to form shocklets. The present results suggest that shocklets are most likely excited by the anisotropy-driven instability. As mentioned in section 2, the nature of the instabilities driven by the beam-ring distribution function in equation (1) is similar to that of a drifting bi-Maxwellian with $T_\perp > T_\parallel$. Thus, we expect the shocklets in the foreshock to be generated by ions with either a beam-ring or a drifting bi-Maxwellian distribution function. Among the various ion beam populations observed in the foreshock, the gyrating ions [*Gosling et al.*, 1982] detected upstream of the quasi-parallel portion of the bow shock seem to be the most likely candidate for the excitation of shocklets. According to *Gosling et al.* [1982], these ions are formed due to specular reflection at the quasi-parallel shock. Upon reflection and near the shock these ions are not gyrotropic. However, further upstream they become gyrotropic, forming a beam-ring type distribution

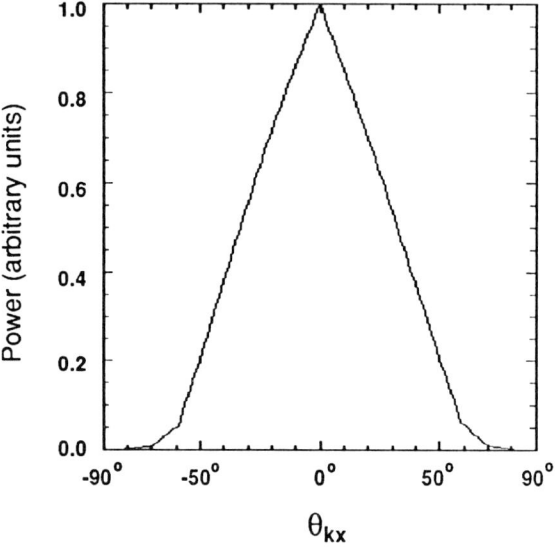

Fig. 6. Power as a function of θ_{kx} for a driven simulation with a beam-ring distribution function. Power peaks at $\theta_{kx} = 0°$ which corresponds to a wave normal angle of $30°$.

with a finite thermal spread. Whether shocklets are indeed excited by the gyrating ions needs to be investigated in more detail through observations and theoretical studies in the future. In our view, such studies are of paramount importance because it seems very likely that the current paradigm on the evolution of the various ion beam populations and magnetosonic waves and the connection between them is not correct.

4. FURTHER EVOLUTION OF SHOCKLETS

One of the important and outstanding issues regarding the nonlinear evolution of hydromagnetic waves is their possible transition to a turbulent state. This transition is often envisioned as the broadening of the spectrum through wave energy cascading into lower and/or higher frequencies via wave-wave interactions (e.g. parametric decay instability). The degree to which such processes may take place in space plasmas has not been fully tested or investigated in the observations. This is despite the fact that large amplitude waves are observed over a broad region of space. For example, the nature of highly nonlinear waves near the quasi-parallel shocks (i.e. pulsations or SLAMS) or at Comet GZ remains obscure. Clearly, one of the obstacles to such an investigation has been a lack of firm theoretical predictions which could be tested in the observations. Given the strong similarity between the observed and simulated shocklets, an obvious next step is to use simulations in order to examine and characterize their further evolution. Such an investigation can shed light on whether and how a transition to a turbulent state can take place and how observations can be used to identify such processes. To this end, we have used the results of 1-D hybrid simulations to investigate the stability and evolution of shocklets. One-dimensional simulations have the advantage that a larger system can be studied for a longer period of time with a higher resolution. Of course eventually it will be necessary to extend the calculations to two-dimensions to allow for the excitation of waves in a range of wave normal angles. The results presented here are preliminary in nature and are from the analysis of a run with parameters similar to the driven simulation in the previous section except that the simulation box is 800 c/ω_p long. Also the background magnetic field is in the X-Z plane making an angle of 30° with the X-axis. As a result, the excited waves have a wave normal angle of 30° similar to the dominant waves in the 2-D simulations.

Figure 7 shows the time evolution of a transverse component of the magnetic field (B_z) from $\Omega_p t = 150$ to $\Omega_p t = 300$ in the form of a surface plot. Also shown are

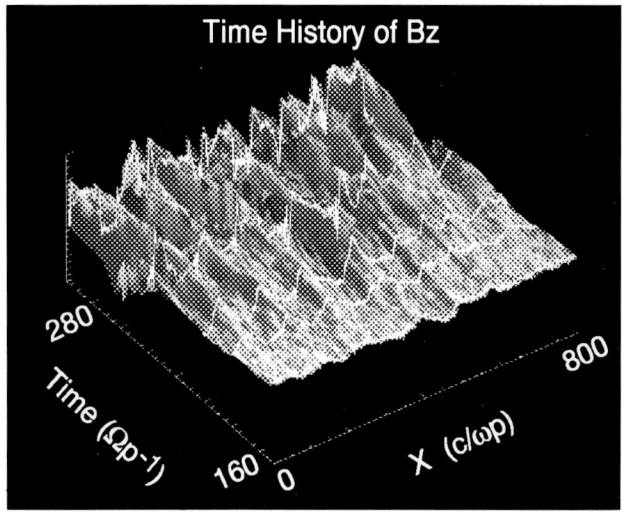

Fig. 7. Surface plot of a transverse component of the magnetic field B_z as a function of X and time. The 5 white lines are cuts in time showing the wave profile as a function of X at a fixed time. These lines show the evolution of small amplitude waves into shocklets which in turn evolve into less coherent structures.

5 cuts in time (white lines) which illustrate the wave profile at various stages of its evolution. At earlier times, the waves have small amplitudes and have a sinusoidal waveform. As the waves grow, they begin to steepen and eventually form shocklets (see e.g. the profile at $\Omega_p t \sim 260$). The shocklets persist for a while but then they evolve into less coherent structures at later times (see e.g. the last profile). This demonstrates that shocklets are not stable entities and that they undergo further change. To characterize the nature of this change, we have used a combination of Fourier and correlation analyses of the magnetic field and the density data. The top panels in Figure 8 show one component of the magnetic field B_y as a function of X at two separate times. The left panel corresponds to a time when the shocklets are in a full developed form. The right panel corresponds to a later time when the waveforms are not as coherent although some remnants of shocklets are still present. The bottom two panels show the Fourier transform of B_y in space. It is evident from the bottom left panel that the spectrum has a peak at mode numbers 6 and 7. The power spectrum at the later time shows that despite the change in the waveform the peak power remains in modes 6 and 7. Thus, while the spectrum has become somewhat broader the overall characteristics of the spectrum remains the same. This suggests that the change in the waveform is not directly tied to the excitation of other modes through nonlinear processes. As

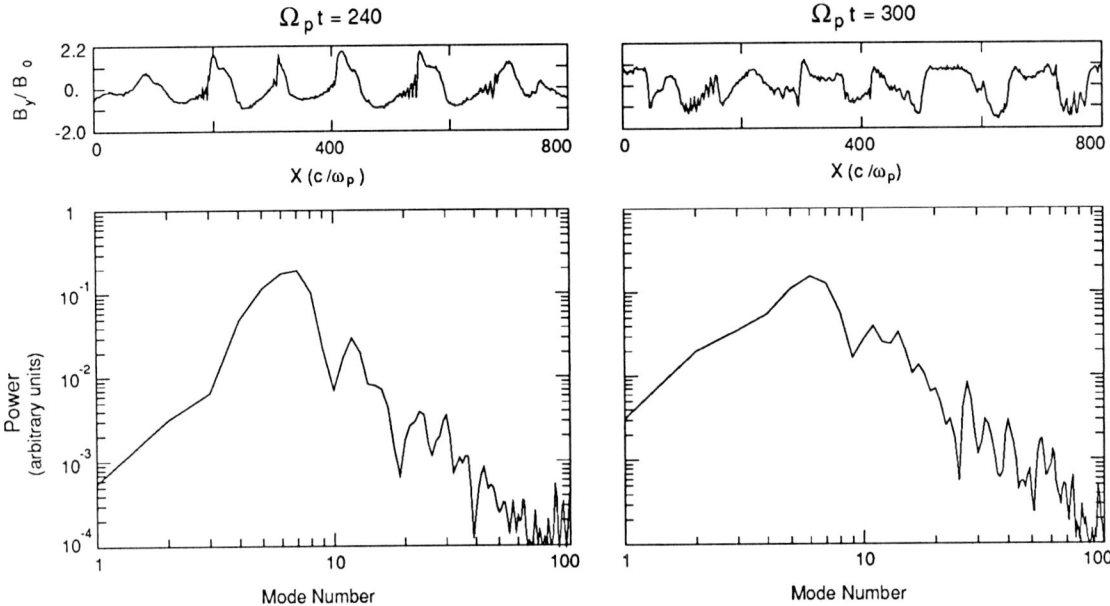

Fig. 8. The profile and power spectrum (in space) of B_y at a time when the shocklets are fully developed (left panels) and at a later time when they are not as coherent (right panels). Note that the two spectra are similar despite a change in the waveform.

a result, utilization of power spectrum in discerning the nature of further evolution of shocklets is of limited use.

Despite this limitation, Fourier analysis can shed some light on this problem. The top two panels in Figure 9 show the total magnetic field at the same times as in the previous figure. The fact that shocklets have evolved into less coherent structures is quite evident in the top right panel. The two middle panels show the helicity (sense of field rotation in space) of the waves where ϕ is the angle between the transverse magnetic field and the Z-axis. Waves with positive helicity (right-hand polarized in the plasma rest frame) correspond to ϕ changing from 2π to 0 while negative helicity waves would have ϕ changing from 0 to 2π. As can be seen in the left middle panel, most of the shocklets show a trend towards positive helicity implying a right-hand polarization which is in agreement with the predictions of linear theory for magnetosonic waves. However, as has been discussed in *Omidi and Winske* [1990], during the steepening process dispersive effects lead to a change in the polarization so that the waves become partially linear. This effect which was detected in the observed shocklets at Comet GZ [*Tsurutani and Smith*, 1986] can also be seen in the middle left panel. Note also that the helicity of the two shocklets at $X \sim 100$ and $X \sim 700$ has reversed implying that shocklets can show left-hand polarization as well. Examination of the helicity at time $\Omega_p t = 300$, when shocklets are no longer coherent, shows that the waves do not exhibit a well defined helicity or polarization. This, of course, implies a lack of ordered phase between the two transverse components of the magnetic field which exists at earlier times. We will come back to this point shortly.

The bottom two panels in Figure 9 show the power spectrum of the total magnetic field at times $\Omega_p t = 240$ (left) and 300 (right). As was the case with B_y, the power spectrum of total B has well defined peaks at $\Omega_p t = 240$. However, at later times (e.g. $\Omega_p t = 300$) the power spectrum of B is essentially flat. Given that the power spectrum of each transverse component of the magnetic field (e.g. the bottom right panel in Figure 8) does not show a similar behavior it is clear that the broadening of the spectrum is not related to the generation of longer and/or shorter wavelength waves. An alternate possibility, which also explains the lack of well defined helicity at later times, is the randomization of the phase between the two transverse components of the magnetic field (i.e. B_y and B_z). Under such a circumstance, the sense of field rotation in space (or time) can change within a wavelength, rendering the notation of helicity (or polarization) ill defined. Similarly, given that the total magnetic field fluctuations are directly tied to the phase relation between the two transverse components of the field, a random phase would result in

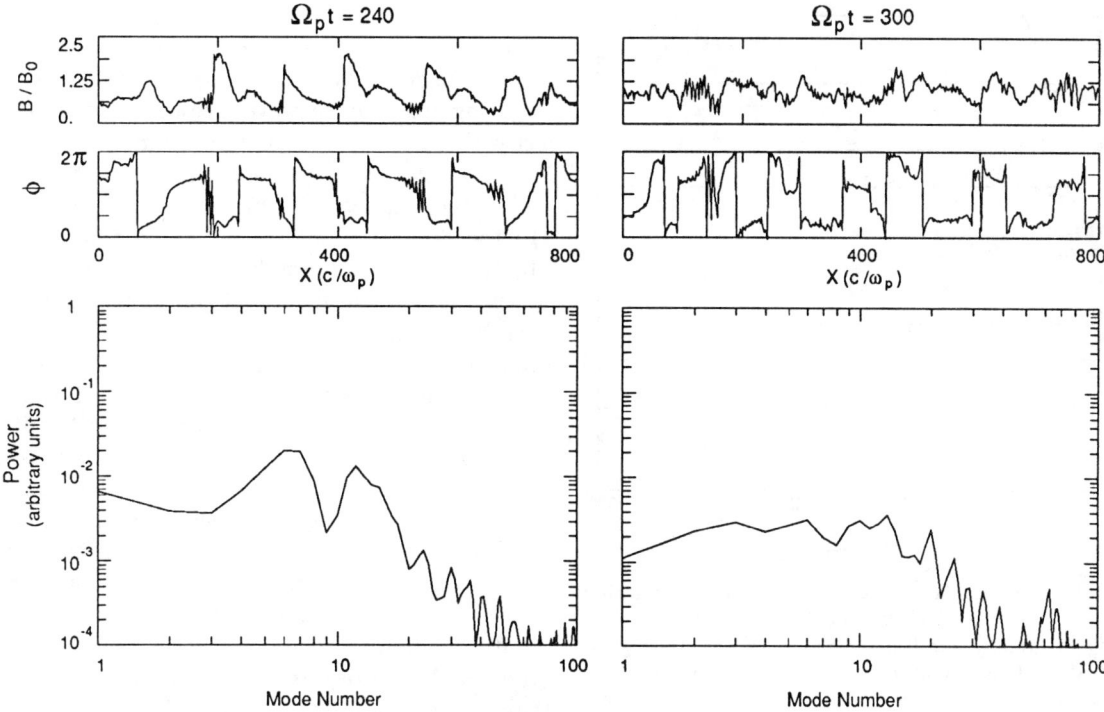

Fig. 9. Plots of total magnetic field (top panels), magnetic helicity (middle panels), and the power spectrum of magnetic field (bottom panels) at the same times as in Figure 8. At the earliest time shown, helicity is well defined (mostly positive) and the magnetic field has sharp peaks. Later, the helicity becomes ill defined and the spectrum is flat due to the randomization of the phase between the two transverse components of the magnetic field.

a similar behavior in the field fluctuations which in turn gives rise to a flat spectrum.

This result shows that one way in which further evolution of shocklets can be characterized is in terms of phase randomization. To consider another signature for this process, Figure 10 shows the results of a correlation analysis between the density and B_z. The top panel shows the correlation as a function of time at zero lag between the density and magnetic field, while the middle and the bottom panels show the maximum correlation and the corresponding lag, respectively. It can be seen in the top panel that the correlation is small (near zero) up to $\Omega_p t \sim 100$ which corresponds to a time interval in which no appreciable wave growth has occurred. Beyond that time the correlation increases, in concert with wave amplitude, reaching a maximum at $\Omega_p t \sim 190$. The correlation remains strong until $\Omega_p t \sim 240$ where it begins to drop to values close to 0. This drop is commensurate with the randomization of the helicity and the total field fluctuations. The middle panel shows a similar behavior except that the drop in correlation at late times is not as severe by virtue of a large lag between density and B_z (see the bottom panel). This suggests that another way in which further evolution of shocklets into less coherent structures can be characterized is in terms of a drop in correlation between the density and magnetic field fluctuations.

We have demonstrated that shocklets are not steady structures and evolve into entities with complex waveforms. This unsteady behavior can be due to a number of physical processes, such as noncoplanarity of the magnetic field, dispersive effects, and shocklet-shocklet collisions. For example, it was shown by *Omidi and Winske* [1988, 1990] that shocklets exhibit many of the behaviors associated with shocks and approximately satisfy the Rankine-Hugoniot relations. However, not all features of shocklets are similar to shocks. For example, in addition to being spatially localized another major difference between shocklets and regular fast shocks is that shocklets are not coplanar, i.e. the upstream and downstream magnetic fields do not lie in the same plane. It is well known, however, that the solutions to the steady state fluid equations (i.e. the Rankine-Hugoniot relations) require the magnetic field in the upstream and downstream to be in the same plane. In other words, non-coplanar shocks cannot be time steady, implying

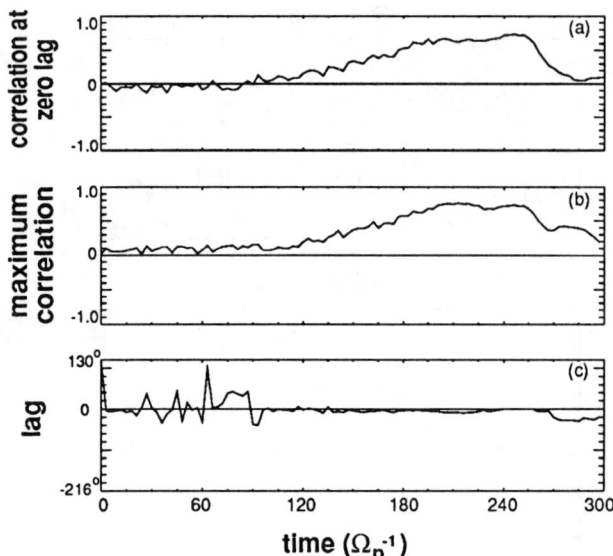

Fig. 10. Plots of correlation (top), maximum correlation (middle), and the corresponding lag (bottom) between density and B_z as a function of time. As the waves begin to grow (at $\Omega_p t \sim 100$) the correlation begins to increase, reaching its maximum value at $\Omega_p t \sim 180$. The drop in correlation occurs at the same time that shocklets breakup into less coherent structures.

that shocklets should evolve with time. This line of reasoning which is based on fluid theory may explain why shocklets are not steady, however, it does not shed light on how the actual evolution takes place. Some preliminary results (not shown here) suggest that dispersive effects could at least be partially responsible for the breakup of shocklets. However, more detailed studies are needed to further characterize the change in the shocklets and determine the nature of the kinetic processes responsible for this change.

5. SUMMARY

In this paper two issues relevant to shocklets namely the generation and the nonlinear evolution of oblique magnetosonic waves were discussed. Using linear theory, it was shown that beam-ring distribution functions can excite fast magnetosonic waves via two separate instabilities. One is driven by the beam part of the distribution with maximum growth along the magnetic field. The other is driven by the ring or the anisotropic ($T_\perp > T_\parallel$) part of the distribution with maximum growth at oblique angles. The results of 2-D hybrid simulations show that field aligned beams lead to the excitation of parallel propagating waves which do not steepen. As such, it is unlikely that the field aligned beams upstream of planetary bow shocks are responsible for the generation of shocklets. Initial value simulations with a beam-ring distribution function show that parallel propagating nonresonant waves grow first due to their larger growth rates. Pitch angle scattering of the ions into a half shell results in the saturation of this instability, while the magnetosonic mode continues to grow, leading to the dominance of the oblique magnetosonic waves. In the 2-D driven simulations where the ions are uniformly injected in space and time, oblique magnetosonic waves are dominant from early times. These waves were shown to steepen and form shocklets. Based on this, we conclude that the shocklets observed at Comet GZ were excited by the ring part of the distribution function. Among the various types of ion beams observed in the foreshock, the most likely population responsible for the generation of shocklets is the gyrating ions, which we expect can excite the anisotropy-driven magnetosonic waves. Further observational and theoretical studies are needed to assess the likelihood of this possibility. A more detailed discussion of linear theory and the simulation results is given by Killen et al. (paper submitted to *J. Geophys. Res.*, 1994).

Using 1-D hybrid simulations, the stability of shocklets was also investigated. It was shown that shocklets evolve into less coherent structures with complex waveforms. This transition can be characterized by the randomization of the phase between the two transverse components of the magnetic field, as well as a drop in correlation between the density and magnetic field oscillations. As a result of the former effect the wave helicity or polarization may not be well defined and also the fluctuations in the total magnetic field (but not each component) may show a broad spectrum. The fact that the power spectrum of the transverse components of the field do not exhibit a significant broadening suggests that wave-wave interactions are not important, at least in the early stages of the transition. On the other hand dispersion may be at least partially responsible for this process. We conclude by noting that these should be viewed as preliminary results and that much more needs to be done before a better understanding of the further evolution of shocklets is at hand. It is hoped that future theoretical studies coupled with observations will enhance our understanding of nonlinear hydromagnetic waves at shocks and comets in particular and space plasmas in general.

Acknowledgements. Useful discussions on the upstream ion populations with S. Fuselier are acknowledged. This research was supported by NASA Grants NAGW-1806 and NAG 5-1492, and IGPP at Los Alamos National Laboratory. N. Omidi's research was performed under the auspices of the California Space Institute. The computations were performed on the CRAY Y-MP at the San Diego Supercomputer Center.

REFERENCES

Akimoto, K., D. Winske, T. B. Onsager, M. F. Thomsen, and S. P. Gary, Steepening of parallel propagating hydromagnetic waves into magnetic pulsations, *J. Geophys. Res.*, *96*, 1991.

Brinca, A. L., and B. T. Tsurutani, Oblique behavior of low-frequency electromagnetic waves excited by newborn cometary ions, *J. Geophys. Res.*, *94*, 3, 1989.

Burgess, D., Cyclic behavior at quasi-parallel collisionless shocks, *Geophys. Res. Lett.*, *16*, 345, 1989.

Fuselier, S. A., Suprathermal ions upstream from the Earth's bow shock, Proc. of the Chapman Conference at Williamsburg, in press, 1993.

Fuselier, S. A., M. F. Thomsen, J. T. Gosling, S. J. Bame, and C. T. Russell, Gyrating and intermediate ion distributions upstream from the Earth's bow shock, *J. Geophys. Res.*, *91*, 91, 1986.

Gary, S. P., and C. D. Madland, Electromagnetic ion instabilities in a cometary environment, *J. Geophys. Res.*, *93*, 235, 1988.

Gary, S. P., J. T. Gosling, and D. W. Forslund, The electromagnetic ion beam instability upstream of the Earth's bow shock, *J. Geophys. Res.*, *86*, 6691, 1981.

Goldstein, M. L., H. K. Wong, K. H. Glassmeier, Generation of low-frequency waves at comet Halley, *J. Geophys. Res.*, *95*, 947, 1990.

Gosling, J. T., J. R. Asbridge, S. J. Bame, G. Paschmann, and N. Sckopke, Observations of two distinct populations of bow shock ions, *Geophys. Res. Lett.*, *5*, 957, 1978.

Gosling, J. T., M. F. Thomsen, S. J. Bame, W. C. Feldman, G. Paschmann, and N. Sckopke, Evidence for specularly reflected ions upstream from the quasi-parallel bow shock, *Geophys. Res. Lett.*, *9*, 1333, 1982.

Gosling, J. T., M. F. Thomsen, S. J. Bame, and C. T. Russell, Ion reflection and downstream thermalization at the quasi-parallel bow shock, *J. Geophys. Res.*, *94*, 10027, 1989.

Greenstadt, E. W., I. M. Green, G. T. Inouye, D. S. Colburn, J. H. Binsack, and E. F. Lyon, Dual satellite observations of the Earth's bow shock I. The thick pulsation shock, *Cosmic Electrodyn.*, *1*, 160, 1970.

Gurgiolo, C., G. K. Parks, B. H. Mauk, C. S. Ling, K. A. Anderson, R. P. Lin, and H. Reme, Non-ExB ordered ion beams upstream of the Earth's bow shock, *J. Geophys. Res.*, *86*, 4415, 1981.

Hada, T., C. F. Kennel, and T. Terasawa, Excitation of compressional waves and the formation of shocklets in the Earth's foreshock, *J. Geophys. Res.*, *92*, 4423, 1987.

Hoppe, M. and C. T. Russell, Plasma rest frame frequencies and polarizations of the low-frequency upstream waves: ISEE 1 and 2 observations, *J. Geophys. Res.*, *88*, 2021, 1983.

Hoppe, M., C. T. Russell, L. A. Frank, T. E. Eastman, and E. W. Greenstadt, Upstream hydromagnetic waves and their association with backstreaming ion populations: ISEE 1 and 2 observations, *J. Geophys. Res.*, *86*, 4471, 1981.

Le, G., Generation of upstream ULF waves in the Earth's foreshock, Ph.D. thesis at UCLA, 1991.

Omidi, N., and D. Winske, Subcritical dispersive shock waves upstream of planetary bow shocks and at comet Giacobini-Zinner, *Geophys. Res. Lett.*, *15*, 1303, 1988.

Omidi, N. and D. Winske, Steepening of kinetic magnetosonic waves into shocklets: Simulations and consequences for planetary shocks and comets, *J. Geophys. Res.*, *95*, 2281, 1990.

Paschmann, G., N. Sckopke, S. J. Bame, J. R. Asbridge, J. T. Gosling, C. T. Russell, and E. W. Greenstadt, Association of low frequency waves with supra-thermal ions in the upstream solar wind, *Geophys. Res. Lett.*, *6*, 209, 1979.

Russell, C. T., and M. Hoppe, Upstream waves and particles, *Space Sci. Revs.*, *34*, 155, 1983.

Scholer, M. and T. Terasawa, Ion reflection and dissipation at quasi-parallel collisionless shocks, *Geophys. Res. Lett.*, *17*, 119, 1990.

Schwartz, S. J., Magnetic field structures and related phenomena at quasi-parallel shocks, *Adv. Space. Res.*, *11(9)*, 231, 1991.

Schwartz, S. J., D. Burgess, W. P. Wilkinson, R. L. Kessel, M. Dunlop, and H. Luhr, Observations of short large-amplitude magnetic structures at a quasi-parallel shock, *J. Geophys. Res.*, *97*, 4209, 1992.

Thomas, V. A., D. Winske, and N. Omidi, Re-forming supercritical quasi-parallel shocks: 1. One and two dimensional simulations, *J. Geophys. Res.*, *95*, 18809, 1990.

Thomsen, M. F., G. T. Gosling, S. J. Bame, and C. T. Russell, Magnetic pulsations at the quasi-parallel shock, *J. Geophys. Res.*, *95*, 957, 1990a.

Thomsen, M. F., J. T. Gosling, S. J. Bame, T. G. Onsager, and C. T. Russell, Two-state ion heating at quasi-parallel shock, *J. Geophys. Res.*, *95*, 6363, 1990b.

Tsurutani, B. T., and E. J. Smith, Hydromagnetic waves and instabilities associated with cometary ion pickup: ICE observations, *Geophys. Res. Lett.*, *13*, 263, 1986.

Tsurutani, B. T., R. M. Thorne, E. J. Smith, J. T. Gosling, and H. Matsumoto, Steepened magnetosonic waves at comet Giacobini-Zinner, *J. Geophys. Res.*, *92*, 11074, 1987.

Vinas, F. A., and M. L. Goldstein, Parametric instabilities of circularly polarized large-amplitude dispersive Alfvén waves: excitation of obliquely-propagating daughter and side-band waves, *J. Plasma Phys.*, *46*, 129, 1991.

Winske, D., and S. P. Gary, Electromagnetic ion beam instabilities driven by cool heavy ion beams, *J. Geophys. Res.*, *91*, 6825, 1986.

Winske, D., and M. M. Leroy, Diffuse ions produced by electromagnetic ion beam instabilities, *J. Geophys. Res.*, *89*, 2673, 1984.

Winske, D., and N. Omidi, Hybrid codes: Methods and applications, in *Computer Space Plasma Physics: Simulation Techniques and Software*, edited by H. Matsumoto and Y. Omura, Terra Sci. Publ., Tokyo, in press, 1992.

Winske, D., C. S. Wu, Y. Y. Li, Z. Z. Mou, and Y. S. Y. Gou, Coupling of newborn ions to the solar wind by electromagnetic instabilities and their interaction with the bow shock, *J. Geophys. Res.*, *90*, 2713, 1985.

Winske, D., N. Omidi, K. B. Quest, and V. A. Thomas, Re-forming supercritical quasi-parallel shocks, 2, Mechanism for wave generation and front re-formation, *J. Geophys. Res.*, *95*, 18821, 1990.

Wu, C. S., and R. C. Davidson, Electromagnetic instabilities produced by neutral particle ionization in interplanetary space, *J. Geophys. Res.*, 77, 5399, 1972.

N. Omidi, Department of Electrical and Computer Engineering and California Space Institute, University of California, San Diego, La Jolla, CA 92093-0407.

H. Karimabadi, Department of Electrical and Computer Engineering, University of California, San Diego, La Jolla, CA 92093-0407.

D. Krauss-Varban, Department of Electrical and Computer Engineering, University of California, San Diego, La Jolla, CA 92093-0407.

K. Killen, Department of Electrical and Computer Engineering, University of California, San Diego, La Jolla, CA 92093-0407.

Magnetospheric and Solar Wind Studies with Co-orbiting Spacecraft

C. T. Russell

*Institute of Geophysics and Planetary Physics, University of California
Los Angeles, USA*

Magnetic field and plasma measurements made by the International Sun Earth Explorer 1 and 2 spacecraft (ISEE-1 and -2) are reviewed as a guide to the operation of future multiple spacecraft missions, such as the ESA Cluster mission. Regions and phenomena examined include interplanetary shocks, upstream waves, the bow shock, magnetosheath waves, the magnetopause, and the magnetotail. No one separation distance is optimum for all phenomena or even all desired studies of a single phenomenon. The need for a variety of different separations over the course of a mission effectively adds a new dimension to the study space and requires longer observing periods in order to obtain the proper combination of plasma conditions and spacecraft configuration to solve any particular problem. Thus, coverage on future multispacecraft missions needs to be more complete and the mission operated for longer periods than on past missions. All regions in and around the magnetosphere can benefit from multiple spacecraft studies.

INTRODUCTION

The plasma environment of the Earth is dynamic. Its boundaries are in constant motion driven ultimately by the solar wind. These boundaries move more rapidly than our spacecraft, often passing over a spacecraft multiple times as the spacecraft travels in the vicinity of the average position of the boundary. In order to compare observations with theory, we would like to be able to convert the temporal profiles that we measure into spatial profiles so that we can compare thicknesses with plasma scale lengths, observed wavelengths with those derived from plasma dispersion theory and periodicities with natural resonances of the plasma. In order to do this we need at least two spacecraft. Four spacecraft are needed for situations in which the natural coordinate system of the phenomena is unknown and five spacecraft or more are needed to test stationarity.

In this review we will examine how we have used multiple spacecraft with ISEE mission for closely spaced studies of discontinuities and boundaries in the vicinity of the Earth. Our emphasis will be on those lessons learned that are applicable to the conduct of the Cluster mission. These lessons have two aspects: information on the nature of the physical process taking place and information about the spacecraft configuration needed to undertake the study. The former lessons will help formulate the objectives of future work; the latter will provide information on how such studies can be undertaken. We will not discuss a second type of multispacecraft correlative study, perhaps best described as widely separated studies, where the spacecraft simultaneously measure separate phenomena such as the variation of the size of the polar cap as the interplanetary magnetic field changes its orientation. Much was learned from the ISEE mission in this regard and this knowledge will be vital to the conduct of the other elements of the ISTP mission, but these other lessons are in general much more obvious and perhaps better appreciated than the lessons to be discussed herein.

Solar System Plasmas in Space and Time
Geophysical Monograph 84
Copyright 1994 by the American Geophysical Union.

The lessons to be learned from ISEE can be presented in almost any order so we will pick the expedience of working from the outside in, starting in the solar wind and moving through the bow shock, magnetosheath and into the magnetosphere. This order also roughly follows the time spent in each region. An orbiting spacecraft will always spend most of its time in the vicinity of the most distant part of its orbit; hence one must choose apogee carefully. If it is chosen to be too low, critical phenomena may be entirely missed and if it is too high, the spacecraft will spend too little time observing prime objectives. This may be considered to be lesson zero of the ISEE program, but it is a rather obvious lesson.

Was the apogee of ISEE 1-and -2 well chosen? The spacecraft did spend about one third of their lives in the solar wind which was not a prime measurement region for ISEE-1 and -2. Interplanetary shocks and discontinuities were, of course, of some interest, but ISEE-1 and -2 were poor monitors of the solar wind either for studies of the solar wind itself or for correlative purposes. On the other hand the 23 Earth radii (R_E) apogee enabled upstream wave and particle studies over a significant and interesting region of space and enabled shock studies over much of the dayside bow shock. On the nightside, one could argue ISEE-1 and -2 did not go far enough. While some would argue that the substorm is initiated between geosynchronous orbit and about 10 R_E, others have shown that quiet and moderately disturbed times the neutral point seems to be seldom inside of 20 R_E [Cattell and Mozer, 1984; Angelopolous, 1993]. To the extent that the physics of the tail neutral point is a key ISEE or Cluster objective, there is an argument for greater apogee than 23 R_E.

We note here an important technical lesson from the ISEE program. Neither ISEE-1 or -2 included a tape recorder preferring additional instruments to complete temporal coverage. This was not a good trade in retrospect. Too much of the time data are missing at one or another of the spacecraft. Furthermore in order to keep ISEE-1 and -2 visible with the same ground station antenna the spacecraft were not allowed to drift too far apart. This restricted some of the science that could be done. With tape recorders the spacecraft could have been separated by arbitrary distances.

Another important technical lesson is that adding multiple spacecraft to a mission increases the time needed for the mission to accomplish its objectives, because it literally adds a new dimension to the study. In many respects it is like the difference in computation time required for running a two dimensional and three dimensional computer simulation. Adding a third dimension does not increase the running time of the computer program 50% as one might very naively believe, but of the order of 100 times since the new dimension will add about 100 parallel planes and communication between them. So too, with multiple spacecraft we have to explore the new dimensions of the separation between the spacecraft and the configuration of the separation vectors. These extra dimensions compound the problem of the scarcity of the critical study periods needed to address the objectives of the mission. For example, quasi-perpendicular bow shock crossings at moderate Mach numbers are very common. However if one wishes to study low beta (less than 0.1) quasi-perpendicular shocks then only 1 in 30 shocks is suitable. If one wishes to study low beta nearly parallel shocks (the so called switch on shock) then we have found only one such shock in 1700 studied [Farris et al., 1993a]. The chance that the Cluster spacecraft will be suitably configured for this extremely interesting observation when it occurs is miniscule. In fact, ISEE-1 and -2 were not suitably configured for optimum studies of this shock when it was encountered. The spacecraft were only 29 km apart at the time which is too close to determine an accurate shock velocity.

INTERPLANETARY SHOCKS

Interplanetary shocks provide an important complement to terrestrial bow shock studies because interplanetary shocks are generally weaker than the bow shock. The Mach number of the bow shock is determined by the flow velocity of the solar wind relative to the Earth as compared to the velocity of the fast magnetosonic wave in the plasma ahead of the shock, the upstream plasma. The Mach number of an interplanetary shock is determined by the velocity of the interplanetary shock relative to the solar wind flow ahead of it. Since the shock and solar wind flow are travelling in the same direction, this difference is often relatively small. It has to exceed the velocity of a magnetosonic wave, of course, to steepen into a shock.

Low Mach number interplanetary shocks occur in the solar wind under all beta conditions because their strength is not correlated with the upstream plasma properties. However, low Mach number bow shocks do have such a correlation and only occur for low beta conditions. Since the solar wind velocity is rather steady, varying at most usually only over a factor of 2, low Mach number bow shocks occur only when the magnetosonic velocity is high. High magnetosonic velocities occur when the interplanetary magnetic field is strong and the solar wind density is low, which are also the conditions for low beta. Thus to completely probe the low Mach number shock requires studies of both interplanetary and bow shocks.

In order to study a shock one needs to know its orientation in space which we specify by quoting the orientation of a vector normal to its surface, the shock normal vector. There are several ways to determine this vector. An important means of doing this for a bow shock is to use the direction perpendicular to the average shock surface. This is called the geometric normal. This estimate is not available for interplanetary shocks and determining their normals becomes a difficult issue, one for which multiple spacecraft are ideally suited.

Figure 1 shows the magnetic field strength measured by 4 spacecraft on August 18, 1978 while ISEE-3 was less than half way to its libration point orbit. Thus ISEE-1, -2 and -3 and IMP-8 were within about 80 R_E if each other. One can see differences between the appearance of the shock at the four spacecraft (except at ISEE-1 and -2 which are identical at this resolution). These differences are due to the different orientation of the magnetic field relative to the shock normal over a distance of 80 R_E. Thus a shock with much the same Mach number at the four sites can provide different examples of the processes acting, so that multiple observations of interplanetary shocks on these scales can be quite interesting. However this observation is not relevant to the Cluster mission per se which will operate at much smaller separations.

The important lesson here concerns the determination of the normal of the shock. When one has four observations, such as here, one can choose one of the spacecraft to serve as the origin of the coordinate system and time and write the following matrix equation:

$$\begin{pmatrix} \delta x_{01} \\ \delta x_{02} \\ \delta x_{03} \end{pmatrix} \cdot N = V_n \begin{pmatrix} \delta t_{01} \\ \delta t_{02} \\ \delta t_{03} \end{pmatrix}$$

where δx_{oi} is the separation vector between spacecraft 'o' and spacecraft 'i'; δt_{oi} is the time separation between the shock passages at the two locations; N is the shock normal and V_n is the speed of the shock along its normal. This equation may be solved for the shock normal and speed [Russell et al., 1983a]. The assumptions are, of course, that the shock normal and speed are constant over the measurement space and the shock is a plane. While a fifth spacecraft can be used to test these assumptions, such a spacecraft is seldom available.

Maxwell's laws and the conservation equations provide some assistance in testing these assumptions or providing information to supplant a missing spacecraft. For example, since the divergence of B is zero, the scalar product of the change in field across the shock and the shock normal direction is zero. This can be added to the equation above as can several other constraints [Russell et al., 1983b] so that satisfactory shock normals can be found with as few as two spacecraft.

In order to test whether our derived normals are in fact satisfactory we can check to see if the change in B_n is zero at each of our shocks. In Figure 2 we show an estimate of the error in our interplanetary shock normals by determining what angular error in the shock normal would be necessary to cause the observed change in the nominal normal component of the field. [Russell and Alexander, 1984]. For our set of interplanetary shocks, the majority of deviations were less than 4°, but one of our shocks appeared to deviate more than 32°.

These deviations are probably an inescapable consequence of the non-uniformity of the solar wind plasma. As sketched in Figure 3, the majority of interplanetary shocks are thought to arise from coronal mass ejections or stream-stream interactions. As the shock propagates away from the CME or the stream interface, it encounters different plasma

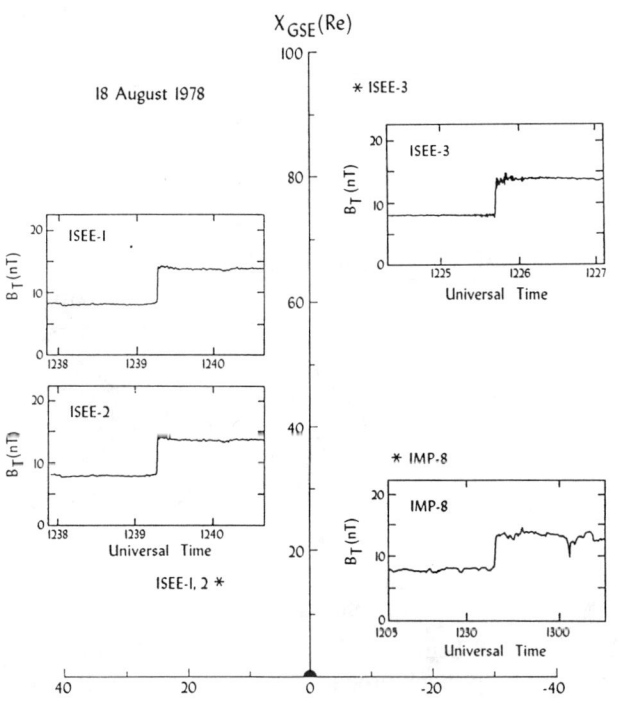

Fig. 1. Magnetic field profile and location of interplanetary shock seen by ISEE-1,-2, and -3 and IMP-8 on August 18, 1978.

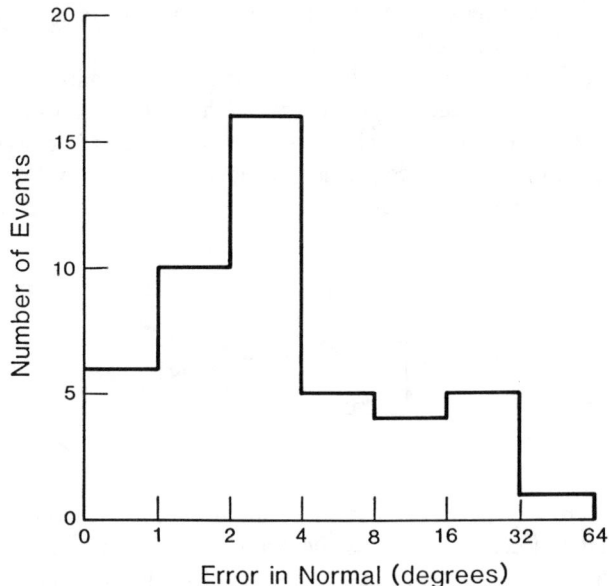

Fig. 2. Local deviation in shock normal direction from average direction of overconstrained multipoint shock normal determinations [Russell, and Alexander, 1984].

magnetic mirror as that field line slips through the shock. Since electrons have thermal velocities much greater than their bulk velocities and gyro radii much smaller than the shock thickness, a portion of the electron distribution is reflected by this moving mirror in a process that has been called fast Fermi acceleration. This process produces an energetic beam of electrons traveling upstream which creates a bump on the tail of the electron distribution which through

INTERPLANETARY SHOCK PROPAGATION

a) Driven Shock

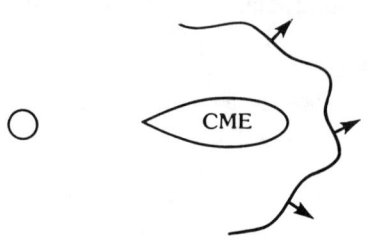

b) Stream-Stream Interaction

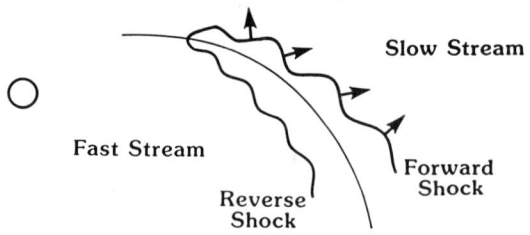

c) Small Scale Fluctuations

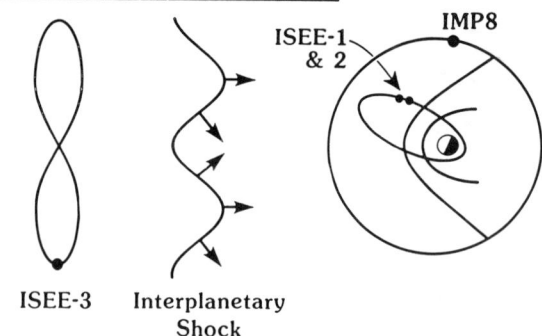

Fig. 3. Schematic diagram showing fluctuations in interplanetary shock propagation directions. a) In front of a shock driven by a coronal mass ejection; b) At a fast-slow stream interface; c) In the local neighborhood of the Earth. The shock direction oscillates as it encounters material of different index of refraction. [Russell, and Alexander, 1984].

conditions which could slow it or speed it up. Thus the shock front can develop a varying normal direction. Even on the smaller scale as illustrated in the bottom panel of Figure 3, the varying magnetic field orientation will affect the local Mach number of the shock and cause a fluctuating normal direction. The lesson here for the Cluster mission is that in studying boundaries the separation of the spacecraft should be kept sufficiently small that the orientation of the structure can be assumed to be constant over the volume enclosed by the spacecraft. Even large scale phenomena, such as interplanetary shocks, can have small scale structure so frequent checks on this assumption are recommended.

UPSTREAM WAVES

In the region upstream from the bow shock, but connected to it by the interplanetary magnetic field, a variety of ULF and VLF waves are found that are not usually observed in interplanetary space. These waves arise from 2 different sources. Some waves are generated at the bow shock and can propagate upstream against the solar wind flow. Other waves are generated in the upstream region by plasma instabilities associated with particle streaming back into the solar wind from the neighborhood of the shock. These latter particles may be ions or electrons and can be accelerated by several different processes. Electrons on the interplanetary field line that first encounter the shock see a rapidly moving

Landau resonance generates Langmuir oscillations at the electron plasma frequency. Thus the leading edge of the foreshock region is relatively easy to detect through its plasma wave signature.

Behind the tangent field line electrons continue to stream back into the solar wind because their thermal velocity exceeds the bulk velocity of the solar wind. These electrons provide a heat flux out of the shock, but this heat flux is not sufficient to make any major change in shock properties. At high Mach numbers ions too become hot enough to stream back upstream. However, instead of fast Fermi acceleration, ions with their larger gyro radii do not reflect along the tangent field line. They can be accelerated by drift along the shock surface so that they cross equipotentials of the upstream electric field and they can be turned around (reflected) by the combined electric and magnetic fields of the shock foot and overshoot. These processes create backstreaming ions along field lines connected to the shock in regions in which the magnetic field makes an angle of less than about 50° [Le and Russell, 1992a]. These backstreaming ions in turn lead to the generation of large amplitude waves. The generation of these waves is not completely understood.

In magnetohydrodynamic (MHD) theory there are four wave modes: a fast magnetosonic mode that compresses the magnetic field and the plasma; an Alfvenic or intermediate mode that twists the magnetic field and flow, but does not compress it; a slow magnetosonic mode in which the plasma pressure and magnetic pressure are anticorrelated so that the total pressure remains almost constant; and an entropy fluctuation which does not propagate and in which the total pressure is constant. Despite the popularity of the MHD treatment of waves and its extension to warm plasmas [Stringer, 1963; Formisano and Kennel, 1969] the MHD treatment does not adequately predict the properties of waves in a warm plasma such as the solar wind [Krauss-Varban et al., 1993; Orlowski et al., 1993]. In order to treat waves in such a plasma a Vlasov or kinetic approach is needed. In the Vlasov treatment there are also four modes: a fast compressional mode; an Alfvenic/ion cyclotron mode; a slow/sound mode; and the non-propagating mirror mode [Krauss-Varban et al., 1993]. The comparison of the properties of these kinetic modes which differ significantly from the MHD modes has only just begun.

The reason that determining the wave modes in the upstream plasma is important is as a means to discovering the underlying physical processes that generate the waves. Figure 4 shows examples of these waves. In the right hand panel are "30-second" waves thought to be generated by the back streaming ions by Doppler-shift ion cyclotron resonance. The bottom panel shows what is believed to be the same waves in an evolved state in which the waves have steepened and produced whistler mode precursors. The left-hand panels show waves at 1 Hz and 0.3 Hz. The former are believed to be generated at the bow shock [Fairfield, 1974; Orlowski and Russell, 1991], but the source of the latter waves is still a mystery [Le et al., 1992b].

With the dual ISEE-1 and -2 spacecraft we can calculate the wave normal directions of these waves by calculating the direction of minimum variance and can calculate the wavelength from the separation along the normal and the phase shift between the two measurements. With knowledge of the plasma velocity along the wave normal we can also calculate the wave velocity in the plasma rest frame. This can then be compared with the expected velocity and wavelength. Other parameters are available such as the polarization of the waves, the ellipticity of the perturbation ellipse and how compressive the wave is.

In order to pursue this wave identification an obvious necessary condition is that the same wave be seen at the two locations. Similar waves at similar frequencies are seen throughout the upstream region, but these may be from different sources and will not provide useful results in any cross comparison. To check to see that the same wavefront is being studied we perform cross correlations between the waves seen at the two spacecraft. Figure 5 shows the cross correlation coefficient as a function of distance perpendicular to the solar wind flow for 30-second waves [Le et al., 1990]. At a distance of 1 Re perpendicular to the flow the correlation coefficient has fallen to 0.5. Thus in general these waves can be studied with spacecraft separations of the order of 1 Re.

Figure 6 shows the cross correlation for three second waves. The waves are more monochromatic than the 30-second waves so that they can be studied at separations that are several wavelengths apart. However, since the wave length of the three-second waves is close to 1500 km, the maximum separation between the spacecraft should be only about 3000 to 4000 km for satisfactory wave analysis.

Figure 7 shows a similar plot of correlation coefficient versus distance for the upstream whistler mode waves at 1 Hz. These waves have very short wave lengths and correspondingly short correlation lengths. For these waves the separation distances should be less than 100 km for adequate investigation. This we believe will be very difficult on the Cluster mission and may be incompatible with other objectives.

In short, in order to study upstream waves a variety of separations is required. Separations of up to 1 Re may be appropriate for 30-second waves, but separations of less than 100 km are needed for upstream whistler wave studies. Since the measurement of time delays and phase lags is not

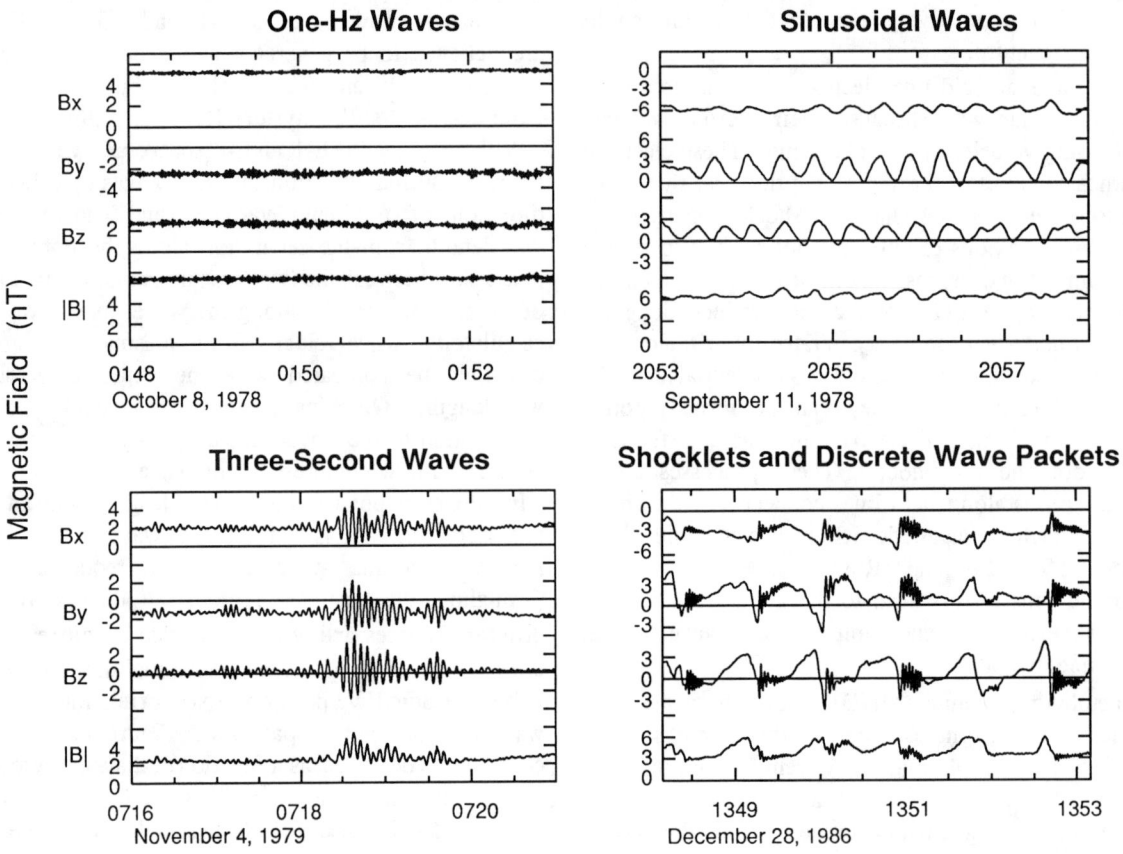

Fig. 4. Magnetic field measurements in solar ecliptic coordinates for the majority of the ULF wave types seen upstream from the Earth's bow shock [Le and Russell, 1993].

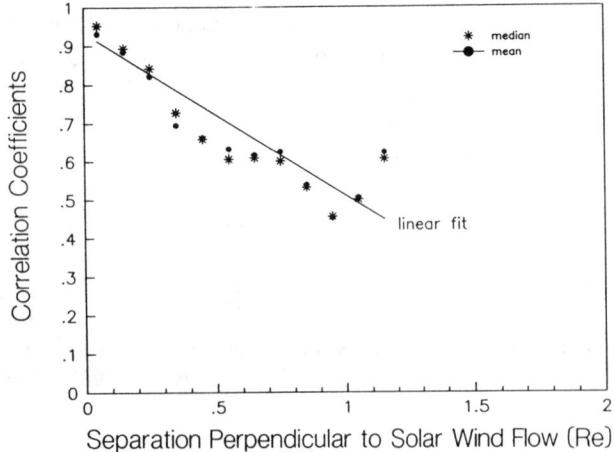

Fig. 5. Correlation coefficient versus distance perpendicular to the solar wind flow for 30-second waves [Le and Russell, 1990]

infinitely accurate a minimum separation is needed to study each wave type. Hence one cannot simply choose the shortest required separation for investigations. One must optimize the separations for each.

THE BOW SHOCK

The function of the bow shock is to slow, deflect and heat the supersonic solar wind so that it can flow around the magnetosphere. The macroscopic behavior of the shock is specified by the Rankine-Hugoniot equations which in turn are derived from the conservation equations and Maxwell's laws. The processes by which occurs the dissipation required by the Rankine-Hugoniot equations is not specified by these equations and the microscopic physics varies with the parameters of the flow. The strength of the shock is important in this regard. Weak shocks are laminar and less turbulent than strong stocks. Shocks that occur under low

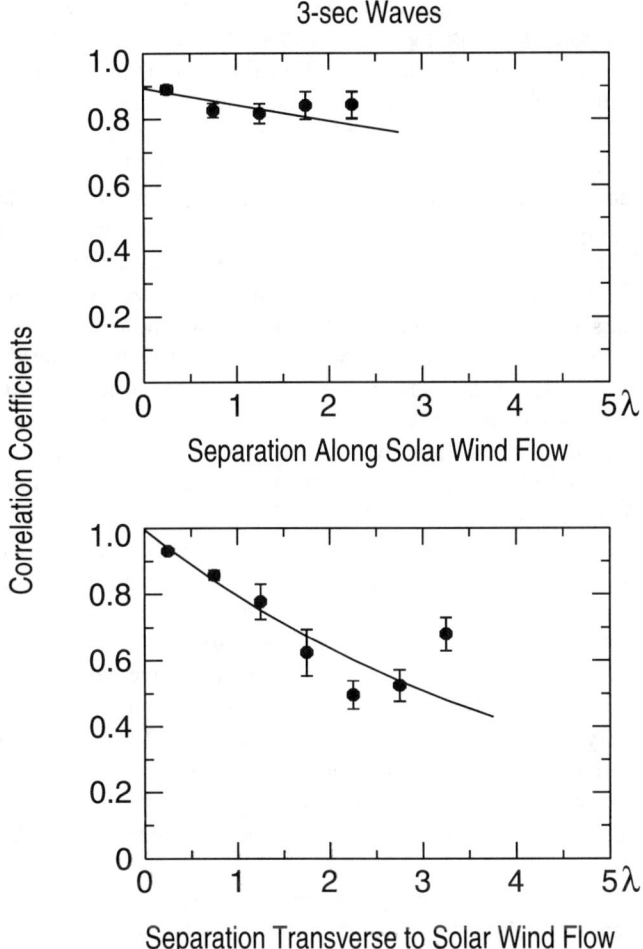

Fig. 6. Correlation coefficient versus distance along the flow and perpendicular to the flow for 3-second waves [Le et al., 1993]

beta conditions when the magnetic pressure is much greater than the thermal plasma pressure are also less turbulent in general. The angle between the upstream magnetic field and the shock normal is also an important factor in controlling these processes. Quasi-perpendicular shocks tend to be more regular than quasi-parallel shocks.

Upstream waves play a very important role in determining shock structure. First there are waves that can propagate far upstream because their group velocity can exceed the incoming solar and velocity. An example of such a wave is the upstream whistler mode discussed in the previous section. These waves can damp in the incoming solar wind and heat the incoming flow. Stronger waves are often seen right at the shock, and co-moving with the shock. These waves are phase standing precursors whose phase velocity exactly matches the solar wind velocity and so appear to be stationary in the shock frame. The group velocity exceeds the phase velocity if these are whistler mode waves so that as before the energy of the wave can move upstream and the solar wind can be heated as it approaches the shock.

The third type of wave that is important is the 30-second wave. These waves are not phase standing. They are propagating toward the sun and steepening with time to form what have been called shocklets [Russell and Hoppe, 1983] because of their similar appearance to weak shocks. They even develop a precursor, phase standing whistler, which grows with time [Le and Russell, 1992b]. Since their group velocity is less than the solar wind velocity they are carried back toward the shock and they most strongly interact with the shock when they reach it. Figure 8 shows an example of ISEE-1 and -2 data obtained very close to the bow shock in front of a high Mach number quasi-parallel shock [Thomsen et al., 1988; 1990]. The steepened waves are being convected by the solar wind flow toward the shock on the left crossing each satellite in turn (ISEE-2 is sunward of ISEE-1 by about 600 km along the Earth-Sun line). Nearer the shock the waves suddenly slow down in the spacecraft frame (which is effectively the shock frame) and the field becomes compressed. The difference in the magnetic field strength between the two spacecraft only 600 km apart is striking when this occurs. Thus a phenomenon that can be studied with a separation distance of 6000 km suddenly switches to a phenomenon that requires separations of less than 600 km, possibly much less than 600 km. Moreover the geometry of these compressed structures may be quite

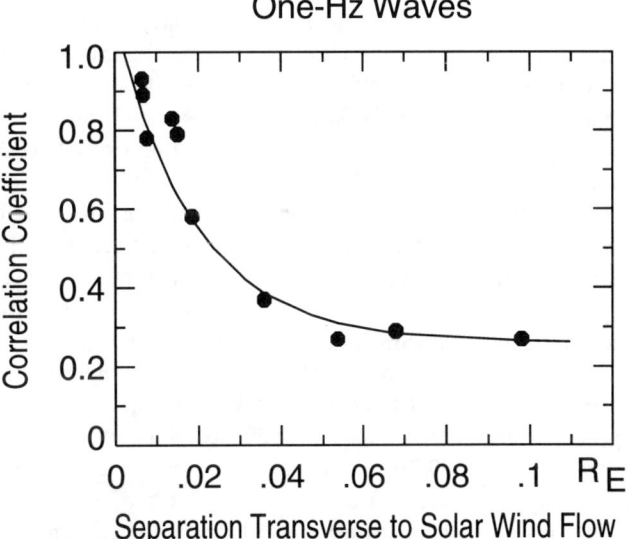

Fig. 7. Correlation coefficient versus distance across the flow for upstream whistler waves (1 Hz waves). [Le et al., 1993].

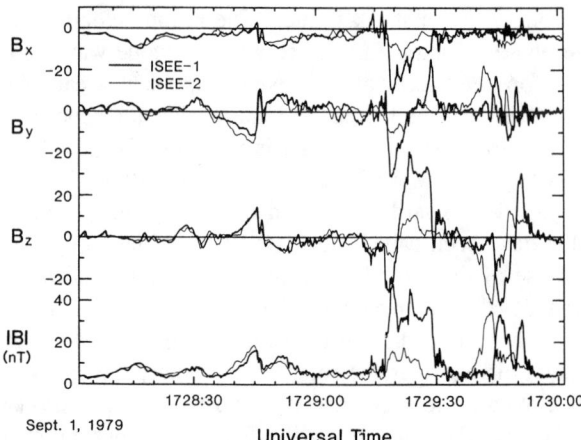

Fig. 8. Magnetic field measurements just upstream of bow shock under conditions of field aligned flow. Just ahead of the bow shock the waves slow down and increase in amplitude as shown on the right of the figure. ISEE-1 and -2 are shown. ISEE-2 is 600 km upstream of ISEE-1.

irregular. It is important for the Cluster mission to study both of these phenomena, but it cannot successfully do so with a single configuration of the 4 spacecraft. We note that recently this same phenomenon has been observed in the AMPTE data and called "short large-amplitude magnetic structures" or SLAMS [Schwartz et al., 1992], and this same phenomenon causes the quasi-parallel shock to periodically become quasi-perpendicular, perhaps interacting with the SLAMS [Greenstadt et al., 1993].

A correlation length of 600 km was found in a cross correlation study of the structure of moderate to high Mach number shocks by Greenstadt et al. [1982]. Figure 9 shows the cross-correlation coefficient between ISEE-1 and -2 for separations of 80 to 6000 km. There is a rapid fall off in the correlation coefficient when the separation exceeds 600 km.

If one restricts oneself to laminar shocks, those with well behaved magnetic profiles, which appear to have a well-defined shock surface, one still finds that rather close separations are required. Figure 10 shows 5 traversals of the bow shock by ISEE when the Mach number and beta were low. The temporal profiles have been converted to spatial profiles by timing the shock crossing at the two spacecraft and determining the velocity along the normal [Farris et al., 1993b]. In order to measure the velocity accurately the spacecraft could not be too close or too far. If the spacecraft were closer than 50 km, the uncertainty in the timing of the shock crossing caused by the fluctuations on the shock profile compared with the transit time at usual shock velocities gives unacceptable errors. A particularly unfortunate example of this effect occurred at the time of the only known crossing of a switch-on shock in the ISEE data [Farris et al., 1993a]. At this time the ISEE-1 and -2 separation along the shock normal was only 29 km and the uncertainty in velocity was well over a factor of 2. On the other end of the scale, the motion of the shock is approximately sinusoidal, and the separation of the spacecraft should be a fraction of the amplitude of the sinusoid for the velocity to be approximately constant. A 500 km separation is usually the largest that can be used for this reason. One important result about these low Mach number shocks obtained by Farris et al. [1993b] is that the thickness of the shock ramp is equal to at least a full wavelength of the upstream precursor wave. One might expect that the shock was merely the non-linear steepening of one half of a wave cycle. Inspection of Figure 10 shows that this is not true. The shock is thicker than expected.

There is one situation in which large spacecraft separations at the shock are useful. This is when one requires a simultaneous measure of the upstream and downstream conditions. An example of such a study is shown in Figure 11 [Farris et al., 1992] where the objective was to test whether the Rankine-Hugoniot equations were satisfied at very high beta shocks, where the plasma and field are very turbulent. In such a situation it is not possible to measure the conditions first upstream and then downstream and assure that there have been no changes during the period that the spacecraft moved from the upstream to the downstream state. One can see by inspection of Figure 11 that some features behind the shock front, such as the overshoot near 1440 UT, are not mirrored in the upstream field (solid line), but others such as the weakening of the downstream field around 1444 UT was directly associated with the weakening of the upstream magnetic field.

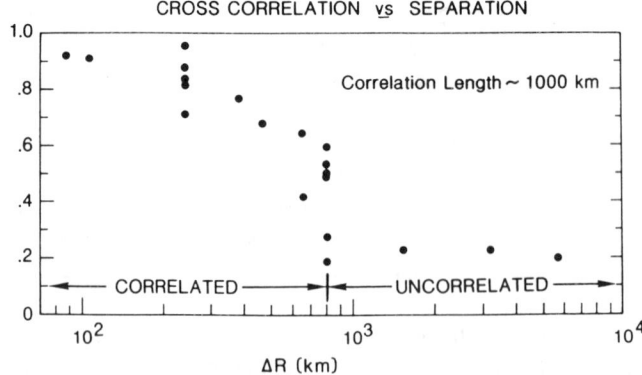

Fig. 9. The cross correlation between magnetic measurements seen at ISEE-1 and -2 in the vicinity of pulsating quasiparallel bow shocks [Greenstadt et al., 1982].

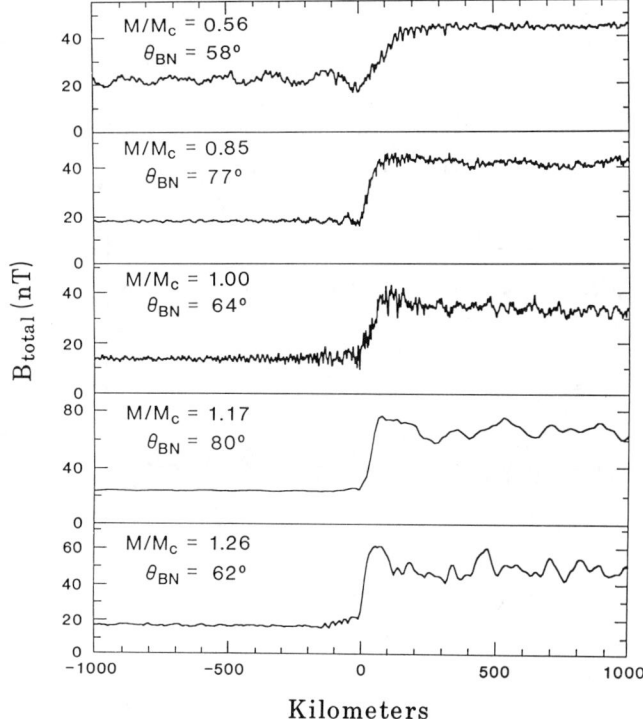

Fig. 10. Magnetic field profiles across 5 low Mach number, low beta shocks. Profiles have been converted from temporal to spatial using the observed stock velocity. The ratio M/Mc in the ratio of the shock's fast magnetosonic Mach number to the critical Mach number [Farris et al., 1993].

In short then, studies of the structure of the shock, whether that be the quasi-parallel shock or the quasi-perpendicular shock and whether that be the low Mach number shock or high, require separations in the range 50 to 500 km. Larger separations are useful for studies of upstream waves for for simultaneous upstream and downstream measurements.

MIRROR MODE WAVES

Many of the problems of studying waves in the magnetosheath are similar to studying waves in the solar wind except that the waves are convected past the observer more slowly and the beta of the plasma is higher. However, there is a change in the relative occurrence of the wave modes. In the solar wind upstream from the shock the ion beams generate fast mode waves. These compressional waves, as we discussed in the previous section, are blown back against the bow shock. We might expect then that the magnetosheath too would be filled with fast magnetosonic waves, perhaps amplified by their passage through the shock. On the contrary, the fast mode appears to play a minor role in the magnetosheath, perhaps because of the mode conversion of the fast mode at the bow shock. What is seen throughout the magnetosheath is the mirror mode wave, the non-propagating mode in a warm plasma [Tsurutani et al., 1982; Hubert et al., 1989a,b; Song et al., 1992; Fazakerley and Southwood, 1993]. These waves are difficult to study with only 2 spacecraft because they are linearly polarized and do not have a well defined direction of minimum variance and because they have a short coherence length. Figure 12 shows mirror mode waves at ISEE-1 and 2 on December 21, 1977. At this time the spacecraft are only 54° km apart along the normal to the magnetopause and 500 km apart along the boundary, but there is almost no correlation at any lag between the two spacecraft. To study this phenomenon with the Cluster mission will require relatively short separations.

Another "new" wave appears in the magnetosheath, the slow/sound oscillation [Song et al., 1992]. This wave is not expected because it damps rapidly in a warm plasma. However in situations in which the wave is driven by some process, or if wave-particle effects can maintain it, such waves may be seen. These waves differ from the mirror mode in that they propagate, albeit slowly, and may be able to stand in slow flows.

THE MAGNETOPAUSE

The structure of the magnetopause is very sensitive to the direction of the interplanetary magnetic field. When the interplanetary field is northward, the magnetopause, at least near the subsolar point, behaves much like a tangential discontinuity with little or no connection between the magnetosheath magnetic field and the magnetospheric magnetic field. When the interplanetary field turns southward, the fields become connected, plasma at the magnetopause becomes accelerated and particles can enter into and leak out of the magnetosphere [Paschmann et al., 1979].

The lack of interconnection between the magnetosphere and magnetosheath when the IMF is northward does not result in a simple boundary. Rather it reduces the dynamics so that the complexity of the boundary can be seen. Figure 13 shows the plasma and magnetic field across the magnetopause on such an occasion [Song et al., 1990]. On the left hand side of the figure the spacecraft is in the magnetosheath. We see the plasma pressure and the magnetic field oscillating out of phase. These are the mirror waves discussed above. At about 1516 UT the mirror waves cease and the plasma density decreases at constant temperature and the magnetic field increases. This is the

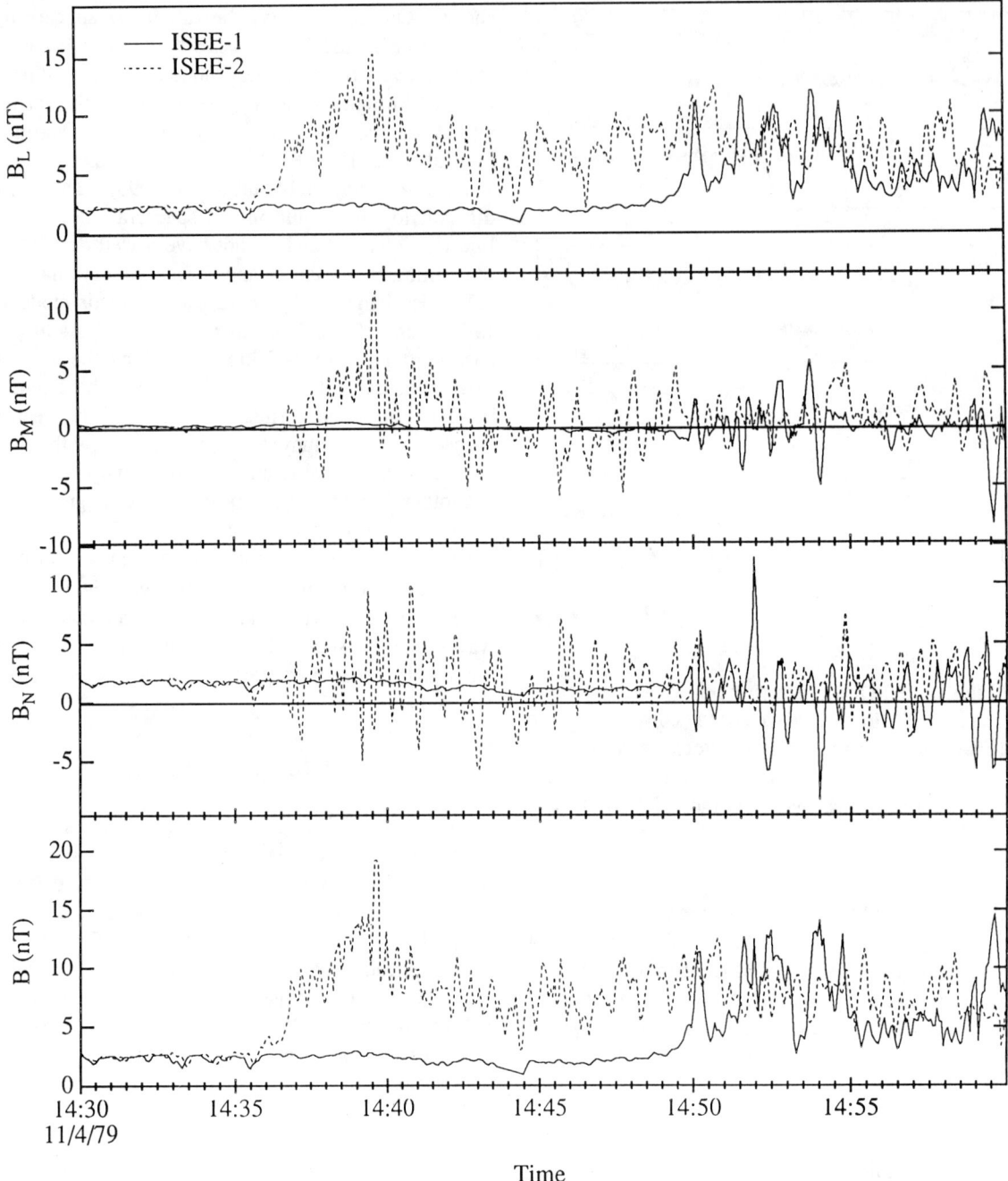

Fig. 11. Magnetic field measurements from ISEE-1 and -2 across a very high beta shock [Farris et al., 1992].

plasma depletion layer. As in the mirror mode waves the sum of the plasma and magnetic pressures are constant in the depletion layer. Since this region is part of the transition from magnetosheath conditions to magnetospheric conditions, we include this region in our definition of the magnetopause and consider it to be a boundary layer of the magnetopause.

Inside the magnetosphere on the right hand side of the figure is another boundary layer with a step-like transition and the final entry into the magnetosphere proper. The important lesson from this figure is the thickness of the layers in comparison with the transitions between them. The depletion layer is typically over 1000 km thick, but the inner

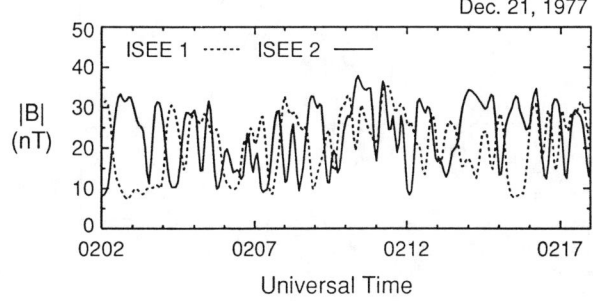

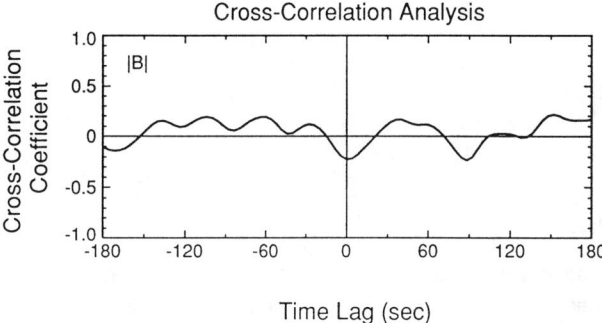

Fig. 12. Time series and cross-correlation coefficient for mirror mode waves in the magnetosheath just outside the magnetopause on Dec. 21, 1977.

These structures have been interpreted in many ways from transient patchy interconnection between the magnetosphere and the magnetosheath to waves on the magnetosheath boundary. It is clear they are not simply magnetopause encounters. The magnetopause encounters here are slow and irregular, and are thin in comparison with the separation of the spacecraft (300 km along the boundary normal). The flux transfer events in contrast appear almost simultaneously at the two locations so they are moving quickly and they have almost identical signatures at the two spacecraft. Thus their scale size is large, approximately 1 R_E. So in order to study FTE's we need rather large separations, about 1 Re, but to study the magnetopause itself we need small separations, perhaps as small as 100 km.

Although FTE's generally have rather large scale sizes, one unusual FTE with significant variation on a very short

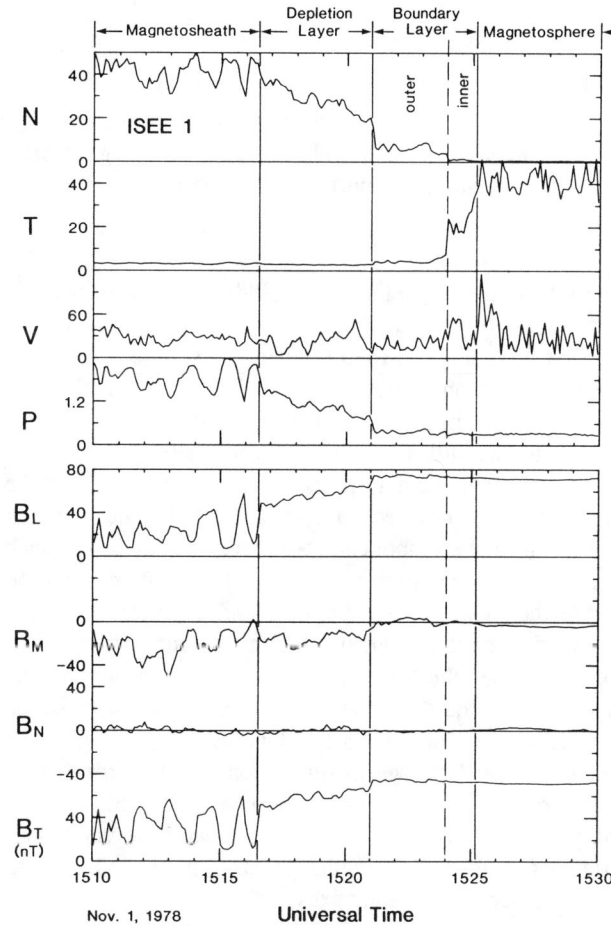

Fig. 13. Ion measurements from the Los Alamos Fast Plasma Experiment on ISEE-1 (top) and magnetic field measurements in boundary normal coordinates [Song et al., 1990].

edge of the depletion layer may be as thick as a few tens of km. We were not able to obtain a more precise thickness here because the two spacecraft were too far apart.

When the IMF turns southward the magnetopause becomes very dynamic and the magnetosheath and magnetosphere magnetic fields become interconnected. Large amplitude surface waves are created that grow with distance away from the subsolar region [Song et al., 1988]. Furthermore a new phenomenon arises known as a Flux Transfer Event [Russell and Elphic, 1978]. An example of a magnetopause with flux transfer events is shown in Figure 14. The main flux transfer events occur at about 0213 UT and 0235 UT. The principal signatures of a flux transfer event as seen in the magnetic field is the bipolar signature in the component of the magnetic field along the direction of the normal to the magnetopause, BN, the strengthening of the magnetic field in the magnetosheath to near magnetospheric levels, the northward turning of the component in the plane of the boundary, but in a direction away from the magnetospheric field. While the BL component moves toward a magnetospheric-like value, the BM component becomes stronger rather than approaching its magnetospheric value of zero.

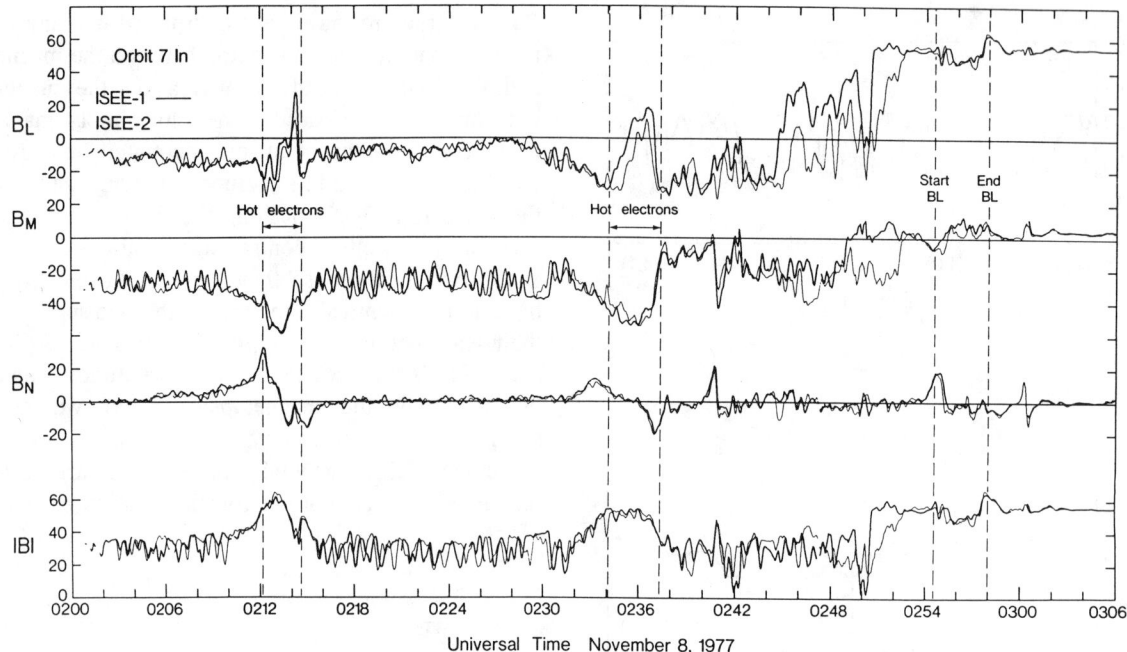

Fig. 14. ISEE-1 and -2 magnetic measurements in boundary normal coordinates across a magnetopause encounter on November 8, 1977 showing examples of flux transfer events. The spacecraft were 300 km apart along the normal at this time.

scale was studied by Zhu et al. [1988]. Their observations are shown in Figure 15 which shows 2 FTE's, both unusual. The first FTE at about 1815 UT shows a similar behavior at both spacecraft only 1000 km apart along the normal. This FTE is unusual in the sense that the field along the normal is close to the full strength of the magnetic field at both spacecraft. The field is pointing almost straight into the magnetopause. The second FTE at 1824 UT has a much smaller normal component, but here the field is much different at the two spacecraft. One spacecraft seems to be outside the main part of the FTE and experiences only a compression of the magnetic field as it wraps around FTE structure. The other spacecraft seems to have penetrated the central part of the FTE. The lesson for Cluster here is that there appear to be two scale lengths for FTE's. The structures may be large in dimension, of the order of an Earth radius across, and they may be very long and have no real gradient along their axes, but they can have very sharp variations with scale lengths much less than 1000 km perpendicular to their long axes.

Understanding flux transfer events has been difficult with the ISEE spacecraft. They are certainly three dimensional structures and have a variety of scale sizes. Thus even the 4 spacecraft of the Cluster mission may have difficulty characterizing them; furthermore if our interpretation of FTE's as a manifestation of time varying reconnection is correct, there may be significant changes in the structures between the times of passing the spacecraft, further complicating the analysis.

In short, the study of the magnetopause even under conditions of northward IMF will challenge the Cluster mission. The magnetopause has a variety of scale lengths from ion gyro radius scales, or possible subgyro radius scales, to fractions of an Earth radius. When the IMF turns southward, it becomes dynamic. Large amplitude surface waves are observed as well as a new phenomenon that is much more than just a surface wave. It has Earth radius scale sizes and moves with magnetosheath velocities and it distorts the magnetic field in unusual directions and to a significant degree; moreover this structure may be time varying on periods comparable to the interspacecraft transit time. These studies will strain the capabilities of the Cluster spacecraft to unfold the geometry of structures and separate spatial from temporal variations.

THE MAGNETOTAIL

The current sheet in the tail lies almost parallel to the ISEE-1 and -2 orbit. Thus generally the spacecraft only slowly cross the current sheet. The slowness of the crossing

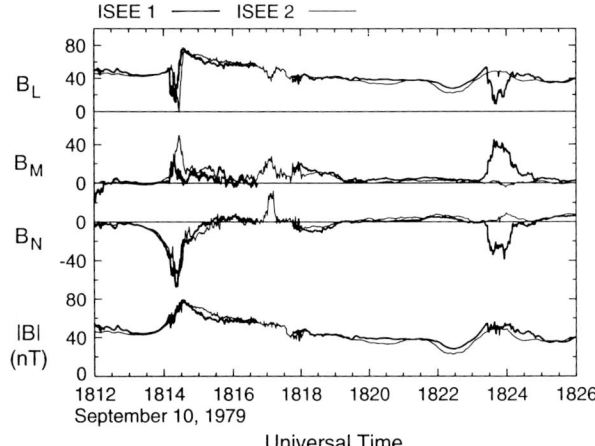

Fig. 15. ISEE-1 and -2 magnetic measurements in boundary normal coordinates across a magnetopause encounter on September 10, 1979 showing two unusual flux transfer events. The spacecraft were 1000 km apart along the normal at this time.

makes the current sheet difficult to study because of the unknown velocity of the current sheet relative to the spacecraft. Another problem in studying the current sheet is to determine its normal. The magnetic fields on either side of the current sheet are nearly antiparallel and the variation of the field through the current sheet is almost linear. In order to find the current density profile of the tail current sheet McComas et al. [1986] searched first for times when the direction of the solar wind changed rapidly so that the tail current sheet was swept over the ISEE spacecraft. They then determined the current sheet geometry by assuming that the direction of maximum variance lay in the sheet (perpendicular to the current), that there was no average field along the direction of the current in the sheet (i.e., J • B = 0), and that the normal direction was perpendicular to the above two directions with a constant magnetic field along it. Figure 16 shows an example of the deduced current structure. The current layer is close to 10,000 km in total thickness with a sharp maximum in current density in the center of the sheet.

The success of Cluster in studying the tail current sheet depends in the final orbit chosen for the mission. An orbit that is perpendicular to the current sheet would cross the sheet faster and provide a better current sheet profile and then 4 spacecraft would allow the orientation of the sheet to be determined without assumptions. However the observations would be limited to a small range of radial distances for such an orbit. We note that since the thickness of the current sheet is over an Earth radius rather large separations of the order of 1 R_E are appropriate for this study. Solar wind directional changes may also be needed to sweep the current sheet across cluster to facilitate this study.

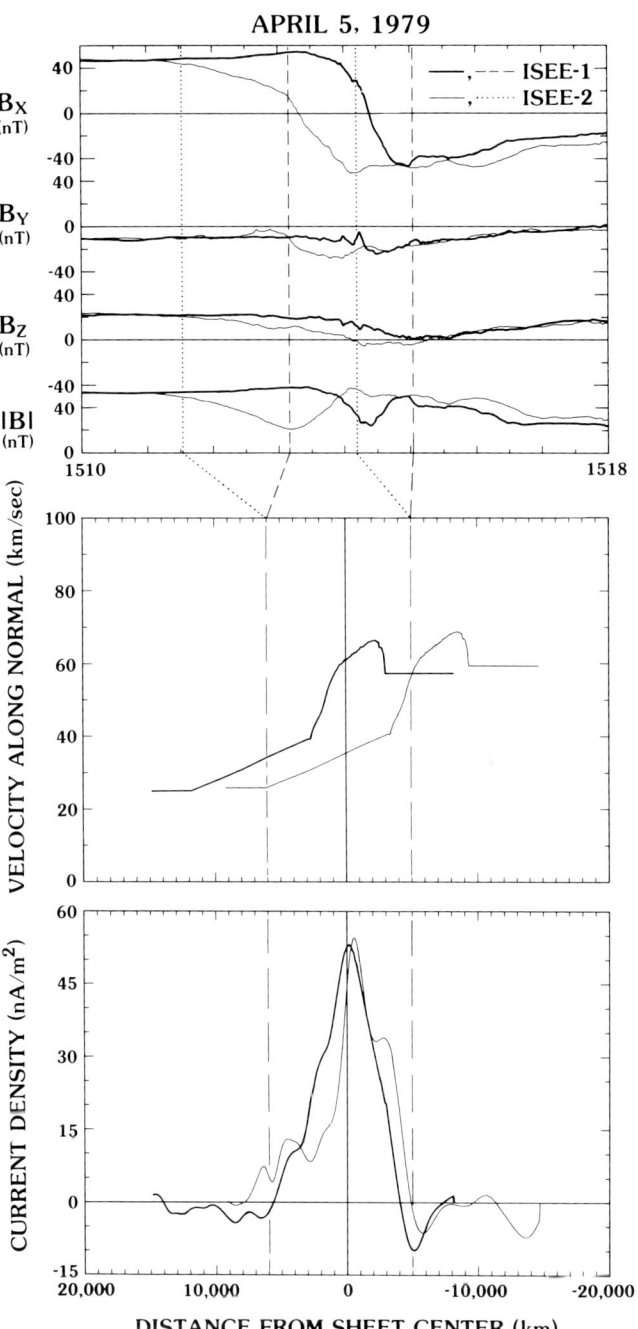

Fig. 16. (Top) ISEE-1 and -2 magnetic measurements in the near tail during a rapid traversal of the current sheet. (Center) Velocity deduced from the two magnetic profiles. (Bottom) Current density derived from magnetic profile and velocity profile. [McComas et al., 1986].

Another important issue to be resolved with multiple spacecraft measurements in the magnetotail is the location of the instability that triggers a substorm. This may or may not be the formation of a neutral point. It is possible that a disturbance in one region later triggers a neutral point formation somewhere else. Thus it is critical to monitor the direction of travel of substorm associated disturbances in the tail. Ohtani et al. [1992] have developed a single spacecraft method based on the phase of changes in the Bx (Sunward) and Bz (poleward) components and then verified this technique with dual ISEE spaced measurements. Since large separations are needed for studies such as this, few ISEE observations are available in this study. ISEE was seldom separated by the 1 Re or more distances needed.

In summary, studies of the steady state structure of the tail current sheet and of the dynamics of the tail were hampered by the geometry of the ISEE orbit and the scarcity of intervals of large separations respectively. The Cluster mission can successfully overcome both those problems by its orbit choice and its separation strategy. However, such a choice will certainly compromise some other objectives.

SUMMARY AND CONCLUSIONS

One obvious lesson from the ISEE mission is that multiple spacecraft are essential for studying the dynamic magnetosphere and the plasma environment. If the orientation of a boundary cannot be determined from means other than spaced measurements at different times, then four such measurements are required. The orientation so obtained will be an average over the region defined by the separation vectors. If this region is large compared to the scale size of irregularities in the medium, the local normals may differ greatly from the average. We have used multiple spacecraft for determining the orientation of interplanetary shocks, using both 4 spacecraft and "mixed mode" techniques, and find that in general these work well for interplanetary shocks. However on occasion there are large deviations of local normals from the average.

A second lesson from the ISEE mission is that a wide variety of spacecraft separations are necessary to study the various phenomena. These varied separations are needed in almost every region of space since phenomena occur over all scales. In the upstream region separations on the order of 1000 to 6000 km are sufficient to study 30 second and 3-second waves, but they are far too large to study the upstream whistler mode waves at about 1 Hz whose coherence length is less than 100 km.

At the bow shock this dilemma is also present. There is no one scale length that is optimum for shock studies. Upstream 30 second waves have rather large scale sizes, about 1 R_E, but at the shock these waves evolve suddenly on scale lengths of a few hundred km. Also shock motion limits the useful separation of the spacecraft. If accurate velocities are to be deduced from the time delays from one spacecraft to the other, the separation between spacecraft should be no longer than 500 km and no shorter than 50 km. It goes without saying that these separations need to be known accurately, at least to several percent. ISEE studies of the low Mach number, low beta laminar shock using time-of-flight measurements of the velocity of the oscillating bow shock have shown that the bow shock is about twice as thick as predicted in one-dimensional computer simulations.

Finally a type of study that is important in shock studies is the truly simultaneous upstream/downstream comparison. This type of study, that requires relatively large separations on the order of an Earth radius or more, allows one to determine the physical processes acting at the shock by allowing one to test predictive equations and to see what "phenomena" are actually features convected through the shock from the solar wind. ISEE measurements for example, have shown that the usual formulation of the Rankine-Hugoniot equations is sufficient for even the very high beta bow shock.

In the magnetosheath many of the same wave phenomena are present as in the solar wind together with one new mode, the mirror mode. The properties of the mirror mode make it very difficult to study. It is linearly polarized and appears to have very small scales. It is possible that the coherence length depends greatly on the direction relative to the wave fronts. Thus the 4 spacecraft of Cluster will be very important in determining the properties of these waves.

The magnetopause again presents us with the problem of multiple scale sizes, even for the most quiet conditions. ISEE shows that, when the IMF is northward, the magnetopause consists of multiple layers each of which plays a role in the transition from magnetosheath conditions to magnetospheric conditions. The outer most depletion layer which reduces the density and increases the field strength is a fraction of an Earth radius thick. Boundary layers inside the magnetosphere which further decrease the density and increase the temperature are of similar dimensions, but transitions between layers may be only tens of kilometers thick or even less. The interior magnetopause boundary layers appear to be dynamic even for strongly northward IMF, indicating that reconnection at high latitudes is important at such times.

When the interplanetary magnetic field is southward ISEE finds the magnetopause to be very dynamic with an amplitude of motion that increases with increasing distance from the subsolar point. The magnetosphere and the magnetosheath become interconnected, plasma and energetic particles flow between them, and accelerated plasma flows appear. Most importantly an entirely new phenomenon occurs for southward IMF, the flux transfer event. It has multiple scale lengths, both along and across the field. It is clearly a three dimensional stucture and may be time varying. Thus the FTE will strain the investigative capabilities of the Cluster mission.

Studies of the current sheet in the Earth's magnetotail have been difficult for ISEE, both because of its orbit and because of the paucity of data obtained at large separations. ISEE has shown that the current sheet at quiet times is quite thick, over 1 Earth radii, but at disturbed times this current sheet can narrow greatly. Cluster can improve upon this in two ways. The current sheet orientation will be determined more accurately with four spacecraft than two and the orbit can be chosen to cross the current sheet at a steeper angle.

ISEE data have also revealed disturbances to be propagating outward from the inner magnetosphere down the tail even though the neutral point is generally thought to be formed near or beyond ISEE apogee at 23 R_E. Cluster can add significantly to this study if large separations can be achieved in the tail. This was not possible on ISEE because of the lack of telemetry antennas to track them independently.

In summary the ISEE has served as an excellent pathfinder for future multispacecraft missions, both the ESA Cluster mission that is now under construction and other multiple spacecraft missions presently under discussion. Many more results are expected from the data in the ISEE archives because these data have not been fully analyzed yet due to lack of research support during the ISEE mission. Some of these results may impact the Cluster science planning, so ISEE funding should now be put at a high priority. Nevertheless the general picture is clear. A wide variety of separations are needed in every region of space because the coherence length of processes is quite varied. Moreover much data is needed at each separation distance because the proper plasma conditions, under which certain processes need to be investigated, are rare. The expected 50% data return from Cluster is particularly unfortunate in this regard. Finally, all regions of space around the magnetosphere will benefit from the multipoint measurements Cluster will provide.

Acknowledgments. The author wishes to thank J. Willett for his efforts at supporting the archiving of Space Physics data sets as well as the NSSDC under both J. Green and J. H. King for their diligence in preserving the archive. These data are critical for planning our future science activities. The author also wishes to thank M. M. Mellott, G. Le, E. W. Greenstadt, M. H. Farris, D. J. McComas, and Paul Song for their assistance in the multiple spacecraft studies described herein. The archiving of the ISEE-1 and -2 magnetometer data is being supported by the National Aeronautics and Space Administration under research grant NAG5-1967.

REFERENCES

Angelopoulos, V., Transport phenomena in the earth's plasma sheet, Ph.D. dissertation, University of California, Los Angeles, 1993.

Cattell, C. A., and F. S. Mozer, Substorm electric fields in the Earth's magnetotail, in *Magnetic Reconnection in Space and Laboratory Plasmas*, pp. 208-215, American Geophysical Union, Washington, D.C., 1984.

Fairfield, D. H., Whistler waves observed upstream of collisionless shocks, *J. Geophys. Res.*, 79, 1368, 1974.

Farris, M. H., C. T. Russell, M. F. Thomsen, and J. T. Gosling, ISEE-1 and -2 observations of the high beta shock, *J. Geophys. Res.*, 97, 19,121-19,127, 1992.

Farris, M. H., C. T. Russell, R. J. Fitzenreiter and K. W. Ogilvie, The subcritical quasi-parallel switch-on shock. *Geophys. Res. Lett., 20*, in press, 1993a.

Farris, M. H., C. T. Russell, and M. F. Thomsen, Magnetic structure of the low beta, quasi-perpendicular shock, *J. Geophys. Res.*, 98, 15,285-15,294, 1993b.

Fazakerley, A. N., and D. J. Southwood, Mirror instability in the magnetosheath, *Adv. Space Res.*, in press, 1993.

Formisano, V., and C. F. Kennel, Small amplitude waves in high beta plasma, *J. Plasma Phys.*, 3, 55-74, 1969.

Greenstadt, E. W., M. M. Hoppe, and C. T. Russell, Large-amplitude magnetic variations in quasi-parallel shocks: Correlation lengths measured by ISEE-1 and -2, *Geophys. Res. Lett.*, 9, 781-784, 1982.

Hubert, D., C. C. Harvey, and C. T. Russell, Observations of magnetohydrodynamic modes in the Earth's magnetosheath at 0600 LT, *J. Geophys. Res.*, 94, 17,305, 1989b.

Hubert, D., C. Perche, C. C. Harvey, C. Lacombe, and C. T. Russell, Observations of mirror waves downstream of a quasi-perpendicular shock, *Geophys. Res. Lett., 16*, 159, 1989a.

Le, G., C. T. Russell, A study of the coherence length of ULF waves in the Earth's foreshock, *J. Geophys. Res.*, 95, 10,703-10,706, 1990.

Le, G., and C. T. Russell, A study of ULF wave foreshock morphology-I: ULF foreshock boundary, *Planet. Space Sci.*, 40, 1203-1213, 1992a.

Le, G., and C. T. Russell, The morphology of ULF waves in the Earth's foreshock, in *Solar Wind Sources of ULF Pulsations*, in press, American Geophysical Union, Washington, D. C. 1993.

Le, G., C. T. Russell, M. F. Thomsen, and J. T. Gosling, Observation of a new class of upstream wave with periods near 3 seconds, *J. Geophys. Res.*, *97*, 2917-2925, 1992.

McComas, D. J., C. T. Russell, R.C. Elphic, and S. J. Bame, The near-Earth cross-tail current sheet: Detailed ISEE-1 and -2 case studies, *J. Geophys. Res.*, *91*, 4287-4301, 1986.

Ohtani, S., S. Kokubun, and C. T. Russell, Radial expansion of the tail current sheet disruption during substorms: A new approach to the substorm onset region, *J. Geophys. Res.*, *97*, 3129-3136, 1992.

Orlowski, D. S., and C. T. Russell, ULF waves upstream of the Venus bow shock: Properties of one-Hertz waves, *J. Geophys. Res.*, *96*, 11,271, 1991.

Paschmann, G., B. U. O. Sonnerup, I. Papamastorakis, N. Sckopke, G. Haerendel, S. J. Bame, J. R. Asbridge, J. T. Gosling, C. T. Russell, and R. C. Elphic, Plasma acceleration at the Earth's magnetopause, *Nature*, *282*, 243-246, 1979.

Russell, C. T., and C.J. Alexander, Multiple spacecraft observations of interplanetary shocks: Shock normal oscillations and their effects, *Adv. Space Res.*, *4*, vol. 2-3, 277-282, 1984.

Russell, C. T., and R. C. Elphic, Initial ISEE magnetometic results: Magnetopause observations, *Space Sci. Rev.*, *22*, 681-715, 1978.

Russell, C. T., and M. M. Hoppe, Upstream waves and particles, *Space Sci. Rev.*, *34*, 155-172, 1983.

Russell, C. T., M. M. Mellott, E. J. Smith, and J. H. King, Multiple spacecraft observations of interplanetary shocks: Four spacecraft determinations of shock normals, *J. Geophys. Res.*, *88*, 4739-4748, 1983a.

Russell, C. T., E. J. Smith, B. T. Tsurutani, J. T. Gosling, and S. J. Bame, Multiple spacecraft observations of interplanetary shocks: Characteristics of the upstream ULF turbulence, in *Solar Wind Five*, (edited by M. Neugebauer), pp. 385-400, NASA Conference Publication 2280, Washington, D.C., 1983b.

Schwartz, S. J., D. Burgess, W. P. Wilkinson, R.L. Kessell, M. Dunlop, and H. Luhr, Observations of short large-amplitude magnetic structures at a quasi-parallel shock, *J. Geophys. Res.*, *97*, 4209, 1992.

Song, P., R. C. Elphic, and C. T. Russell, ISEE-1 and ISEE-2 observations of the oscillating magnetopause, *Geophys. Res. Lett.*, *15*, 744-747, 1988.

Song, P., R. C. Elphic, C. T. Russell, J. T. Gosling, and C. A. Cattell, Structure and properties of the subsolar magnetopause for northward IMF: ISEE observations, *J. Geophys. Res.*, *95*, 6375-6387, 1990.

Song, P., C. T. Russell, and M. F. Thomsen, Waves in the inner magnetosheath: A case study, *Geophys. Res. Lett.*, *19*, 2191-2194, 1992.

Stringer, T. E., Low frequency waves in an unbounded plasma, *Plasma Phys. (J. Nucl.Energy)*, *5*, 89-117, 1963.

Thomsen, M. F., J. T. Gosling, S. J. Bame, and C. T. Russell, Magnetic pulsations at the quasi-parallel shock, *J. Geophys. Res.*, *95*, 957-966, 1990.

Thomsen, M. F., J. T. Gosling, and C. T. Russell, ISEE studies of the quasi-parallel bow shock, *Adv. Space Res.*, *8*, (9)175-(9)178, 1988.

Tsurutani, B. T., E. J. Smith, R. R. Anderson, K. W. Ogilvie, J. D. Scudder, D. N. Baker, and S. J. Bame, Lion roars and non-oscillatory drift mirror waves in the magnetosheath, *J. Geophys. Res.*, *87*, 6060, 1982.

Zhu, X. M., M. G. Kivelson, R. J. Walker, C. T. Russell, M. F. Thomsen, and D. J. McComas, An ISEE 1/2 spacecraft study of an unusual flux transfer event, *Adv. Space Res.*, *8*, (9)259-(9)262, 1988.

C. T. Russell, Institute of Geophysics and Planetary Physics, University of California, Los Angeles, CA 90024-1567, USA.

Multi-Spacecraft Study of a Substorm Growth and Expansion Phase Features Using a Time-Evolving Field Model

D. N. Baker,[1] T. I. Pulkkinen,[2] R. L. McPherron,[3] and C. R. Clauer[4]

A large substorm commenced at 0111 UT on 3 May 1986 and had a peak AE index value of ~1500 nT. The event was observed by a variety of satellites including ISEE-1, -2, and IMP-8 in the magnetotail plus SCATHA, GOES, GMS, and LANL spacecraft at or near geostationary orbit. An important feature of the 0111 UT substorm was the simultaneous imaging of the southern auroral oval by DE-1 and of the northern auroral oval by Viking. The constellation of spacecraft near local midnight in the radial range 5-9 R_E made it possible to study the strong cross-tail current development during the substorm growth phase and the current disruption and current wedge development during the expansion phase. A time-evolving magnetic field model was used to map observed auroral features out into the magnetospheric equatorial plane. There was both a dominant eastward and a weaker westward progression of activity following the expansion phase. A substantial latitudinal separation ($\geq 10°$) of the initial region of auroral brightening and the region of intense westward electrojet current was identified. The combined ground, near-tail, and imaging data for this event provided a good opportunity to separate spatial and temporal effects during tail current development and substorm onset. Evidence is found for current redistribution within the equatorial plasma sheet during the late growth phase and rapid current disruption and field-aligned current formation from deeper in the tail at substorm onset. It is concluded that these results are consistent with a model of magnetic neutral line formation in the late growth phase which causes plasma sheet current redistribution before the substorm onset. The expansion phase onset occurs considerably later due to reconnection of lobe-like magnetic field lines and roughly concurrent cross-tail disruption in the inner plasma sheet region.

INTRODUCTION

Key outstanding problems in the study of magnetospheric substorms include the determination of where major instabilities are initiated, what physical processes are involved, and how these instabilities evolve to eventually constitute a global magnetospheric disturbance. New tools to resolve these issues include multispacecraft observations, global auroral images, and realistic magnetic field mapping capabilities. Coordinated Data Analysis Workshops (CDAWs) have made readily available a wide array of ground-based measurements, high-altitude particle and field data from numerous spacecraft, and auroral imaging sequences from such polar-orbiting satellites as Dynamics Explorer (DE-1) and Viking.

A period of particularly extensive data collection was the PROMIS (Polar Region Outer Magnetosphere International Study) interval of March-June 1986. The PROMIS period designated as the CDAW-9 Event C (0000 to 1200 UT on 3 May 1986] had several discrete substorm onsets contained within it. Here we will focus on the first substorm of 3 May 1986 which had an identified expansion phase onset time of 0111 UT. Because of the very disturbed nature of the magnetosphere at this time, the substorm effects in the magnetotail were exceptionally close to the Earth. This allowed numerous geostationary spacecraft to observe directly the substorm initiation processes. In this paper we describe the localized nature in space and time of the substorm onset and we map this initiation from ionospheric to magnetospheric altitudes using realistic, temporally-evolved magnetic field models.

An intriguing aspect of this event is the very taillike stretching of the magnetic field very close to the Earth late in the substorm growth phase. This feature of strong substorms has received recent attention in the literature [Kaufmann, 1987; Baker and McPherron, 1990; Baker and Pulkkinen, 1991]. It is clear that

[1]NASA/Goddard Space Flight Center, Greenbelt, MD 20771.
[2]Finnish Meteorological Institute, Box 503, SF-00101 Helsinki, Finland.
[3]University of California, Los Angeles, CA 90024
[4]University of Michigan, Ann Arbor, MI 49109

Solar System Plasmas in Space and Time
Geophysical Monograph 84
Copyright 1994 by the American Geophysical Union.

late growth phase field distortions require a quite thin, radially localized cross-tail current sheet extending inward at least to the vicinity of geosynchronous orbit. As shown by the modeling presented in recent work [Sergeev et al., 1990; Pulkkinen et al., 1991, 1992], the inclusion of such thin current sheets can replicate the observations very well. The question left unanswered by such empirical modeling, of course, is what physical processes lead to such an extreme configurational situation.

Baker and McPherron [1990] suggested that a modest departure from the "standard" substorm model could possibly help address late growth phase observations. They proposed that rather than having a magnetic neutral line form suddenly right at substorm expansion phase onset, instead the neutral line could form at a geocentric distance of ~ 20 R_E in the nightside magnetotail (as in the standard model), but they speculated that in this early stage the neutral line would form on the closed field lines of the central plasma sheet. It has been suggested that reconnection within such a relatively thick plasma sheet configuration would be slow and plasma flow would be weak because the $J \times B$ stresses would be quite small [Coroniti, 1985].

An important effect of the near-Earth neutral line in the standard substorm model is to "disrupt" and divert cross-tail current from the plasma sheet into the ionosphere to form the current wedge [Atkinson, 1967; McPherron et al., 1973]. Baker and McPherron [1990] suggested that the neutral line would initially divert cross-tail current earthward within the plasma sheet because of the relatively low ionospheric conductivities at that stage of the growth phase and because of the large inductances required to set up field aligned currents (FACs) to the ionosphere. Thus, as implied in Fig. 1, this suggests that for some extended period, the initial neutral line would reconnect closed plasma sheet field lines rather slowly (hence not changing the basic field configuration drastically) and would divert cross-tail current toward the inner edge of the plasma sheet (without yet forming a substorm current wedge or strong FACs). A principal effect in this scenario would be for the weak, growth-phase neutral line to drive the formation of the intense cross-tail current sheet that is observed so prominently near the Earth toward the end of the growth phase. We explore the observational support for this scenario in this paper.

OBSERVATIONS

Definition of the substorm timing is possible using individual ground magnetograms as shown in Figure 2. Panel (a) shows the H-component records for the period 0000-0400 UT on 3 May 1986 from the Greenland array [E. Friis-Christensen, private communication]. Data from the western coast of Greenland ranging from Thule (THL at 85.8° invariant latitude) down to Narssarsuaq (NAQ at 67.1° INVL) all support the identification of an isolated substorm with maximum development shortly before 0200 UT. In the data records from the auroral latitude stations of Godthab (GHB), Frederikshab (FHB), and Narssarsuaq, the substorm appears as a single episode of activity with the negative bay commencing at ~ 0125 UT. The higher latitude stations such as Sondre Stromfjord (STF, 73.7° INVL)

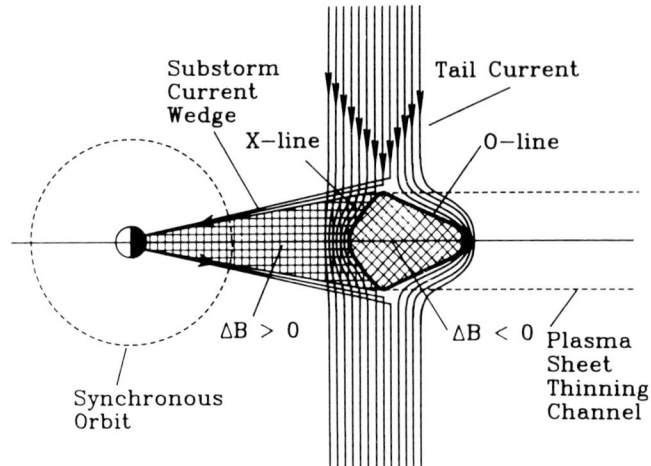

Fig. 1. A schematic illustration of the formation of a connected pair of X- and O-type neutral lines in the central plasma sheet through current diversion. The cross-hatched region is the channel within which the plasma sheet collapses during loss of the plasmoid.

show a first onset at ~ 0140 UT and a second intensification near ~ 0200 UT. We would, however, interpret this feature as being related to the substorm recovery phase and its associated high-latitude current systems [Pellinen et al., 1992].

Magnetometer data from the Canadian sector are presented in panels (b) and (c) of Fig. 2. Panel (b) shows the X-component records for the selected stations for the period 0000-0400 UT. Poste de la Baleine (66.2° CML [corrected geomagnetic latitude]) and Ottawa (56° CML) are in the same magnetic meridian and were at ~ 2100 MLT at 0100 UT on 3 May. As seen in panel (c) the much larger X and Z components of the perturbations seen at PDB compared to Ottawa demonstrate that electrojet currents were flowing much closer to magnetic latitude of ~ 66° than 56°[see, also, Baker et al., 1993].

The DE-1 spacecraft was positioned over the south polar region during the period of the 3 May 1986 event. Complete views of the southern auroral oval were obtained by the imaging experiment [Frank et al., 1981] on DE-1 throughout the period 2352 UT on 2 May to 0347 UT on 3 May. Images were acquired with 8-min resolution at 117-170 nm. A broad, disturbed auroral oval was seen in the image obtained at 0000 UT [see Baker et al., 1993]. Bright auroral regions were seen extending from ~1900 MLT to local midnight. In the postmidnight section, the auroral oval extended from ~60° latitude to ~ 78° latitude. Similar features were seen in later images, but the identifiable auroral oval tended to reduce in latitudinal width compared to the earlier times.

Quieting of the auroral oval occurred during the period 0032 to 0105 UT as seen in DE-1 images. The bright auroras in the premidnight sector lessened in latitudinal and longitudinal width, and the entire oval became substantially narrower in latitudinal extent with time. By 0105 UT, the detectable oval extended from ~ 58° latitude to ~ 66° latitude. Thus, during this general period

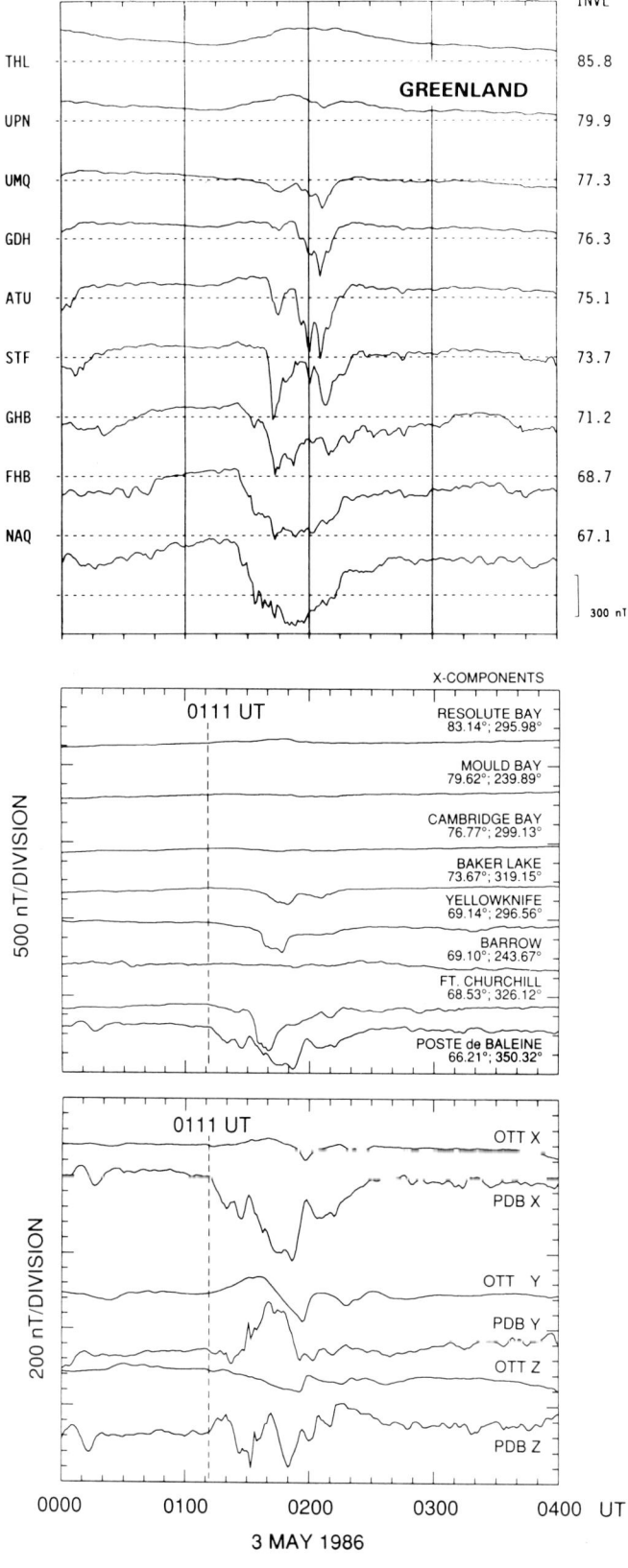

(0000-0105 UT) there was some equatorward motion of the auroral oval and the polar cap size increased substantially. In the image that was assembled starting at 0113 UT, the auroras brightened markedly in the premidnight sector demonstrating an obvious substorm onset at about that time [Baker et al., 1993]. The auroras remained bright and there were expansions of activity both eastward and westward until at least 0129 UT.

Recently, tools have been developed to allow computer analysis of auroral images [Samadani et al., 1990]. These methods have been applied to the DE-1 auroral images and the computer algorithm locates rapidly the auroral luminosity belt at all local times. By determining the region poleward of the auroral intensity belt, it is possible to specify the approximate boundary of the polar cap and thereby numerically determine the polar cap area for each DE-1 image. In Figure 3 we show the southern polar cap area determined by this method for the period 0000 - 0250 UT on 3 May. The variation of the area is quite interesting in that the polar cap grows monotonically and systematically in size during the period 0000 ~ 0115 UT and then begins abruptly to decrease in area after ~ 0120 UT. This sequence would seem to be strongly supportive of the tail lobe "loading-unloading" model of substorms.

The Viking spacecraft was near apogee over the northern polar region early on 3 May 1986. The imaging experiment [Anger et al., 1987] onboard the spacecraft returned 1-sec snapshot images of the auroral oval at 40-second resolution from 0040:58 UT until 0126:35 UT. At 0048:02 UT, the northern auroral oval was relatively broad, extending from ~ 56° to ~ 68° magnetic latitude in the premidnight sector [Baker et al., 1993]. There was a distinct equatorward arc system at ~ 60° latitude, but clear auroral luminosity was evident far poleward of this arc. This is a typical configuration observed during active intervals and the recovery phase of a previous substorm [Elphinstone and Hearn, 1992]. This "double oval" configuration allows important inferences to be made about the location of the optical substorm onset.

Figure 4 is a sequence of detailed images from Viking. It shows the 0112:10 UT, 0112:50 UT, and 0118:32 UT images of the brightening and expanding auroras. The ground data indicated a substorm onset at ~ 0111 UT, i.e., just before the first image in Figure 4. These images make clear that the initial auroral brightening occurred at quite low latitude, while there were bright auroral features more poleward that were eventually enveloped by the spreading auroral expansion. There also were diffuse auroral luminosity signatures in the premidnight sector extending even further poleward. The discrete structure at the poleward edge of the auroral feature at ~ 0113 UT is close to the latitude of PDB and, hence, to the region where the strongest electrojet currents were inferred to flow. We conclude that the onset of this substorm occurred quite close to the Earth since the aurora brightened so far equatorward [see also Murphree et al., 1991; Murphree et al., 1992].

Fig. 2. (a) Greenland magnetometer chain data (H-component) for 0000-0400 UT on 3 May 1986. (b) Selected Canadian-sector magnetometer data (X-component). (c) Detailed magnetometer data from Ottawa (OTT) and Poste de la Baleine (PDB) [from Baker et al., 1993].

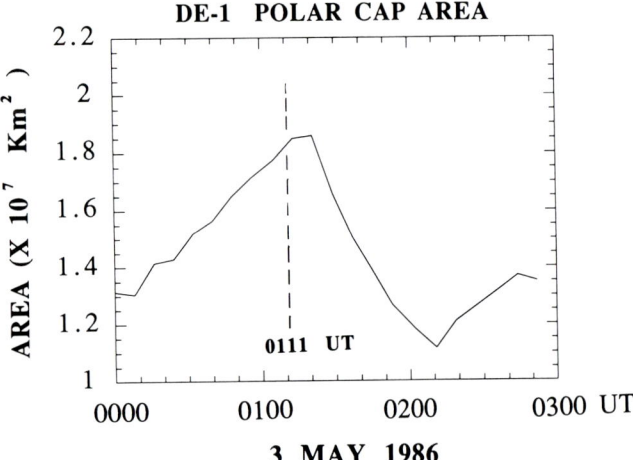

Fig. 3. The time development of the polar cap area inferred from the sequence of DE-1 auroral images between 0000 and 0300 UT on 3 May 1986.

There was a progressive taillike stretching of the magnetic field (0000 ~ 0110 UT) and also a progressive decrease in the energetic particle flux [see Baker et al., 1993]. Thus, the 70-min period prior to ~ 0110 UT would be taken as the energy storage

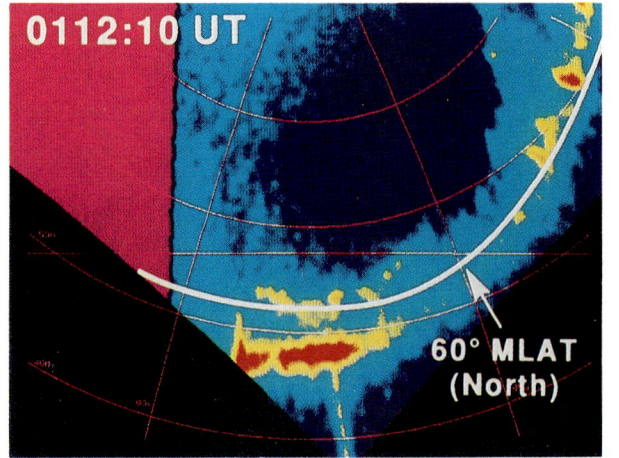

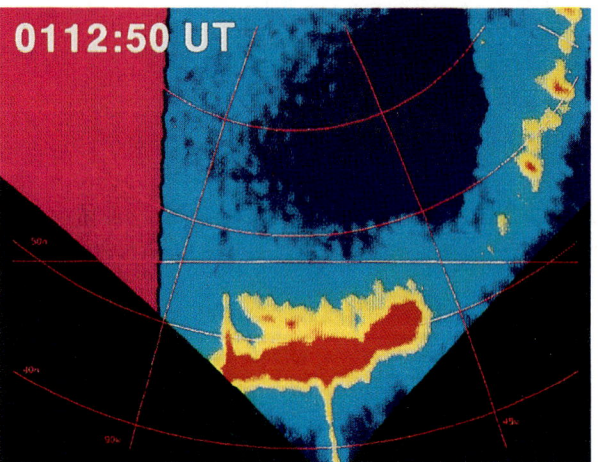

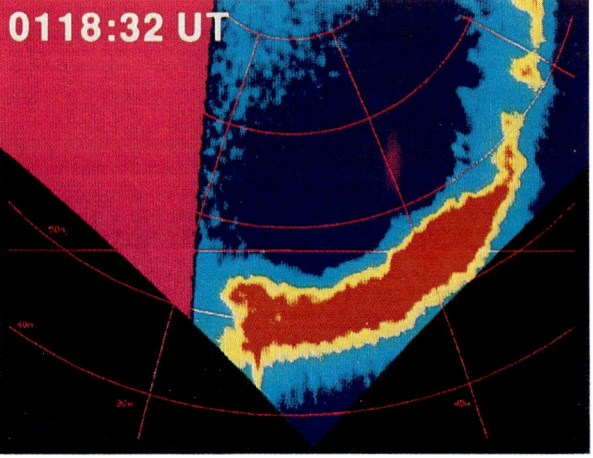

In addition to the imaging data described above, there were a number of other satellite data sets on 3 May 1986 available for analysis. The inset in Figure 5 provides the locations of several spacecraft that have been examined. The constellation of spacecraft at, or near, geostationary orbit are particularly important in the midnight and premidnight sector: SCATHA, S/C 1982-019 (S19, hereafter), ISEE-1 (and -2) and GOES-5. These spacecraft were in the vicinity of the substorm initiation region and they provided key information concerning the growth phase of the substorm as well. Evidence of the growth phase development is provided by geostationary orbit spacecraft such as S19 near ~ 2300 LT. As shown in Figure 5, the 92 keV to ~ 150 keV protons measured by S19 showed a systematic decrease in flux from 0000 UT until at least 0045 UT. The higher energy particle fluxes showed a complete "dropout" and were at background until the substorm onset time. The S19 data also revealed a progressive development of a "cigarlike" anisotropy of the energetic electrons (data not shown here) during this same interval. These flux and anisotropy features at geosynchronous orbit are traditional, highly repetitive features of the substorm growth phase [Baker et al., 1978].

The SCATHA spacecraft [Fennell, 1982] also made measurements slightly inside geostationary orbit (r ~ 5 R_E) during this interval. The SCATHA magnetic field data for the period 0000 to 0300 UT are shown in the lower part of Figure 5.

Fig. 4. A detail of the Viking auroral images acquired at 0112:10 UT, 0112:50 UT, and 0118;32 UT on 3 May 1986. The isointensity contours delineating the auroral arc regions associated with the initial substorm onset were used to map features out to the magnetic equator. In Fig. 4a, a trace at MLAT = 60° is included for reference and for comparison with the geographic latitudes used in these projections [from Baker et al., 1993].

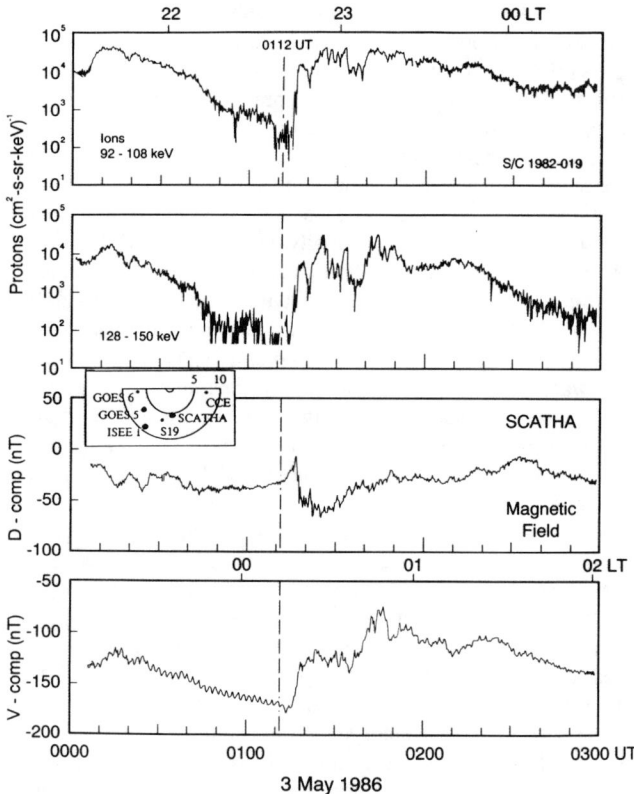

Fig. 5. Energetic ion (top panels) from spacecraft 1981-019 and magnetic field (lower panels) data from the SCATHA spacecraft. The curves show the growth and expansion phase features observed near geostationary orbit early on 3 May 1986. The small inset shows the various spacecraft locations.

interval [e.g., Baker et al., 1984] in which tail energy was loaded prior to the unloading phase commencing at ~ 0111 UT. Such results are very consistent with the polar cap changes shown in Fig. 3.

From the imaging and ground-based measurements presented above, we estimate that at ~ 0100 UT ISEE-1 and GOES-5 were in the general longitude sector of the substorm initiation (i.e., ~ 21 LT). Figure 6 shows in the top panel magnetic field data [Russell et al., 1978] from the ISEE-1 spacecraft. During this time, ISEE-1 was outbound at ~ 2200 LT and ~ 9 R_E geocentric distance. A very evident feature of the magnetic measurements is the systematic and progressive "taillike" stretching of the field [see Pulkkinen et al., 1991]. From essentially the beginning of the day until 0111 UT, the field inclination substantially increased and B generally diminished. This is a classic signature of the substorm growth phase [McPherron, 1970]. The middle panel of Figure 6 shows GOES-5 magnetic field data for the period 0000 - 0300 UT on 3 May. The spacecraft was near 2000 LT early in this interval and it observed a progressive and substantial stretching of the field (inclination and B both increased) on a time scale very similar to that seen by ISEE-1.

The lower panel shows similar information for SCATHA (in a different format than Fig. 5) for the period 0000 to 0300 UT.

The growth phase effects for the period 0000-0110 UT on 3 May have been included in the basic Tsyganenko [1989] model by varying the model parameters and by modification of the functional form of the field components [Pulkkinen, 1991; Pulkkinen et al., 1991]. The current sheet was locally thinned by modification of the function representing the current sheet thickness for X and Y positions in the model. Pulkkinen et al. [1991] included the minimum thickness, location of the minimum thickness, and size of the thinned region as free parameters whose values were set to give an optimal agreement with actual field observations. The enhanced tail magnetic flux was represented by increasing the model tail current intensity, which produced appropriate enhancement of the field terms. Pulkkinen et al. noted that because the usual [Tsyganenko, 1989] model tail current peaks beyond 10 R_E, its enhancement does not account adequately for stretching of the field in the region between 5 and 10 R_E. In order to further increase the taillike stretching close to the Earth, a new current sheet was added to the model. The functional form of the new current sheet was chosen to be similar

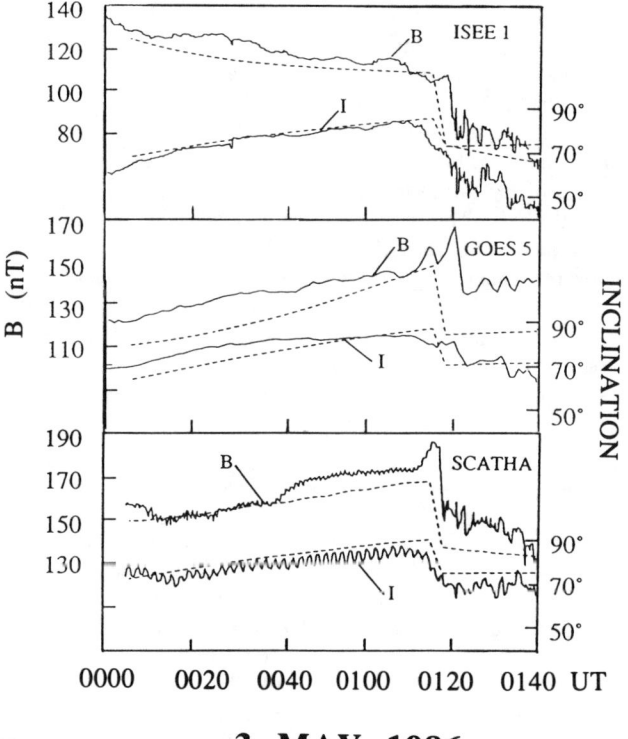

Fig. 6. Magnetic field magnitude and inclination variations observed by ISEE-1, GOES-5, and SCATHA spacecraft on May 3, 1986. Dashed lines show the field computed using the modified field model during the growth phase. After substorm onset the model field is computed using the T89 model with Kp ≥ 5-.

to the ring current. The peak intensity, location of the current maximum, and the thickness of the added current sheet was set according to the observed degree of field stretching. Thus, the modifications of the Tsyganenko [1989] model were represented by a group of parameters whose values were set by a least-squares type of comparison with the field observations [see also Pulkkinen et al., 1992; Baker et al., 1993]. The parameters describing the current intensity and current sheet thickness were increased linearly with time from zero at the beginning of the growth phase (field represented by the Tsyganenko model) to maximal values at substorm onset.

To obtain the optimal correspondence with the observations from all spacecraft in this case, the current sheet was thinned until it reached a minimum thickness which was only 10% of the original value (which, at X ~ -10 R_E, was about 4 R_E). The tail flux was enhanced by a factor of 1.2. The additional thin current sheet added to the model had a peak intensity which was 55% of the ring current intensity and a current maximum located at 4.7 R_E. The thickness of this current sheet was 10% of the thickness of the ring current sheet at that location.

The dashed lines in Fig. 6 show the model fits to the measured near-tail magnetic field and inclination between 0000 UT and 0112 UT. At 0112 UT the model was allowed to transition back (over a 2-min interval) to the Tsyganenko [1989] representation. The dashed line in each case between ~ 0115 and 0140 UT simply shows the unmodified Tsyganenko model representation for the location of the respective spacecraft. These comparisons show that, using the methods of Pulkkinen [1991] and Pulkkinen et at. [1991], it is possible to use multi-spacecraft observations throughout the (pre)midnight sector to arrive at a reasonable model of the time-varying magnetic field during the substorm growth phase. This model provides a global representation of the nightside region at any given time. As would be suggested by the large temporal changes in the field components between 0000 UT and 0110 UT in Fig. 6, the modified model is quite important if one wishes to have a realistic "snapshot" of the magnetic field at some arbitrary point during the growth phase sequence.

EXPANSION PHASE ONSET

Ground magnetic and auroral imaging data discussed above allow us to specify with considerable confidence that the substorm expansion phase onset occurred at 0111 UT. As shown in Fig. 6, the ISEE-1 and GOES-5 data tended to support this as the earliest probably onset time. Careful examination of the ISEE and GOES magnetic data show that, in reality, it was the B_Z component at each spacecraft that showed the most prompt signature, with the B_X, B_y, and plasma signatures being somewhat more delayed and variable in their response [Baker et al., 1993]. The B_X component at ISEE-1 did not diminish in magnitude until ~ 0119 UT. At GOES-5, the field inclination increased slightly right at 0111 UT. However, Bx actually became larger until nearly 0115 UT. A more complete "dipolarization" of the field at GOES did not occur until nearly 0120 UT. In Fig. 5, it is seen that the recovery of energetic proton fluxes at S19 above background levels was delayed by several minutes relative to the substorm onset at 0111 UT.

SCATHA is interesting also in its substorm timing sequence properties. As noted for ISEE and GOES, the field inclination at SCATHA showed the most prompt response to the substorm onset. Because of the spacecraft spin-modulation signal in the data it is hard to be precise, but it appears that SCATHA measured an increase in B_Z at ~ 0112 UT. As noted for the other spacecraft, however, B_X continued to increase at SCATHA (in fact, the rate of increase was enhanced for a time) after 0111 UT and B did not decrease substantially until at least 0117 UT. Thus, the indications are that the north-south component strengthened promptly, but weakly, throughout most of the (pre)midnight sector shortly after the 0111 UT auroral brightening observed by Viking and DE-1. There were also prompt signatures seen in the eastward components measured at various near-tail spacecraft, but B_X components were large and persisted for several minutes after the onset time at the various spacecraft locations. This suggests that there was not an immediate, fully-developed current wedge formation throughout the midnight sector, but rather, there was a more gradual and progressive development of the current wedge system. Observation of the current wedge at a particular time depended sensitively on the specific spacecraft position.

Using the modified Tsyganenko model described above, we have demonstrated that a realistic global representation of the near-tail field is possible for the entire growth phase of the 3 May substorm event. In particular, the model provides a good description of the magnetospheric field configuration right at the time of the expansion phase onset (~ 0111 UT). The top panel of Figure 7 shows the projections of selected model field lines onto the equatorial (X-Y) plane at the end of the growth phase. The field lines in Fig. 7 were chosen to be at constant magnetic latitudes and traces were done at one-hour longitude intervals in the night sector. Two sets of field lines were mapped, one set originating from 59° latitude and the second set from 61°latitude. As is evident, the 59° field lines cross the equator near synchronous orbit at ~ 21 LT and ~ 03 LT, but extend to ≥ 15 R_E nearer local midnight. The 61° field lines extend to ~ 15 R_E away from midnight, but extend to ~ 30 R_E (i.e., well outside the figure) near the midnight meridian.

The auroral imaging data showed that the onset took place between 0110:29 and 0112:10 UT [Baker et al., 1993] and ground-based magnetic records show that the substorm onset occurred in a narrow region near 21 MLT. The positions of several relevant near-tail spacecraft at this time are shown projected onto the X-Y plane in Fig. 7 as is the meridian of substorm onset at 0111 UT: This line was seen to be between ISEE-1 and GOES-5, but was closest to the ISEE meridian. The analysis of substorm expansion features seen at the various near-tail spacecraft shows that the substorm disturbances expanded eastward and westward in a progressive way. By 0112 UT ISEE-1 and GOES-5 had definitely begun to observe evidence of the substorm current wedge (SCW). This is illustrated in Fig. 7b. (The dashed lines in each panel of Fig. 7 are the field line projections calculated at the end of the growth phase.) By 0115 UT, S19 had seen a dipolarization of the magnetic field and a recovery of energetic electron fluxes. This suggests that the current wedge had, by ~ 0116 UT, spread beyond the longitudes of both S19 and GOES-5; this is shown in Fig. 7c. By 0120 UT

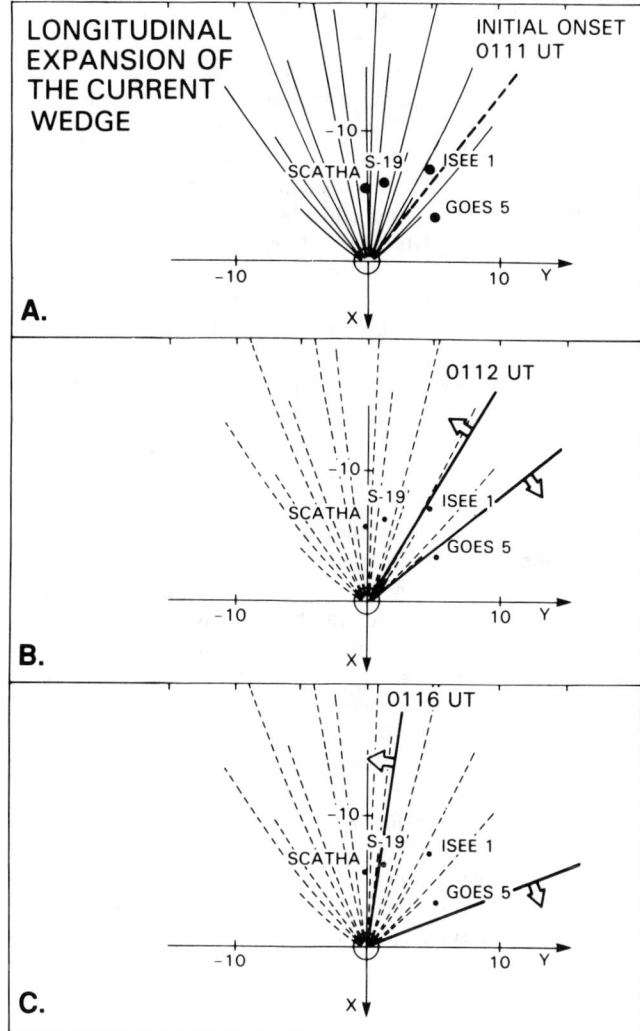

hemisphere. These geographic contours were determined for both the poleward and equatorward edges of the brightened auroral features at ~ 1° intervals of longitude. (The DE-1 image [acquired over an 8-min interval] represented a much more expanded auroral configuration than did the 1-s Viking snapshot.)

Figure 8a shows the mapping of the three Viking images taken at 0112 UT, 0113 UT, and 0118 UT (as shown in Figure 4). The three different shadings denote the mappings at different times, with the darkest shaded region corresponding to the initial auroral brightening seen by Viking at 0112 UT. Note that it maps to well before midnight and quite close to the Earth. The larger and lighter shaded regions in Figure 8a show the subsequent spreading of the aurora seen by Viking. The lightest shading covers the largest area corresponding to the mapping of bright aurora for the latest time at 0118 UT. The spreading is largely eastward (toward midnight) which is quite consistent with the results shown in Figure 7. There is also an evident spreading outward in radial distance [see Lopez and Lui, 1990; Ohtani et al., 1992]. Note that the Viking projected auroral region for 0118 UT in Figure 8a is rather comparable in character to the DE-1 projected auroral image in Figure 8b where we explicitly overlay the ~ 0115-0117 UT DE auroral projection with the 0118 UT Viking projection. Except in the region of local midnight for $-5 \lesssim x \lesssim -15\ R_E$, there is rather good correspondence between the two projections.

DISCUSSION AND INTERPRETATION

Careful examination of many auroral images at substorm onset has revealed that auroral brightenings often occur relatively far

all of the inner tail spacecraft, including SCATHA (~ 0117 UT), had observed the current wedge formation. The development of the substorm current wedge described here showing both eastward and westward progressive spreading, is consistent with that determined for many substorms using midlatitude magnetic measurements by Clauer and McPherron [1974].

Using the modified magnetic model described previously, we can utilize the magnetic field configuration inferred right at the end of the growth phase to map the auroral features out to the magnetic equatorial plane. Obviously, there is some limitation to the correctness of this mapping since after substorm onset the global field begins rapidly to reconfigure itself. However, for a brief period right at the expansion phase onset time, this approximation should be quite valid. For both the DE-1 image right at onset and Viking image sequence (Fig. 4) we have determined the geographic contour of the brightened auroral regions. This intensity contouring traces the outer fringes of the respective auroral features at constant intensity as measured by Viking in the northern hemisphere and by DE-1 in the southern

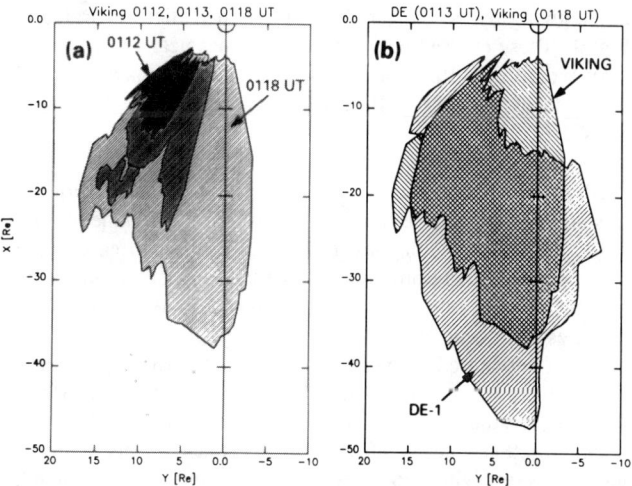

Fig. 8. (a) A mapping of the Viking auroral onset image to the equatorial plane using the model of Pulkkinen et al. [1991] using the Viking image sequence (Fig. 4) corresponding to the three Viking images taken at 0112 UT, 0113 UT, and 0118 UT. The three different shadings denote mappings at different times with the darkest shaded region corresponding to the initial auroral brightening at 0112 UT. The lightest shading covers the largest area corresponding to the mapping of the aurora seen at the latest time of 0118 UT. (b) An overlay of the DE-1 and Viking (0118 UT) mapped images [adapted from Baker et al., 1993]..

equatorward of the putative open-closed field line boundary between the polar cap and the low-altitude extention of the plasma sheet [Murphree et al., 1991; 1992]. The case study here clearly shows such behavior (see Fig. 4) since we see that the auroral intensification at ~ 0111 UT occurred most prominently at MLAT ≲ 60°, while the auroral oval region extended rather evidently to MLAT > 65°. Thus, the brightening aurora clearly could not map directly to a region of the plasma sheet in which open field lines were reconnecting. Our interpretation is that the initially brightening aurora mapped to a region quite near the Earth (~ 8-12 R_E), while the most intense westward electrojet currents mapped to a region deeper (15-20 R_E) in the tail. We argue that the abrupt onset of the substorm (~ 0111 UT) corresponded to explosive reconnection of low -β flux tubes. Evidently, the region of brightening aurora spreads poleward implying that there is a tailward motion of disturbance in the equatorial plane at substorm onset [see Jacquey et al., 1991; Ohtani et al., 1992].

In Figure 9 we illustrate the situation that appears to apply in this case right at the end of the substorm growth phase and just at the beginning of the expansion phase. Following the suggestion of Baker and McPherron [1990], it is proposed that an initial X-line (and accompanying O-line) begins to form (well tailward of the geosynchronous orbit distance) in the hot central plasma sheet. This would begin weakly reconnecting closed flux tubes [see, also, Coroniti, 1985] and would lead to tailward "magnetic islands", but these would be confined and restrained by the overlying (still-closed) plasma sheet flux tubes. Earthward of the X-line, the current that had been flowing within the neutral line region would be diverted into an increasingly confined radial range giving rise to the very high current densities illustrated in Fig. 9. These concentrated current flow lines near geostationary orbit would constitute the thin current sheet we have invoked above to explain the taillike stretching at GOES, ISEE, and SCATHA.

The extreme taillike field stretching that occurs close to the Earth implies that the current sheet (and field reversal region) in this area is very thin in the north-south sense [Pulkkinen et al., 1991, 1992]. Such thin current sheets have strong effects on the particle confinement properties of the tail magnetic fields [e.g., Mitchell et al., 1990] and, in fact, considerable scattering of particles can occur in these thin current sheets as the field line radius of curvature becomes comparable to the particle gyroradii [Buchner and Zelenyi, 1987; 1989]. Pulkkinen et al. [1991; 1992] have studied the ion and electron "chaotization" which occurs in these cases of very thin current sheets using a temporally evolving global magnetic field model. The model results suggest that in most substorm cases studied, including the 0111 UT substorm under investigation here, the current sheet became so thin that the motion of even thermal (≈ 1 keV) electrons became strongly chaotic right at the end of the growth phase. Moreover, Pulkkinen et al. showed that the region of initial and maximum electron chaotic motion mapped closely to the region of first auroral brightening.

We show in Figure 9 the regions where Pulkkinen et al. [1992] found the electron motion to be strongly chaotic by plotting the constant-κ contours. (The κ parameter is basically the ratio of the field line radius of curvature in the current sheet to the particle gyroradius.) The centrally embedded contour would correspond to κ = 1.6 which would be the region of extreme thermal electron scattering. As discussed by Pulkkinen et al. the chaotic electron region occurs remarkably close to the Earth and it maps closely to the region of initial auroral brightening seen by Viking at 56° MLAT and ~ 21 MLT. This relationship is illustrated in Fig. 9. Chaotization of the electron orbits has been suggested to trigger the tearing instability [Buchner and Zelenyi, 1987], which in turn would lead to partial disruption of the cross-tail current. This would feed electrons into the ionospheric loss cone which, together with parallel acceleration at lower altitudes, may account for the auroral features observed.

The 0111· UT substorm on 3 May showed that the intense westward electrojet currents at the expansion phase onset did not flow at the low latitudes where the first auroral brightenings occur (56° MLAT). Rather, as shown by the ground magnetic records, the westward electrojet currents flowed close to PDB at a latitude of ~ 66°. It is evident from the Ottawa data (latitude ~ 56°) that the westward electrojet was far northward of that station until at least 0130 or 0140 UT. Thus, this provides a demonstration that the auroral brightening occurred in the ionosphere above Ottawa but the wedge electrojet currents flowed nearly 10° in latitude poleward of this auroral brightening.

In Figure 9, a potential explanation of this set of observations is offered. By ~ 0111 UT on 3 May the magnetic reconnection at the initial X-line had progressed so as to reconnect most of the closed, central-plasma sheet field lines with no obvious auroral signature. Then, the X-line would suddenly begin reconnecting open lobe flux tubes. At this point, the reconnection rate would increase significantly: the Alfven speed in the inflow region would jump to much higher values (1000-2000 km/s) than for the central plasma sheet and, consequently, the electric fields at the

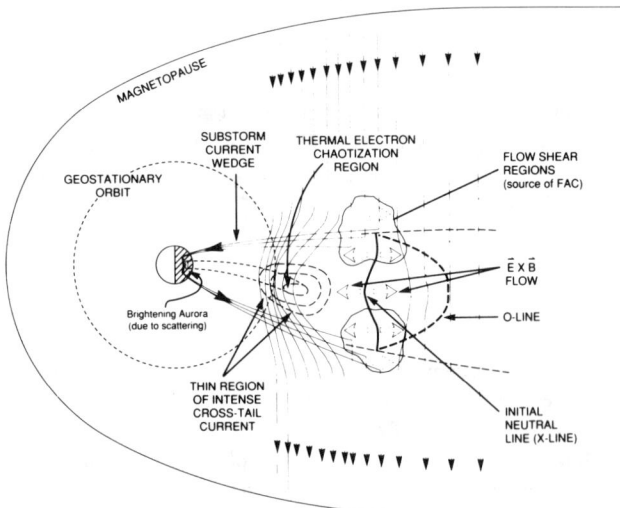

Fig. 9. A schematic diagram of the near-Earth plasma sheet and cross-tail current sheet region during the late growth phase (and very early expansion phase) of the 0111 UT substorm on 3 May 1986 [Baker et al., 1993]..

X-line would be much larger. Such a large change in the reconnection rates could have an almost explosive character [Baker and Belian, 1986] leading to very strong energetic particle production. This would also lead to much stronger jetting (ExB flow) away from the neutral line region [Coroniti, 1985].

As shown in Figure 9, we estimate that the neutral line extends over only a limited width of the magnetotail. At the eastward and westward extremes of the X-line, there would be strong plasma sheet shear flows. These shears would drive strong vorticity development which, in turn, could drive large field aligned currents [Birn and Hesse, 1991; Hesse and Birn, 1991]. Thus, we would suggest that it is the more tailward region, near the original neutral line location, that is the initial and strongest source of the FACs. These would map to much higher magnetic latitudes than would the auroral brightening, as shown in Fig. 9. This could account for the large latitudinal separation of the westward electrojet and the auroral brightening in this very disturbed case.

In summary, we find evidence of a narrowly localized substorm onset region in the near-Earth tail. This region spread rapidly eastward and poleward after the 0111 UT onset. Our approach in this study was novel in that:

- An ideally located constellation of four spacecraft allowed detailed observations of the substorm growth phase in the near-tail region;
- A realistic time-evolving magnetic field model provided a global representation of the field configuration throughout the growth and early expansion phase of the substorm;
- The substorm initiation was well-observed in both the northern and southern hemisphere by global auroral imagers;
- Auroral features and their spreading in longitude and latitude were studied in a conjugate sense using the realistic 3-D magnetic field model;
- Auroral structures were mapped to the equatorial plane and the time-evolution was directly compared with multispacecraft particle and field data;
- An apparent latitudinal separation (≥10°) of the initial region of the auroral brightening and the region of most intense westward electrojet current was identified;
- We developed further the notion [Pulkkinen et al., 1991] that the region of initial auroral brightening mapped into the near-tail equatorial plane in a region of strong thermal electron chaotization;
- A comprehensive interpretation of this event was presented in terms of a modification of the standard near-Earth neutral line model (Fig. 9).

We conclude that these results are consistent with a model of late growth phase formation of a magnetic neutral line. This reconnection region caused plasma sheet current diversion before the substorm onset and eventually led to cross-tail current disruption at the time of the substorm onset. This event and the tools used allowed us to separate remarkably well the temporal and spatial effects in a large substorm. We hope that the analysis methods may be prototypical of the tools that can be regularly employed in the upcoming ISTP program.

Acknowledgments. The authors thank the staff of the National Space Science Data Center for the excellent support provided to the CDAW-9 effort. They also thank R. H. Manka for his organizational work and many useful discussions. Important contributions by J. Craven, R. Elphinstone, J. Fennell, L. Frank, E. Friis-Christensen, R. Lopez, A. Lui, S. Murphree, T. Nagai, and G. Rostoker are gratefully acknowledged. Support for auroral image analysis at Stanford University and the University of Michigan has been provided by the Center for Excellence in Space Data Information Systems.

REFERENCES

Anger, C. D., S. K. Babey, A. L. Broadfoot, R. G. Brown, L. L. Cogger, R. Gattinger, J. W. Haslett, R. A. King, D. J. McEwen, J. S. Murphree, E. H. Richardson, B. R. Sandel, K. Smith, and A. Vallance Jones, An ultraviolet auroral imager for the Viking spacecraft, *Geophys. Res. Lett.*, *14*, 387-390, 1987.

Atkinson, G., An approximate flow equation for geomagnetic flux tubes and its application to polar substorms, *J. Geophys. Res.*, *72*, 5373, 1967.

Baker, D. N., and R. D. Belian, Impulsive ion acceleration Earth's outer magnetosphere, in *Ion Acceleration in the Magnetosphere and Ionosphere, Geophys. Monogr. Ser.*, vol. 38, edited by T. Chang, p. 375, AGU, Washington, DC, 1986.

Baker, D. N., and R. L. McPherron, Extreme energetic particle decreases near geostationary orbit: A manifestation of current diversion within the inner plasma sheet, *J. Geophys. Res.*, *95*, 6591, 1990.

Baker, D. N., and T. I. Pulkkinen, The Earthward edge of the plasma sheet in magnetospheric substorms, in *Magnetospheric Substorms, Geophys. Monogr.*, vol. 64, p. 147, AGU, Washington, DC, 1991.

Baker, D. N., P. R. Higbie, E. W. Hones, Jr., and R. D. Belian, High-resolution energetic particle measurements at 6.6 R_E. Low-energy electron anisotropies and short-term substorm predictions, *J. Geophys. Res.*, *83*, 4863, 1978.

Baker, D. N., E. W. Hones, Jr., P. R. Higbie, R. D. Belian, and P. Stauning, Global properties of the magnetosphere during a substorm growth phase: A case study, *J. Geophys. Res.*, *86*, 8941, 1981.

Baker, D. N., S.-I. Akasofu, W. Baumjohann, J. W. Bieber, D. H. Fairfield, E. W. Hones, Jr., B. H. Mauk, R. L. McPherron, and T. E. Moore, "Substorms in the magnetosphere," Chapter 8 of *Solar Terrestrial Physics - Present and Future*, NASA Pub. 1120, Washington, DC, 1984.

Baker, D. N., T. I. Pulkkinen, R. L. McPherron, J. D. Craven, L. A. Frank, R. D. Elphinstone, J. S. Murphree, J. F. Fennell, R. E. Lopez, and T. Nagai, CDAW-9 analysis of magnetospheric events on 3 May 1986: Event C, *J. Geophys. Res.*, *98*, 3815, 1993.

Birn, J., and M. Hesse, The substorm current wedge and field-aligned currents in MHD simulations of magnetotail reconnection, *J. Geophys. Res.*, *96*, 13456, 1987.

Büchner, J., and L. M. Zelenyi, Chaotization of the electron motion as the cause of an internal magnetotail instability and substorm onset, *J. Geophys. Res.*, *92*, 13456, 1987.

Büchner, J. and L. M. Zelenyi, Regular and chaotic charged particle motion in magnetotaillike field reversals 1. Basic theory of trapped motion, *J. Geophys. Res., 94*, 11821, 1989.

Clauer, C. R., and R. L. McPherron, Mapping the local-time-universal time development of magnetopsheric substorms at midlatitudes, *J. Geophys. Res., 79*, 2811, 1974.

Coroniti, F. V., Explosive tail reconnection: The growth and expansion phases of magnetospheric substorms, *J. Geophys. Res., 90*, 7427, 1985.

Elphinstone, R. D. and D. J. Hearn, Mapping of the auroral distribution during quiet times and substorm recovery, ESA Sp. Publ. SP-335, 13, 1992.

Fennell, J. F., Description of P78-2 (SCATHA) satellite and experiments, in *The IMS Source Book* (eds. C. T. Russell and D. J. Southwood), p. 65, AGU, Washington, DC, 1982.

Frank, L. A., J. D. Craven, K. L. Ackerson, M. R. English, R. H. Eather, and R. L. Carovillano, Global auroral imaging instrumentation for the Dynamics Explorer Mission, *Space Sci. Instrum., 5*, 369, 1981.

Hesse, M. and J. Birn, On dipolarization and its relation to the substorm current wedge, J. Geophys. Res., 96, 19417, 1991.

Jacquey, C., J. A. Sauvaud, and J. Dandouras, Location and propagation of the magnetotail current disruption during substorm expansion: Analysis and simulation of an ISEE multi-onset event, *Geophys. Res. Lett., 18*, 389, 1991.

Kaufmann, R. L. Substorm currents: Growth phase and onset, *J. Geophys. Res., 92*, 7471, 1987.

Lopez, R. E., and A. T. Y. Lui, A multisatellite case study of the expansion of a substorm current wedge in the near-Earth magnetotail, *J. Geophys. Res., 95*, 8009, 1990.

McPherron, R. L., Growth phase of magnetospheric substorms, *J. Geophys. Res., 75*, 5592, 1970.

McPherron, R. L., C. T. Russell, and M. A. Aubry, Satellite studies of magnetospheric substorms on August 15, 1968, A phenomenological model for substorms, *J. Geophys. Res., 78*, 3131, 1973.

Mitchell, D. G., D. J. Williams, C. Y. Huang, L. A. Frank, and C. T. Russell, Current carriers in the near-Earth cross-tail current sheet during substorm growth phase, *Geophys. Res. Lett., 17*, 583, 1990.

Murphree, J. S., R. D. Elphinstone, D. Hearn, and L. L. Cogger, Interpretation of optical substorm onset observations, J. Atmos. Terr. Phys., in press, 1992.

Murphree, J. S., R. D. Elphinstone, L. L. Cogger, and D. Hearn, Viking optical substorm signatures, in Magnetospheric Substorms, AGU Monograph, Vol. 64, edited by J. R. Kan, T. A. Potemra, S. Kokubun, and T. Ijima, pg. 241, 1991.

Nagai, T., D. N. Baker, and P. R. Higbie, Development of substorm activity in multiple-onset substorms at synchronous orbit, *J. Geophys. Res., 88*, 6994, 1983.

Ohtani, S., S. Kokubun, and C. T. Russell, Radial expansion of the tail current disruption during substorms. A new approach to the substorm onset region, *J. Geophys. Res., 97*, 3129, 1992.

Pellinen, R. J., H. J. Opgenoorth, T. I. Pulkkinen, Substorm recovery phase: Relationship to next activation, in Substorms 1, ESA SP-335, p. 469, European Space Agency, Paris, 1992.

Pulkkinen, T. I., A study of magnetic field and current configurations in the magnetotail at time of a substorm onset, *Planet. Space Sci., 39*, 883, 1991.

Pulkkinen, T. I., D. N. Baker, D. H. Fairfield, R. J. Pellinen, J. S. Murphree, R. D. Elphinstone, R. L. McPherron, J. F. Fennell, R. E. Lopez, and T. Nagai, Modeling the growth phase of a substorm using the Tsyganenko model and multi-spacecraft observations: CDAW-9, *Geophys. Res. Lett., 18*, 1963, 1991.

Pulkkinen, T. I., D. N. Baker, R. J. Pellinen, J. Büchner, H. E. J. Koskinen, R. E. Lopez, R. L. Dyson, and L. A. Frank, Particle scattering and current sheet stability in the geomagnetic tail during the substorm growth phase, *J. Geophys. Res., 97*, 19283, 1992.

Russell, C. T., The ISEE-1 and -2 fluxgate magnetometers, *IEEE Trans. Geosc. Electr., GE-16*, 239, 1978.

Samadani, R., D. Mihovilovic, C. R. Clauer, J. D. Craven, and L. A. Frank, Evaluation of an elastic curve technique for automatically finding the auroral oval from satellite images, *IEEE Trans. Geosc. Rem. Sens., 28*, 590, 1990.

Sergeev, V. A., P. Tanskanen, K. Mursala, A. Korth, and R. C. Elphic, Current sheet thickness in the near earth plasma sheet during substorm growth phase, J. Geophys. Res., 95, 3819, 1990.

Tsyganenko, N. A., Magnetospheric magnetic field model with a warped tail current sheet, *Planet. Space Sci., 37*, 5, 1989.

D. N. Baker, NASA/GSFC, Code 690, Greenbelt, MD 20771.

Ion Anisotropy-Driven Waves in the Earth's Magnetosheath and Plasma Depletion Layer

Richard E. Denton,[1] Brian J. Anderson,[2]
Stephen A. Fuselier,[3] S. Peter Gary,[4] and Mary K. Hudson[1]

Abstract.
Recent studies of low-frequency waves ($\omega_r \leq \Omega_p$, where Ω_p is the proton gyrofrequency) observed by AMPTE/CCE in the plasma depletion layer and magnetosheath proper are reviewed. These waves are shown to be well identified with ion cyclotron and mirror mode waves. By statistically analyzing the transitions between the magnetopause and time intervals with ion cyclotron and mirror mode waves, it is established that the regions in which ion cyclotron waves occur are between the magnetopause and the regions where the mirror mode is observed. This result is shown to follow from the fact that the wave spectral properties are ordered with respect to the proton parallel beta, $\beta_{\|p}$. The later result is predicted by linear Vlasov theory using a simple model for the magnetosheath and plasma depletion layer. Thus, the observed spectral type can be associated with relative distance from the magnetopause. The anisotropy–beta relation, $A_p \equiv (T_\perp/T_\|)_p - 1 = 0.85\beta_{\|p}^{-0.48}$, results from the fact that the waves pitch angle scatter the particles so that the plasma is near marginal stability, and is a fundamental constraint on the plasma.

Introduction

The plasma depletion layer [*Crooker et al.*, 1979, *Midgley and Davis*, 1963, *Lees*, 1964, *Zwan and Wolf*, 1976, *Wu*, 1992] is a relatively thin layer ($\approx 0.5\mathcal{R}_E$ thick, where $\mathcal{R}_E$ is the radius of the Earth) which forms on the sunward side of the magnetopause in the subsolar region when the Interplanetary Magnetic Field (IMF) is quasi-perpendicular (at a large angle) to the Earth–Sun line, that is, roughly tangent to the subsolar magnetopause. Magnetic flux from the magnetosheath is compressed as it convects through the plasma depletion layer and presses against the magnetopause; at the same time, the plasma density is depleted. The combined effect of increased magnetic field strength, B_0, and depleted density is that the plasma beta is greatly reduced at the magnetopause edge of the plasma depletion layer compared to that in the magnetosheath proper (that region of the magnetosheath excluding the plasma depletion layer).

In the magnetosheath downstream of the quasi-perpendicular bow shock, the proton and doubly ionized helium species (alpha particles) often may be characterized approximately by bi-Maxwellian distributions with $T_\perp > T_\|$ [*Sckopke et al.*, 1990, *Anderson et al.*, 1991, *Fuselier*, 1992]. Linear Vlasov theory predicts in this case the existence of three instabilities. The first two are the proton cyclotron and alpha cyclotron insta-

---

[1] Physics and Astronomy Department, Dartmouth College, Hanover, New Hampshire.

[2] Applied Physics Laboratory, Johns Hopkins University, Laurel, Maryland.

[3] Lockheed Palo Alto Research Laboratory, Palo Alto, California.

[4] Los Alamos National Laboratory, Los Alamos, New Mexico.

Solar System Plasmas in Space and Time
Geophysical Monograph 84
Copyright 1994 by the American Geophysical Union.

bilities driven by the temperature anisotropies of the protons and alphas respectively and with real frequencies, ω_r, below their respective gyrofrequencies [*Gary et al.*, 1993a, *Denton et al.*, 1993]. The third instability is the mirror mode at much lower ω_r [*Gary et al.*, 1976]. The ion cyclotron waves are transverse, that is, the fluctuating magnetic field, $\tilde{\mathbf{B}}$, is perpendicular to $\mathbf{B_0}$, while the mirror mode is compressional, with $\tilde{\mathbf{B}} \parallel \mathbf{B_0}$.

The AMPTE/CCE spacecraft passes through the magnetosheath under conditions such that the solar wind dynamic pressure is high, so that the magnetosheath beta is high, the magnetosphere is compressed, and the magnetopause lies within the AMPTE/CCE apogee of 8.8 $\mathcal{R}_E$. Using the AMPTE/CCE data set, *Anderson and Fuselier* [1993] and *Anderson et al.* [1994] have shown that the spectral properties of observed low-frequency waves (0.1 - 0.4 Hz, corresponding to ω_r less than the proton gyrofrequency, $\Omega_p \equiv eB_0/m_pc$) are correlated with the variation of plasma parameters across the plasma depletion layer and into the magnetosheath proper. An example illustrating this correlation is shown in Figure 1. On October 18 (day 292), 1984, the AMPTE/CCE spacecraft crossed the subsolar magnetopause at a time during which the companion spacecraft AMPTE/IRM and UKS were upstream of the bow shock in the solar wind. Because of this, it was possible to determine conclusively that changes in density and field strength in the vicinity of the magnetopause resulted from a crossing of the plasma depletion layer, as distinct from temporal changes in solar wind conditions [*Fuselier et al.*, 1991]. Figure 1 (adapted from *Anderson and Fuselier* [1993] and *Fuselier et al.* [1991]) shows dynamic spectra of AMPTE/CCE magnetic field data together with CCE and IRM density and magnetic field data. The top panel (labeled "XY") shows the transverse (to $\mathbf{B_0}$) spectral power from 0 to 2 Hz and the second panel (labeled "Z") shows the parallel power over the same frequency range. The third panel shows the electron density at CCE, calculated as the sum of the proton density and twice the He^{2+} density measured by the Hot Plasma Composition Experiment on CCE. The IRM electron density, measured by the electron instrument was multiplied by 4.92, determined as the ratio of the IRM and CCE densities from 1340 to 1400 UT. The IRM time has been shifted by 3.5 minutes to account for the plasma convection from IRM to CCE. This time was determined as discussed by *Fuselier et al.* [1991]. The bottom panel shows the CCE and IRM magnetic field strength, B_0. The IRM magnetic field magnitude was multiplied by 4.81, the ratio of the average IRM and CCE field magnitudes from 1340–1400 UT. The absolute density and magnetic field values are quite different upstream of the bow shock and close to the magnetopause; the shift in IRM values is made so that a relative comparison can be made. It is clear that the CCE and IRM densities are correlated during the interval 1340–1400, but not beforehand.

From 1300 to 1340 UT the CCE density is low and the CCE field strength is high relative to the values at IRM immediately upstream in the solar wind. As discussed by *Fuselier et al.* [1991] and *Anderson et al.* [1994], the CCE spacecraft was generally located in the magnetosheath after a magnetopause crossing at $\sim$ 1300 UT. Comparison with the upstream solar wind monitor indicates that the interval from 1303 to 1340 UT has characteristics of a plasma depletion layer since the density is depleted and the field strength enhanced at CCE relative to IRM. The correspondence of this interval with the transverse magnetic signals (top panel of Figure 1) indicative of ion cyclotron waves is clear as is the progressive intensification of low-frequency compressional fluctuations (second panel) after 1330 UT. We identify the later as mirror mode fluctuations. We will show below that these mode identifications are well established theoretically. Thus the region in which the ion cyclotron waves are observed is the plasma depletion layer (1303–1340 UT), which is between the magnetopause ($\sim$ 1300 UT) and the magnetosheath proper (1340–1400 UT), where mirror fluctuations are primarily observed.

In order to demonstrate that the spatial ordering magnetopause–ion cyclotron waves–mirror mode holds in general for the high magnetosheath beta conditions during which AMPTE/CCE observes the magnetosheath *Anderson and Fuselier* [1993] made a statistical study of transitions from magnetopause crossings and from time intervals with ion cyclotron or mirror mode fluctuations. The results are shown in Table 1 (adapted from *Anderson and Fuselier* [1993]). Transitions are shown to magnetopause crossings and to time intervals with ion cyclotron, mirror mode, or "other sheath" waves. "Other sheath" indicates intervals in which the waves are broadband (structureless) fluctuations extending above Ω_p, or some combination of ion cyclotron fluctuations and mirror mode fluctuations not considered in *Anderson and Fuselier* [1993] (see below). Most of the magnetopause transitions are to ion cyclotron intervals. About half of the transitions from ion cyclotron waves are to the magnetopause, while the other half are to intervals with mirror mode or "other sheath" waves. None of the mirror mode transitions are to the magnetopause. Therefore, it is clear that there is a spatial ordering with the positions corresponding to ion cyclotron fluctuations close to the magnetopause while the positions corresponding to the mirror mode are farther away.

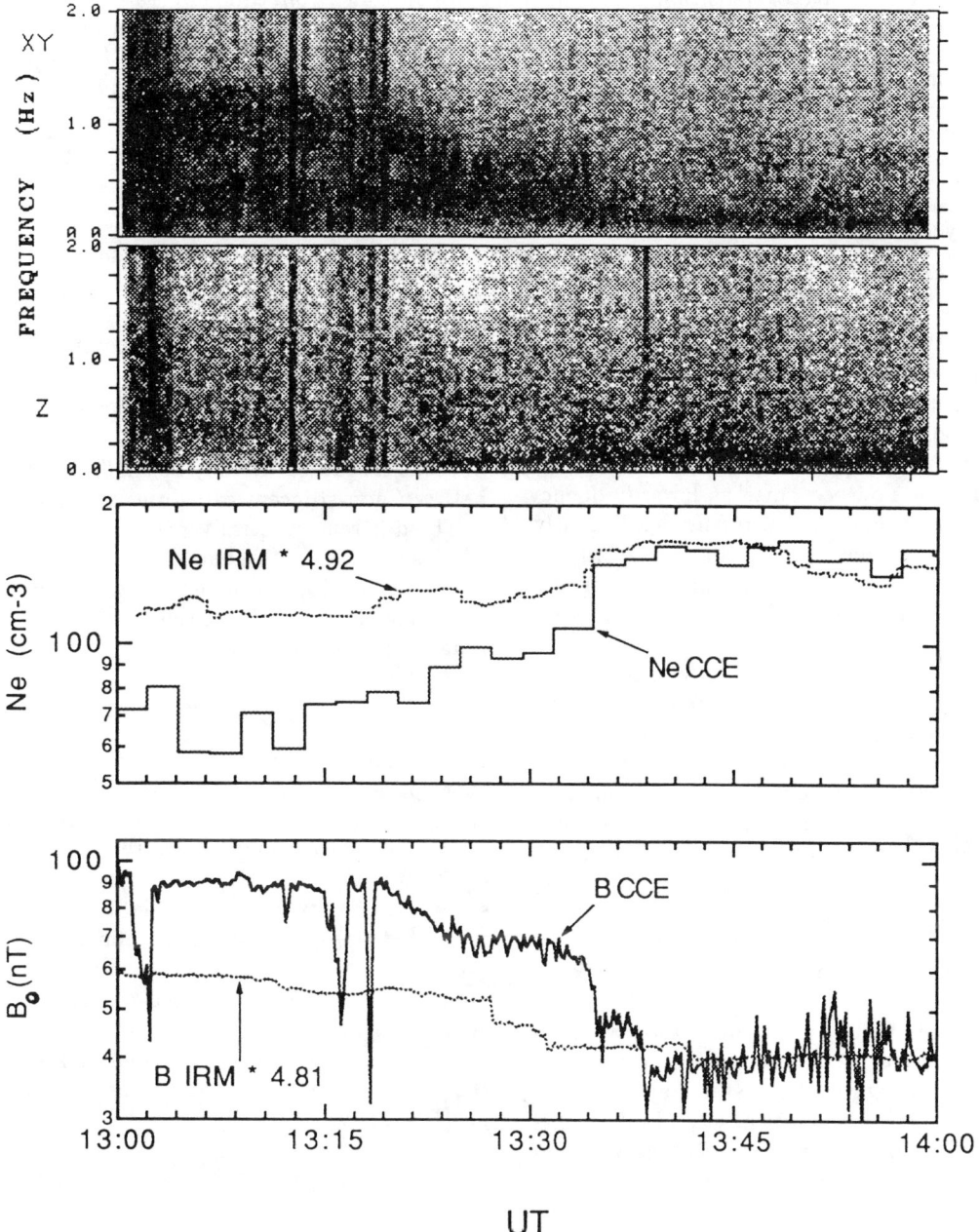

Figure 1. Dynamic spectra of AMPTE/CCE magnetic field data together with CCE and IRM density and magnetic field data. The top panel (labeled "XY") shows the transverse (to $\mathbf{B_0}$) spectral power from 0 to 2 Hz and the second panel (labeled "Z") shows the parallel power over the same frequency range. The third panel shows the electron density at CCE, calculated as the sum of the proton density and twice the He2+ density measured by the HPCE instrument. The IRM electron density, measured by the electron instrument was multiplied by 4.92, determined as the ratio between the densities from 1340 to 1400 UT. The IRM time has been shifted by 3.5 minutes to account for the plasma convection from IRM to CCE. The bottom panel shows the CCE and IRM magnetic field strength, B_0. The IRM field magnitude was multiplied by 4.81, the ratio obtained from the 1340–1400 UT interval.

Table 1. Percentage of transitions from the magnetopause and from time intervals during which ion cyclotron and mirror mode waves were observed, to the magnetopause or to time intervals with ion cyclotron, mirror mode, or "other sheath" (see text) waves. For each of the categories from which transitions occur, the number of total transitions is indicated as "No. Events".

Transitions From:	No. Events	to MgPause	to Ion Cyc.	to Mirror	to Other Sheath
MgPause	30	—	77%	0%	23%
Ion Cyc.	36	64%	—	19%	17%
Mirror	18	0%	50%	—	50%

Anderson and Fuselier [1993] distinguished three types of magnetic fluctuation spectra. These were ion cyclotron waves with transverse fluctuations and with ω_r extending above the alpha particle gyrofrequency $\Omega_\alpha = 0.5\Omega_p$, low-frequency transverse fluctuations with $\omega_r < \Omega_\alpha$ (which we now recognize as lower-frequency ion cyclotron waves), and the mirror mode. Recently, *Anderson et al.* [1994] have provided a more complete categorization by distinguishing five different categories of fluctuation spectra bounded above by the proton cyclotron frequency.

Examples of spectral power density versus normalized wave frequency, ω_r/Ω_p, are shown for each of these spectral categories in Figure 2 (which is taken from *Anderson et al.* [1994]). In the Figure, the left hand polarized transverse (to $\mathbf{B_0}$) power density is indicated by the solid curve, while the right hand transverse and parallel power density are indicated by the dotted and dashed lines respectively. (Note that the peak at very low frequency, $\omega_r \leq 0.05\Omega_p$ which occurs in the BIF, CON, and LOW spectra is a numerical artifact, and should be ignored.) The characteristics of the spectral categories are:

1. Bifurcated (BIF, or B — the three letter abbreviations are used in Figure 2, while the one letter abbreviations are used in Figure 4) transverse spectra with two peaks in wave power, and a clearly resolved gap between the two. The higher-frequency peak may extend above the alpha particle gyrofrequency, Ω_α, while the lower-frequency peak is entirely below Ω_α.

2. Continuous (CON, or C) transverse spectra having a single peak in wave power with frequency width extending above Ω_α.

3. Low-frequency transverse (Low, or L) spectra with a single maximum at a frequency below Ω_α and with no significant energy at frequencies above Ω_α.

4. Mirror plus low-frequency transverse (MRL, or L and M) spectra in which the mirror-like longitudinal fluctuations and low-frequency transverse fluctuations are both present at different frequencies.

5. Mirror (MIR, or M) spectra with predominantly longitudinal magnetic fluctuations, spectral maxima at $\omega_r \ll \Omega_\alpha$, and no transverse components above the noise level.

There is in fact a continuous transition between these types, so the differences between the observed spectra in adjacent categories may be small, but the differences between non-adjacent categories are significant.

The different spectral types occur at different values of plasma beta. We characterize the plasma beta using the parallel proton beta, $\beta_{\|p}$, which we define in this paper as the ratio of the parallel pressure and the magnetic pressure,

$$\beta_{\|p} \equiv 8\pi n_p T_{\|p}/B_0^2, \qquad (1)$$

with n_p and $T_{\|p}$ being the proton density and temperature parallel to $\mathbf{B_0}$ respectively. Figure 3, taken from *Anderson et al.* [1994], shows 102 wave events plotted in A_p–$\beta_{\|p}$ space, where A_p is the proton temperature anisotropy, $A_p \equiv (T_\perp/T_\|)_p - 1$. Each wave event represents from 5 to 30 minutes of data (see *Anderson et al.* [1994] for more details). In the upper left panel, all the events are plotted together, while in the five remaining panels, the events are plotted separately by category (BIF, CON, etc).

Note from the upper left panel of Figure 3 that all the events lie along a straight line, which *Anderson et al.* [1994] have fit to

$$A_p \equiv (T_\perp/T_\|)_p - 1 = 0.85\beta_{\|p}^{-0.48} \qquad (2)$$

(the thin solid line in the Figure). The events lie along this line because it is close to the marginal stability condition for ion cyclotron and mirror mode waves [*Gary et al.*, 1993a, *Denton et al.*, 1993, *Gary and Winske*, 1993, *Gary et al.*, 1993b]. Though the magnetosheath plasma convects through the plasma depletion layer, and the plasma parameters corresponding to a particular moving flux tube are continually changing, *Denton et al.* [1993] have shown that the local instability has sufficient time to saturate nonlinearly and reduce the anisotropy, so that the plasma is near marginal stability. Therefore, the plasma depletion layer is in a driven,

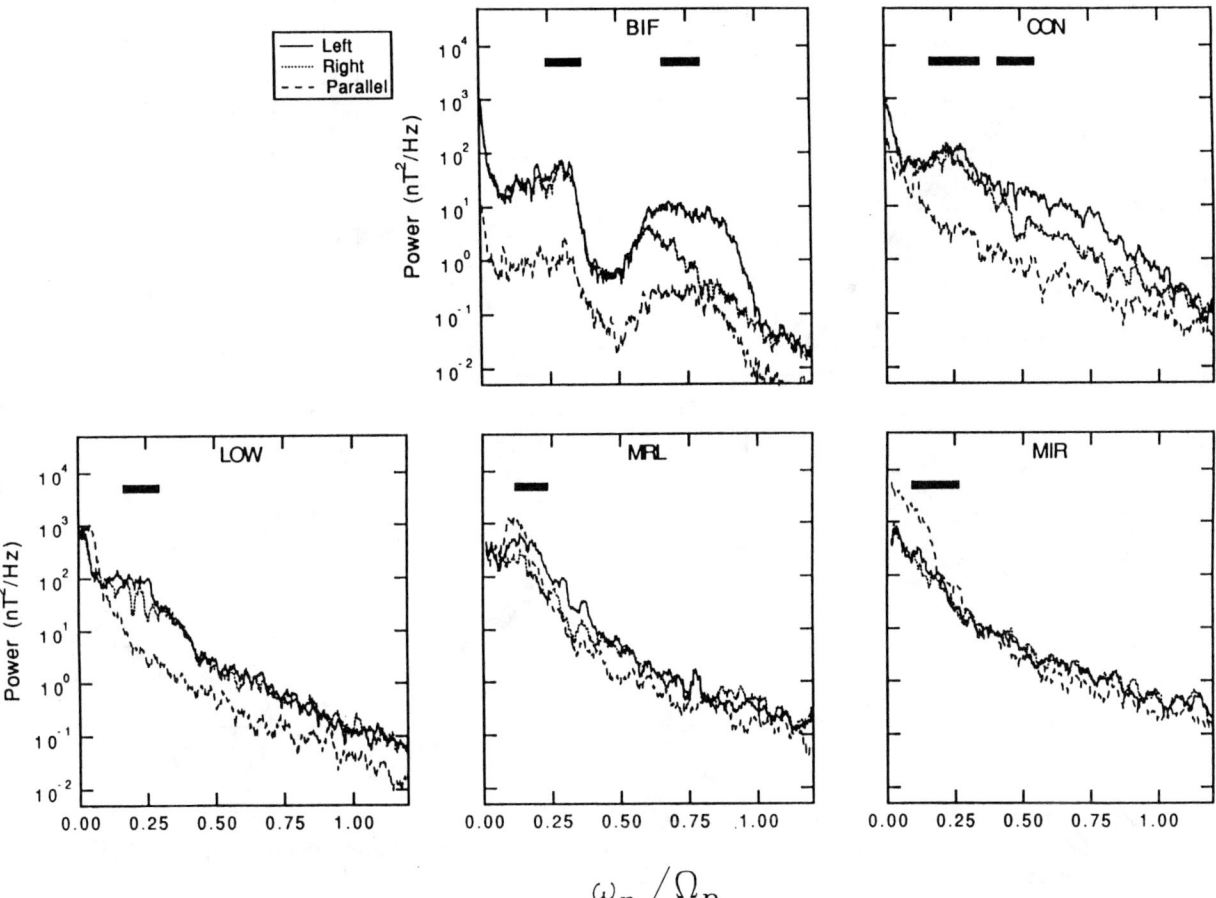

Figure 2. Example power spectra of magnetic field data for the five spectral categories of *Anderson et al.* [1994] defined in the text. Left and right hand polarized power in the plane perpendicular to $\mathbf{B}_0$ are indicated by the solid and dotted curves respectively, while the power of fluctuations parallel to $\mathbf{B}_0$ is indicated by the dashed curve. The horizontal bars above the spectra show the frequency range of linear instability predicted by Vlasov theory.

time independent state. The dotted curve in the upper left panel of Figure 3 shows the contour in A_p–$\beta_{\|p}$ space for which the proton cyclotron (driven by the proton anisotropy) growth rate is equal to $0.01\Omega_p$, while the bold line shows the corresponding curve for the mirror mode. Note how these contours parallel the data values.

The remaining panels showing wave events separated by spectral category demonstrate clearly that the spectral type of the observed waves is ordered by $\beta_{\|p}$, with the lowest values of $\beta_{\|p}$ corresponding to bifurcated (BIF) events, while the largest values of $\beta_{\|p}$ correspond to mirror mode (MIR) events. The events with the mirror mode exclusively (MIR) generally have values of $\beta_{\|p}$ for which the contour of constant growth rate for the mirror mode (the bold line) lies below that for the proton cyclotron mode (the dotted line) [*Gary et al.*, 1993a], so that the mirror mode growth rate is larger.

Since variation of $\beta_{\|p}$ is strongly related to the spatial structure across the plasma depletion layer, the fact that the spectral wave types are ordered by $\beta_{\|p}$ indicates that the variation of spectral type relates to the spatial variation of $\beta_{\|p}$ across the plasma depletion layer and into the magnetosheath proper. Thus the bifurcated (BIF) category occurs close to the magnetopause, while the mirror mode (MIR) occurs more often in the magnetosheath proper.

Denton et al. [1993] showed theoretically that the detailed frequency spectrum of waves could be explained

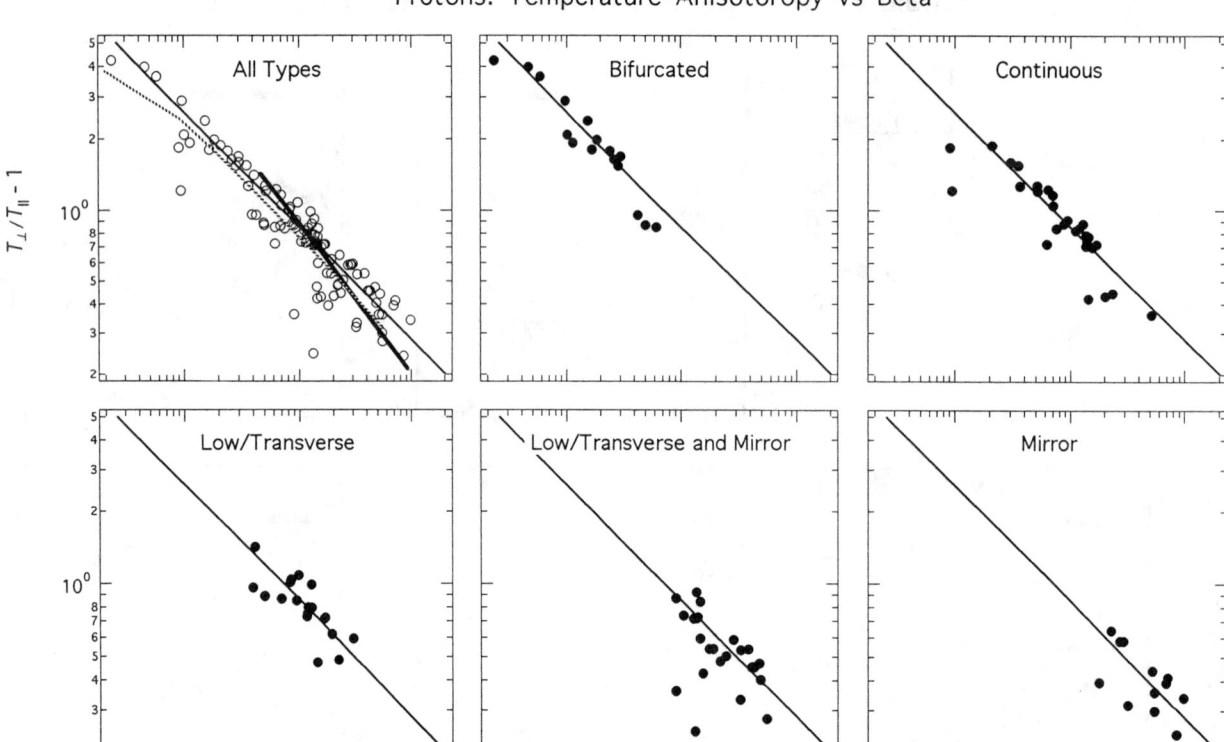

Figure 3. Plot of proton temperature anisotropy, $A_p \equiv (T_\perp/T_\parallel)_p - 1$ versus parallel proton beta, $\beta_{\parallel p}$, for all the events in the study of *Anderson et al.* [1994] (upper left panel), and for the events in each spectral category individually. The solid line in the upper left panel shows the least squares fit relation: $A_p = 0.85\beta_{\parallel p}^{-0.48}$. The dotted curve shows the contour of proton cyclotron growth rate of $0.01\Omega_p$ and the bold curve shows the corresponding contour for the mirror mode assuming 4% He^{2+}.

in the case of one event with a bifurcated (BIF) spectrum. Subsequently, *Denton et al.* [1994] extended their analysis of the frequency spectrum to the full range of plasma depletion layer and magnetosheath conditions. They introduced a model for the plasma variation across the plasma depletion layer and into the magnetosheath proper (the model is similar to that of *Gary et al.* [1993a], *Gary and Winske* [1993] and *Gary et al.* [1993b]). The model incorporates anisotropic protons and alpha particles (at 4% of the proton density with $(T_\perp/T_\parallel)_\alpha = 1.3(T_\perp/T_\parallel)_p$) and isotropic electrons (see *Denton et al.* [1994] for details). The central feature of the model is the anisotropy-beta relation discussed above, $A_p = 0.85\beta_{\parallel p}^{-0.48}$.

Figure 4 (taken from *Denton et al.* [1994]) shows how the variation of the observed fluctuation spectra follows naturally from the magnetosheath and plasma depletion layer model. At each value of $\beta_{\parallel p}$ (plotted on the horizontal axis), the model has been used to generate the plasma parameters. Figure 4a shows the normalized frequency range (ω_r/Ω_p) of ion cyclotron instability using the full model, while Figure 4b shows the result with A_α lowered to zero, and Figure 4c shows the result with A_p lowered to zero. The solid lines indicate frequency values corresponding to marginal stability, while the long and short dashed lines are plotted at frequencies corresponding to maxima in the growth rate. We note from Figure 4a that at low $\beta_{\parallel p}$ there are two regions of instability, one with frequency above Ω_α, and one with frequency below. These regions of instability merge at $\beta_{\parallel p}$ slightly less than unity. Figure 4b and 4c show clearly that the higher-frequency mode is driven by the anisotropy of the protons, while the lower-frequency mode is driven by the anisotropy of the alpha particles.

Thus we identify the higher-frequency mode as the proton cyclotron mode and the lower-frequency mode as the alpha cyclotron mode (indicated by the letters p and α in the Figure). Figure 4d shows the normalized temporal growth rate, γ/Ω_p, for the higher-frequency proton cyclotron mode (long dashes), lower-frequency alpha cyclotron mode (shorter dashes), and the mirror mode (solid line), using the full model. Figure 4e and 4f have $A_\alpha = 0$ and $A_p = 0$ respectively. Note from Figure 4d that the mirror mode growth rate becomes larger than the ion cyclotron mode growth rates at $\beta_{\|p} \sim 6$.

The letters B, C, L, L over M, and M in Figure 4a refer to the spectral categories discussed above. Each one of these letters is plotted at a horizontal position which corresponds to the average value of $\beta_{\|p}$ for the events in that category in the study of *Anderson et al.* [1994]. We see that B is plotted where the model predicts a bifurcated spectrum of ion cyclotron waves, C and L are plotted where the model predicts a merged spectrum, L over M is plotted where the mirror mode growth rate starts to become significant, and M is plotted where the mirror mode growth rate becomes greater than that of the ion cyclotron modes. Thus the $\beta_{\|p}$ variation of spectral properties is well predicted by the theory. In fact, there is a remarkable resemblance between Figure 1 and Figure 4 because the passage of the satellite shown in Figure 1 corresponds to passage into the plasma depletion layer and into the magnetosheath where $\beta_{\|p}$ becomes larger. As further evidence of the adequacy of linear Vlasov theory to predict the wave properties, the unstable frequency bands for the example cases are shown in Figure 2 as horizontal bars; again there is good agreement, except for the mirror mode (MIR) case, for which the theory predicts ion cyclotron waves. Possibly, presence of the mirror mode inhibits the ion cyclotron mode in this case.

The fact that the proton and alpha cyclotron instabilities approach their respective gyrofrequencies at low $\beta_{\|p}$ is easy to understand. Instability occurs for particles which are Doppler shifted up to their gyrofrequency. At low $\beta_{\|p}$, the thermal velocity is small compared to the Alfvén speed, which is roughly equal to the phase velocity of the waves. So at low $\beta_{\|p}$, the wave frequency

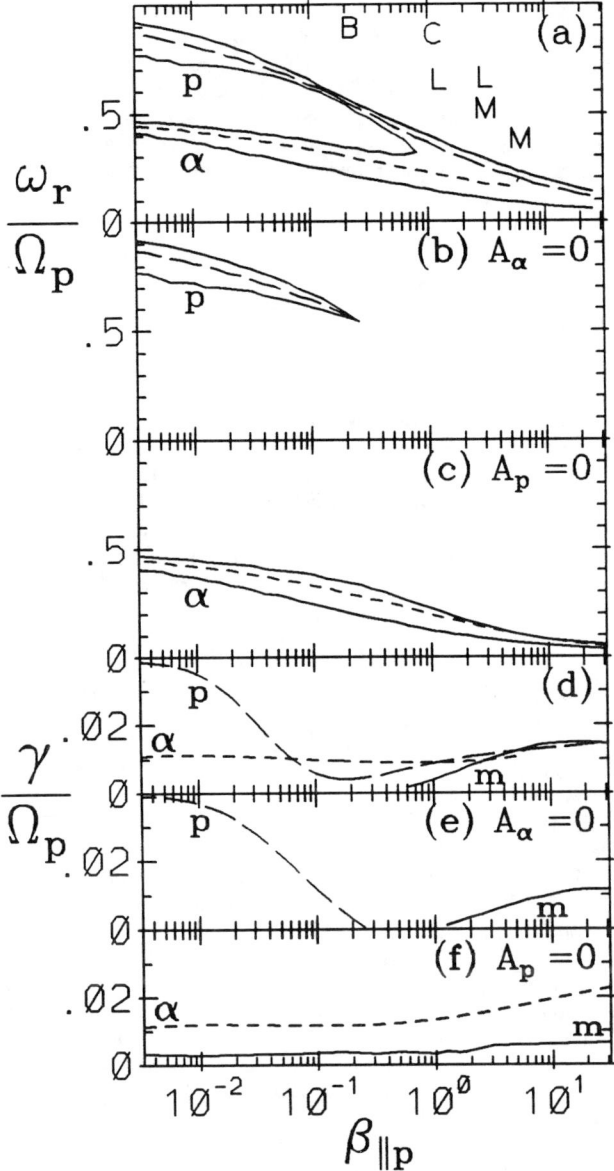

Figure 4. Properties of the ion cyclotron anisotropy instabilities and the mirror instability as functions of $\beta_{\|p}$ under the parameters of the model of *Denton et al.*, [1994]. The top three panels illustrate frequencies ω_r which are relevant to the ion cyclotron anisotropy instabilities; the solid lines indicate the frequencies which bound the instability regimes, $\gamma/\Omega_p > 1 \times 10^{-4}$, and the dashed lines indicate the frequencies which correspond to a local maximum (with respect to ω_r) in temporal growth rate, γ. The bottom three panels illustrate the maximum growth rates for the proton cyclotron instability (the long dashed lines), the helium cyclotron instability (the short dashed lines), and the mirror instability (the solid lines). Panels (a) and (d) correspond to the full model. Panels (b) and (e) represent results from the model except that $(T_\perp/T_\|)_\alpha = 1.0$, and panels (c) and (f) represent results from the model except that $(T_\perp/T_\|)_p = 1.0$. The letters B, C, L, L over M, and M in panel (a) indicate the spectral categories of *Anderson et al.* [1994] defined in the text. They are plotted at a horizontal position corresponding to the average value of $\beta_{\|p}$ for the events in each category.

must approach the gyrofrequency. At higher $\beta_{\|p}$, the Doppler shift is greater, hence the wave frequency is lower.

The data presented here pertain only to periods with high solar wind dynamic pressure, i.e. high magnetosheath beta. For low dynamic pressure conditions the beta just downstream of the bow shock may be quite low, < 1. In this case, the magnetosheath plasma will evolve along the same A_p–$\beta_{\|p}$ relation obtained here as it convects toward the magnetopause, but will start on the curve at a lower value of $\beta_{\|p}$ and higher value of A_p than the high beta cases studied to date. Hence, under these conditions it is possible that ion cyclotron waves could dominate the observed spectrum throughout the magnetosheath [*Sckopke et al.*, 1990], so that the mirror mode is not observed. Such a result is entirely consistent with the results stated here provided that the low beta plasma in the magnetosheath is highly anisotropic (see Figure 3). Indeed, highly anisotropic proton distributions have been observed downstream of low beta shocks [*Thomsen et al.*, 1985, *Sckopke et al.*, 1990]. While distributions downstream of low beta shocks are much more anisotropic than the distributions observed in the high beta magnetosheath proper by AMPTE/CCE, it is important to realize that they still only represent a plasma that is marginally stable to the growth of ion cyclotron waves.

We have shown that the low-frequency waves ($\omega_r \leq \Omega_p$) observed by AMPTE/CCE in the plasma depletion layer and magnetosheath proper are well identified with proton cyclotron waves (driven by A_p), alpha cyclotron waves (driven by A_α), and the mirror mode. The wave spectral properties are well ordered with respect to the proton parallel beta, $\beta_{\|p}$. Such a result is predicted by linear Vlasov theory using a simple model for the plasma depletion layer. Because of this, the observed spectral type is related to relative distance from the magnetopause, since $\beta_{\|p}$ increases sharply away from the magnetopause within the plasma depletion layer and into the magnetosheath proper. The anisotropy-beta relation, $A_p = 0.85\beta_{\|p}^{-0.48}$, results from the fact that the waves pitch angle scatter the particles so that the plasma is near marginal stability. This relation supplies a fundamental constraint on the plasma, and should be taken into account in fluid descriptions of the plasma depletion layer and magnetosheath proper.

Acknowledgments. The work at Dartmouth has been supported with funding from NASA under grants NAG-1652 and NAG-5-1098. Work at the Johns Hopkins University Applied Physics Laboratory was supported by NASA under the AMPTE Missions Operations and Data Analysis program. Research at Lockheed Palo Alto Research Laboratory was funded through NASA contract NAS5-30565 and the NASA Guest Investigator program NAS5-31213. The Los Alamos portion of this work was performed under the auspices of the U.S. Department of Energy (DOE) and was supported by the DOE Office of Basic Energy Sciences, Division of Engineering and Geosciences, and the SR&T Program of the National Aeronautics and Space Administration (NASA). We also acknowledge a grant of computer time from NCSA.

References

Anderson, B. J., and S. A. Fuselier, Magnetic pulsations from 0.1 to 4.0 Hz and associated plasma properties in the Earth's subsolar magnetosheath and plasma depletion layer, *J. Geophys. Res.*, 98, 1461–1479, 1993.

Anderson, B. J., S. A. Fuselier, and D. Murr, Electromagnetic ion cyclotron waves observed in the plasma depletion layer, *Geophys. Res. Lett.*, 18, 1955–1958, 1991.

Anderson, B. J., S. A. Fuselier, S. P. Gary, and R. E. Denton, Magnetic spectral signatures in the Earth's magnetosheath and plasma depletion layer, *J. Geophys. Res.*, in press, 1994.

Crooker, N.U., T.E. Eastman, and G.S. Stiles, Observations of plasma depletion in the magnetosheath at the dayside magnetopause, *J. Geophys. Res.*, 84, 869–874, 1979.

Denton, R. E., M. K. Hudson, S. A. Fuselier, and B. J. Anderson, Electromagnetic ion cyclotron waves in the plasma depletion layer, *J. Geophys. Res.*, 98, 13,477, 1993.

Denton, R.E. S.P. Gary, B.J. Anderson, S.A. Fuselier, and M.K. Hudson, Low–frequency magnetic fluctuation spectra in the magnetosheath and plasma depletion layer, *J. Geophys. Res.*, in press, 1994.

Fuselier, S. A., Energetic magnetospheric protons in the plasma depletion layer, *J. Geophys. Res.*, 97, 13759, 1992.

Fuselier, S. A., D. M. Klumpar, E. G. Shelley, B. J. Anderson, and A. J. Coates, He^{2+} and H^+ dynamics in the subsolar magnetosheath and plasma depletion layer, *J. Geophys. Res.*, 96, 21,095, 1991.

Gary, S. P., and D. Winske, Simulations of ion cyclotron anisotropy instabilities in the terrestrial magnetosheath, *J. Geophys. Res.*, 98, 9171–9179, 1993.

Gary, S.P., M.D. Montgomery, W.C. Feldman, and D.W. Forslund, Proton temperature anisotropy instabilities in the solar wind, *J. Geophys. Res.*, 81, 1241–1246, 1976.

Gary, S. P., S. A. Fuselier, and B. J. Anderson, Ion anisotropy instabilities in the magnetosheath, *J. Geophys. Res.*, 98, 1481–1488, 1993a.

Gary, S. P., B. J. Anderson, R. E. Denton, S. F. Fuselier, M. E. McKean, and D. Winske, Ion anisotropies in the magnetosheath, *Geophys. Res. Lett.*, 20, 1767, 1993b.

Lees, L.C., Interaction between the solar plasma wind and the geomagnetic cavity, *Amer. Inst. Aero. Astro. J.*, 2, 2065, 1964.

Midgley, J.E., and L. Davis, Calculation by a moment technique of the perturbation of the geomagnetic field by the solar wind, *J. Geophys. Res.*, 68, 5111–5123, 1963.

Sckopke, N., G. Paschmann, A. L. Brinca, C. W. Carlson, and H. Luhr, Ion thermalization in quasi-perpendicular shocks involving reflected ions, *J. Geophys. Res.*, *95*, 6337, 1990.

Thomsen, M.F., J.T. Gosling, S.J. Bame, and M.M. Mellott, Ion and electron heating at collisionless shocks near the critical mach number, *J. Geophys. Res.*, *90*, 137–148, 1985.

Wu, C.C., MHD flow past an obstacle: large-scale flow in the magnetosheath, *Geophys. Res. Lett.*, *19* (2), 87–90, 1992.

Zwan, B.J., and R.A. Wolf, Depletion of solar wind plasma near a planetary boundary, *J. Geophys. Res.*, *81*, 1636–1648, 1976.

B. J. Anderson, Applied Physics Laboratory, Johns Hopkins Univ., Laurel, MD 20723-6099. (e-mail: SPAN.ampte::anderson)

R. E. Denton and M. K. Hudson, Physics and Astronomy Dept., Dartmouth College, Hanover, NH 03755-3528. (e-mail: Internet.richard.denton@dartmouth.edu; Internet.mary.hudson@dartmouth.edu)

S. A. Fuselier, Lockheed Palo Alto Research Laboratory, Palo Alto, CA 94304. (e-mail: SPAN.lockhd::fuselier)

S. P. Gary, Los Alamos National Laboratory, Los Alamos, NM 87545. (e-mail: Internet.pgary@lanl.gov)

Multidimensional Fourier Analysis of a Whistler Pulse Excited by a Loop Antenna

C. L. ROUSCULP, J. M. URRUTIA, AND R. L. STENZEL

Department of Physics, University of California, Los Angeles

Large magnetic antennas (loops) are of interest in active space plasma experiments. In a scaled laboratory experiment, the perturbed magnetic field, $\mathbf{B}(\mathbf{r},t)$, of a pulsed, ($\omega_{ci} \ll 2\pi/\Delta t_{pulse} \ll \omega_{ce}$) dipole has been mapped from the near zone ($r < \lambda$) to the radiation zone. From the Fourier time transform, $\mathbf{B}(\mathbf{r},\omega)$, single frequencies are examined and the radiation pattern of an antenna operating in CW mode is displayed. From the full Fourier transform, $\mathbf{B}(\mathbf{k},\omega)$, contours of constant index of refraction for $k_\parallel$ and $k_\perp$, including magnitude and direction of the phase and group velocities, are extracted and compared to whistler plane wave theory.

INTRODUCTION

Loop antennas are of interest in active space plasma experiments [*Shapiro et al.*, 1990]. Loops also represent a simple physical structure and have been theoretically analyzed in plasma for the near zone [*Karpman*, 1985] and radiation zone [*Wang and Bell*, 1972]. Presented here is an experimental study of the excitation of a packet of whistler waves from a pulsed loop antenna using Fourier analysis. This allows decomposition of the wave packet into plane waves which are difficult to excite directly. The study resolves this process in time and space with greater detail than is possible from active space experiments.

EXPERIMENT AND ANALYSIS TECHNIQUES

Data are taken in a large magnetized afterglow plasma column (2.5 m length, 1.0 m diameter, $kT_e = 1.3$ eV, $n_e = 6.0 \times 10^{11}$ cm^{-3}, $B_0 = 20$ G, $\Delta t_{pulse} = 100$ ns). A probe with three orthogonal loops, moveable in three dimensions, is used to take time series of the perturbed magnetic field near a 5 cm diameter insulated loop antenna whose dipole moment is parallel to $\mathbf{B}_0$ [*Stenzel et al.*, 1993]. The spatial volume of interest is approximately a 0.1 m^3 with the loop located at one end. The duration of the experiment is approximately 1 μs during which the plasma parameters can be considered constant and uniform. The data at each time step are interpolated from an irregular grid onto a (35 cm x 39 cm x 75 cm) regular rectangular grid with a time resolution of 10 ns so that a Fast Fourier Transformation (FFT)

can be used [*Press et al.*, 1991]. Figure 1 shows contours of B_z at $t = 410$ ns after the start of the pulse. The near field is no longer present and the axially symmetric wave field has propagated from the antenna, located at $z = 0$ cm, approximately 50 cm along the ambient field.

FOURIER TIME TRANSFORMATION

Since the pulsed data contains a spectrum of frequencies, Fourier transformation can be utilized to examine a single mode of excitation that lies within this spectrum. An FFT is performed on each component of $\mathbf{B}(\mathbf{r},t)$ in time at each point in space. Next, a particular frequency is selected out of each spectrum and combined into a single spatial volume. Since the FFT yields both real and imaginary parts, phase information exists for the wave at each point in space. Using the formula

$$\mathbf{B}_{\omega_0}(\mathbf{r},t) = \frac{Re\mathbf{B}(\mathbf{r},\omega_0)\cos(\omega_0 t) + Im\mathbf{B}(\mathbf{r},\omega_0)\sin(\omega_0 t)}{2} \quad (1)$$

the wave emitted by an antenna operating in CW mode is extracted. Figure 2 shows the propagation of B_z in a $y-z$ plane that is centered on the antenna axis for two different frequencies. The higher frequency has more spatial oscillations indicating a shorter wavelength. Also, the wave amplitude decreases with distance from the loop. This may be caused by either collisions ($\nu_{ei} \approx 10^7$) or simple spreading due to the finite size of the loop. Thus, by using this technique the time consuming task of taking volume data for each frequency is bypassed.

FOURIER TIME AND SPACE TRANSFORMATION

Further Fourier transformation of the the data set in space gives $\mathbf{B}(\mathbf{k},\omega)$. The dispersion characteristics of the pulse can

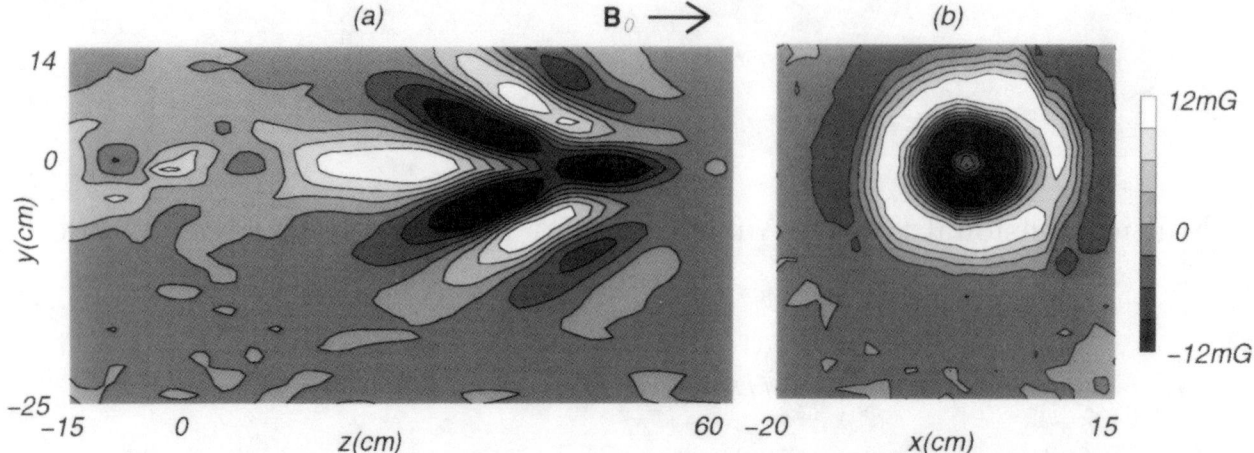

Fig. 1. The perturbed magnetic field, $B_z(\mathbf{r}, t)$, of a pulsed loop located at $(0,0,0)$ at $t = 410$ ns after switch-on of the pulse. (a) Contours in a $y - z$ plane on the loop axis ($x = 0$ cm). (b) Contours in a $x - y$ plane at $z = 40$ cm from the loop.

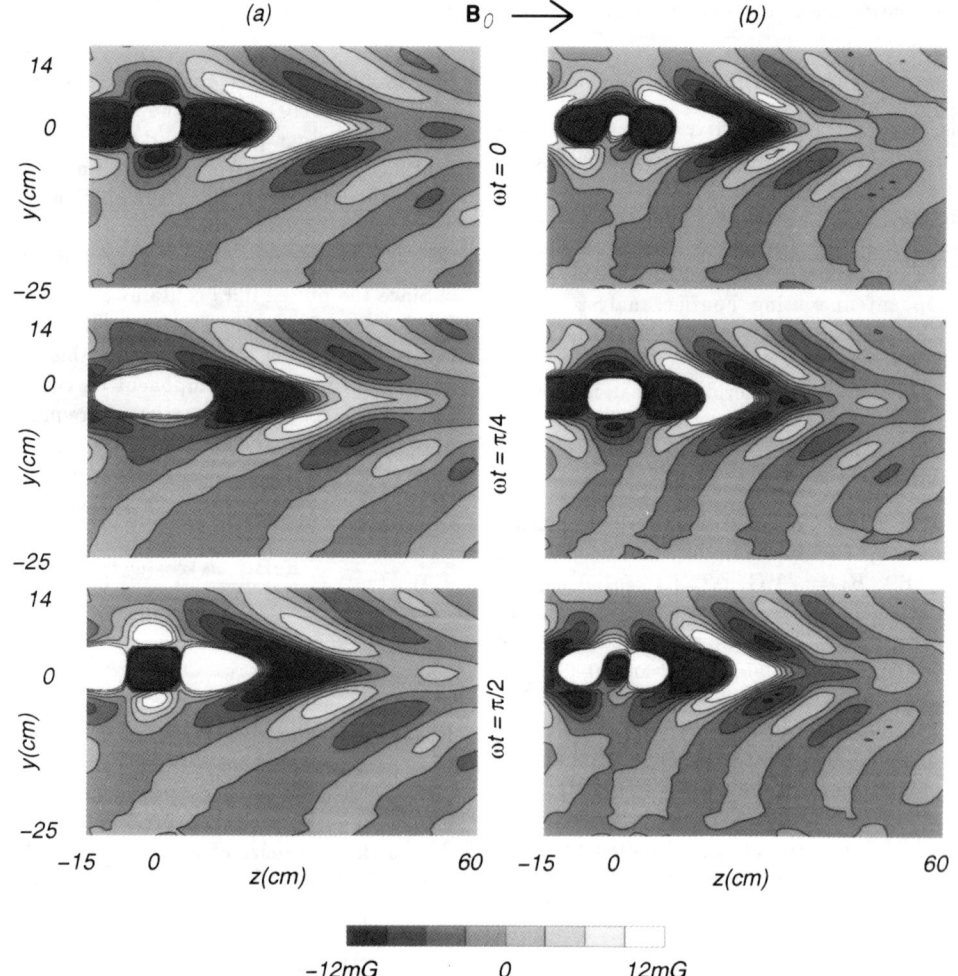

Fig. 2. Radiation field of the antenna operating in CW mode. Contours of B_z for (a) $\omega/\omega_{ce} = 0.042$, (b) $\omega/\omega_{ce} = 0.084$ at different phase angles ωt.

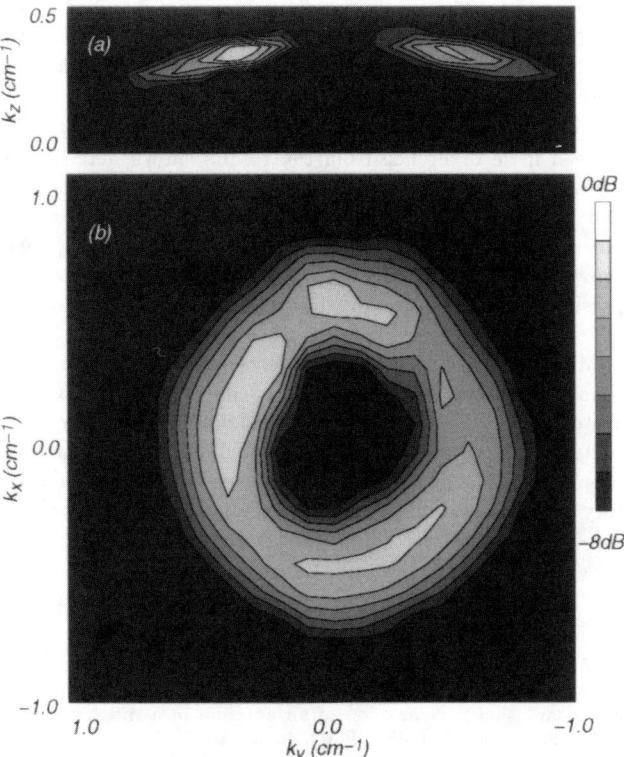

Fig. 3. Power spectrum, $|\mathbf{B}(\mathbf{k},\omega)|^2$, of the pulse for $\omega/\omega_{ce} = 0.084$. (a) Contours in the $k_y - k_z$ plane at $k_x = 0.0$ cm^{-1}. (b) Contours in the $k_x - k_y$ plane at $k_z = 0.3$ cm^{-1}.

then be analyzed. Figure 3 shows two different cuts of the power spectrum, $|\mathbf{B}(\mathbf{k},\omega)|^2$, of the pulse. The spectrum is roughly symmetric about $\mathbf{B}_0$ (the k_z axis) and peaks sharply in the radial direction. The radius at which the spectrum peaks appears to be a function of the angle, θ, between $\mathbf{k}$ and $\mathbf{B}_0$. In Figure 4, maxima of $|\mathbf{B}(\mathbf{k},\omega)|^2$ are converted into index of refraction, and plotted as a function of θ for three different frequencies (note the change in scale). For comparison the theoretical index of refraction surfaces have also been plotted [Helliwell, 1965]

$$n(\theta) = \sqrt{\omega_{pe}^2/\omega(\omega_{ce}\cos(\theta) - \omega)} \qquad (2)$$

for plane whistler waves with $\mathbf{k}$ at an angle θ from $\mathbf{B}_0$.

Good agreement with theory can be seen. The error bars come from converting the maximum in $|\mathbf{B}(\mathbf{k},\omega)|^2$ into index of refraction and represent the uncertainty due to one grid step in the FFT. The discrepancy between experiment and theory in Figure 4(a) is due to the finite size in x and y of the data volume. Part of the packet, unavoidably, gets outside of the data volume which leads to errors because the FFT assumes a repeated signal outside the boundary. When the signal has a discontinuity at the boundary, spurious high $\mathbf{k}_\perp$ values are introduced and show up as errors in $n_\perp$. Despite this error, the decomposed pulse agrees well over a range

of frequencies and waves are excited with the direction of the phase velocity extending from 25° to 70°. Note that these angles are well within the phase velocity resonance cone angle, $\theta_c = \cos^{-1}(\omega/\omega_{ce}) \approx 87°$.

The angle of the group velocity, $\theta_g = 13°$, is calculated by looking at the normal to the index of refraction data for the central frequency, $\omega/\omega_{ce} = 0.084$. This compares to the theoretical maximum ray angle $\theta_{max} \approx 19°$ for $\omega \ll \omega_{ce}$ [Helliwell, 1965].

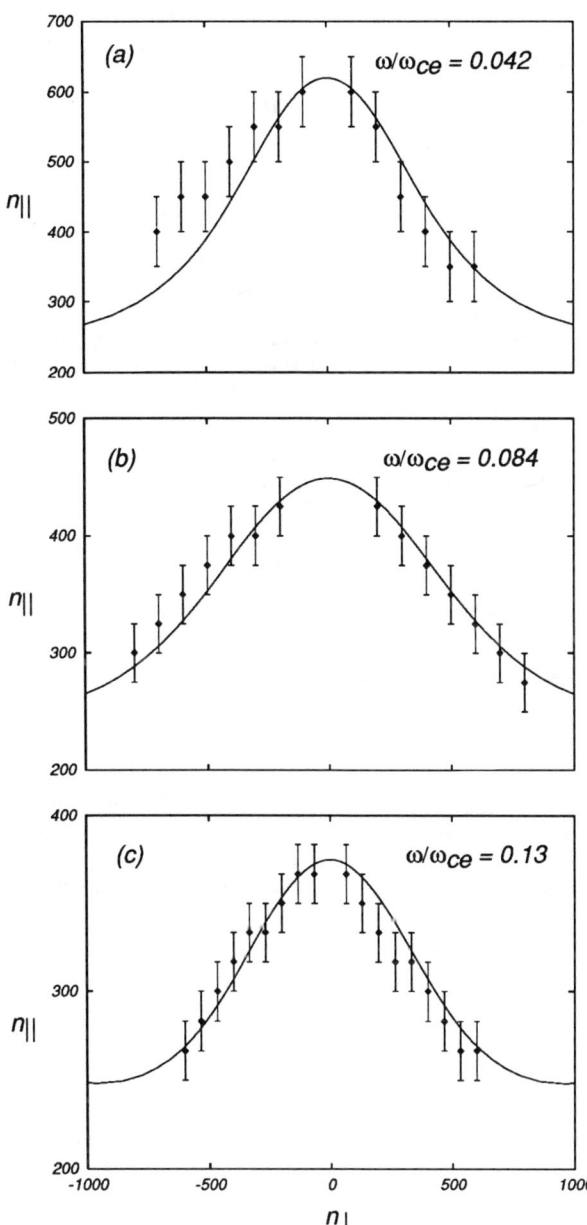

Fig. 4. The index of refraction from experiment and theory (solid curves) for (a) $\omega/\omega_{ce} = 0.042$, (b) $\omega/\omega_{ce} = 0.084$, and (c) $\omega/\omega_{ce} = 0.130$.

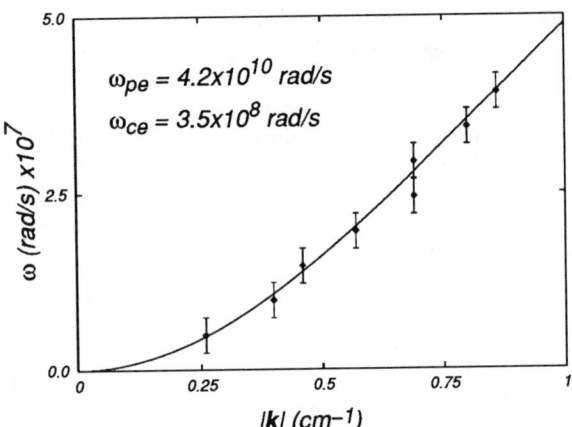

Fig. 5. Dispersion relation ω versus $|\mathbf{k}|$ from experiment and theory (solid curve) for $\theta = 64°$.

DISPERSION RELATION

By looking at the maximum in $|\mathbf{B}(\mathbf{k},\omega)|^2$ at a particular θ for different frequencies, ω versus $|\mathbf{k}|$ plots can be made to compare the dispersion of the wave packet with theory. $\theta = 64°$ is chosen because $|\mathbf{B}(\mathbf{k},\omega)|^2$ peaks in θ at roughly this angle for the frequency spectrum. Figure 5 is a plot of ω versus $|\mathbf{k}|$ with the data superimposed on a plot of the theoretical dispersion curve [Helliwell, 1965]

$$\omega = \omega_{ce}\cos(\theta)\frac{|\mathbf{k}|^2}{(\omega_{pe}/c)^2 + |\mathbf{k}|^2} \quad (3)$$

Once again, excellent agreement between the data and whistler plane wave theory is seen. The slope of the data at $\omega/\omega_{ce} = 0.084$ is the group velocity of the packet and has a value of 5.74×10^7 cm s^{-1}.

CONCLUSIONS

From the above analyses, it has been found that a pulsed antenna launches a packet of whistler waves with a spectrum in frequency and wavenumber. The v_{phase} ranges in angle from 25° to 70° with a maximum at approximately 64° from $\mathbf{B}_0$. The v_{group} is 5.74×10^7 cm s^{-1} and is at an angle of 13° from $\mathbf{B}_0$. Near zone ($r < \lambda$) contribution to the FFT seems negligible since it occupies a small fraction of the data volume. Additionally, similar Fourier transformations performed on data volumes excluding the near zone showed little difference from the results shown here. Even though this analysis is performed on low amplitude waves and are not subject to nonlinear affects, laboratory experiments with high amplitude pulses [Stenzel et al., 1993] have shown radiation patterns similar to Figure 1. Therefore, dispersion characteristics should also be similar for high amplitude pulses.

Most theory is expressed in terms of plane waves even though they are very difficult to excite directly in the laboratory. Active space experiments cannot adequately map perturbed fields to determine the predominant direction of propagation. Multidimensional Fourier analysis of laboratory data elegantly decomposes a wave packet into plane waves allowing direct comparison with theory.

REFERENCES

Helliwell, R. A., *Whistlers and Related Ionospheric Phenomena*. Stanford University Press, Stanford, CA, 1965.

Karpman, V. I., Near zone of an antenna in a magnetoactive plasma. *Sov. Phys.-JETP*, 62, 40–47, 1985.

Press, W. H., et al., *Numerical Recipes in C, The Art of Scientific Computing*. Cambridge University Press, Cambridge, 1991.

Shapiro, V. D., Shevchenko, V. I., Sotnikov, V. I., Fiala, V., and Trivka, P., Plasma heating near a VLF antenna. *Plasma Phys. Controlled Fusion*, 32, 221–224, 1990.

Stenzel, R. L., et al., Pulsed currents carried by whistlers part I: excitation by magnetic antennas. *Phys. Fluids B*, 5, 325–338, 1993.

Wang, T. N. C. and Bell, T. F., VLF/ELF radiation patterns of arbitrary oriented electric and magnetic dipoles in a cold lossless multicomponent manetoplasma. *J. Geophys. Res.*, 77, 1174–1189, 1972.

C. L. Rousculp, J. M. Urrutia, and R. L. Stenzel, Department of Physics, University of California Los Angeles, Los Angeles, California 90024-1547.

Thermal Magnetic Fluctuations in Maxwellian and Non-Maxwellian Plasmas at Whistler and Electron Cyclotron Harmonic Frequencies

R.L. Stenzel, G. Golubyatnikov and J.M. Urrutia

Department of Physics, University of California, Los Angeles, CA.

Thermal magnetic fluctuations have been measured with a magnetic loop antenna inside a large magnetoplasma in the frequency regime of whistlers and electron cyclotron harmonics ($f_{ce} \approx 30$ MHz, $f_{pe} \approx 3000$ MHz, discharge plasma 1 m diam., 2.5 m length). In a Maxwellian afterglow plasma the fluctuations $\tilde{B}(\omega)$ exhibit a 1/f-like spectrum in the whistler wave regime ($f < f_{ce}$), no resonant enhancement at the cyclotron frequency and a flat spectrum in the evanescent regime ($f_{ce} < f \ll f_{pe}$). The whistler noise consists of obliquely propagating waves which are thought to be excited by Cherenkov ($\omega = k_\parallel v_\parallel$) and cyclotron ($\omega - \omega_{ce} = k_\parallel v_\parallel$) wave-particle interactions. In the non-Maxwellian discharge plasma, which contains energetic primary electrons (½ m $v_p^2 \approx$ e $V_{dis} \approx 40$ eV, $kT_e \approx 3$ eV, $n_p/n_e < 1\%$), enhanced lines at the cyclotron frequency and harmonics are observed. Up to 15 cyclotron harmonic lines are produced when spiralling electrons are injected into an afterglow plasma with a small beam source. The emission process is identified to be thermal fluctuations of the spiralling electrons whose broadband spectrum is filtered by interference effects, i.e. at $\omega = n\omega_{ce}$ the ballistic beam modes form a long [$k_\parallel = (\omega - n\omega_{ce})/v_\parallel \approx 0$] solenoidal rf magnetic field while at $\omega \neq n\omega_{ce}$ the fields destructively interfere ($k_\parallel \neq 0$). Thus, the observed fluctuations are dominated by particle effects rather than plasma eigenmodes excited or destabilized by kinetic effects. These results should be relevant to the interpretation of noise spectra from satellite-borne magnetic antennas.

INTRODUCTION

Magnetic fluctuations are of general interest in plasma physics and of particular importance to space plasmas for communication, properties of antennas, interpretation of active experiments and instabilities. While thermal noise on electric dipole antennas has received attention [Meyer-Vernet, 1979; Stenzel, 1989] less is known about thermal noise levels on magnetic loop antennas, in particular near the electron cyclotron frequency [Cable and Tajima, 1992; Golubyatnikov and Stenzel, 1993a]. In non-Maxwellian plasmas emissions at cyclotron harmonics of ions [Kintner et al., 1986] and electrons [Landauer, 1962] have been observed and explained by a kinetic instability involving particles with excess perpendicular energy interacting with cyclotron harmonic waves. In the present work

Solar System Plasmas in Space and Time
Geophysical Monograph 84
Copyright 1994 by the American Geophysical Union.

[Golubyatnikov and Stenzel, 1993b] we observe cyclotron harmonic lines in the thermal fluctuation spectrum of spiralling electrons in a dense magnetoplasma. No instabilities of plasma eigenmodes or nonlinear effects are involved and the fluctuations are only due to ballistic particle effects. A new physical model for generating line emissions from broadband thermal noise is presented.

EXPERIMENTAL SETUP

The experiments are performed in a large (1 m diam. × 2.5 m length) pulsed discharge plasma ($n_e \leq 5 \times 10^{11}$ cm^{-3}, $kT_e < 3$ eV, $B_o \leq 20$ G, Argon p $\approx 1.3 \times 10^{-4}$ Torr) shown in Figure 1. Fluctuation measurements are performed both during the discharge which contains a small tail of energetic primary electrons (½ m $v_p^2 \approx$ e $V_{dis} \approx 40$ eV, $n_p/n_e \leq 1\%$) and in the Maxwellian afterglow plasma [n/(dn/dt) ≥ 2 ms, $T_e/(dT_e/dt) \leq 1$ ms]. A pulsed test electron beam of variable parameters ($0 < V_b < 100$ V, $0 < I_b < 100$ mA, 8 mm

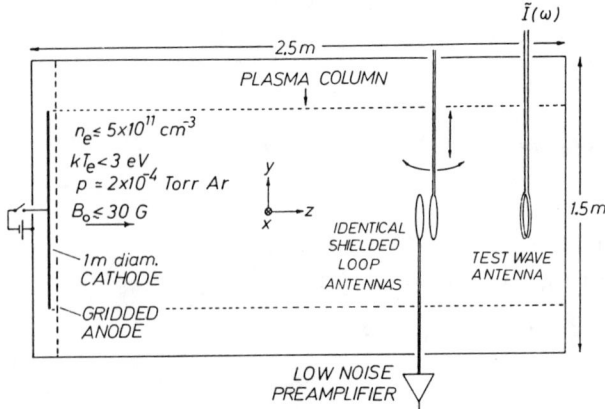

Fig. 1. Schematic view of the experimental setup to measure magnetic fluctuations in a Maxwellian afterglow plasma.

diam., $0 < \theta_{pitch} < 90°$, $t_{pulse} \sim 5$ μs) is injected into the afterglow plasma in order to identify the harmonic emission mechanism under controlled conditions. Magnetic fluctuations are detected with two electrostatically shielded magnetic loop antennas ($r_{major} \sim 2.2$ cm, $r_{minor} \sim 1$ mm) connected via low-noise broadband preamplifiers (NF = 1.4 dB, 1-500 MHz) to a digital oscilloscope (Le Croy 7200, 400 MHz, 1 Gs/S) which is used to analyze the fluctuations in time, frequency, and to perform statistical averages such as cross-correlations.

THERMAL MAGNETIC NOISE IN MAXWELLIAN PLASMAS

Starting with the conceptually simplest case, a Maxwellian plasma, we have measured the spectral density of magnetic fluctuations, $B_{z\,rms}/(\Delta f)^{1/2}$ and displayed it vs. frequency for three different dc magnetic fields in Figure 2. In the whistler wave regime, the amplitude decay can be approximated by $B \propto \omega^{-\alpha}$ where $1 < \alpha < 1.4$. There is no resonant enhancement at the cyclotron frequency or its harmonics but for $\omega > \omega_{ce}$ the noise spectrum is essentially flat and can be accounted for by the thermal noise $P = kT_e \Delta f$ at the electron temperature ($kT_e \sim 1$ eV), antenna rf resistance ($R_a \sim 1$ Ω), antenna inductance ($L \sim 0.25$ μH) and transmission line impedance ($Z = 50$ Ω), $B_{rms}/(\Delta f)^{1/2} \sim 2 R_a^{1/2} kT_e [Z^2 + (\omega L)^2]^{1/2}/(Z \pi r^2 \omega) \sim 10^{-11}$ G/Hz$^{1/2}$. The enhancement of the magnetic fluctuation amplitude with decreasing frequency for $\omega < \omega_{ce}$ must be the result of the collective whistler mode. From cross-spectral measurements it has been verified that the magnetic fluctuations are obliquely propagating whistlers which are highly correlated along $\mathbf{B}_0$ but poorly across $\mathbf{B}_0$ ($\lambda_\perp \sim \ell_{corr.\perp} \sim 3$ cm, $\lambda_\parallel \sim$ 7 cm $< \ell_{corr.\parallel} \gtrsim 25$ cm @ $\omega/\omega_{ce} = 0.4$). For $\omega > \omega_{ce}$ the correlation length in all directions is a few collisionless skin depths ($\ell_{corr.} \lesssim 3\,c/\omega_{pe} \sim 5$ cm). The polarization of the magnetic fluctuations is essentially isotropic ($B_\perp \sim B_\parallel$).

In the evanescent regime ($\omega > \omega_{ce}$) the fluctuations can only be due to particles whose transit through the antenna near-zone induces shot-like noise. Electron transit-time damping also accounts for the measured real part of the antenna impedance R_a at $\omega > \omega_{ce}$. However, in the whistler wave regime propagating waves also contribute to the observed noise. These waves can be excited by wave-particle interactions, i.e. Cherenkov radiation ($\omega = k_\parallel v_\parallel$) predominantly below $\omega = 0.5\,\omega_{ce}$, and cyclotron radiation ($\omega - \omega_{ce} = k_\parallel v_\parallel$) predominantly above $\omega = 0.5\,\omega_{ce}$. The former involves only few fast electrons in a Maxwellian, (e.g. for $f = 0.4\,f_{ce} = 26$ MHz, $\lambda_\parallel \sim 7$ cm, $v_\parallel = 1.8 \times 10^8$ cm/s $\sim 3\,v_{th}$) while the latter should involve the bulk electron population (e.g. for $f = 0.63\,f_{ce} \sim 16$ MHz, $\lambda_\parallel \sim 5$ cm, $v_\parallel = (f - f_{ce})\lambda_\parallel = 4.7 \times 10^7$ cm/s $\sim 0.8\,v_{th}$). The enhancement

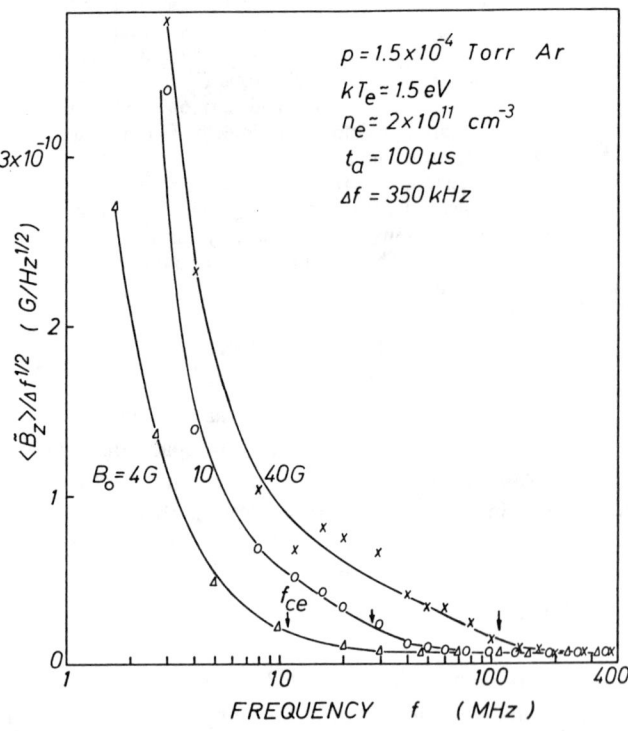

Fig. 2. Spectral noise density of axial magnetic field fluctuations $\tilde{B}_z/(\Delta f)^{1/2}$ vs. frequency in a Maxwellian afterglow plasma at different dc magnetic fields B_0. In the whistler wave regime ($f < f_{ce}$) the fluctuation amplitude decreases with frequency, exhibits no enhancement at the cyclotron resonance, and goes over into a white noise spectrum in the evanescent regime $\omega_{ce} < \omega \ll \omega_p$. The spectral characteristics are the same over an order of magnitude variation in B_0.

of the magnetic fluctuations above the thermal level at decreasing whistler wave frequencies may result from the larger volume (longer axial correlation length) from which the antenna receives weakly damped whistlers. A theoretical analysis by Cable and Tajima [1992] has predicted similar fluctuation spectra.

CYCLOTRON HARMONIC LINES IN NON-MAXWELLIAN PLASMAS

Enhanced emissions at the electron cyclotron frequency and its harmonics are observed in plasmas with non-Maxwellian electron distributions. These include (i) the active discharge plasma containing tails of energetic primary electrons (½ $mv_p^2 \lesssim eV_{dis} \approx 45$ V, $n_p/n_e \lesssim 1\%$) which have been scattered in energy and pitch-angle by a cold beam-plasma instability near the cathode ($\Delta z < 5$ cm) and (ii), a pulsed spiralling electron beam ($V_b < 100$ V, $I_b < 100$ mA, $t_{pulse} \approx 5$ μs, beam radius $r_b \approx 4$ mm < cyclotron radius $r_c \lesssim 2$ cm, $0 < \theta_{pitch} < 90°$) injected from a small cathode into a Maxwellian afterglow plasma. The second configuration is used to study under controlled conditions the mechanism of line emissions by spiralling electrons.

Figure 3 displays the fluctuation spectrum of $B_z(\omega)$ detected in the flux tube of a spiralling dc beam. Up to 15 cyclotron harmonic lines are superimposed on the thermal background noise spectrum. The line frequencies are exact harmonics and a least-square fit yields the dc magnetic field to with $\lesssim 0.1\%$ uncertainty. The line amplitude scales linearly with beam current as does the background thermal noise with electron temperature.

In order to determine the emission mechanism spatially resolved measurements have been performed. It is observed

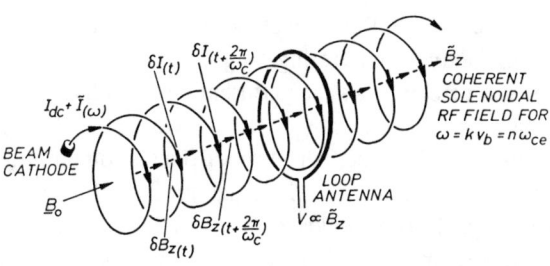

Fig. 4. A new physical model explaining the generation of cyclotron harmonic lines by a spiralling electron beam. At the thermal level the injected beam current contains broadband fluctuations $\tilde{I}(\omega)$ ($0 < f < f_{pb} \approx 500$ MHz). Those frequency components which rotate synchronously with the cyclotron motion ($\omega = n\omega_{ce}$) produce identical magnetic perturbations for each orbit which add up to a coherent solenoidal rf field structure observable with a linked magnetic loop antenna. Off-resonance, the perturbations δI, δB slip from orbit to orbit and interfere destructively which explains the observed filtering effect. No collective plasma mode or instability is required to explain the observed line spectrum.

that the line spectrum evolves axially without growth or decay from a broadband spectrum ($0 < f < f_{pb} \approx 500$ MHz) near the injection point. Thus, no convective instability is excited. Radially, the fluctuation amplitude is peaked in or near the beam flux tube. Cross-spectral measurements show that the fluctuations at $n\omega_{ce}$ propagate azimuthally with the same angular velocity ω_{ce} as the spiralling electrons. Thus, at the fundamental ($\omega = \omega_{ce}$) the azimuthal mode number is n = 1 (2π phase shift per revolution), at the second harmonic n = 2 ($\Delta\phi = 4\pi$), etc. Axially, there is no phase shift ($k_\parallel = 0$) but a finite group velocity equal to the parallel beam velocity. When the beam current is modulated an axial wavenumber is observed consistent with the Doppler-shifted cyclotron resonance $\omega = n\omega_{ce} + k_\parallel v_\parallel$. The modulation enhances the magnetic field $\tilde{B}_z$ strongly when tuned to a resonance ($\omega = n\omega_{ce}$) but not off resonance. The lines are linear, independent resonances.

Based on these observations, a physical model of the emission process is proposed. As shown in Figure 4 the injected electron beam carries both a dc current I_{dc} and broadband fluctuations $\tilde{I}(\omega)$ up to the beam plasma frequencies, $0 < \omega < \omega_{pb} \approx 2\pi \times 500$ MHz ($n_b \approx 3 \times 10^9$ cm^{-3}). The current fluctuations can arise from various sources such as shot noise emission, velocity modulations by sheath potential fluctuations, wave-particle scattering off plasma waves etc. The current perturbations convect along the spiralling beam as a free-streaming, ballistic perturbation of dispersion $\omega = \mathbf{k} \cdot \mathbf{v}_b$ where $\mathbf{k}$ is along $\mathbf{v}_b$. For those frequency components which are synchronous with the cyclotron rotation ($\omega = n\omega_{ce}$) the current or magnetic field

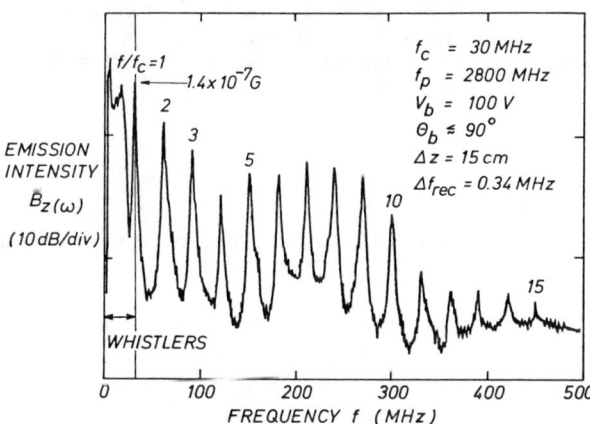

Fig. 3. Typical fluctuation spectrum of $\tilde{B}_z$ in the flux tube of a spiralling electron beam injected into an afterglow plasma, exhibiting whistlers ($f < f_{ce}$) and cyclotron harmonics ($f/f_{ce} = 1 \ldots 15$).

perturbations (δI_θ, δB_z) of each orbit are in phase resulting in a constructively interfering long solenoidal rf field. For off-resonant frequencies the perturbations on adjacent orbits are phase shifted ($\omega - n\omega_{ce} = k_\| v_\|$) which can lead to a destructive interference, hence small fluctuation amplitudes between the harmonics. The axial group velocity of the ballistic signal corresponds to the parallel beam velocity. The ballistic mode differs from extraordinary cyclotron harmonic modes by its large symmetric bandwidth around each harmonic with $k_\perp \rho_e < 1$, non-vanishing group velocities on resonance, and absence of wave propagation into the flux tube behind the beam source. No convective or absolute instabilities are excited. Nonlinear effects (e.g. bunching) can be excluded for waves at the thermal level, non-circular cyclotron orbits due to collisions with sheaths/boundaries [Simon and Rosenbluth, 1963] cannot occur in a large uniform plasma, and both mechanisms would lead to a phase relation between harmonics which is not observed.

In the present overdense plasma ($\omega_{pe} > \omega \sim n\omega_{ce}$) the particle fluctuations are shielded and cannot be observed externally. However, by lowering the density the harmonics have been observed with the loop antenna outside the plasma. Thus, it is possible that previous related observations in underdense plasmas using external microwave antennas [Landauer, 1962; Ikegami and Crawford, 1965] have a similar explanation rather than the presently accepted model of cyclotron harmonic wave instabilities [Canobbio and Croci, 1964]. The present physical model calls for a quantitative theory to explain the line amplitudes, widths, and mode structures at high harmonics.

Although large magnetic loop antennas have not yet been successfully flown in space [Shapirov et al., 1990] it should be possible to detect cyclotron harmonic line emissions in regions of energetic spiralling electrons such as auroral arcs or active beam injection experiments [Frank et al., 1989]. With a suitable theoretical model it may be possible to infer the pitch angle distributions from the fluctuation spectrum. In any case the cyclotron harmonic lines yield a precise diagnostics for the local dc magnetic field.

Acknowledgments. The authors would like to acknowledge helpful discussions with Dr. G. Morales. This work was supported by NSF grant PHY 91-02132 and NASA grant NAGW 1570.

REFERENCES

Cable, S. and T. Tajima, Low frequency fluctuations in plasma magnetic fields, *Phys. Rev. A*, 46, 3413-3441, 1992.

Canobbio, E. and R. Croci, Harmonics of the electron cyclotron frequency in a PIG discharge, *Proc. VIth. Intl. Conf. on Ionization Phenomena in Gases*, Vol.III, p.269, 1964.

Frank, L. A., W. R. Paterson, M. Ashour-Abdalla, D. Schriver, W. S. Kurth, D. A. Gurnett, N. Omidi, P. M. Banks, R. I. Bush, and W. J. Raitt, Electron velocity distributions and plasma waves associated with the injection of an electron beam into the ionosphere, *J. Geophys. Res.*, 94, 6995, 1989.

Golubyatnikov, G. and R. L. Stenzel, Thermal magnetic fluctuations of whistlers in a Maxwellian plasma, *Phys. Fluids*, September 1993a.

Golubyatnikov, G., and R. L. Stenzel, Cyclotron harmonic lines in magnetic fluctuations of spiralling electrons in plasmas, *Phys. Rev. Lett.*, 70, 940-942, 1993b; also *Phys. Fluids* (Oct. 1993).

Ikegami, H., and F. W. Crawford, Noise radiation from a warm magnetoplasma, *Proc. VIIth Intl. Conf. on Ionization Phenomena in Gases*, Belgrade, Vol. II, p. 503, 1965.

Kintner, P. M., I. LaBelle, W. Scales, R. Erlandson, and L. J. Cahill, Jr., A comparison of plasma waves produced by ion accelerators in the F-region ionosphere, in *Ion Acceleration in the Magnetosphere*, T. Chang, editor, *Geophys. Monograph*, 38, 206-208, AGU, Washington, DC, 1986.

Landauer, G., Generation of harmonics at the electro-gyro frequency in a Penning discharge, Plasma Phys. (J. Nucl. Energy C) 4, 395, 1962.

Meyer-Vernet, N., On natural noises detected by antennas in plasmas, *J. Geophys. Res.*, 84, 5373, 1979.

Shapirov, V. D., V. I. Shevchenko, V. I. Sotnikov, V. Fiala, and P. Triska, Plasma heating near a VLF antenna, *Plasma Phys. Contr. Fus.*, 32, 221-224, 1990.

Simon, A., and M. N. Rosenbluth, Single particle cyclotron radiation near walls and sheaths, *Phys. Fluids*, 6, 1566, 1963.

Stenzel, R. L., High-frequency noise on antennas, *Phys. Fluids B*, 1, 1369-1380, 1989.

R.L. Stenzel and J.M. Urrutia, Department of Physics, University of California, Los Angeles, CA 90024-1547.

Magnetic Dipole Antennas in Moving Plasmas: a Laboratory Simulation

J. M. URRUTIA, C. L. ROUSCULP, AND R. L. STENZEL

Department of Physics, University of California, Los Angeles

The magnetic perturbation to a large laboratory plasma due to a pulsed ($1/\omega_{ci} \gg \Delta t \gg 1/\omega_{ce}$) magnetic dipole is repeatedly superposed in space and time to simulate the motion of a dc-driven dipole. "Shock"/"wing" structures are formed when the dipole moves along/across the ambient magnetic field, $\mathbf{B}_0$.

INTRODUCTION

The effect created by a moving magnetic field in a plasma has long been an important topic of space physics research. The best known example is the study of the interaction of Earth's dipolar magnetic field with the solar wind. Proposals to create artificial magnetospheres in space [*Lane et al.*, 1987; *Birykov et al.*, 1992] are made from time to time while laboratory works can be traced to Birkeland's famous "terrella" experiments [*Egeland*, 1984]. Similarly, modulation of magnetic sources (e.g., a dipole) is thought to produce VLF/ELF waves [*Armand et al.*, 1989; *Triska and Shevchenko*, 1990], important for advanced communications. More recently, magnetic signatures observed during the flyby of 951 Gaspra by Galileo imply that asteroids may carry a magnetosphere [*Kivelson et al.*, 1993]. Although these perturbation sources do not, in general, exhibit a variation in time, the plasma in which they are immersed sees them as time-dependent and, consequently, responds differently to each. Thus, a slow moving, large source induces waves at or below the Alfvén frequency while a fast, small one produces R, X, and/or O waves. The intermediate regime of whistler waves, believed to apply to objects such as 951 Gaspra [*Kivelson et al.*, 1993], is the subject of our work. We approach the problem from a novel direction: instead of creating a moving plasma that buffets a stationary source and then mapping the perturbation, we create a wave packet by pulsing the source in a stationary plasma, measure it, and then digitally superpose the observations to study motion along and across the ambient magnetic field $\mathbf{B}_0$. In this way, a single, detailed data set allows us to study various ratios of source-to-plasma speed. The analysis reveals that structures that resemble the wake of boats in water ("wings" in three dimensions) are produced when the source moves across $\mathbf{B}_0$. Motion along $\mathbf{B}_0$ produces something reminiscent of a shock, albeit without the nonlinear features (e.g., particle reflections or plasma parameter modification).

EXPERIMENTAL ARRANGEMENT

The experiments are conducted in the afterglow ($t_{decay} \approx$ 2.5 ms, $t_{afterglow} = 150$ μs) of a large, cylindrical (1 m diam., 2.25 m length), magnetized ($B = 20$ G), plasma column ($n_e = 6 \times 10^{11}$ cm^{-3}, $kT_e = 1.3$ eV) [*Stenzel et al.*, 1993]. The magnetic perturbation is launched by applying a short current pulse ($I_{max} \approx 0.1$ A, $\Delta t = 0.1 \mu$s) to a two-turn, insulated magnetic dipole antenna (4.5 cm diam., $B_z \approx 50$ mG in vacuum at its center) located at the center of the plasma column. The perturbation is measured with a magnetic probe containing three orthogonal loops at many points ($\geq 15,000$) surrounding the source and averaged over ten highly repeatable discharges. The stored measurements of $\partial \mathbf{B}(\mathbf{r}, t)/\partial t$ are then processed to yield $\mathbf{B}(\mathbf{r}, t)$ over a regularly spaced grid. The current density, $\mathbf{J}(\mathbf{r}, t)$, is then available from Ampere's law, $\nabla \times \mathbf{B}(\mathbf{r}, t) = \mu_0 \mathbf{J}(\mathbf{r}, t)$.

EXPERIMENTAL RESULTS

Previous work has shown that, for our experimental parameters, magnetic perturbations are carried via whistler waves [*Urrutia and Stenzel*, 1989]. This conclusion was reached on the basis of interferometric studies of pulse trains that yielded the appropriate dispersion and polarization. Spectral analysis employing Fast Fourier Transformations in both $\mathbf{k}$ and ω space (see *Rousculp et al.*, in this volume) of a single wave packet has now confirmed this. In this paper, however, we restrict the discussion to the temporal and spatial evolution of the single pulse wave packet and its superposition. Figure 1 displays records of $B_x(t)$ at various points along a line parallel to $\mathbf{B_0}$ ($x \approx 0, y \approx -3, z$; antenna is at origin). The dashed lines tag similar points in the packet and make a $z - t$ graph of its motion along the chosen line in z. Since the lines have different slopes (equivalent to speeds), the packet disperses. Hence, it may be deduced that it consists of waves that propagate at different group velocities. The spatial evolution of B_x is depicted in Figure 2a for a plane bisecting the dipole and containing the points displayed in Figure 1. Inspection of the propagating packet shows that it travels across and along $\mathbf{B_0}$. Tracking the minimum and maximum of B_x, as described by *Stenzel et al.* [1993], reveals that the packet moves within a cone of angle $\theta < 19°$ consistent with classical, oblique whistler propagation [*Helliwell*, 1965]. This angular spread has been confirmed by *Rousculp et al.* (this volume). The packet's group velocity, v_{group}, along $\mathbf{B_0}$ is roughly 10^8 cm/s. For completeness, we show in Figure 2b a three-dimensional display of $B = 2$ mG at $t = 0.4$ μs. The symmetry of the wave packet about $\mathbf{B_0}$, expected from whistler propagation (i.e., within the resonance cone) as well as source symmetry, is very much in evidence.

For efficient radiation of whistlers, it is generally assumed that the source's dipole moment should be oriented perpendicular to $\mathbf{B_0}$ in order to excite a wave with $\mathbf{k}$ predominantly along $\mathbf{B_0}$. We have found this not to be a necessary condition since the antenna excites equally well when oriented along (Figures 1 and 2) and across $\mathbf{B_0}$ (not shown). The explanation is that it is the plasma response to the perturbation that allows for the excitation of whistler waves. In

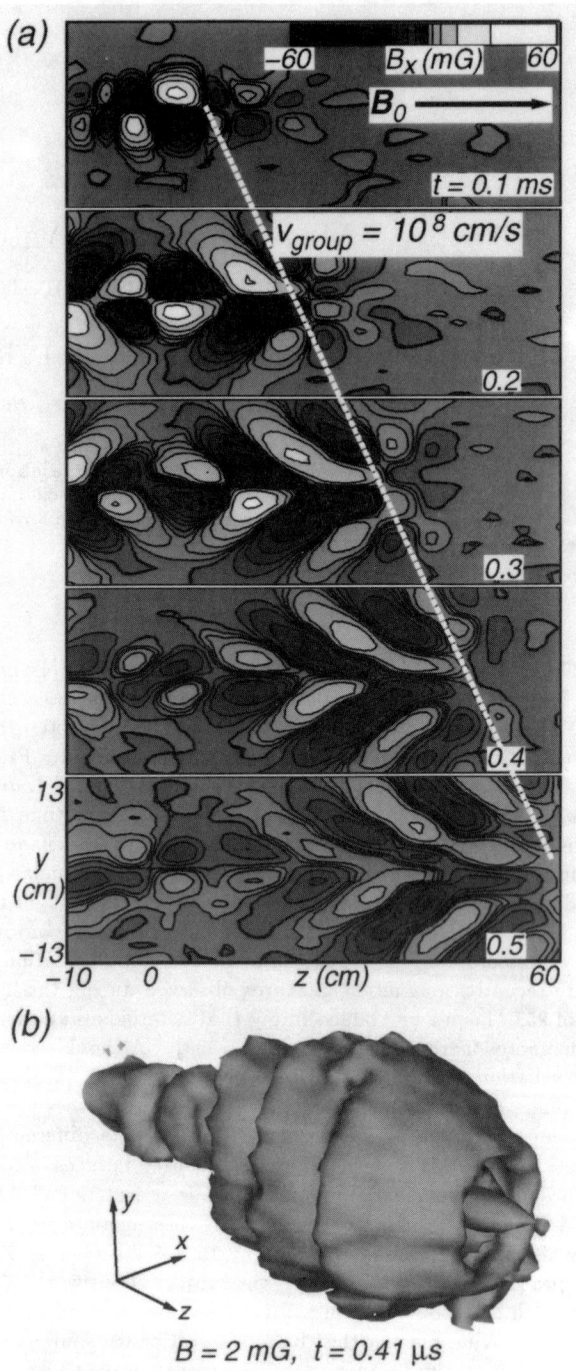

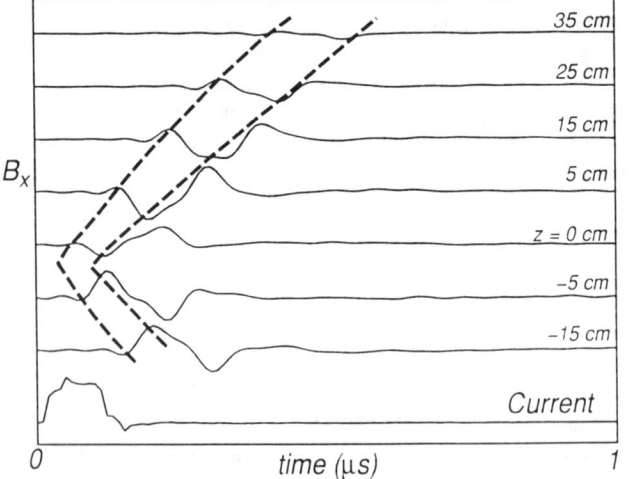

Fig. 1. Temporal evolution of one of the components of the perturbed magnetic field, $B_x \approx B_\theta$, at different axial distances, z, from a pulsed loop antenna with axis along $\mathbf{B_0}$ (bottom trace is the externally measured current, $\Delta t = 0.1$ μs, $I_{max} = 0.1$ A). The traces show that the perturbation propagates dispersively since the packet front moves at a different speed than its middle.

Fig. 2. (a) Topology of $B_x \approx B_\theta$ in the y-z plane ($x \approx 0$) at different times t after applying a 0.1 μs current pulse to an insulated magnetic loop antenna whose normal is parallel to $\mathbf{B_0}$. Image currents and their associated magentic fields are induced which propagate in the whistler mode away from the exciter structure. (b) Constant magnitude surface of B at 2 mG (at $t = 0.4$ μs) displaying the cylindrical symmetry of the wave packet consistent with whistler wave propagation.

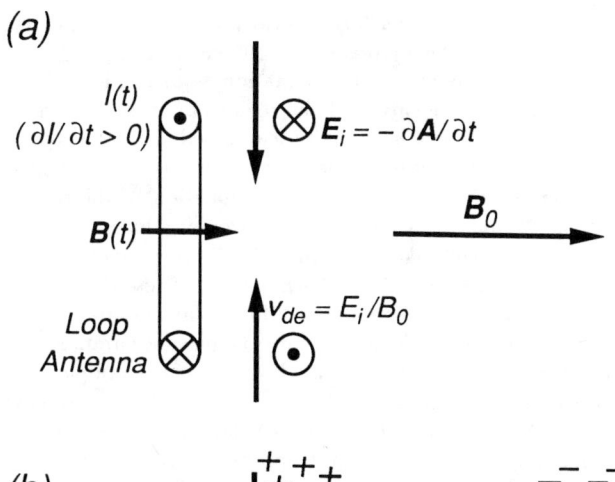

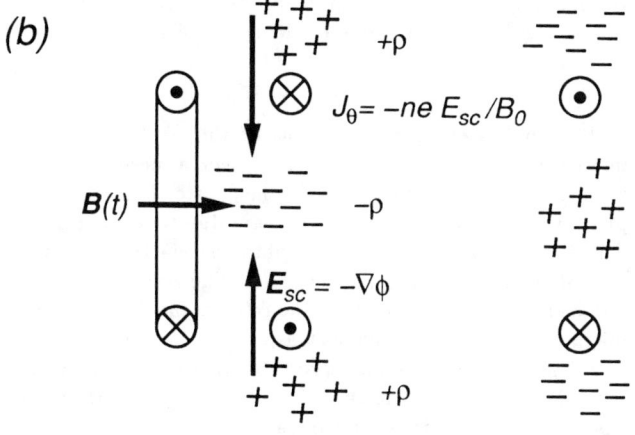

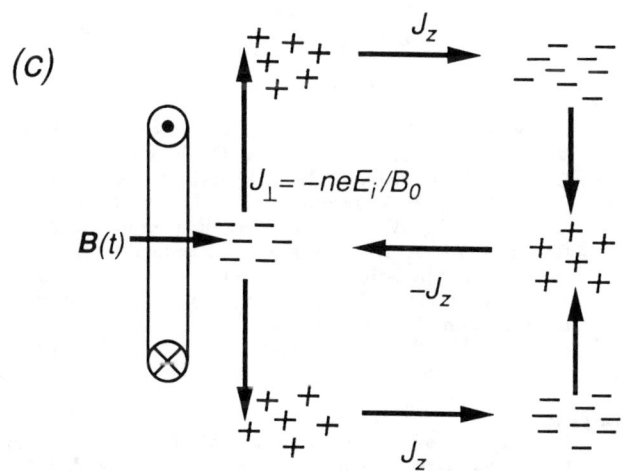

Fig. 3. Model of whistler excitation by a pulsed loop antenna with normal parallel to $\mathbf{B}_0$. (a) The inductive electric field due to $\partial I/\partial t > 0$ gives rise to a radial electron drift which, for unmagnetized ions, results in a space charge separation. (b) The space charge separation produces a radial electric field that drives Hall currents in the $-\theta$ direction opposing the antenna's current. The inductive electric field of the image current then creates an opposing current system. (c) Parallel and transverse current system induced by image currents.

the simplest term, it is Lenz's law. Nevertheless, it is instructive to see how physically this comes about. We offer a model for the mechanism in Figure 3 when the antenna's normal is parallel to $\mathbf{B}_0$. Please note that the model applies to both sides while the figure discusses only one side of the antenna ($z > 0$). As the current begins to flow in the loop, an oppositely directed, inductive electric field, $\mathbf{E}_i$, is imposed on the plasma (Figure 3a). This induces an $\mathbf{E}_i \times \mathbf{B}_0$ drift which produces a radial space charge imbalance. The excess charge is due to a shift of the electron population since the ions move along $\mathbf{E}_i$ during the time scale of the experiment ($r_{ci} \gg r_{ce}$). The space charge electric field, $\mathbf{E}_{sc}$, which has been experimentally confirmed by *Stenzel et al.* [1993], in turn drives the shielding current, $\mathbf{J}_\theta = -ne\mathbf{E}_{sc} \times \mathbf{B}_0/B_0^2$ (Figure 3b). This perturbation produces an opposite system further into the plasma which begets another one and so on (Figure 3b). The charge imbalances also lead to radial and axial currents as depicted in Figure 3c. Thus, a full three-dimensional magnetic field/current system carried by whistler waves is induced in the plasma. It is worth noting that such a model strongly depends on the orientation chosen as well as on the unmagnetized ions. Furthermore, there may be no waves that can carry the signals under other parameter regimes (e.g., $\Delta t < 1/\omega_{ce}$). Under those conditions, other mechanisms must be considered (see Stenzel et al. in this volume).

Because we cannot move the perturbation source or the plasma, we simulate the process by re-constructing their relative motion in space. The process is sketched in Figure 4 and has been discussed elsewhere in greater detail [*Stenzel and Urrutia*, 1990]. At n discrete points along the source trajectory, we assume that a packet was emitted at some time $t = -n\delta t = -nD/v$, where v is the source velocity

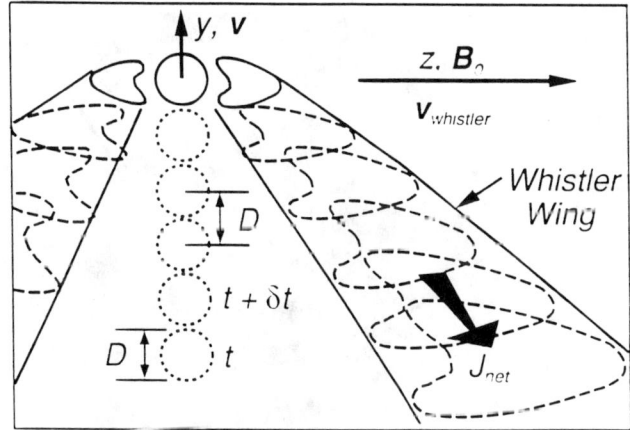

Fig. 4. Schematic diagram for the construction of whistler wings of a dc-magnetic field source moving with velocity $\mathbf{v} \perp \mathbf{B}_0$. A whistler packet is emitted at each position since the source resides in the flux tube of size D only for a short duration $\delta t = D/v$. The superposition of wavelets forms a whistler wing that co-moves with the source. Motion along $\mathbf{B}_0$ follows a similar construction.

and D is the source extent along the direction of motion as well as the distance between discrete points. The data corresponding to that time is put into the volume surrounding this location. Owing to the wave propagation, each point in the volume can have contributions from different trajectory points. All contributions are then summed and the overall pattern is deduced. For motion across $\mathbf{B}_0$, a structure resembling a wing results. Its inclination with respect to $\mathbf{B}_0$ is given by $\theta_{wing} = \tan^{-1}(v/v_{group})$. It is important to note that the structure co-moves with the source which approximates a dc-driven antenna. Thus, the perturbed magnetic field pattern does not change in time relative to the loop and should not be considered a classical radiation pattern. Nevertheless, it is formed because the time rate of change of magnetic flux experienced by the plasma gives rise to the local excitation of whistler waves. This process is similar to the creation of eddy currents in a metal as a magnetic field source moves in its vicinity.

Figures 5a and 5b display the magnitude of the magnetic perturbation structures resulting from motion across and along $\mathbf{B}_0$. For illustration purposes, a lower limit for B has been chosen (50 mG in 5a) and 90 mG in 5b)). The simulated speed of the source is $v = 10^8$ cm/s, comparable to the packet speed; hence, the inclination angle in Figure 5a is close to 45°. The location of the regions of maximum magnetic energy are not necessarily associated with the source location but are a result of constructive interference between the discrete steps employed in the simulation of the source motion. It must be kept in mind, however, that some spatial "aliasing" will take place since the discrete current is not a perfect square pulse. Furthermore, all effects due to the structures must be describable via linear superposition (e.g., the packet front should not modify the plasma parameters). The structures can then be analyzed to, for example, obtain the instantaneous directions of $\mathbf{B}(\mathbf{r})$ and $\mathbf{J}(\mathbf{r})$. A slow ($f^{-1} \gg \delta t = D_{source}/v_{source}$) modulation can also be applied to the antenna current to produce VLF waves. Such topics are, however, outside of the scope of this article.

Conclusions

The spatio-temporal evolution of the field of a pulsed magnetic loop antenna is measured over a large volume in a quiescent magnetoplasma. The fields are then computer-processed and the magnetic structure due to a moving, dc-driven magnetic loop antenna is obtained via the superposition of the single current pulse data. The results should be applicable to situations where the relative motion between a magnetic field source, such as Gaspra [*Kivelson et al., 1993*], and a magnetoplasma may produce whistler waves. Other possible applications are the interaction of spacecraft antennas with the solar wind or auroral arcs.

Acknowledgements. The authors gratefully acknowledge support for this work by grants from NSF PHY, ATM, and NASA.

References

Armand, N. A., Yu. P. Semenov, B Ye. Chertok, V. V. Migulin, V. V. Akindinov, V. I. Aksenov, G. V. Bashilov, P. M. Belousov, V. A. Blinov, E. P. Vyatkin, S. A. Gorbunov, S. M. Yermin, I. V. Lishin, D. S. Lukin, A. V. Moshkov, L. I. Nezhinskiy, V. G. Osipov, V. B. Presnyakov, A. E. Reznikov, Ye. A. Rudenchik, P. P. Savchenko, G. K. Sosulin, S. V. Starostin, N. P. Chubinskiy, A. V. Shabanov, V. A. Shlykov, and I. N. Shugalev, Experimental study of the radiation of a frame antenna, placed on *Mir-Progress-28-Soyuz TM-2* orbiter complex, at very low frequencies in the Earth's ionosphere, *Sov. J. Comm. Tech. and Electron.*, *34*, 50-57, 1989.

Biryukov, A. S., I. S. Veselovsky, O. R. Grigoryan, A. D. Koval, S. N. Kuznetsov, A. P. Kropotkin, M. I. Panasyuk, S. B. Ryabukha, A. A. Us, and V. A. Shuvalov, Investigation of magnetospheric processes with the use of a source of strong magnetic field in the ionosphere, *Adv. Space Res.*, *12*, 135-141, 1992.

Egeland, A., Kristian Birkeland, the man and the scientist, in *Magnetospheric Currents*, edited by T. A. Potemra, pp. 1-16, American Geophysical Union, Washington, DC, 1984.

Helliwell, R. A., *Whistlers and Related Ionospheric Phenomena*, Stanford University Press, Stanford, CA, 1965.

Kivelson, M. G., L. F. Bargatze, K. K. Khurana, D. J. Southwood, R. J. Walker, and P. J. Coleman, Jr., Magnetic sig-

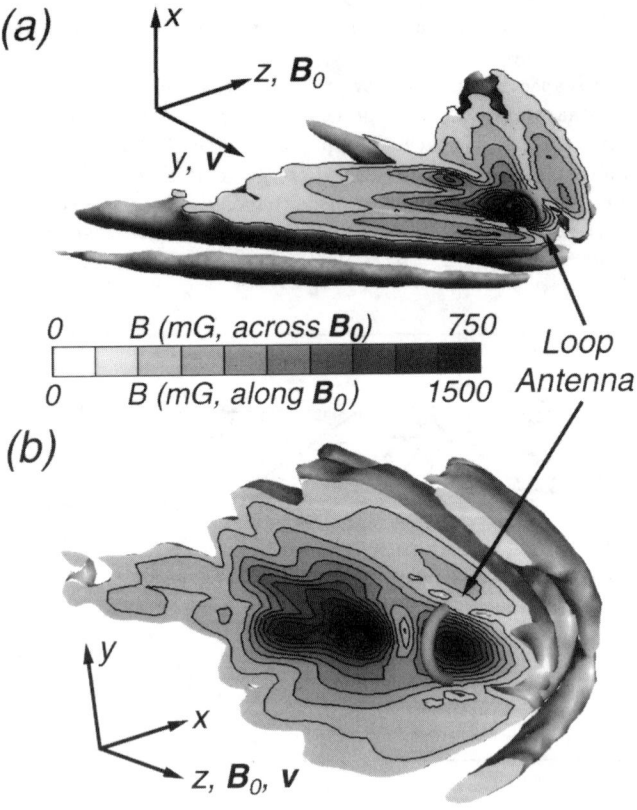

Fig. 5. Magnetic field structure of dc-driven sources moving (a) across and (b) along $\mathbf{B}_0$. The $B = constant$ surfaces and contours are constructed following the scheme presented in Figure 4. Simulated speeds are comparable to the wave packet overall group velocity parallel to $\mathbf{B}_0$.

natures near Galileo's closest approach to Gaspra, *Science*, *261*, 331-334, 1993.

Lane, B. G., R. S. Post, and J. D. Sullivan, Physics of manmade extended magnetic structures in low Earth orbit, *EOS Transactions*, *68*, 398, 1987.

Rousculp, C. L., J. M. Urrutia, and R. L. Stenzel, Multidimensional Fourier analysis of a whistler pulse excited by a loop antenna, this volume.

Stenzel, R. L., and J. M. Urrutia, Currents between tethered electrodes in a magnetized laboratory plasma, *J. Geophys. Res.*, *95*, 6209-6226, 1990.

Stenzel, R. L., J. M. Urrutia, and C. L. Rousculp, Pulsed currents carried by whistlers. Part I: Excitation by magnetic antennas, *Phys. Fluids, B*, *5*, 325-338, 1993.

Stenzel, R. L., G. Yu. Golubyatnikov, and J. M. Urrutia, Thermal magnetic fluctuations in Maxwellian and non-Maxwellian plasmas at whistler and electron cyclotron harmonic frequencies, this volume.

Triska, P., and V. I. Shevchenko, "Active" project satellites, Abstract S 12.1.1, XXVII COSPAR, The Hague, The Netherlands, p. 78, 1990.

Urrutia, J. M., and R. L. Stenzel, Transport of current by whistler waves, *Phys. Rev. Lett.*, *62*, 272-275, 1989.

J. M. Urrutia, C. L. Rousculp, and R. L. Stenzel, Department of Physics, University of California, Los Angeles, CA 90024-1547.

Interaction Between Global MHD and Kinetic Processes in the Magnetotail

G. GANGULI, H. ROMERO[1] AND J. FEDDER

Space Plasma Branch, Plasma Physics Division, Naval Research Laboratory, Washington, DC 20375-5000

Progress in global MHD modelling of the magnetosphere for the substorm growth phase indicates a dramatic thinning and possible compression of the plasmasheet over an extended region of the near earth tail. Since the MHD approximation becomes invalid, and finite Larmor radius corrections are little better, for length scales which approach a few ion gyroradii; it is important to study details of both the localized equilibrium conditions and possible instabilities for a highly stressed magnetotail using a kinetic approach. We have recently investigated the structure of a stressed PSBL (which is intimately related to the stressed magnetotail) using a Vlasov formalism, and its nonlinear dynamics using a 2 1/2 D PIC code. In this situation, we find a hierarchy of plasma oscillations, having both small and large temporal and spatial scales, which mediate the dynamical equilibrium. Depending on the degree of stress, i.e., the thinness of the layer, one or more of these modes can be triggered leading to relaxation. From our preliminary studies we find that the strong inhomogeneities in a stressed PSBL can affect its nonlinear evolution significantly. In particular, it appears that the width of the PSBL is determined by a balance between Poynting flux transported by global convection and the dissipation in the layer. Moreover, the micro instabilities can lead to plasma energization through both anomalous viscosity and resistivity. These types of plasma micro processes may play an important role maintaining dynamical equilibrium during the substorm growth phase, as well as providing plasma thermalizatiom during the substorm expansion.

INTRODUCTION

The structure of the Earth's magnetotail is strongly dependent on the solar wind parameters. This is particularly true during the magnetospheric substorm process. When the coupling of solar wind energy and momentum to the magnetosphere is strong, during periods of strongly southward interplanetary magnetic field, the quantity of magnetic flux and the field strength in the tail lobes increases (see for instance Stern [1991], Lui [1991], and references therein). As the tail lobes grow, increasing stress is transmitted to the near Earth plasmasheet with the dawn to dusk currents increasing in order to maintain momentum and energy equilibrium. Recent results from global MHD simulations are contributing to this picture [*Fedder et al.*, 1992; Lyon and Fedder, manuscript in preparation]. During this magnetotail growth phase, the increase in the lobe magnetic field and the convection of the lobe plasma towards the tail midplane compress the plasmasheet causing the plasmasheet plasma to drift tailward. Since the lobe convection electric potential field is directed from dawn to dusk, but the plasmasheet induction field, caused by the increasing lobe field strength and the tailward plasma drift, is dusk to dawn; a substantial increase in the magnetic shear and the resultant current density takes place at the boundary between the plasmasheet and the tail lobes. This profound steepening of the plasmasheet boundary layer (PSBL) during magnetotail growth must be limited by a local dissipation process. In the case of MHD numerical simulations, the dissipation is provided by the discrete cell size associated with the numerical algorithm. In the magnetosphere, the dissipation is expected to be provided by kinetic plasma processes, and in particular by processes which occur at cyclotron radius scale sizes. In this paper we provide a first assessment of the micro processes which may be important in a highly steepened PSBL. These processes provide electromagnetic dissipation, thermalization, and energization of the plasma; and thereby contribute to the maintainance of the global equilibrium during the magnetotail growth phase. They may also play an important role in the loss of equilibrium and plasma energization at substorm expansion onset.

The dynamical state of the plasma sheet boundary layer (PSBL) is closely related to that of the magnetotail and its degree of coupling with the solar wind. For instance, when the magnetotail is highly stressed, the plasma sheet and its boundary layer become narrow [*Mitchell et al.*, 1990] and the micro- processes discussed here would be most active. During these times particles originating from different regions of

Solar System Plasmas in Space and Time
Geophysical Monograph 84
This paper is not subject to U.S. copyright. Published in 1994 by the American Geophysical Union.

the magnetosphere are vigorously mixed and energized by the dissipative processes taking place within the PSBL. This in turn gives rise to unique distributions which can be accelerated at lower altitudes in the auroral zone and provide the spectacular auroral displays. During highly stressed conditions the entire plasma sheet thins to about a few ion gyroradii and in such situations the difference between the PSBL and the plasma sheet is not clear. As the plasma sheet thins beyond a certain critical width, it is observed [*Mitchell, et al.*, 1990] that there is a large enhancement of the cross-field ion current in the dawn-dusk direction which is closely related to the expansion onset. Following the substorm there is a relaxation. As the magnetosphere relaxes, leading to thicker boundaries, the PSBL will likely become less active. The boundary plasma distributions and the plasma wave activity can therefore be considered as a diagnostic of the global magnetospheric conditions. Thus, it is important to understand the dynamics of the PSBL and how it is affected by global changes in the magnetosphere. Specifically, it is desirable to quantify such observables as composition, structure, width of the PSBL, and its activity level in relation to its dynamical state. In doing so, the subject of wave generation and the nonlinear evolution of strongly inhomogeneous boundary layers will have to be addressed. With the availability of high-time-resolution detectors in the upcoming Cluster and Geotail missions, small scale-sizes will be resolved more accurately and hence microscopic analysis of stress relief mechanisms is quite topical.

Sufficient observational evidence has now been gathered to warrant a comprehensive and systematic modelling effort of the PSBL and a stressed magnetotail. To date, most previous studies of this region have focused on the influence of counterstreaming uniform beams on the generation of broadband electrostatic noise (BEN). Virtually no attention has been given to the study of the origin and dynamical evolution of the strong inhomogeneities inherent to this region. Also, while effects due to parallel beams have been extensively studied, no attention has been given to the origin and the role played by the observed shear in the beam velocities which occur throughout the PSBL [*Takahashi and Hones*, 1988; *Lui*, 1987]. We find that inhomogeneities can profoundly affect the nonlinear evolution of a stressed layer. In particular, it appears that the PSBL width is determined from a balance between the stress generated by the solar wind-magnetosphere coupling and that dissipated in the PSBL. During active periods the rate of solar wind energy input can exceed the rate of dissipation in the PSBL. This leads to narrower PSBL widths and ultimately to plasma sheet thinning [*Mitchell et al.*, 1990] with their own unique properties and signatures, while the quiet time broad PSBL follows a different evolution scheme and is typified by a different signature.

During active periods when the magnetosphere is stressed, the PSBL may contain structures of small scale sizes of a few hundred kms (approximately, 1 to 2 ρ_i, the ion gyroradius) but during moderate to low activity periods, the scale sizes may be a few thousand kms (around 6 to 10 ρ_i) [*Dandouras et al.*, 1986, *Parks et al.*, 1992]. The highly active nature of boundary layers offers considerable challenge as far as modelling this region is concerned. Major outstanding issues concerning this region are: (1) the origin and nature of the north-south component of d.c. electric fields and inhomogeneous flows [*Cattell et al.*, 1982; *Orsini et al.*, 1984; *Takahashi and Hones*, 1988], (2) the origin, nature and evolution of stressed ($L \leq 2\rho_i$) boundary layers [*Parks et al.*, 1992], (3) the condition for the rapid enhancement of the dawn-dusk ion current prior to the expansion phase, (4) the nature of dissipation and transport processes in the PSBL and their influence on its structure (i.e., thickness), and (5) the nature and origin of the broadband electrostatic noise and its relationship to the dynamical evolution of the PSBL. In the following we describe our preliminary results addressing some of these issues.

CURRENT MODELS OF MAGNETOSPHERIC BOUNDARY LAYERS

While some attention has been devoted to developing models of the dayside boundary layer (i.e., the magnetopause) [*Okuda*, 1991; *Cargill and Eastman*, 1991; *Winske et al.* [1990], *Lee* [1990] and references therein], comparatively little work is available concerning models of the nightside boundary layer (i.e., the PSBL) which is intimately related to the plasma sheet conditions and hence relevant to substorm events. Nonetheless, a great number of studies concerning wave generation in the PSBL region are available [e.g., Ashour-Abdalla and Schriver, 1989, and references therein]. Recent works of Romero et al. [1990] and Onsager et al., [1990] are the first attempts to systematically model the inhomogeneous equilibrium properties of the PSBL. The Onsager et al., model uses time of flight concepts to obtain the plasma density variation in the PSBL. It treats the PSBL as a time stationary region while observations indicate that it is a very dynamic structure. Also, Ashour-Abdalla et al., [1991], have developed a model, based on non-self-consistent single particle orbits, which exhibits the existence of field-aligned beams in the PSBL. Test particle models preclude plasma effects and hence their validity is restricted primarily to non-dynamic cases where collective effects are non-existent. Such models cannot, therefore, be applied under stressed conditions where plasma collective effects are likely to play a crucial role. We have taken the initial steps to understand the structure and stability of the PSBL and their relationship to the global plasma conditions [*Romero et al.*, 1990], a nonlocal theory to study its linear stability properties [*Romero et al.*, 1992a], and have developed a 2-1/2 dimension Particle In Cell (PIC) code to study its nonlinear evolution, dissipation and transport properties [*Romero et al.*, 1992b; *Romero and Ganguli*, 1993]. In the following we provide a description of our model and demonstrate its relation to global MHD models.

The PSBL Model

Observational Evidence of Strong Inhomogeneities

To illustrate the nature of strong nonuniformities present in the PSBL, we note here the existence of thin electron layers on its outer edge. These layers were first identified in the ISEE particle data by Parks et al. [1979], and were later shown to be a persistent feature of the PSBL by Parks et al. [1984; 1992] and by Takahashi and Hones [1988]. Observations have shown that the width of these layers is typically of the order of 100 km and at times may even be smaller [G. Parks, private communication, 1992]. Fig. 1 (courtesy of G. Parks) illustrates these features by showing 1/4 second averages of the electron and ion fluxes from the two lowest energy channels (2 keV and 6 keV, respectively) of the fixed energy detector on ISEE 1 [Anderson et al., 1978]. The particles measured by this instrument have pitch angles in the range 70-90 degrees. It is seen that electron fluxes in both energy channels increase substantially at 23:17 UT. This is followed about 30 seconds later by an ion flux increase. The width of the electron layer can be estimated to be $\sim$ 100 km but smaller distances cannot be ruled out because the motion of the PSBL is difficult to determine accurately. Since the ion gyroradius (ρ_i) for keV range protons in a 40 nT magnetic field is around 125 km, Fig. 1 reveals that significant variations in plasma parameters may occur at the PSBL-lobe interface on a scale length smaller than ρ_i. At

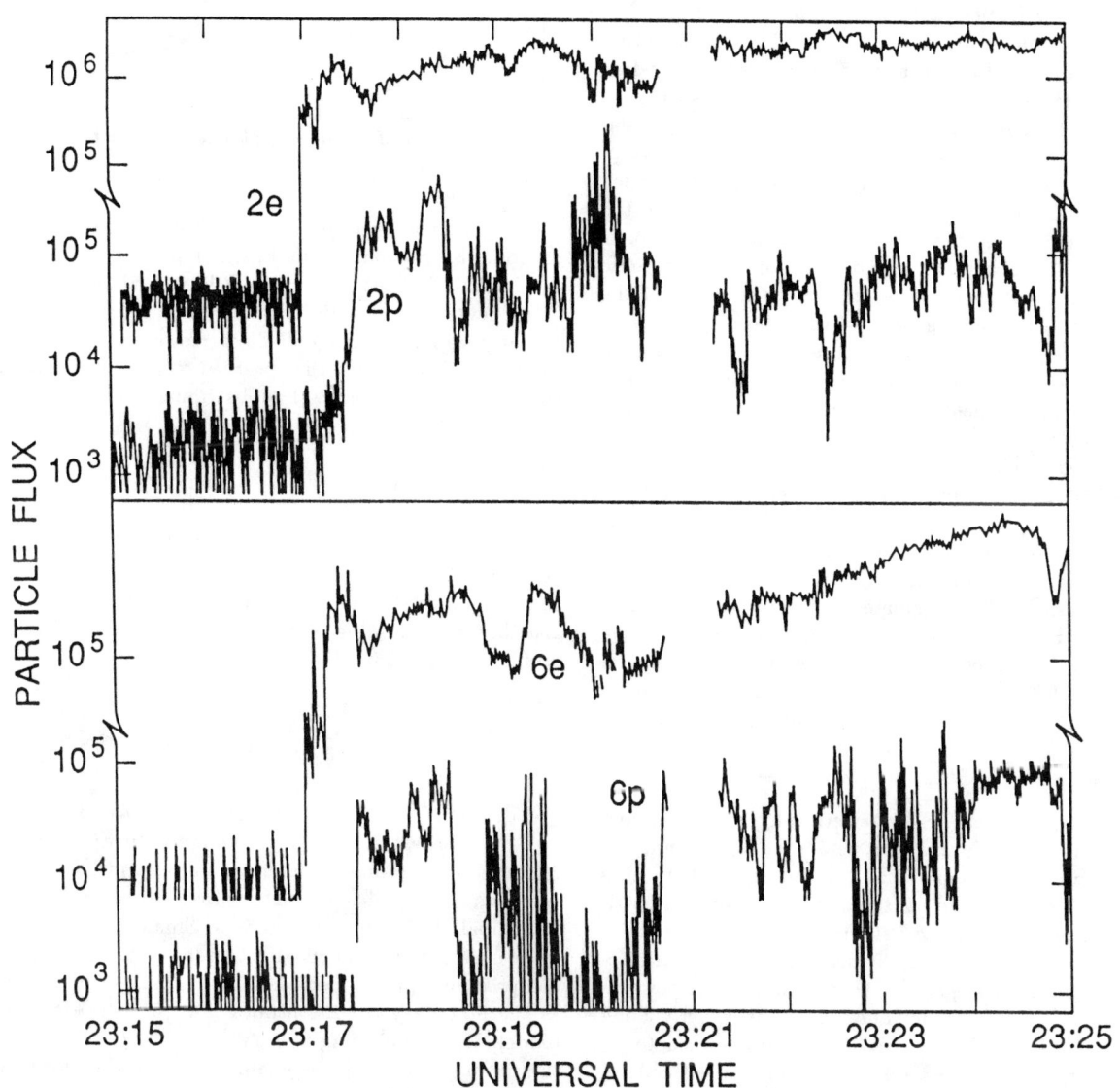

Fig. 1. Ion (p) and electron (e) flux data from ISEE 1 (March 31, 1979) versus UT for two energy channels (2 keV and 6 keV) [courtesy, G. Parks].

least 30 more cases with similar features have been recently identified and are currently under investigation [G. Parks, private communication, 1992]. The origin of these layers, their contribution to PSBL dynamics, and their relaxation mechanism(s) may have a direct bearing on the substorm onset but yet have not been explored. These aspects and the tendency of these electron layers to be more prominent during strong geomagnetic activity are important issues to be addressed in this article.

In order to place our studies of the micro-processes in context, we postulate a one dimensional phenomenological model for the plasma density, magnetic field, and current density in the near Earth plasmasheet. A cartoon of this model is sketched in Fig. 2, which incorporates many of the observed features [Mitchell et al., 1990; Lui, 1991] and some of the new numerical simulation features of the plasmasheet during the magnetotail growth phase. Momentum balance is maintained between the plasma sheet energy density and the tail lobe magnetic energy density. As the lobe field strength is increased, the increasing plasma energy density in the sheet causes the sheet plasma to drift tailward, and induces a dusk to dawn electric field in the sheet which is consistent with the field strength increase and opposes the convection field in the tail. As a consequence, steep gradients in both the plasma energy density and the magnetic field strength tend to be concentrated at the edges of the plasma sheet. Continued growth of the lobe magnetic flux compresses the plasmasheet and causes the steep magnetic gradients to converge toward each other. We postulate that the steep plasma and field gradients are the primary location of kinetic plasma dissipation processes and the gradients are themselves limited to a few ion cyclotron radii. When the steep gradients finally approach each other, any attempt to continue the compression will lead to a burst of kinetic dissipation, a loss of macro scale equilibrium, possibly a burst of magnetic reconnection, and the consequent substorm explosive expansion. This phenomenological model motivates our further studies of kinetic processes.

In the following sections (i) we develop a Vlasov model for the PSBL-lobe system, (ii) investigate its stability properties, and (iii) follow its nonlinear evolution.

Vlasov Formalism for Stressed Plasma Sheet

We assume an ideal PSBL (electron-hydrogen plasma, one temperature, no beams) but retain its highly inhomogeneous nature. Fig. 3 gives the equilibrium configuration under consideration. We construct an equilibrium distribution function using the constants of motion [Romero et al., 1990], which are: (1) $E_j = m_j v^2/2 + q_j \phi(x)$, the total energy, (2) $X_g = x + v_y/\Omega_j$, the guiding center position where Ω_j is the gyrofrequency and (3) v_z, the velocity along the magnetic field. The magnetic field is assumed to be nearly uniform and in the z (earthward, or x in a GSM coordinate system) direction and the strong inhomogeneities are assumed to be in the x (north-south, or z in a GSM coordinate system) direction. Unlike the magnetopause where

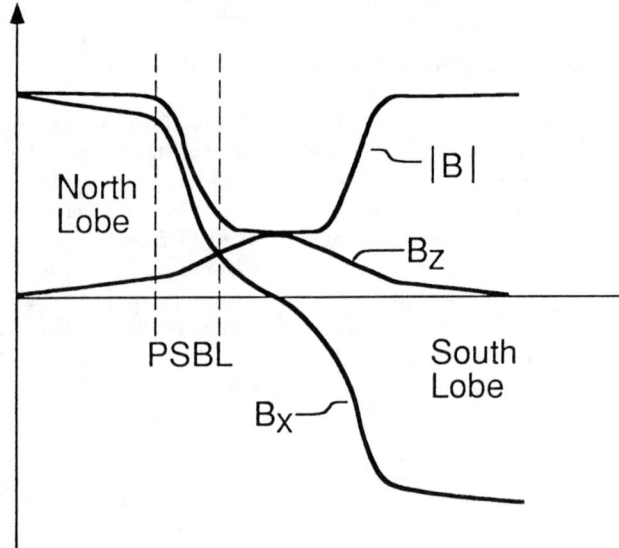

Fig. 2. A sketch of steepening plasma sheet (not necessarily to scale).

the magnetic field variation is very strong and can change direction, observations in the PSBL [Parks et al., 1992; Frank et al., 1981] indicate that $\Delta B/B$ here is about 0.5 and the fields never change sign. Hence, for simplicity, we assume a nearly uniform average B field. However, this model can be extended to include the observed magnetic field variation without much difficulty. The distribution function is given by,

$$f_0(E_j, X_g, v_z) = \left(\pi v_j^2\right)^{-3/2} N_j Q_j(\zeta_j) \exp\left(-\frac{E_j}{T_j}\right), \quad (1)$$

where $\zeta_j = C_j P_j - D_j v_z$, $P_j = \Omega_j X_g$, v_j is the thermal velocity, C_j and D_j are constants, N_j is a normalization constant, and 'j' denotes the species. The function Q_j is the distribution of guiding centers which determines the density profile and is defined by,

$$Q_j(\zeta) = \begin{cases} R_j, & \zeta < \zeta_1 \\ R_j + (S_j - R_j)\left(\frac{\zeta - \zeta_1}{\zeta_2 - \zeta_1}\right), & \zeta_1 < \zeta < \zeta_2 \\ S_j, & \zeta > \zeta_2 \end{cases} \quad (2)$$

We define $N_j R_j = n_{sj}$ the density in the PSBL, while $N_j S_j = n_{Lj}$ is the density in the lobe. In Eq. (2), ζ_1 and ζ_2 are input parameters determining the width of the structure and can be obtained either from observations or from the global MHD simulation model. Smaller Δ_ζ ($= |\zeta_1 - \zeta_2|$) implies stronger gradient and therefore represents stronger compression (active periods). Hence, Δ_ζ can be viewed as the boundary condition obtained from the global MHD model. In the following we shall show that Δ_ζ is crucial to the nature of the micro-dynamics. Thus, the global MHD process strongly influences the micro- processes that ensues in a stressed layer. Other inputs are the magnetic field strength B_0 and the ion and electron temperatures. In terms

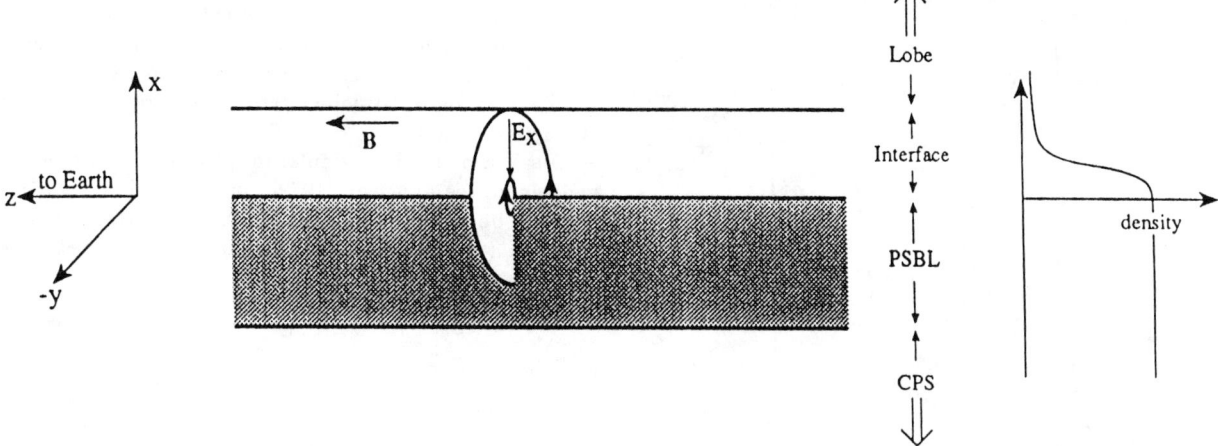

Fig. 3. A sketch of the equilibrium configuration at the PSBL.

of these input quantities f_0 can be integrated over velocity space to obtain the density profiles across the boundary layer. Imposing quasi-neutrality, we solve for the electrostatic potential $\phi(x)$ self-consistently. With $\phi(x)$ determined, f_0 is fully specified and the flows (therefore, currents) and pressures can be readily calculated by taking higher moments of f_0.

Here we assume that the variations in the PSBL region are much stronger across (in the north-south direction) than along (earthward direction) the ambient magnetic field. To study the cross-field variations and its consequences we set $C_j = 1$ and $D_j = 0$.

Using numbers pertinent to Fig. 1 (described in more detail by Romero et al. [1990]) the model predicts the existence of a d.c. electric field in the north-south direction. A rough average value of this d.c. electric field is estimated to be around 5 - 10 mV/m which compares favourably with observed north-south electric fields [*Cattell et al.*, 1982 and *Orsini et al.*, 1984]. Fig. 4a is the density profile and Fig. 4b is the corresponding self-consistent electrostatic potential at the PSBL-lobe interface.

We also find highly localized ion and electron flows and current perpendicular to the magnetic field. In Fig. 5 we show the equilibrium transverse ion and electron flows nor-

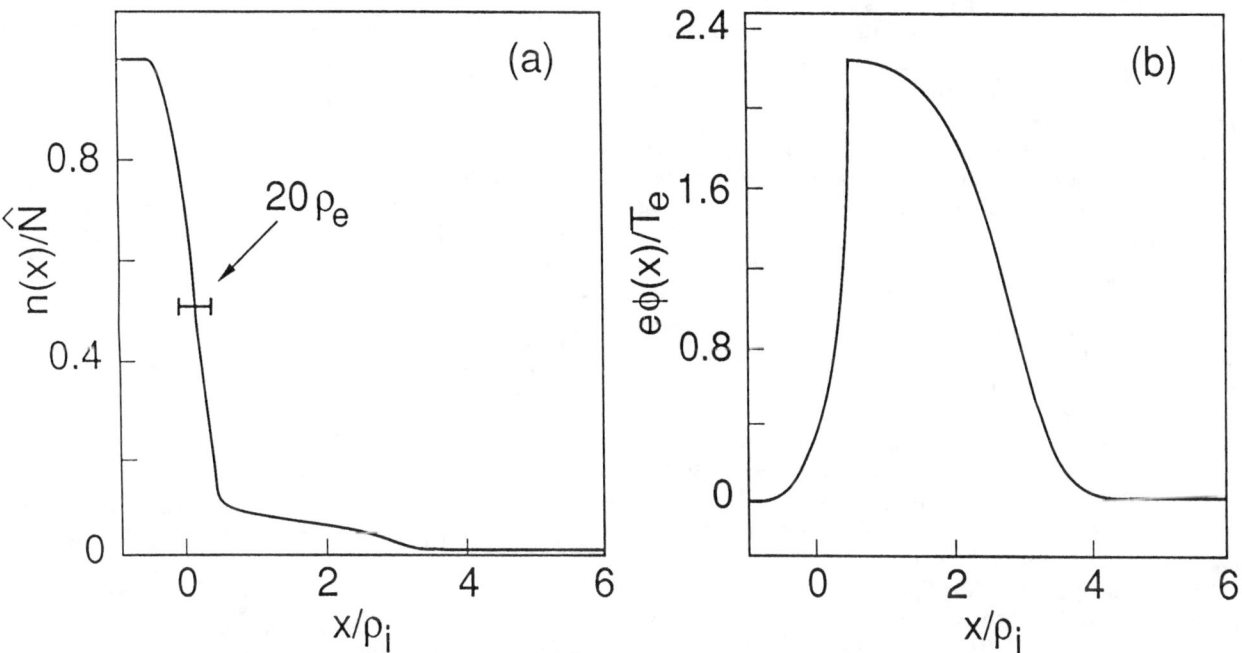

Fig. 4. Self-consistent density (a) and the electrostatic potential (b) at the boundary layer. Note $x < 0$ represents the PSBL and $x > 0$ represents the lobe.

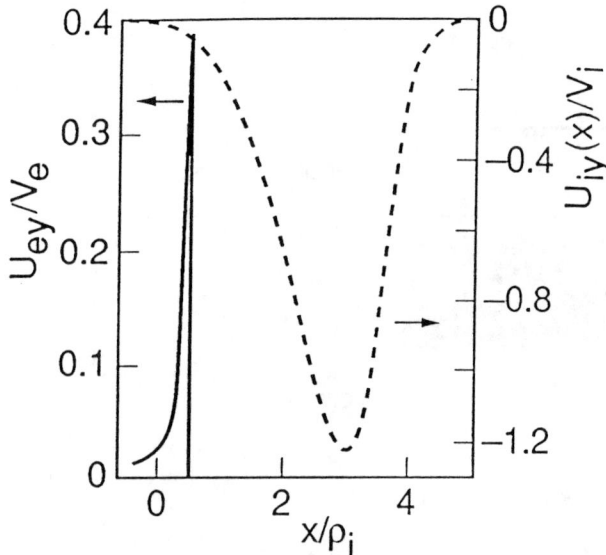

Fig. 5. Self-consistent ion (dotted) and electron (solid) flows normalized by their respective thermal velocities as a function of position.

stronger flow with larger shear in it. Hence, the thin energetic electron current layers observed by Parks et al. [1984; 1992] and Takahashi and Hones [1988] at the outer edge of the PSBL may be a consequence of the density (or pressure) gradient at the PSBL-lobe interface and are intensified during active periods.

Our Vlasov model is similar in spirit to the magnetopause models of Lee and Kan [1979], and Whipple et al., [1984]. It must be pointed out that the choice of the distribution function (1) is not unique. Nonuniqueness arises because

malized by their respective thermal velocities. Similar flows parallel to the magnetic field are also predicted [Romero et al., 1990]. Smaller $\Delta\zeta$ leads to larger flow magnitudes.

In Fig. 6 we plot the gradient in these flows normalized by the species gyrofrequency (i.e., $(dV/dx)/\Omega$). The peak value of this dimensionless number is defined as the shear parameter $\alpha_j = \omega_{sj}/\Omega_j$, where $\omega_{sj} = V_j^0/L_j$ is the shear frequency, V_j^0 is the peak flow, L_j is the scale-size of this flow, and j is the species. It is found that smaller $\Delta\zeta$ leads to larger α_j. For the numbers chosen here the electrons shear parameter (α_e) is less than unity. This implies that the electron gyrofrequency dominates its shear frequency and hence the electrons are magnetized. On the other hand the ion shear frequency is larger than its gyrofrequency, implying that the ions are essentially unmagnetized. During quiet (relaxed) times the shear frequency ω_s is smaller than Ω_i and Ω_e and consequently both ions and the electrons are magnetized. As stress builds up, the magnitude of ω_s increases and first exceeds Ω_i. At this stage the ions become demagnetized. For $\omega_s < \Omega_e$, the electrons remain magnetized and are tied to the magnetic field lines. Consequently, the ions can respond to the large-scale convection electric field in the dawn-dusk direction and constitute an ion current in this direction. This ion current has been observed prior to the onset of the expansion phase [Mitchell et al., 1990]. Thus, the condition for the rapid enhancement of the dawn-dusk ion current is $\omega_s > \Omega_i$.

Thus, the magnitude of the flows and their spatial gradients (i.e., α_j) are found to be closely related to the gradient in the density at the PSBL-lobe interface which in turn is directly related to the magnetotail compression by the solar wind. A smaller density gradient scale-size leads to a

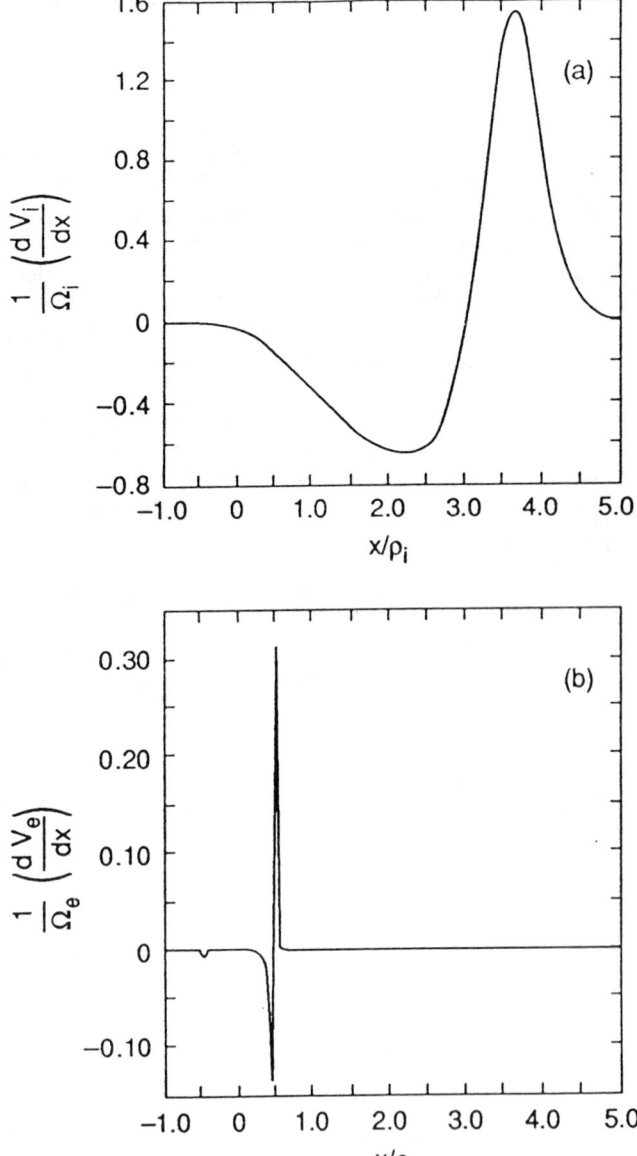

Fig. 6. The spatial gradients of the ion (a) and electron (b) flows as a function of position. Note the peak value of the ion flow gradient exceeds its gyrofrequency while the corresponding parameter for the electrons does not.

the choice of the function Q_j is arbitrary, and is chosen primarily for convenience. We note however, that many of the qualitative features of our Vlasov model can also be derived from fluid theory where this ambiguity is not a problem. For example, consider a stressed PSBL where the ions are unmagnetized. Then the ion pressure gradient is balanced by a d.c. electric field, E, so that the ion momentum balance equation is given by,

$$eE - \frac{T_i}{n}\left(\frac{dn}{dx}\right) = 0. \quad (3)$$

Electrons are magnetized and hence the electron momentum balance equation is

$$eE - \frac{e\mathbf{V} \times \mathbf{B}}{c} - \frac{T_e}{n}\frac{dn}{dx} = 0, \quad (4)$$

leading to

$$V_{\perp e} = -\frac{cE}{B} - \frac{cT_e}{eB}\left(\frac{1}{n}\frac{dn}{dx}\right) = -\frac{cE}{B}\left(1 + \frac{T_e}{T_i}\right). \quad (5)$$

Thus, we see that the electrostatic potential as obtained from the Vlasov model (Fig. 4b) corresponds to the electric field given by Eq. (3), and the electron cross-field flow and the shear in it (corresponding to Figs. 5 and 6) are given by Eq. (5). We shall show in the following that the wave properties, and hence the relaxation mechanism of a stressed layer, are influenced and controlled by these quantities.

Stress Relief Mechanism (Linear)

We now examine the stability of the equilibrium described in the previous section. One of the important results emerging from our Vlasov model is the self-consistent generation of sheared flows both along and across the magnetic field and their direct relation to the stress build-up in the magnetotail. This velocity shear is capable of inducing turbulence in a broad frequency band which leads to the depletion of the velocity shear and hence to stress relief. From the linear stability analysis of sheared flows we find that the shear parameter α_j determines the character of the waves.

The stability of sheared flow layers, perpendicular to a magnetic field, has been extensively discussed in the published literature and need not be repeated here. In general, we find that for $\alpha_i < 1$ and $\alpha_e < 1$, both electrons and ions are magnetized. This is the likely scenario in a less stressed PSBL, whose width could be $(6-10)\rho_i$. The resulting shear driven waves are of low frequency ($\omega \sim \Omega_i$ or less) [*Ganguli et al.*, 1985a,b; 1988a; 1989a; 1991; *Nishikawa et al.*, 1988]. As discussed in Ganguli et al. [1988a, 1991], these modes, along with the well known Kelvin-Helmholtz (KH) modes, form the two branches of a magnetized plasma with cross-field sheared flow. Since the $\omega \sim \Omega_i$ modes are generated by an inhomogeneity in the energy density introduced by the localized nature of the cross-field flow, they have been referred to as the Inhomogeneous Energy Density Driven Instability (IEDDI). While the KH branch is characterized by low frequency ($\omega \ll \Omega_i$) and long wavelengths ($k_\perp \rho_i \ll 1, k_\parallel \sim 0$), the IEDDI branch is characterized by higher frequency ($\omega \sim \Omega_i$) and shorter wavelengths ($k_\perp \rho_i \sim 1, k_\parallel$ can be non zero and larger). The strong density gradients, such as those found in the PSBL-lobe interface and at the magnetopause, have a damping effect on the KH branch [*Satyanarayana et al.*, 1987], while they can enhance the IEDDI branch. PIC simulations of Nishikawa et al. [1988; 1990] have shown that the IEDDI is capable of contributing significantly to particle diffusion and viscosity.

For $\alpha_i > 1$ and $\alpha_e < 1$, electrons are magnetized but the ions become unmagnetized. This is the likely case in a stressed PSBL during highly active periods, whose widths could be $(1-2)\rho_i$. The resulting shear driven waves are in the intermediate frequency range ($\omega \sim \omega_{LH}$, the lower hybrid frequency to Ω_e) [*Ganguli et al.*, 1988b; 1989c; *Romero et al.*, 1992a,b; *Romero and Ganguli*, 1993]. These modes have been referred to as the Electron-Ion-Hybrid (EIH) modes. In Section 3.1.4 we describe the nonlinear evolution of this mode and its relevance to stressed boundary layers. We have recently examined the effects of the observed $\Delta B/B (\sim 0.5)$ in the PSBL on the EIH modes. It is found that the growth rate is enhanced by a factor of 2 without major qualitative changes. Thus, the observed magnitude of ΔB is not expected to affect the dynamics in a significant way.

For $\alpha_i > 1$ and $\alpha_e > 1$, both ions and electrons are unmagnetized and the resulting waves are of high frequency ($\omega \sim \omega_{pe}$, the plasma frequency) [*Mikhailovskii*, 1974; *Romero et al.*, 1992a].

A similar hierarchy of parallel flow shear driven modes also exist and are relevant to the PSBL since field-aligned beams [*Eastman et al.*, 1984; *Takahashi and Hones*, 1988] and currents [*Frank et al.*, 1981, *Frank* 1985] are important constituents of the PSBL. Accordingly, there are a great number of theoretical studies in the published literature concerning the generation of BEN by PSBL beams [for example see Ashour-Abdalla and Schriver, 1989, and references therein]. These studies neglected the effects of transverse shear in the beam and current velocities which may be a drastic simplification for a stressed PSBL.

Velocity shear can significantly affect the properties of parallel flow and current driven waves. The threshold for some modes, for example the low frequency D'Angelo waves [*D'Angelo*, 1965], which are driven by shear in the parallel ion flow, is significantly increased by transverse velocity shears [*Ganguli et al.*, 1989b]; while the threshold for the current driven ion-cyclotron instability (CDICI) is lowered [*Ganguli et al.*, 1989a; 1991; *Nishikawa et al.*, 1990]. This prediction has recently been verified in a series of well diagnosed laboratory experiments [*Koepke et al.*, 1992; *Koepke and Amatucci*, 1992]. In view of this, it is important to ask whether and how the observed shear in beam velocities in the PSBL will alter the conclusions of previous investigations and affect the PSBL's nonlinear state.

Our preliminary investigation of the effect of velocity shear on parallel current driven modes is encouraging. For example, significant changes take place in the well known dispersion relation of the Current Driven Ion-Cyclotron Instability [*Drummond and Rosenbluth*, 1962; *Kindel and Kennel*, 1971] when shear in the field-aligned current (FAC) is included. By including shear in V_d, the field-aligned electron drift, the dispersion relation for the CDICI in the local limit becomes [*Ganguli et al*, 1991],

$$1 + \sum_n \Gamma_n(b) \left(\frac{\omega}{\sqrt{2}|k_z|v_i}\right) Z\left(\frac{\omega - n\Omega}{\sqrt{2}|k_z|v_i}\right)$$

$$+ \tau'\left(1 + \left(\frac{\omega - k_z V_d}{\sqrt{2}|k_z|v_e}\right)\left(\frac{\omega - k_z V_d}{\sqrt{2}|k_z|v_e}\right)\right) = 0. \quad (6)$$

where v_i and v_e are ion and electron thermal velocities, $\tau' = \tau(1 - V_d'/u\Omega_e)$, $\tau = T_i/T_e$, $u = k_z/k_\perp$, $b = k_\perp^2 \rho_i^2$ and $\Gamma_n(b) = \exp(-b)I_n(b)$ where $I_n(b)$ are the modified Bessel function. Thus, if shear in the parallel flow $V_d' = dV_d/dx = 0$, then $\tau' = \tau$, and Eq. (6) reduces to the dispersion relation for the CDICI, otherwise shear affects the system by reducing the temperature ratio and thereby reducing the threshold for this instability. The effective temperature ratio τ', can even become negative. This is an interesting domain never explored before. For $\tau = 10$, Kindel and Kennel predict a threshold velocity at $V_d \sim 50v_i$, where v_i is the ion thermal velocity. However, for the same τ we find that there is a strongly growing ion cyclotron root ($\omega = (1.03 + 0.60i)\Omega_i$) for $V_d = 15v_i$, where $k_y\rho_i = 44.7$, $u = 0.0015$, and $\mu = 1837$, even if we assume a small $V_d' = 0.0019\Omega_e$. Thus, shear in the parallel drift velocity can substantially reduce the threshold drift for the CDICI.

When ions are unmagnetized but electrons are magnetized, FAC will give rise to higher frequency ($\omega_r \sim (\omega_{LH} \to \Omega_e)$) instabilities. The nonlocal differential dispersion relation for these modes including density and velocity gradients is currently under investigation. Preliminary results indicate that the threshold for the electron acoustic and two stream branches is lowered due to velocity shear. For a homogeneous plasma the electron acoustic branch is stabilized when the thermal velocity approaches the drift velocity, i.e., $v_e \sim V_d$, making it less relevant to PSBL conditions where the flow speed can be of the order of the thermal velocity. However, by including velocity shear effects, the thermal velocity can be raised to 5 times the drift velocity. Even for such strong thermal effects there is growth around $\omega_r \sim 0.2\Omega_e$ and $\gamma \sim \omega_{LH}$. Thus, this mode may contribute substantially to the dynamics of the PSBL and deserves further attention. A detailed analysis of these modes is deferred to a future article.

Mikhailovskii, [1974] has shown that inhomogeneous electron flows lead to an instability around the electron plasma frequency, ω_{pe}, in an unmagnetized plasma. He considered the ions to be stationary with a relative drift between ions and electrons. We have extended this to the case where both ions and electrons are drifting with no relative velocity between them. We find that an instability near ω_{pe} can still be sustained by shear in the flow velocity [*Romero et al.*, 1992a]. From our preliminary work we see that shear-driven instabilities around ω_{pe} need very high values of velocity shear which may be attainable during the final moments of the growth phase. Further investigation of the properties of this branch is needed to examine whether the expected velocity shear in the PSBL can sustain this instability and, if so, can it explain the observed power around the plasma frequency?

We see that the instabilities discussed above can be triggered by increasing α_j due to stress build-up self-consistently. As they develop the instabilities can contribute to substantial viscosity which can reduce the magnitude of the shear, i.e. decrease α_j, and thereby lower the stress in the magnetotail. Also, they can contribute to resistivity and can energize the plasma. In the following we quantify this by following the nonlinear evolution of one such instability (the EIH instability). We discuss its role in stress-relief in the magnetotail and its contributions to anomalous viscosity and resistivity.

Stress Relief Mechanism (Nonlinear)

In this section, we consider the dynamical evolution of a stressed PSBL by an electrostatic, 2-1/2D PIC code [*Romero et al.*, 1992b; *Romero and Ganguli*, 1993]. In its current version, 800,000 particles are used in the simulation with, typically, 40 particles per cell. The electric field is decomposed into two constituents: the first is doubly periodic in the two spatial dimensions (hence, it is represented in a double Fourier series and its solution is facilitated by the use of fast Fourier transform techniques using a grid employing 64 × 128 nodes), and the second one is a time-independent, externally imposed field. This latter component models the effects of a constant driver in the system. It represents the effects of compression of the magnetotail owing to coupling with the solar wind. Increased coupling, especially during a substorm growth phase, will tend to steepen the PSBL as well as cause the well known plasma sheet thinning prior to substorm onset. This effect is modelled by intensifying the driving electric field. To begin the simulation, the particle loading is accomplished in a way such that both the electrons and ions are in force balance with the externally imposed electric field. This is achieved by a density gradient in the system similar in form to that which occurs at the PSBL-lobe interface (or at the magnetopause). As predicted by our Vlasov model we find a cross-field electron flow with respect to the ions. The grid spacing in both directions is chosen to be 1.4 to 2 times smaller than the Debye length. The time evolution of the system is then observed through the use of various diagnostic tools.

The value of the shear parameter is obtained from the Vlasov analysis of the equilibrium. From a number of simulation runs with different α_e (from 0.25 to 0.01) we find that if the shear frequency is larger than the wave frequency (in this case the lower hybrid frequency), shear effects are substantial and the EIH instability dominates the onset and character of the ensuing wave turbulence as well as the corresponding transport properties of the system [Romero et al., 1992b, Romero and Ganguli, 1993]. Fig. 7 shows the time evolution of a typical stressed boundary layer simulation. The plasma parameters corresponding to this simulation are as follows: $L_x/\lambda_D = 32, L_y/\lambda_D = 76.8, M_i/m_e = 400, T_i/T_e = 1$, and $\omega_{pe}/\Omega_e = 0.5$. Here λ_D is the Debye length and ω_{pe} is the electron plasma frequency. In more recent simulations we have increased the value of ω_{pe}/Ω_e to larger than unity without any appreciable modification to these results. In this example the electric field is chosen such that $\alpha_e = 0.25$. It is found that 6 vortices are formed after 5 lower hybrid times. This number is in close accord with the fastest growing mode in the system as derived from linear theory for the EIH mode [Romero et al., 1992a]. The wavelength of this mode is much larger than that of the lower hybrid drift instability.

Fig. 8 shows the depletion of the initial sheared flow layer as a function of time. Large anomalous viscosity and resistivity are generated by the waves. Within 20 lower hybrid times (less than 2 ion gyro periods) we see a significant reduction in the magnitude of the flow accompanied by a broadening of its scale-size. This leads to a reduction in the shear parameter, α_e, and therefore to stress relief. From a

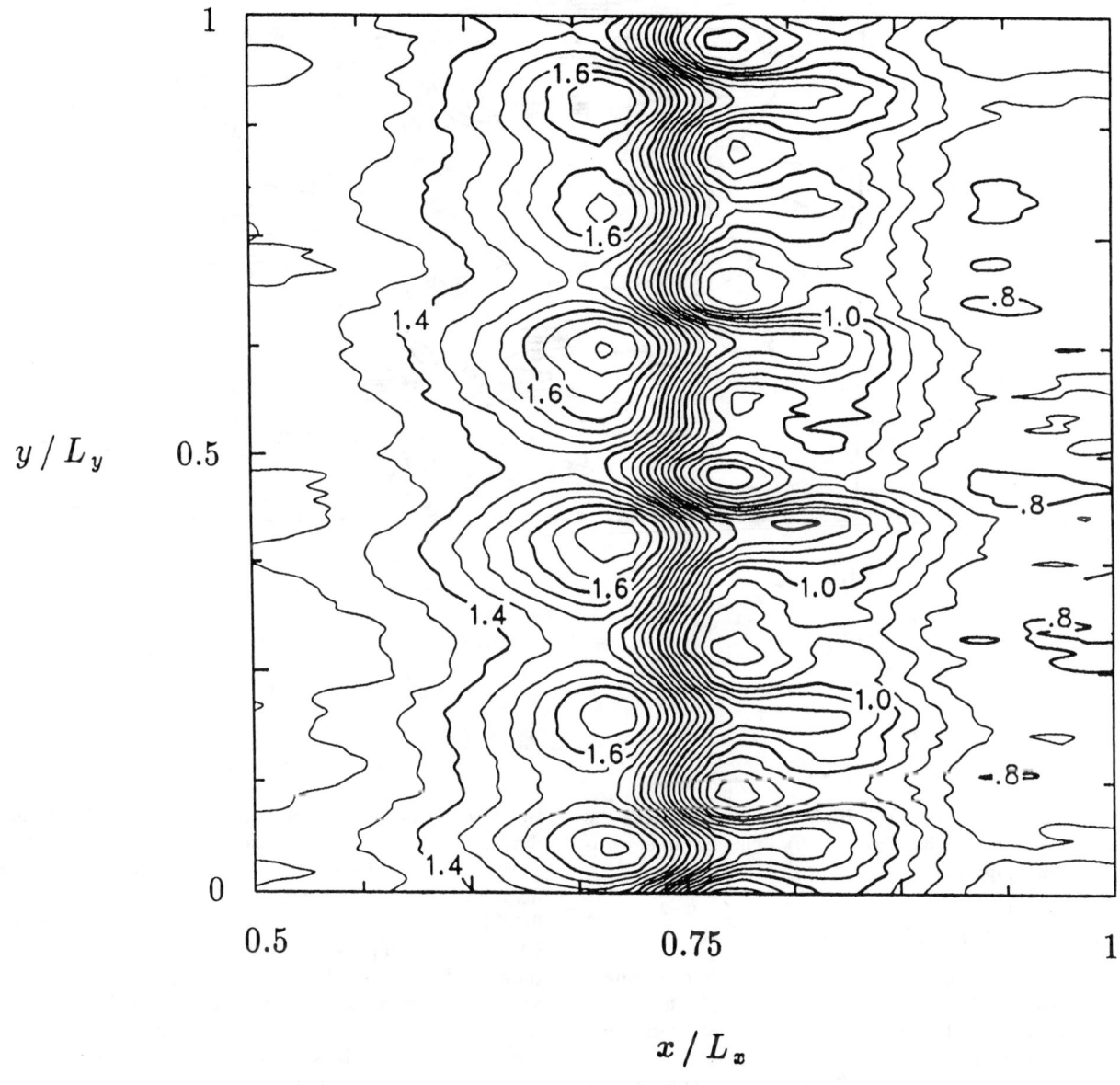

Fig. 7. Time sequence of the nonlinear evolution of a stressed boundary layer. The electrostatic potential is shown at times (a) $\omega_{LH}t = 7.2$, and (b) 19.8, respectively.

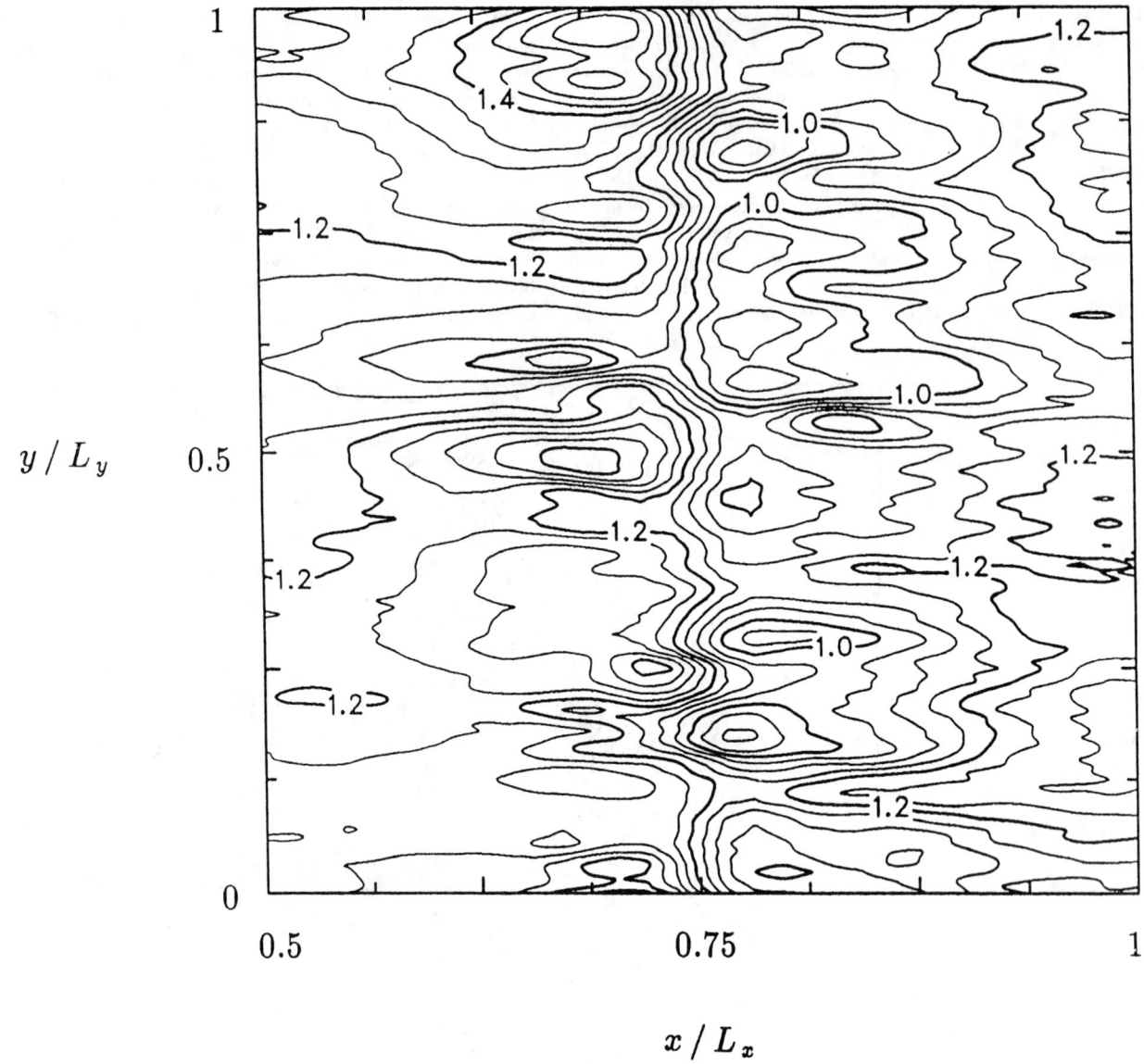

Fig. 7 (b.)

number of simulation runs we obtain a relation between the shear parameter and the viscosity (μ). This parameterization is given (approximately) by,

$$\mu = \begin{bmatrix} 0 & \alpha_e \leq 0.08 \\ 0.33 \left(\rho_i^2 \Omega_i \right) [\alpha_e - 0.08] ; & \alpha_e \geq 0.08 \end{bmatrix} \quad (7)$$

Thus, we see that while the MHD model leads to steepening and hence to a stress build-up (increase α_e), the microinstabilities lead to broadening and hence to stress relief (decrease α_e). It is envisioned that in nature these two opposing tendencies will lead to a balance which will prevent the indefinite steepening observed in MHD or fluid codes. This also emphasizes the crucial role of micro-processes and the obvious need for coupling the micro- and the macrodynamics for a realistic representation of natural systems.

We have also parametrized the resistivity due to the EIH instability by determining the value of the electron crossfield flow at the end of the simulation [Romero and Ganguli, 1993]. This value is found to be oscillatory in time and roughly of the order of the ion thermal velocity. As a result, an accurate estimate of the anomalous resistivity is difficult to obtain. However, by time averaging the plasma current in the latter phase of the simulation, the anomalous resistivity is obtained by dividing the resulting time-averaged current by the externally imposed value of the electric field. Using a linear fit to the data, we find the following representation

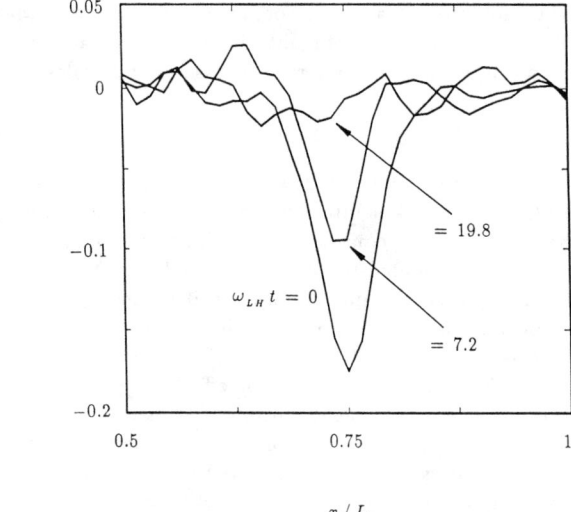

Fig. 8. The transverse flow layer at $\omega_{LH}t = 0, 7.2,$ and 19.8. Substantial anomalous viscosity leads to the depletion of cross-field electron flow within 20 lower hybrid times.

for the anomalous resistivity,

$$\eta_R = 26 \left(\frac{\Omega_e}{\epsilon_0 \omega_{pe}^2} \right) \alpha_e. \tag{8}$$

Here ϵ_0 is the permitivity in free space.

As shown in Fig. 9 considerable transverse ion acceleration is produced due to the EIH instability. (A much larger ion acceleration and heating is expected from the IEDDI since these waves oscillate around the ion gyrofrequency [*Ganguli et al.*, 1985b].) As ions are accelerated, energy is tapped and transferred from the velocity shear. In these runs we have maintained $k_\parallel = 0$. However, when this condition is removed we see parallel electron heating due to Landau damping mechanisms. Thus, these instabilities which feed on the available velocity shear in the transverse direction can transfer some energy to the parallel direction via dissipation due to Landau damping. While the ions are primarily energized in the transverse direction, the electrons are energized in the parallel direction.

The power spectrum for this simulation is given in Fig. 10. The spectrum is broadband in frequency, extending from below ω_{LH} to around ω_{pe} with more power in the lower frequency portion. This is consistent with observations [*Cattell and Mozer*, 1986]. Inclusion of electron beams in the model is expected to extend the spectrum toward higher frequency and possibly provide an enhancement near ω_{pe}, while the inclusion of ion beams and lower frequency ($\omega \sim \Omega_i$) IEDDI contributions is expected to reinforce the lower frequency portion of the spectrum. However, even in this preliminary simulation the spectrum is similar in character to that seen by satellites. A remarkable feature is that the broadband nature is a direct consequence of the inherent inhomogeneity of the PSBL and is not critically dependent on the existence of a large amount of cold plasma. Nevertheless, it may be interesting to investigate the effects of a cold plasma component on these wave modes.

As shown in Fig. 8, the electron cross field flow is reduced by a factor of 3 in 17 lower hybrid times while its width broadens. Concomitant with this decrease, there is

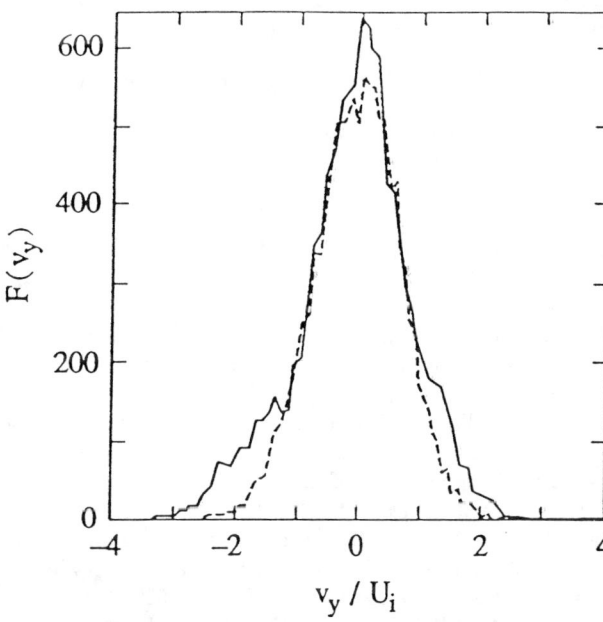

Fig. 9. Ion acceleration due to the EIH modes. Initial (dotted) and final (solid) ion distribution functions after only 20 lower hybrid times.

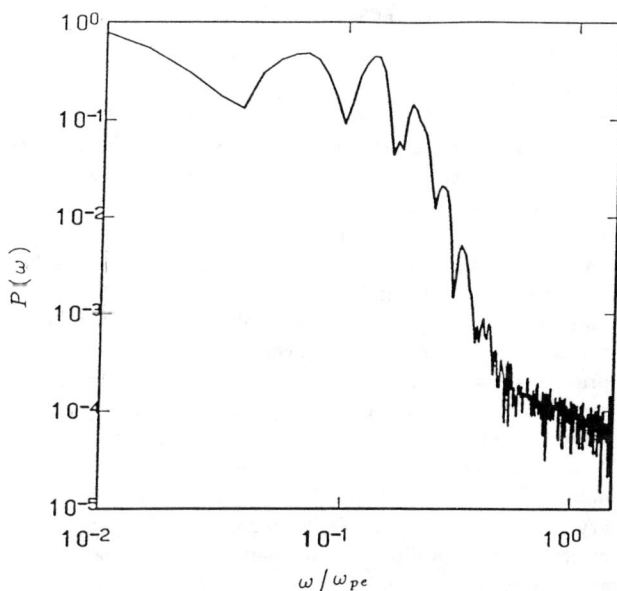

Fig. 10. Wave power spectrum from the PIC simulation of a stressed boundary layer. Note that the broadband character is due to the inherent inhomogeneities of the PSBL.

a reduction in the source of free energy for the EIH mode and the intensity of this instability is reduced. As the scale-size of the flow broadens to become larger than ρ_i, ω_s becomes smaller than Ω_i and the ions become magnetized. The conditions are now met for cascading toward lower frequencies and the longer wavelength IEDDI [*Ganguli et al.*, 1988a; *Nishikawa et al.*, 1988]. As seen in the simulations of Nishikawa et al. [1988; 1990], a shift towards lower frequency and longer wavelength is also observed in this simulation during the nonlinear evolution. Since this run was limited to 20 lower hybrid times, the full development of the downward spreading of the wave power remains to be investigated. Spreading of wave power from high ($> \omega_{LH}$) to low ($\ll \Omega_i$) frequency and the attendant consequences and nonlinear signatures are extremely relevant to the boundary layer dynamics, especially with regard to the relaxation of highly stressed layers.

As discussed earlier, during the compression phase enhanced stress can lead to strong gradients in the background plasma parameters in the PSBL. From the simulations discussed here, we find that there is a strong tendency to relax the stress via generation of plasma waves. In the evolution of a typical stressed PSBL discussed here, it took only 20 lower hybrid times to relax a flow layer of $\rho_i/2$ to $2\rho_i$. In order to maintain the layer width at its initial value the solar wind stress will have to be sufficient to overcome the strong dissipative agents acting on the system. This situation may be attained only during highly active periods, which is in keeping with observed characteristics that (1) the energetic electron layers as described by Parks et al., [1992] are infrequently observed and (2) they are generally observed during active periods.

Micro-Macro Coupling

This study constitutes a first step towards merging a complete theory of dissipation scale processes with global dynamics. On the global scale the solar wind, via reconnection on the bow magnetopause, is responsible for the magnetic and flow structures throughout the magnetosphere, but especially in the magnetotail. When the solar wind - magnetosphere coupling is strong both convection and magnetic flux are enhanced in the magnetotail. This circumstance leads to increased stresses in the tail and particularly to the plasmasheet which separates the tail lobes. The stresses are ultimately expressed in a thin layer where micro-scale plasma processes determine the physics of the stress balance. Here we have considered a preliminary model of the PSBL and a number of micro-scale processes which may occur there. These micro-scale dissipation processes are responsible for determining the physical scales associated with the boundary layers. In any case the thickness of the layer is determined by a dynamical balance between the globally driven steepening and the microscopic dissipation, i.e., viscosity, resistivity, etc.. As the resolution of the global simulation models is increased towards the dissipation scale the micro-plasma physics can be incorporated into the global models

via anomalous transport coefficients. Ongoing studies like that reported here hold out the promise of a final understanding of magnetospheric processes at all scales.

Conclusions

We have presented a new model for the dynamics of the PSBL which takes into account the inherent inhomogeneity of this region caused by the solar wind - magnetosphere coupling. Although the model reveals interesting global features of the macro-micro coupling, it is incomplete in many respects. The key purpose is to demonstrate the importance of velocity shear and other nonuniformities in the PSBL (ignored thus far) and their unique effects on particle energization and the nonlinear evolution of stressed layers. We find that the boundary layer inhomogeneity relaxes by giving rise to instabilities generated by velocity shear. These instabilities influence the nonlinear behaviour of the PSBL significantly. They can also determine the width of the boundary layer by balancing the micro-scale dissipation with the global steepening. We find that the ensuing turbulence can be broadband in frequency and is not necessarily dependent on a cold plasma component. Because the dissipation due to micro-processes balances the steepening effects due to solar wind compression, different instabilities of the hierarchy are effective for different globally driven steepening rates. Although these early results are encouraging, the model needs to be improved in a number of ways in order to make it more realistic. For example, multi-temperature plasma components and counter streaming beams are now being added to the model. This will allow us to investigate any additional role played by cold electron and ion components on the waves generated in a spatially inhomogeneous boundary layer.

Acknowledgments. Stimulating discussions with G. Parks, A.T. Lui, D. Baker, and D. Winske are gratefully acknowledged. This work is supported by ONR.

References

Anderson, K.A., R.P. Lin, R.J. Paoli, G.K. Parks, C.S. Lin, H. Reme, J.S. Bosqued, F. Martel, F. Cotin, *IEEE Trans. Geosci, GE-16*, 213, 1978.

Ashour-Abdalla, M. and D. Scriver, Acceleration and transport in the plasma sheet boundary layer, *Geophysical Monograph*, 54, 305, 1989.

Ashour-Abdalla, M., J. Berchem, J. Buchner, and L.M. Zelenyi, Large and small scale structures in the plasma sheet: A signature of chaotic motion and resonance effects, *Geophys Res Lett.*, 18, 1603, 1991.

Cattell, C.A., M. Kim, R.P. Lin, and F.S. Mozer, Observations of large electric fields near the plasma sheet boundary by ISEE 1, *Geophys. Res. Lett.*, 9, 539, 1982.

Cattell, C.A. and F.S. Mozer, Experimental determination of the dominant wave mode in the active near-earth magnetotail, *Geophys. Res. Lett.*, 13, 221, 1986.

Cargill, P.J. and T.E. Eastman, The structure of tangential discontinuities 1. Results of hybrid simulations, *J. Geophys Res.*, *13* 763, 1991.

D'Angelo, N., Kelvin-Helmholtz instability in a fully ionized plasma in a magnetic field, *Phys. Fluids.*, *8*, 1748, 1965.

Dandours, J., A. Saint-Marc, H. Reme, J. Sanvaud and G. Parks, *Adv Sp Res.*, *6*, 159, 1986.

Drummond, W.E. and M.N. Rosenbluth, Anomalous diffusion arising from microinstabilities in a plasma, *Phys. Fluids.*, *5*, 1507 (1962).

Eastman, T.E., L.A. Frank, W.K. Peterson and W. Lennartsson, The plasma sheet boundary layer, *J. Geophys. Res.*, *89*, 1553, 1984.

Fedder, J.A., J.G. Lyon, and S.P. Slinker, Numerical simulations of geomagnetic substorms?, *EOS Trans. Amer. Geophys. Union*, *73*, 457, 1992.

Frank, L.A., R.L. McPherron, R.J. DeCoster, B.G. Burek, K.L. Ackerson, and C.T. Russel, Field-aligned currents in the Earth's magnetotail, *J. Geophys. Res.*, *86*, 687, 1981.

Frank, L.A., Plasmas in the Earth's magnetotail, in *Space Plasma Simulations*, edited by M. Ashour-Abdalla and D.A. Dutton, p. 211, D. Reidel Publ. Co., Dordrecht, Holland, 1985.

Ganguli, G., Y.C. Lee and P. Palmadesso, Electrostatic ion cyclotron instability due to a nonuniform electric field perpendicular to the external magnetic field, *Phys. Fluids.*, *28*, 761, (1985a).

Ganguli, G., P. Palmadesso and Y.C. Lee, Electrostatic ion cyclotron instability due to a nonuniform electric field perpendicular to the external magnetic field, *Geophys. Res. Lett.*, *12*, 643, (1985b).

Ganguli G., Y.C. Lee, and P.J. Palmadesso, A new mechanism for excitation of waves in a magnetoplasma I. Linear theory, *Geophysical Monograph 38*, Proc. Chapman Meeting on Ion Acceleration, Boston, p 297, (1985c).

Ganguli, G. and P.J. Palmadesso, Electrostatic ion instabilities in the presence of parallel currents and transverse electric fields, *Geophys. Res. Lett.*, *15*, 103, 1988.

Ganguli, G., Y.C. Lee and P.J. Palmadesso, Kinetic theory for electrostatic waves due to transverse velocity shears, *Phys. Fluids*, *31*, 823, 1988a.

Ganguli, G., Y.C. Lee and P.J. Palmadesso, Electron-Ion hybrid mode due to transverse velocity shear, *Phys. Fluids.*, *31*, 2753, 1988b.

Ganguli, G., Y.C. Lee, P.J. Palmadesso and S.L. Ossakow, Oscillations in a plasma with parallel currents and transverse velocity shears, in *Physics of Space Plasmas (1988)*, SPI Conference Proceedings and Reprint Series, edited by T. Chang, G.B. Crew and J.R. Jasperse, 8, pp. 231, Scintific Publishers, Inc., Cambridge, MA, 1989a.

Ganguli, G., Y.C. Lee, P.J. Palmadesso and S.L. Ossakow, D.C. electric field stabilization of plasma fluctuations due to a velocity shear in the parallel ion flow, *Geophys. Res. Lett.*, *16*, 735, 1989b.

Ganguli, G., P.J. Palmadesso, Y.C. Lee and J.D. Huba, High Frequency Waves due to Velocity Shear in the Electron Flow, in *Proceedings of the 1989 International Conference on Plasma Physics*, New Delhi, India, p 197, 1989c.

Ganguli G., Y.C. Lee, and P.J. Palmadesso, Role of small scale processes in global plasma modelling, *AGU Monograph 62*, The second Huntsville Workshop on Magnetosphere/Ionosphere Plasma Models, Huntsville, Alabama, 17, 1991.

Ganguli, S.B. and P.J. Palmadesso, Plasma transport in the auroral return current region, *J. Geophys. Res.*, *92*, 8673, 1987.

Gurnett, D.A., L.A. Frank and R.P. Lepping, Plasma waves in distant magnetotail, *J. Geophys. Res.*, *81*, 6059, 1976.

Koepke, M.E., M.J. Alport, T.E. Sheriden, W.E. Amatucci, and J.J. Carroll III, Waves driven by strong transverse potential structures, in *Proc. of International Conference on Plasma Physics*, Jun-Jul 1992, Innsbruck, Austria; (to appear).

Koepke, M.E. and W.E. Amatucci, Electrostatic ion cyclotron wave experiments in the WVU Q-machine, *IEEE Trans. Plasma Sci.*, *20*, 631, 1992.

Kindel, J.M. and C.F. Kennel, Topside current instabilities, *J. Geophys. Res.*, *76*, 3055, 1971.

Lee, L.C., The Magnetopause: A tutorial review, in Physics of Space Plasmas (1990), *SPI Conference Reprint Series*, *10*, T. Chang, G.B. Crew, and J. Jasperes, eds., (Scientific Publishers Inc., Cambridge, MA, 1991), p.33.

Lee, L.C. and J.R. Kan, A unified kinetic model of the tangential magnetopause structure, *J. Geophys Res.*, *84*, 6417, 1979.

Lui, A.T.Y., Road map to magnetotail domains, in *Magnetotail Physics*, edited by A.T.Y. Lui, John Hopkins Press, Baltimore and London, p 3, 1987.

Lui, A.T., Extended consideration of a synthesis model for magnetospheric substorms, *Magnetospheric Substorms*, Geophysical Monograph 64 (American Geophysical Union, Washington D.C., 1991), p. 43.

Mikhailovskii, A.B., *Theory of Plasma Instabilities*, Vol. I, p. 18, (Consultants Bureau, New York, 1974).

Mitchell, D.G., D.J. Williams, C.Y. Huang, L.A. Frank, and C.T. russell, Current carriers in the near-earth cross-tail current sheet during substorm growth phase, *Geophys. Res. Lett.*, *17*, 583, 1990.

Nishikawa, K.-I., G. Ganguli, Y.C. Lee and P.J. Palmadesso, Simulation of ion- cyclotron-like modes in a magnetoplasma with transverse inhomogeneous electric field, *Phys. Fluids.*, *31*, 1568, 1988.

Nishikawa, K.-I., G. Ganguli, Y.C. Lee and P.J. Palmadesso, Simulation of electrostatic turbulence due to sheared flows parallel and transverse to the magnetic field, *J. Geophys. Res*, *95*, 1029, 1990.

Okuda, H., Numerical simulations on the magnetopause current layer, *AGU Monograph 62*, The second Huntsville Workshop on Magnetosphere/Ionosphere Plasma Models, Huntsville, Alabama, 9, 1991.

Onsager, T., M.F. Thomsen, J.T. Gosling, and S.J. Bame,

Electron distributions in the plasma sheet boundary layer: Time-of-flight effects, *Geophys. Res. Lett., 17*, 1837, 1990.

Orsini, S., M. Candidi, V. Formisano, H. Balsiger, A. Ghielmetti and K.W. Ogilvie, The structure of plasma sheet-lobe boundary in the Earth's magnetotail, *J. Geophys. Res., 89*, 1573, 1984.

Parks, G.K., C.S. Lin, K.A. Anderson, R.P. Lin, and H. Reme, ISEE-1 and -2 observations of the outer plasma sheet boundary, *J. Geophys. Res., 84*, 6471, 1979.

Parks, G.K., M. McCarthy, R.J. Fitzenreiter, J. Etcheto, K.A. Anderson, R.R. Anderson, T.E. Eastman, L.A. Frank, D.A. Gurnett, C. Huang, R.P. Lin, A.T.Y. Lui, K.W. Ogilvie, A. Pedersen, H. Reme and D.J. Williams, Particle and field characteristics of the high-latitude plasma sheet boundary layer, *J. Geophys. Res., 89*, 8885, 1984.

Parks, G.K., R. Fitzenreiter, K.W. Ogilvie, C. Huang, K.A. Anderson, J. Dandouras, L. Frank, R.P. Lin, M. McCarthy, H. Reme, J.A. Sauvaud, and S. Werden, Low-energy particle layer outside of the plasma sheet boundary, *J. Geophys Res., 97*, 2943, 1992.

Romero, H., G. Ganguli, P.B. Dusenbery and P.J. Palmadesso, Equilibrium structure of the plasma sheet boundary layer-lobe interface, *Geophys. Res. Lett., 17*, 2313, 1990.

Romero, H., G. Ganguli, Y.C. Lee and P.J. Palmadesso, Electron-ion hybrid instabilities driven by velocity shear in a magnetized plasma, *Phys. Fluids. B, 4*, 1708, 1992a.

Romero, H., G. Ganguli, and Y.C., Lee, Ion acceleration and coherent structures generated by lower hybrid shear-driven instabilities, *Phys. Rev. Lett., 69*, 3503, 1992b.

Romero, H. and G. Ganguli, Nonlinear evolution of a strongly sheared cross-field plasma flow, Phys. Fluids B, *5*, 3163, 1993.

Satyanarayana, P., Y.C. Lee, and J.D. Huba, Stability of a stratified shear layer, *Phys Fluids., 30*, 81, 1987.

Stern, D.P., The begining of substorm research, *Magnetospheric Substorms*, Geophysical Monograph 64 (American Geophysical Union, Washington D.C., 1991), p. 11.

Takahashi, K. and E.W. Hones, Jr., ISEE 1 and 2 observations of ion distributions at the plasma sheet-tail lobe boundary, *J. Geophys. Res., 93*, 8558, 1988.

Whipple, E. C., J.R. Hill and J.D. Nichols, Magnetopause structures and the question of particle accessibility, *J. Geophys. Res., 89*, 1508, 1984.

Winske, D., S.P. Gary, and D.S. Lemmons, Diffusive transport at the magnetopause, in Physics of Space Plasmas (1990), *SPI Conference Reprint Series*, **10**, T. Chang, G.B. Crew, and J. Jasperes, eds., (Scientific Publishers Inc., Cambridge, MA, 1991), p. 397.

The Structure and Dynamics of the Plasma Sheet During the Galileo Earth-1 Flyby

G. D. Reeves, T. A. Fritz, and R. D. Belian,[1] R. W. McEntire, D. J. Williams, and E. C. Roelof,[2]
M. G. Kivelson,[3] and B. Wilken[4]

The flyby of the Earth by the Galileo spacecraft on December 8, 1990 provided an excellent opportunity to study the spatial structure and temporal dynamics of the near-Earth plasma sheet. From 1700 to 2000 UT, when Galileo was within $R < 20\ R_E$, the magnetotail was in a stretched configuration. Galileo's trajectory was ideal for investigating spatial structure in the tail. We identified periods in which Galileo was in the plasma sheet, in the trapping boundary for keV particles, and in the stable radiation belts. Geosynchronous spacecraft 1984-129 monitored the temporal dynamics at fixed radius. We compared temporal dynamics at Galileo and at geosynchronous orbit. Both spacecraft saw a long-term decline in energetic particle fluxes which appears to have been a purely temporal change. Only 1984-129 observed two more traditional, and dramatic, growth phase dropouts illustrating temporal changes confined to one spatial region. In the plasma sheet Galileo did not observe flux variations due to radial gradients but did observe temporal variations caused primarily by flapping of the tail or magnetic reconfiguration. One sequence of such variations caused the spacecraft to enter and exit the trapping boundary region quite close to the earth. When Galileo was near spacecraft 1984-129 the two spacecraft observed very different fluxes which implies the presence of very strong spatial gradients in that region of space.

INTRODUCTION

It is a well-known quandary in space plasma physics that it is generally impossible to distinguish between spatial structures and temporal processes using single-point measurements. Indeed, even with measurements from multiple spacecraft one is rarely fortunate enough to have the spacecraft in the right places at the right time. The

[1]Los Alamos National Laboratory, Los Alamos, NM 87545
[2]The Johns Hopkins University, Applied Physics Laboratory, Laurel, MD
[3]University of California, Los Angeles, CA
[4]Max Plack Institut fur Aeronomie, Katlenburg-Lindau, Germany

Solar System Plasmas in Space and Time
Geophysical Monograph 84
Copyright 1994 by the American Geophysical Union.

Galileo Earth-1 flyby on December 8, 1990 was one of those serendipitous occasions. Galileo's trajectory was designed to use the Earth's gravity to propel it on its way to Jupiter. Fortunately for our purposes this trajectory also allowed Galileo to rapidly sample spatial structures in the magnetotail. At the same time spacecraft 1984-129 monitored the midnight local time sector at geosynchronous altitudes. It's orbit, at a fixed radius, is well suited to observe temporal variations. Galileo's passage through the magnetotail required several hours and thus was too slow to provide a "snapshot" of the magnetosphere. This is particularly true since the flyby occurred during an active period. However the combined data from Galileo and from 1984-129 allowed us to separate spatial and temporal effects and to study the structure and dynamics of the plasma sheet when it was in a stretched, growth phase configuration. (See *Kivelson et al.* [1993] for an overview of the Earth-1 flyby.)

150 GALILEO EARTH 1 FLYBY

Instruments

The Galileo spacecraft carries a complement of field and particle instruments. In this study we have used data from the fluxgate magnetometer (described in detail by *Kivelson et al.* [1992]) and the energetic particle detector (EPD) [*Williams et al.*, 1992]. The low-energy magnetospheric measurement system (LEMMS) of the Galileo energetic particle detector measures electrons and protons with energies of tens of keV. The channels used for this study are the A0–A5 channels that measure protons in differential energy bins with energies of 22–42, 42–65, 65–120, 120–280, 280–515, and 515–825 keV; and the E0–E3 and F0 channels that measure electrons in differential energy bins with energies of 15–29, 29–42, 42–55, 55–93, and 93–188 keV. In this study we use spin-averaged LEMMS data. The spin period of Galileo is approximately 20 s. In order to obtain full unit sphere coverage the EPD is also articulated with respect to the spin plane. A stepper motor moves the sensors in the plane containing the spin axis after each spin. When the particle distribution is not isotropic this adds a modulation to the EPD data that has not been removed. Although the number of steps is controllable, during the Earth-1 flyby the modulation period is approximately 200 s.

Spacecraft 1984-129 was one of a continuously operating constellation of three geosynchronous spacecraft that carried Los Alamos charged particle analyzer (CPA) instruments from 1976 to the present. The CPA includes two instrument sub-systems. The LoE sub-system measures electrons in six nested energy channels with low-energy thresholds of 30, 45, 65, 95, 140, and 200 keV. All share a common high energy limit of 300 keV. The LoP subsystem has ten channels with thresholds of 72, 91, 104, 125, 153, 190, 235, 292, 365, and 475 keV. The common upper energy limit is 573 keV. Although both instruments have 256 ms resolution, throughout this paper we use data that were averaged over all telescopes, 6 spins (approximately 1 minute), and therefore over the unit sphere. Differential energy measurements are obtained from the nested energy measurements by subtracting adjacent channels. More information on the CPA detectors can be found in the paper by *Higbie et al.* [1978].

OBSERVATIONS

Figure 1 shows the trajectories of Galileo and spacecraft 1984-129. The top plot shows the trajectories projected into the equatorial plane for times between 1700 and 2000 UT. Both spacecraft were in the midnight sector. Galileo's trajectory was nearly radial and 1984-129's was azimuthal. Galileo crossed the geosynchronous drift shell at

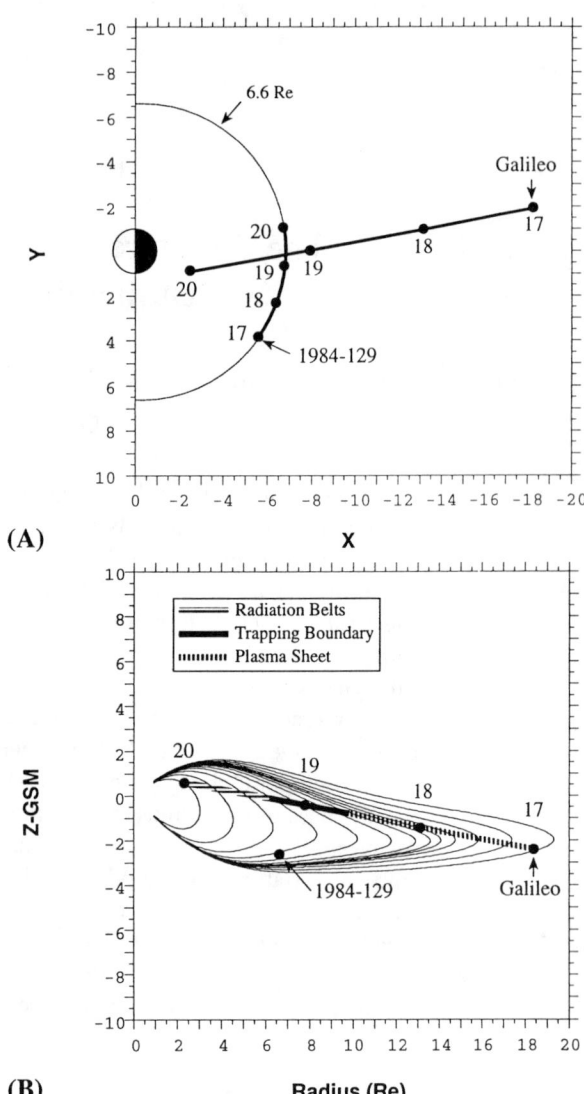

Fig. 1. The trajectories of Galileo and geosynchronous spacecraft 1984-129. Positions are marked for each hour from 1700 to 2000 UT. (A) The trajectories in the X-Y plane. Galileo made a nearly radial pass through the midnight sector while 1984-129 made an azimuthal pass. At 1917:30 UT Galileo crossed the geosynchronous orbit within 2.5° longitude of the position of 1984-129. (B) The Galileo trajectory plotted as a function of Z-GSM and Radius, $(X^2+Y^2+Z^2)^{1/2}$. In this coordinate system 1984-129 is nearly motionless over this 3-hour period. Also shown are the magnetic field lines connected to Galileo at each 15 minute interval as predicted by the *Tsyganenko* [1989] model for Kp>4+. The trajectory of Galileo is plotted with shaded lines representing the magnetotail regions measured by Galileo (see text).

1917:30 UT, at which time the spacecraft were separated by approximately 2.5° in azimuth. The bottom plot shows the trajectories as a function of Z_{GSM} and radius. In this coordinate system spacecraft 1984-129 is nearly motionless over this 3-hour period. Also shown are field line traces according to the *Tsyganenko* [1989] field model for its most stretched configuration (Kp > 4+). The field lines are traced starting at the position of the Galileo spacecraft at 15 minute intervals. The shading of the Galileo trajectory shows the magnetotail regions measured at different locations along the trajectory and is discussed in the following section.

We show the fluxes of energetic particles measured by Galileo and 1984-129 between 1700 and 2000 UT in Figure 2. The top panel shows electrons and the bottom panel shows protons. We have only plotted one energy range for each species and chose similar energy ranges for both spacecraft. The fluxes in other energy ranges show similar features and are plotted in *Reeves et al.* [1993]. The universal time and the position of the two spacecraft are indicated along the bottom of the plot and the fluxes measured at the two spacecraft are plotted on the same scale.

Between 1700 and 2000 UT the fluxes of both protons and electrons at geosynchronous orbit declined by almost two orders of magnitude. Superimposed upon that long-term decline were two dropouts of approximately 30 min duration. The first dropout began at approximately 1730 UT. The proton and electron fluxes both returned at 1811 UT but did not attain their previous levels. The second dropout was observed primarily in the electron fluxes. This is because the proton fluxes measured by 1984-129 were already near the 1-count level and lower fluxes could not be recorded. The second dropout began at approximately 1912 UT. It intensified at 1927 UT and recovery was at 1941 UT. The final recovery (and injection) of energetic particles at geosynchronous orbit was not recorded until 2053 UT, after the period of interest for this study.

The Galileo energetic particle fluxes were quite different. The only feature that was similar to the geosynchronous energetic particle fluxes was a gradual decline in flux levels observed between 1700 and 1841 UT. The similarity is particularly apparent in the proton fluxes in the lower panel of Figure 2. The Galileo and geosynchronous fluxes track each other quite closely in that interval except during the first geosynchronous dropout when the Galileo electron fluxes increased. The Galileo proton fluxes were unaffected. At 1811 UT when the geosynchronous particle fluxes recovered there was little change at Galileo except a small injection of electrons. Between 1825 and 1841 UT three narrow spikes were observed in the Galileo electron fluxes. They are the result of modulation caused by the LEMMS stepper motor combined with a very anisotropic pitch angle distribution. A more dramatic change in flux levels was recorded by Galileo at 1841 UT when the flux levels increased by approximately an order of magnitude and remained high. Between 1841 and 1917 UT the Galileo fluxes were highly variable. Peak Galileo electron and proton fluxes were comparable to the fluxes measured by 1984-129 at geosynchronous orbit. After 1917:30 UT the Galileo fluxes were higher than those measured by 1984-129 and the fluxes were not highly variable.

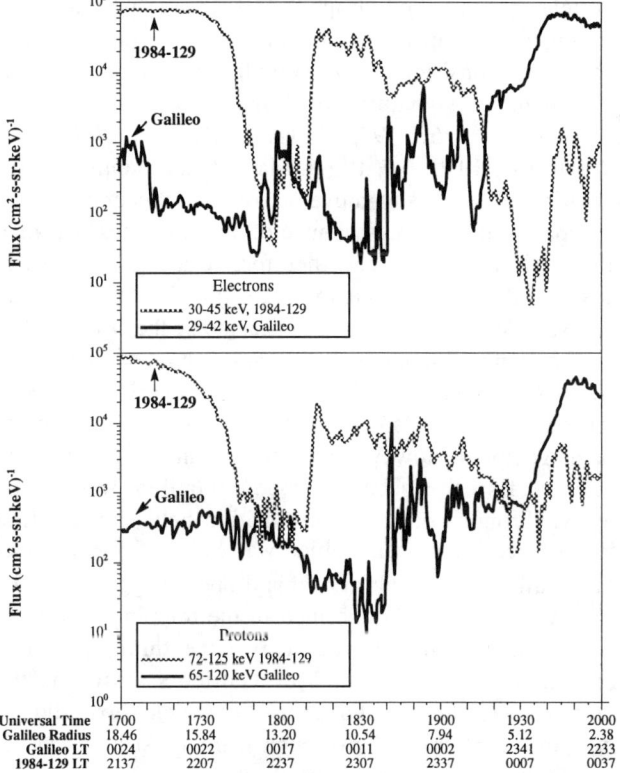

Fig. 2. A comparison of geosynchronous (shaded line) and Galileo (solid line) fluxes. Electrons are shown in the top panel and protons are shown in the bottom panel. For electrons, the Galileo LEMMS and 1984-129 CPA instruments had very similar energy bands so a direct flux comparison can be made. For protons, three energy bands from the 1984-129 CPA instrument have been added together to show the flux from 72 to 125 keV which can be compared with the Galileo LEMMS fluxes from 65 to 120 keV.

Figure 3 shows the Galileo magnetometer measurements [*Kivelson et al.*, 1992, 1993] along with the Galileo electron fluxes. From 1700 to about 1747 UT B_x was quite small and varied in sign indicating that Galileo was very close to the neutral sheet, as expected from its orbit (Figure 1). After about 1747 UT when B_x was uniformly positive, the inclination of the field (θ) was close to 90°, indicating that the field was in a highly stressed, tail-like configuration. The increasing field magnitude was a

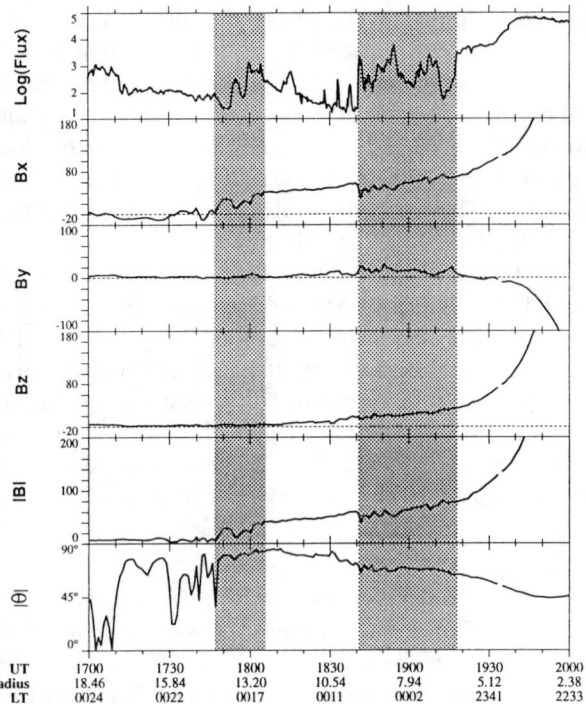

Fig. 3. A comparison of energetic electron fluxes with data from the UCLA magnetometer on Galileo. The top panel shows 29–42 keV electron fluxes (in counts/cm^2/s/sr/keV) and the lower panel shows all three components of the magnetic field (in nT), its magnitude (in nT), and the inclination angle of the field ($\theta = \tan^{-1}(B_x/B_z)$). The shaded areas show periods of particle and magnetic field variability.

result of Galileo's motion toward the Earth. Two intervals have been shaded. During those times both the field components and the particle fluxes were variable. The first of these intervals occurred during the first geosynchronous dropout and the second spans the interval between 1841 and 1917 UT discussed above. Close examination shows that the peak particle fluxes are correlated with minimums in the field intensity suggesting that the flux variations may be due to apparent motion of the spacecraft across magnetic flux surfaces. (The three energetic electron peaks between 1825 and 1841 UT were not included in the shaded portion of Figure 3 because they are caused by instrumental effects, not temporal variability.) Interestingly the recovery of fluxes at geosynchronous orbit and the small injection of electrons at Galileo at 1811 UT were not accompanied by any variation in the magnetic field magnitude or direction. In particular, no "dipolarization" of the field was observed.

ANALYSIS

The Galileo energetic particle signatures during the three hours analyzed in this study can be used to identify three spatial regions: the plasma sheet, the trapping boundary, and the radiation belts. At the beginning of the interval Galileo was clearly in the plasma sheet. (In this study we do not distinguish between the plasma sheet and the plasma sheet boundary layer.) At the end of the interval it is equally clear that Galileo was in the radiation belts. Below we identify the interval between 1841 and 1917 UT as the period of interaction with the trapping boundary. In Figure 1b we use shading of the Galileo trajectory to indicate where Galileo was when it measured each of these regions. A vertical-striped line indicates the plasma sheet (before 1841 UT), the black line indicates the trapping boundary (1841-1917:30 UT), and the horizontal-striped line indicates the stable radiation belts (after 1917:30).

1700–1841 UT

Between 1700 and 1841 UT, while Galileo was in the plasma sheet, 1984-129 recorded a growth phase energetic particle dropout. This dropout of energetic particles was examined in detail by *Reeves et al.* [1993] who concluded that it was consistent with the traditional interpretation of growth phase signatures of magnetotail thinning [e.g., *Hones et al.*, 1967, 1973; *Baker et al.*, 1981; *Baker and McPherron*, 1990; *Lui*, 1991]. Magnetotail thinning causes magnetic flux tubes to move across the spacecraft. Hence the spacecraft becomes connected to field lines that map further down the tail and therefore to a region of lower fluxes. If the thinning is extreme the trapping boundary can move across the spacecraft as appears to be the case here.

Magnetotail thinning is caused by an intensification of the cross-tail current and is therefore primarily a temporal phenomenon. It is also expected to affect a large portion of the magnetotail. However, neither dropout was observed by Galileo. For the second dropout Galileo was inside geosynchronous orbit in the radiation belts, but the first dropout occurred when Galileo was at 13–15 R_E. The fact that Galileo did not observe this dropout suggests that its effects were probably confined to the trapping region. In other words, while the tail may have thinned both at geosynchronous orbit and at Galileo's position, the necessary radial gradient of energetic particle fluxes did not exist in the plasma sheet though it did exist between the trapping region and the plasma sheet.

Comparison of the fluxes measured by Galileo and 1984-129 provide direct evidence of the gradient between geosynchronous orbit and the plasma sheet (Figure 2) but, we can only infer the lack of such a gradient in the plasma sheet proper. Supporting evidence is seen during the deepest part of the dropout. From approximately 1745 to 1811 UT the fluxes measured by Galileo and 1984-129 were nearly equal (Figure 2). This was true for both protons and electrons at all measured energies (data not shown). Even some of the temporal variations in the electron fluxes during the dropout were seen by both spacecraft. While it is likely that the field line connected to 1984-129 mapped to the plasma sheet it is highly unlikely that it mapped to the same radius or the same local time as Galileo (Figure 1a). Therefore the equality of the fluxes during the dropout suggests some uniformity in the level of energetic particle fluxes in the plasma sheet during this time. The temporal variations seen by both spacecraft during the deepest part of the dropout were correlated with magnetic field variations (Figure 3) and hence may be the result of flapping of the magnetotail.

1841–1917:30 UT

At 1841 UT an abrupt change in the Galileo energetic particle fluxes was observed. The 1841 UT transition clearly had a temporal as well as a spatial character. At that time there was a marked decrease in B_x and $|B|$ and an increase in B_y indicating both a rotation and a relaxation of the field. From 1841 to 1917 UT both the particle fluxes and magnetic field measured by Galileo were highly variable (Figure 3). The highest fluxes were measured when B_x (and therefore $|B|$) was lowest. At those times the field was slightly less tail-like. Also at those times the Galileo energetic particle fluxes were comparable in magnitude to the fluxes measured by 1984-129. Therefore, during times of peak fluxes, Galileo was probably in the same flux region as 1984-129, namely the trapping region for keV particles. At other times the Galileo fluxes were significantly lower than those measured by 1984-129. Therefore we suggest that between 1841 and 1917 UT Galileo skimmed along the trapping boundary and that "flapping" of the magnetotail moved that boundary moved across the spacecraft causing the variations in the measured magnetic field and particle fluxes.

The interval during which Galileo was in the trapping boundary is particularly interesting because of its implications for the structure of the magnetotail. We have seen that Galileo moved in and out of the trapping boundary and therefore conclude that the trapping boundary was nearly parallel to the Galileo trajectory. This, in turn, implies that the magnetic flux surfaces were nearly parallel to the Galileo trajectory. While the Tsyganenko model predicts that Galileo flew nearly along the magnetic field from 1800 to about 1830 UT, by 1900 UT the Tsyganenko model predicts that Galileo should have been rapidly crossing flux surfaces. One can see from Figure 1, though, that if the magnetotail were "squeezed" in the vicinity of 6–10 R_E that Galileo's trajectory could have been more parallel to the flux surfaces. The conclusion that the Tsyganenko model is not sufficiently stretched at distances of 6–10 R_E has been reached before based on magnetic field measurements alone [*Kaufmann,* 1988], but here we see direct evidence of the effect of magnetotail thinning on the structure of energetic particle drift shells.

1917:30–2000 UT

Reeves et al. [1993] determined that Galileo crossed the geosynchronous drift shell at 1917:30 UT. At that time the fluxes of electrons were equal at the two spacecraft for all measured energies. After that time Galileo measured fluxes which were higher than those measured by 1984-129. The fluxes measured after 1917:30 showed little variability other than that expected as Galileo moved from high to low altitudes. Therefore we identify the region sampled between 1917:30 and 2000 UT as the stable radiation belts.

It is interesting to contrast the two transitions at 1841 UT and at 1917:30 UT. At 1917:30, as Galileo crossed the geosynchronous drift shell, the electron fluxes increased abruptly. Unlike the 1841 UT transition, the abruptness does not appear to have been due to a temporal change. Spacecraft 1984-129 remained at fixed radius and observed a continuing growth phase dropout as the Galileo fluxes increased. Since the two spacecraft were in such close proximity one would expect a temporal change to be apparent in both data sets. Furthermore, the Galileo protons did not experience a similar abrupt increase and there was no signature in the Galileo magnetometer data. Therefore we attribute this transition to the presence of a strong gradient in the energetic electron fluxes in the vicinity of geosynchronous orbit. (*Reeves et al.* [1993] have examined this gradient in more detail.) Hence the Galileo data show that the region in which growth phase dropouts are observed is limited to a region with both an inner boundary and an outer boundary. The outer boundary is the trapping boundary. The inner boundary must be defined by where magnetotail thinning does or does not cause particle gradients to move across the spacecraft. These results suggest that boundary may have been very near geosynchronous orbit and was highly localized in azimuth, radius, or latitude.

CONCLUSIONS

We have examined the structure and dynamics of the magnetotail during the Galileo Earth-1 flyby. We determined the times in which Galileo was in the plasma sheet, the trapping boundary, and the radiation belts. The measurements suggest that Galileo's trajectory skimmed the trapping boundary and that variations in the magnetic field caused that boundary to pass back and forth across the spacecraft. A comparison of the shape of the magnetic field predicted by the Tsyganenko model with the measurements of the trapping boundary show that the actual magnetic field was much more tail-like than the model field, especially at distances of 6–10 R_E.

A long-duration decline of energetic particle fluxes was observed both by 1984-129 at geosynchronous orbit and by Galileo in the plasma sheet. Superimposed on that feature were two growth phase dropouts of energetic particle fluxes at geosynchronous orbit. The first occurred when Galileo was in the plasma sheet and the second occurred when Galileo was entering or within the radiation belts. Since neither dropout was observed by Galileo we concluded that the effects magnetotail thinning that produced the dropouts at geosynchronous orbit were limited to the trapping region. We also found that during the first dropout the fluxes measured by 1984-129 became nearly equal to those measured by Galileo even though it is highly unlikely that the field connected to 1984-129 mapped to the vicinity of Galileo. Therefore we concluded that the energetic particle gradients in the plasma sheet during this time were relatively small and that the variations in energetic electron fluxes which were observed at that time were probably related to a flapping motion of the magnetotail. We compared the two boundary crossings identified in the Galileo data and concluded that the transition from the plasma sheet to the trapping boundary was caused by a reconfiguration of the magnetic field while the transition from the trapping boundary to the radiation belts occurred due to Galileo's motion alone, without temporal variation.

The combination of data from a fast-moving spacecraft in the plasma sheet and a slow-moving geosynchronous spacecraft near local midnight was a vital asset for this study. We look forward with anticipation to similar studies which should be possible utilizing Los Alamos geosynchronous and Geotail data.

Acknowledgments: This work was supported by the U.S. Department of Energy Office of Basic Energy Sciences. Work at UCLA was partially supported by the Jet Propulsion Laboratory under contract JPL-958694. The authors would like to thank Tom Armstrong, Vassilis Angelopoulos, Alan Roux, Lou Frank, Tom Cayton, and Bob McPherron for their insights and enthusiasm.

REFERENCES

Baker, D. N. and R. L. McPherron, Extreme energetic particle decreases near geostationary orbit: A manifestation of current diversion within the inner plasma sheet, *Adv. Space Res., 10,* 131, 1990.

Baker, D. N., E. W. Hones, P. R. Higbie, R. D. Belian, and P. Stauning, Global properties of the magnetosphere during a substorm growth phase: A case study, *J. Geophys. Res., 86,* 8941, 1981.

Higbie, P. R., R. D. Belian, and D. N. Baker, High-resolution energetic particle measurements at 6.6 R_E 1, Electron micropulsations, *J. Geophys. Res., 83,* 4851, 1978.

Hones, E. W., J. R. Asbridge, S. J. Bame, and I. B. Strong, Outward flow of plasma in the magnetotail following geomagnetic bays, *J. Geophys. Res., 72,* 5879, 1967.

Hones, E. W., J. R. Asbridge, S. J. Bame, and S. Singer, Substorm variations of the magnetotail plasma sheet from X-sm≈-6 R_E to X-sm≈60 R_E, *J. Geophys. Res., 78,* 109, 1973.

Kaufmann, R. L., Substorm currents: growth phase and onset, *J. Geophys. Res., 92,* 7471-7486, 1987.

Kivelson, M. G., K. K. Khurana, J. D. Means, C. T. Russell, and R. C. Snare, The Galileo magnetic field investigation, *Space Sci. Rev., 60,* 357, 1992.

Kivelson, M. G., C. F. Kennel, R. L. McPherron, C. T. Russell, D. J. Southwood, R. J. Walker, K. K. Khurana, P. J. Coleman, C. M. Hammond, V. Angelopoulos, A. J. Lazarus, R. P. Lepping, and T. J. Hughes, The Galileo Earth encounter: Magnetometer and allied measurements, *J. Geophys. Res., 98,* 11,299, 1993.

Lui, A. T. Y., A synthesis of magnetospheric substorm models, *J. Geophys. Res., 96,* 1849, 1991.

Reeves, G. D., T. A. Fritz, R. D. Belian, R. W. McEntire, D. J. Williams, E. C. Roelof, M. G. Kivelson, and B. Wilken, Structured plasma sheet thinning observed by Galileo and 1984-129, *J. Geophys. Res., 98,* 21,323, 1993.

Tsyganenko, N. A., A Magnetospheric magnetic field model with a warped tail current sheet, *Planet. Space Sci, 37,* 5, 1989.

Williams, D. J., R. W. McEntire, S. Jaskuler, and B. Wilken, The Galileo energetic particle detector, *Space Sci. Rev., 60,* 385, 1992.

R. D. Belian, T. A. Fritz, and G. D. Reeves, Los Alamos National Laboratory, NIS-2, Mail Stop D-436, Los Alamos, NM 87545.

M. G. Kivelson, University of California, Los Angeles, CA 90024.

R. W. McEntire, E. C. Roelof, and D. J. Williams, The Johns Hopkins University Applied Physics Laboratory, Laurel, MD 20723.

B. Wilken, Max-Planck-Institut für Aeronomie, Postfach 20, D-3411 Katlenburg-Lindau, Germany

Initial Observations of the Medium Distance Magnetotail Plasma by GEOTAIL: Cold Ion Beams

T. Mukai[1], M. Hirahara[2], S. Machida[3], Y. Saito[1],
T. Terasawa[2], and A. Nishida[1]

[1] *The Institute of Space and Astronautical Science, Sagamihara, Japan*
[2] *Department of Earth and Planetary Physics, University of Tokyo, Tokyo, Japan*
[3] *Geophysical Institute, Kyoto University, Kyoto, Japan*

We observed cold ion beams in the magnetotail lobe at $X_{GSM} \sim -42$ Re in the initial operations of the Low Energy Particle (LEP) experiment onboard the GEOTAIL satellite on August 22, 1992, when multiple onsets of substorms took place. These ion beams generally consisted of protons and singly-charged oxygen ions (O^+), flowing tailward with nearly the same velocities $\sim 100-200$ km/s. The H^+ number density was generally of order 10^{-2} cm^{-3}, while the O^+ density was of order 10^{-3} cm^{-3} but at times increased sporadically by an order of magnitude. These ions presumably would be transported along magnetic field lines through the polar mantle from the dayside polar ionosphere and convected to the mid-magnetotail where the observation was made. That different ion species had equal velocities is consistent with the velocity filter effect due to $\mathbf{E} \times \mathbf{B}$ convection. Three-dimensional determinations of their velocity distributions have revealed off-ecliptic angles of the flow directions as large as several tens of degrees. We discuss this feature in terms of the enhanced convection velocity that is associated with the plasma-sheet thinning confined to a limited region in the Y direction. Characteristic changes in the flow velocity are also found in association with a substorm plasmoid passage.

INTRODUCTION

One of the outstanding problems in magnetospheric physics is to determine the physical processes of refilling of the plasma sheet against its losses. It is now well known that both solar wind and ionospheric plasmas contribute to the plasma sheet, and these sources vary according to the global magnetospheric activity [*Peterson et al.*, 1981; *Lennartsson et al.*, 1981; *Sharp et al.*, 1982]. One open question is how the plasma is transported to the plasma sheet from the high-latitude topside ionosphere where ionospheric ions are thought to be accelerated and solar wind ions are mirrored. Based on the DE-1 observations, *Chappell* [1988] argued the importance of the ionospheric contribution and presented a schematic drawing in which the low-energy particles outflowing from the polar ionosphere could be transported through the lobe region and convected into the plasma sheet. Recently *Delcourt et al.*[1993] have also demonstrated their role quantitatively by means of three-dimensional particle codes.

As for observations in the magnetotail lobe region and in the plasma mantle at distances up to ~ 20 Re from the earth, the ion composition and plasma experiments onboard ISEE 1 and 2 satellites have revealed that both O^+ and H^+ are streaming tailward roughly along the magnetic field lines and are injected into the plasma sheet [*Sharp et al.*, 1981; *Candidi et al.*, 1982; *Orsini et al.*, 1985; *Orsini et al.*, 1990; *Akinrimisi et al.*, 1990]. In the region further downtail, there have been only two papers, to the authors' knowledge, reporting similar observations at distances of ~ 35 Re by the Imp-7 satellite [*Frank et al.*, 1977] and on the lunar surface [*Hardy et al.*, 1977]. Both observations showed occasional occurrence of cold ions of ionospheric origin streaming tailward.

Previous observations were two-dimensional measurements of the ion distribution functions with time resolution of ~1.5 min. During the past decade, the capability of plasma instruments onboard spacecraft has been so much improved as to provide three-dimensional velocity distributions of charged particles with finer time resolution. In the present paper we report on the GEOTAIL/LEP observation of the cold ion streams of H^+ and O^+ in the tail lobe region at geocentric distances of ~42 Re. This is a case study obtained during the initial operations of the LEP instrument, which has unfortunately become latched up soon after the present observation. The observation was made in the course of a substorm, and the cold ion streams showed interesting characteristics, some of which may be associated with the passage of a substorm plasmoid [*Hones et al.*, 1984; *Baker et al.*, 1987].

INSTRUMENTATION

The GEOTAIL spacecraft was launched on July 24, 1992 into a translunar orbit, and injected into a nightside distant-tail orbit on September 8. The present observation was made on August 22 in a series of initial spacecraft operations during the translunar orbit. During the period of interest, the spacecraft spin axis was nearly perpendicular (~ 87.1°) to the ecliptic plane, and the spin period was 3.276 sec. The LEP experiment onboard GEOTAIL consists of three sensors which we call EA, SW and MS [*Mukai et al.*, 1993]. We will here mention only those features of EA (Energy-per-charge Analyzer) which are essential for the understanding of the result to be presented. The EA sensor consists of two nested sets of quadrispherical electrostatic analyzers to measure three-dimensional velocity distributions of electrons (with EA-e) and ions (with EA-i), simultaneously and separately, over the energy-per-charge range of several eV/q to 43 keV/q, The energy range is selectable independently for EA-e and EA-i, and the selected range is divided into 32 bins. In the present observation, EA-i covers the energy range of 32 eV/q to 39 keV/q divided into 32 bins, in which 24 bins are equally spaced on a logarithmic scale in energies higher than 630 eV/q and have width of ±9.4 % of the center energy, while the lower-energy 8 bins are spaced linearly with width of ±40 eV/q (±20 eV/q for the lowest-energy bin). The full energy range is swept in a time which is 1/16 of a spin period (synchronized with the spacecraft spin motion). The field of view is fan-shaped with ~10° × 145°, in which the longer dimension is perpendicular to the spin plane and divided into seven directions centered at elevation angles of 0°, ±22.5°, ±45° and ±67.5° with each of width 10°. Thus, count rate data of dimension $32(E) \times 16(Az) \times 7(El)$ are generated in one spin period. While the velocity moments are calculated onboard every spin period, the complete three-dimensional velocity distributions can only be obtained in a period of four spins owing to the telemetry constraints; the count data are accumulated during the four-spin period. The MS sensor (energetic ion Mass Spectrometer) was operated in a two-dimensional mode with a limited field of view for a series of examinations of the instrument performance. The magnetometer was not yet operational during the present observation, which was made prior to extension of the magnetometer mast.

OBSERVATION

Plate 1 shows energy-time spectrograms of electrons and ions observed at $X_{GSM} \sim -42$ Re, $Y_{GSM} \sim 5$ Re and $Z_{GSM} \sim 2$ Re on August 22, 1992. The spacecraft was located in the plasma sheet from the beginning of the time period shown, 1200 UT, up to ~1235 UT. Hot plasma with keV-energy ions was observed continuously, though the energy density was variable. Electron data also exhibited similar features. After ~1235 UT, the hot plasma density decreased drastically, and the spacecraft remained in the lobe region until the end of the observation period. The geomagnetic conditions were quite active during the observation interval. In Plate 1 three bursty events of high-speed tailward plasma flows with duration of ~1 minute are evident around 1212 UT, 1225 UT and 1300 UT. The first and the second events were observed in the plasma sheet, while the third event took place in the lobe region. Detailed descriptions of these events are given in *Machida et al.*[preprint, 1993]. As shown in Figure 1, the GMS-4 geosynchronous satellite also detected multiple injections of energetic particles roughly coinciding with the three events observed in GEOTAIL [Nagai, personal communication]. The mid-latitude magnetogram indicated the Pi-2 onset at 1208 UT [Yumoto, personal communication]. Thus the present observation was made in the course of a substorm.

In the lobe region after ~1235 UT, beam-like distributions at two discrete energies of ~100 eV/q and 1–2 keV/q are detected instead of the previously-observed hot component of the plasma sheet proper. The beams appear intermittently and in some cases alternatively.

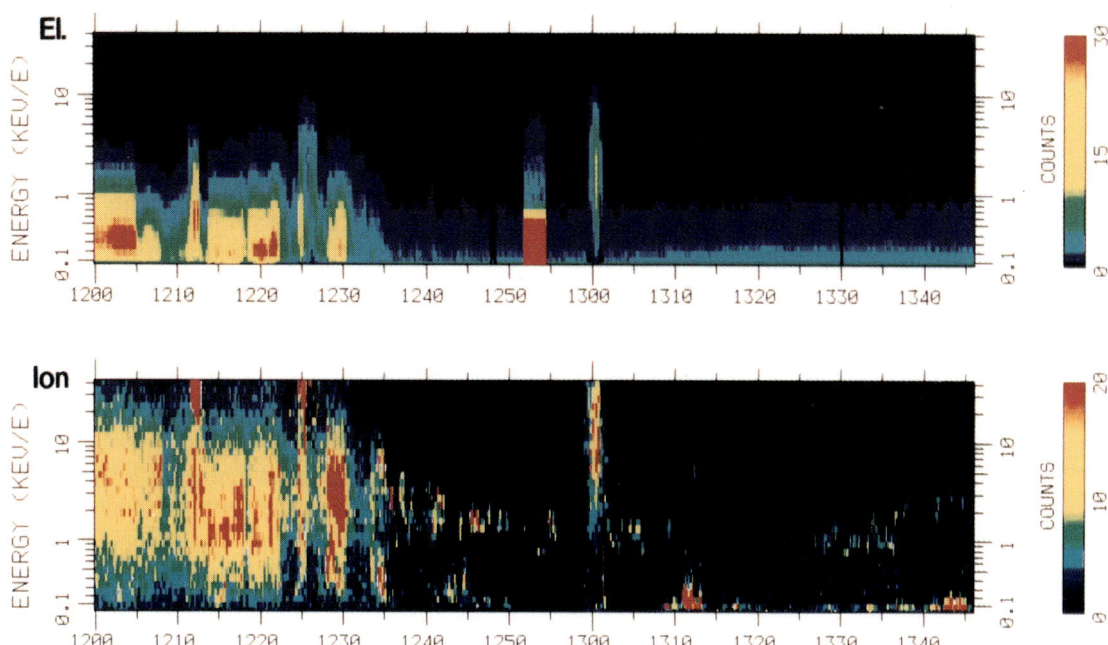

Plate 1. A summary of GEOTAIL/LEP electron (upper panel) and ion (lower panel) energy-time spectrograms on August 22, 1992. The abscissa is Universal Time from 1200 to 1346, and the ordinate is the ion energy on a logarithmic scale, except for lower energies below 0.4 keV/q where the scale is linear. Color-coded intensity shows the maximum of the 112 count data measured in seven elevation angles and 16 azimuthal sectors at a given energy and time so as to identify an event easily if it exists. It should be noted that the electron energy during the time interval of 1251:46−1254:22 UT is not scaled by the ordinate, since the electron energy range was changed to measure the detailed spectra in lower energies as shown in Plate 3.

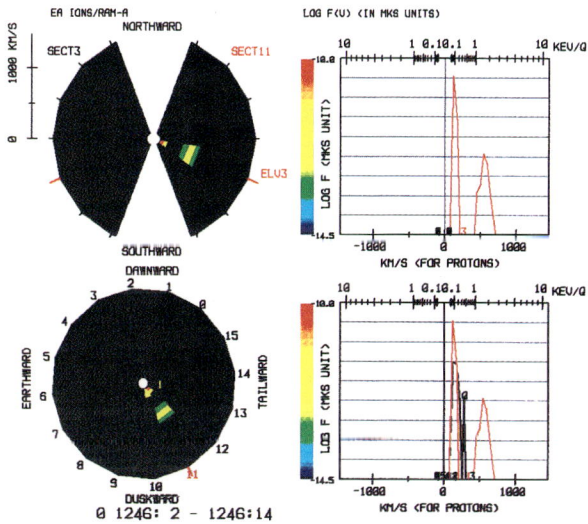

Plate 2. An example of the three-dimensional distribution of ions during the time interval of 1246:02−1246:14 UT in the lobe region. The radial scale in the left-hand panels and the abscissa scale in the right-hand panels are equal, and represent the ion velocity in km/s. Here the data in a lower energy range below 11.2 keV/q are only displayed to make the nature of the cold ion streams clear. The phase space density on a logarithmic scale are coded by color in the left-hand panels and are shown by the ordinates in the right-hand panels. Here all ions are assumed to be H^+.

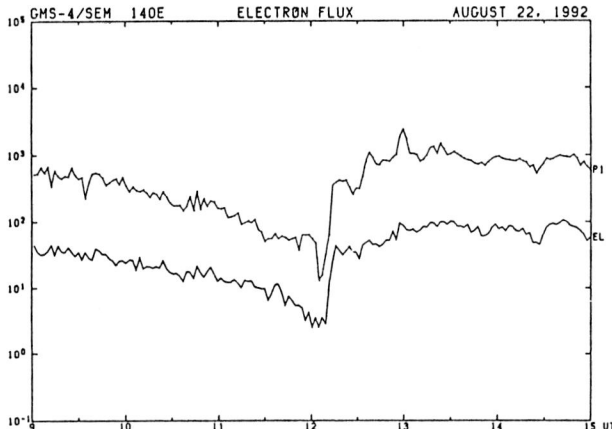

Fig. 1 Energetic particle data observed by the GMS-4 geosynchronous satellite.

They are what we call 'cold ion streams', since their velocities are narrowly confined around the bulk flow velocity, as shown in Plate 2. Plate 2 shows an example of the three-dimensional phase space distributions, where the scaling is for H^+ ions. For other ion species, the velocity scale should be divided by $(m/q)^{1/2}$ while the phase space density should be multiplied by $(m/q)^2$, where m and q are the mass number and the charge state of the ions, respectively. On the left-hand side in Plate 2, the three-dimensional nature of the distribution function is revealed by two panels; latitudinal (upper) and azimuthal (lower) distributions. The upper left panel shows the distribution in a meridional plane with an azimuthal angle of 112.5° relative to the GSE-X direction as indicated by a red mark (No. 11) in the lower panel, while the lower left panel shows the azimuthal distribution on the conical surface with an elevation angle of 22.5° southward of the spin plane (roughly, the ecliptic plane) as indicated by a red mark in the upper panel. In the upper and lower panels on the right, the red curves show the distribution functions along the lines indicated by the red marks in the corresponding left-hand panels. Black curves in the lower panel show distribution functions in other azimuthal angles. For example, a black line coexisting with the main peak at 80 eV/q (the red curve for Azimuth No. 11 as indicated in the lower left panel) shows the distribution function in Azimuth No. 10.

In Plate 2 two distinct peaks are evident at the energies of 80 eV/q and 1.6–1.9 keV/q at nearly the same elevation and azimuth. These energies, however, should be corrected by taking the spacecraft potential into account. In the present case the spacecraft potential can be estimated as +40 volts by the apparent cutoff of spacecraft photoelectrons in the electron data measured simultaneously with the EA-e sensor, as shown in Plate 3. Thus the lower energy peak is at 120 eV/q in ambient plasma, while the correction at the higher energy peak is negligible. Hence the ratio of the two peak energies is nearly 16. The energetic ion mass spectrometer (LEP-MS) also identified the mass-per-charge ratio of the higher energy component as 16 when the beam entered its field of view. We therefore conclude that the lower energy component corresponds to protons and the higher energy component to singly-charged oxygen ions. Both ion species appear to be streaming with nearly equal velocities.

The flow parameters of each ion species can be independently obtained by taking the velocity moments in the separate phase space volumes. Plate 4 shows temporal variation of the flow parameters for H^+ and O^+ from 1235 to 1346 UT, in which the spacecraft stayed in the lobe region. (These calculations are done assuming a constant spacecraft potential of +40 volts.) The H^+ number density is of order 10^{-2} cm^{-3}, while the O^+ density is of order 10^{-3} cm^{-3} but at times is enhanced sporadically by order of magnitude reaching 1×10^{-2} cm^{-3}; see the data in 1240–1250 UT. These values are comparable to the ion densities previously observed in the lobe region closer to the earth [Sharp et al., 1981; Eastman et al., 1984]. The H^+ density increased after 1308 UT, and the electron density profile [not shown] also shows the similar enhancement. It should be noted that Plate 1 shows the corresponding enhancement in the low-energy electron flux, and hence they represent not the spacecraft photoelectrons but ambient electrons in the lobe region. In Plate 4 both the H^+ and O^+ densities show significant temporal fluctuations which are consistent with their intermittent appearance in the E-t diagram of Plate 1. However, a similar feature could be observed if the flow directions of a very narrow cold beam fluctuated on a time scale comparable to the measurement cycle, even if the density fluctuations were small. Actually the cold ions were sometimes seen as a sharp and intense spot in a very narrow region of phase space, and in other cases as diffuse in a wider region of phase space. Detailed studies of such distribution functions and their temporal evolution will be pursued in the future.

It is reasonable to interpret these cold ions as having originated from the polar ionosphere and having been transported along magnetic field lines to the lobe region in the mid-magnetotail. In Plate 4 the speeds and directions of H^+ and O^+ flows agree well most of the time, though occasionally the flow directions of the two ions

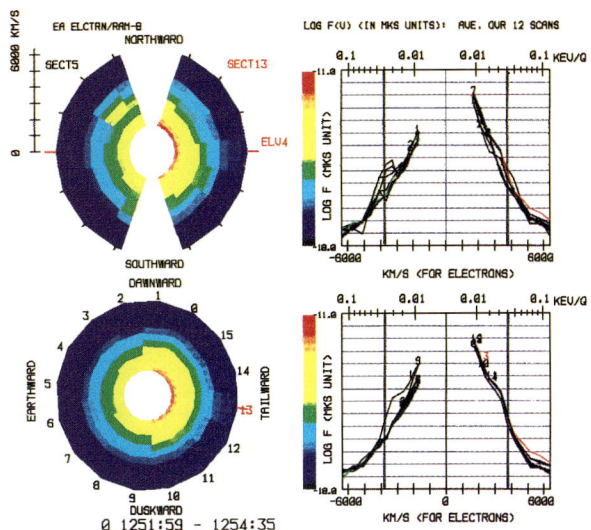

Plate 3. Three-dimensional distribution of electrons in the lobe region. The format is similar to that in Fig.3, except that the number flux is displayed instead of the phase space density.

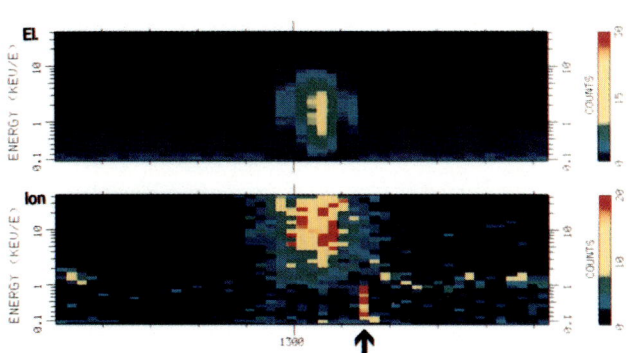

Plate 5. A blow-up of the electron and ion energy-time spectrograms during the time interval of 1255–1305 UT.

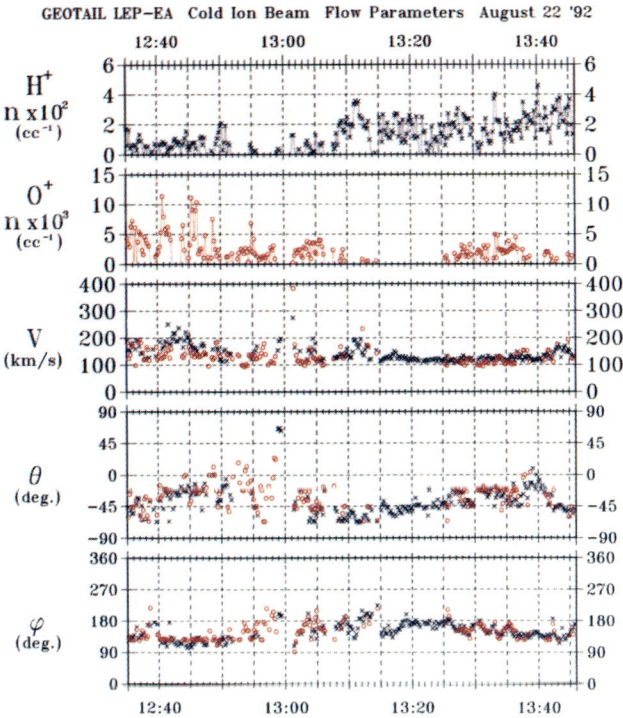

Plate 4. Flow parameters of cold ion streams in the lobe region; from top, H^+ densities, O^+ densities, flow velocities, and their elevation and azimuthal angles in the spacecraft coordinate system which is nearly identical to GSE. Blue marks indicate the H^+ data, while red one are the O^+ data.

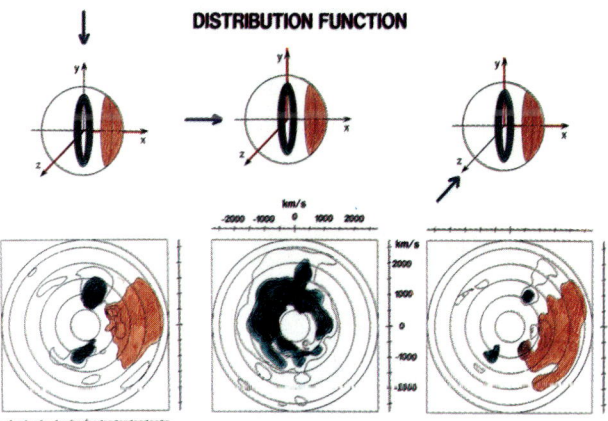

Plate 6. Distribution function of ions observed in a plasmoid at 1300:03 UT. Lower three panels show contour plots of phase space density on the three orthogonal planes as indicated in the corresponding top panels. As shown schematically in top panel, a red part of the distribution shows a fast ion flow in the tailward direction, while a green distribution indicates a ring distribution.

appear to be slightly different. This is consistent with a velocity filter effect due to **E** × **B** convection. The flow direction is generally tailward ($\phi \sim 110° - 210°$) and southward ($\theta < 0$) and the latter is qualitatively consistent with the expected direction of global convection in the northern lobe region. It is however noted that the azimuthal direction of the streaming is more or less duskward. This duskward flow is opposite to the direction expected from the ISEE results in regions closer to the earth [*Orsini et al.*, 1990]. In their paper the convection was observed to be directed toward the central axis of the plasma sheet as is consistent with a simple convection model proposed by Rostoker and Bostrom [1976]. Here we should note that the GEOTAIL spacecraft was located in the duskside lobe at $Y_{GSM} \sim 5$ Re, where dawnward streaming is expected.

Temporal variations in the flow velocities and directions around 1300 UT are also interesting, since the third event of the high-speed plasma flows, which seem to represent plasmoids [*Hones et al.*, 1984; *Baker et al.*, 1987; *Machida et al.*, preprint, 1993], was seen by the spacecraft during the interval of 1259:37 UT to 1301:09, as shown in Plate 5. Both of the cold ion species, H^+ and O^+ could not always be identified simultaneously, but in Plate 4 the combined data set clearly indicates that the elevation angle of the flow was changed from southward ($\theta \sim -30°$) to northward ($\theta \sim +67°$) for about one minute prior to the plasmoid arrival. It is most probable that this directional change was caused by fast MHD waves traveling ahead of the plasmoid. The fast waves would generate a TCR (Traveling Compressional Region) [*Slavin et al.*, 1984] where the plasma tends to be pushed aside (northward in the present case). This signature is also revealed by a MHD simulation related to the magnetic reconnection, as shown in Figure 2 [Maezawa, personal communication]. Inside the plasmoid, the cold ions could not simply be identified as the beams, but they are found out to form a ring-like distribution in the velocity space, as shown in Plate 6. The detailed description of this ring distribution will be published elsewhere [*Saito et al.*, in preparation]. Just after the plasmoid passage the cold ions were not only accelerated but also widely spread in energies, as shown by a black arrow in Plate 5. The acceleration might be caused by a slow shock transition behind the plasmoid passage [*Feldman et al.*, 1985]. The energy spreading may represent heating of the ions as well, but at present we cannot exclude another possibility that the apparent spreading of the energies was caused by faster temporal variation in the flow velocity than the measurement cycle (~ 13.1 sec.).

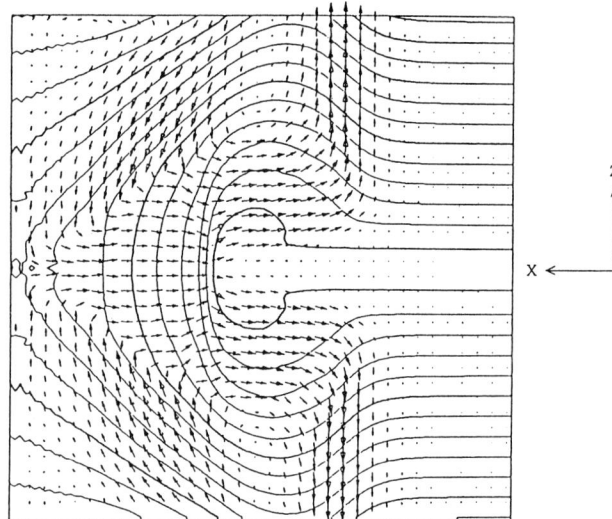

Fig. 2 MHD simulation result of flow velocities and magnetic fields related to the magnetic reconnection and the associated plasmoid formation. The scale in the X-direction is compressed by factor 5. [by courtesy of Maezawa]

DISCUSSION

We have observed tailward streaming of cold ions in the mid-tail lobe region at $X_{GSM} \sim -42$ Re, $Y_{GSM} \sim 5$ Re and $Z_{GSM} \sim 2$ Re in the course of a substorm on August 22, 1992. The ions consisted of protons and singly-charged oxygen ions flowing with nearly the same velocities. They are likely to have originated from the polar ionosphere. Upflowing ion conics and beams in the dayside cusp/cleft region are the most probable source, since the upflowing ions leaving the nightside auroral region would probably have been convected into the plasma sheet closer to the earth. Upflowing ion conics and beams are very frequently observed at altitudes of 1-2 Re in the dayside cusp/cleft region [*e.g., Lockwood et al.*, 1985; *Miyake et al.*, 1993]. They would be transported along magnetic field lines through the polar mantle and convected to the lobe region in the mid-magnetotail where the observation was made, as suggested by *Chappell* [1988] and *Delcourt et al.*[1993]. However, the existence of the heavy, cold ions like O^+ at this downtail distance of ~ 42 Re cannot be simply expected from the suggestion of *Chappell* [1988] and *Delcourt et al.*[1993]. The observed flow speed was generally in the 100-200 km/s range. After 1305 UT the velocity was rather stable at ~ 120 km/s, which corresponds to energy-per-charge of 75 eV/q for H^+ and 1.2

keV/q for O^+, respectively. This O^+ energy is much higher than that of upflowing ions usually observed at altitudes of 1–2 Re in the dayside cusp/cleft region. The H^+ energy is reasonable for this source region. Processes that heat the O^+ ions selectively may operate at higher altitudes; for example, multi-step heating, such as proposed by *Singh and Schunk* [1985] and/or ion-ion (H^+–O^+) two-stream instability [*Winglee et at.*, 1989].

It takes about one hour for ions to flow from the ionosphere to the observation site (X $\sim$ −42 Re) with a velocity of 100 km/s along the field line. In Plate 4, the H^+ density increased after 1308 UT, and this might be related to the first substorm onset around 1210 UT (that is, about one hour before the enhancement) in a sense that the ion outflow might be intensified after the substorm onset. This might also be due to some change in the magnetotail configuration and convection in relation to the plasmoid passage around 1300 UT.

The elevation angles of the flow directions observed in the northern lobe were significantly negative, that is, southward toward the plasma sheet. That the different ion species had equal velocities is consistent with the velocity filter effect due to $\mathbf{E} \times \mathbf{B}$ convection. However, the observed off-ecliptic angles as large as several tens degrees are difficult to understand if the cold ions stream along the magnetic field line with a small velocity component across the field line, as previously observed in regions closer to the earth [*e.g.*, *Candidi et al.*, 1982]. (It is however noted that the previous observations were limited to two-dimensional distributions in the spin plane, that is, the ecliptic plane.) The velocity component across the magnetic field line should correspond to the convection velocity. Since the elevation angle of the magnetic fields is typically below 10° in the lobe region at distances of $\sim$40 Re away from the earth, a significant part of the off-ecliptic angles must correspond to enhanced convection. If the elevation angle of 45° (seen mostly after 1305 UT in Plate 4) is regarded as being due to convection, we can estimate the convection speed as $\sim$ 100 km/s and the convection electric field as 1.5 mV/m (assuming a magnetic field of 15 nT). This is unreasonably large if the convection electric field is uniform along the dawn-dusk line (GSM-y direction). However, we should recall that the present observation was made during a plasma-sheet thinning after the second event of the high-speed plasma flow, and the third plasmoid event was also detected in the midst of this observation [*Machida et al.*, preprint, 1993]. The observed features seem to suggest that the lobe plasma was drawn toward the thinned plasma sheet by the enhanced convection. The enhanced convection would therefore be localized in the y coordinate. The region of the plasma sheet thinning would also be localized in the y direction and related to near-earth three-dimensional reconnection, and the associated plasmoid formation.

Temporal variations in the flow speed and its direction observed around 1300 UT are also quite interesting, as mentioned in the previous section. These variations are most likely to be related to the plasmoid passage. The three-dimensional distribution functions of ions in the plasmoid have also revealed interesting features, such as a ring distribution, the details of which will be described elsewhere [*Saito et al., in preparation*].

In summary, the initial GEOTAIL/LEP observations provided us with quite interesting information on the cold ion dynamics in the mid-magnetotail. However, a definite conclusion on the above discussion cannot be drawn by the present observations, since simultaneous data of the magnetic field were not available unfortunately. We hope that the GEOTAIL/LEP observations will be restarted in the near future and give further clue to the problem on the dynamics of ionospheric ions presented in this paper.

Acknowledgments. We thank all members of GEOTAIL project team in Japan as well as in the U.S.A for their extensive efforts to the success of GEOTAIL, especially K. T. Uesugi, the Project Manager in ISAS. We appreciate K. Maezawa for providing his unpublished result (Figure 2) of the MHD simulation and useful discussions on the dynamics of the plasma flow in relation to the plasmoid passage. We also thank T. Onsager and W. K. Peterson for discussions on the cold ion dynamics in the magnetotail lobe. The GMS-4 particle data were provided by Meteorological Satellite Center, Japan.

References

Akinrimisi, J., S. Orsini, M. Candidi, and H. Balsiger, Ion dynamics in the plasma mantle, *Ann. Geophys.*, 8, 739-754, 1990.

Baker, D. N., R. C. Anderson, and R. D. Zwickl, Average plasma and magnetic field variations in the distant magnetotail associated with near-earth substorm effects, *J. Geophys. Res.*, 92, 71-81, 1987.

Candidi, M., S. Orsini, and V. Formisano, The properties of ionospheric O^+ ions as observed in the magnetotail boundary layer and northern plasma lobe, *J. Geophys. Res.*, 87, 9097-9106, 1982.

Chappell, The terrestrial plasma source: a new perspective in solar-terrestrial processes from Dynamic Explorer, *Rev. Geophys.*, 26, 229-248, 1988.

Delcourt, D.C., J. A. Sauvaud, and T. E. Moore, Polar wind ion dynamics in the magnetotail, *J. Geophys. Res.*, 98, 9155-9169, 1993.

Eastman, T. E., L. A. Frank, W. K. Peterson, and W.

Lennartsson, The plasma sheet boundary layer, *J. Geophys. Res.*, *89*, 1553-1572, 1984.

Feldman, W. C., D. N. Baker, S. J. Bame, J. Birn, J. T. Gosling, E. W. Hones, Jr., and S. J. Schwartz, Slow-mode shocks: a semipermanent feature of the distant geomagnetic tail, *J. Geophys. Res.*, *90*, 233-240, 1985.

Frank, L. A., K. L. Ackerson, and D. M. Yeager, Observations of atomic oxygen (O^+) in the earth's magnetotail, *J. Geophys. Res.*, *82*, 129-134, 1977.

Hardy, D. A., J. F. freeman, and H. K. Hills, Double-peaked ion spectra in the lobe plasma: evidence for massive ions?, *J. Geophys. Res.*, *82*, 5529-5540, 1977.

Hones, E. W., Jr., D. N. Baker, S. J. Bame, W. C. Feldman, J. T. Gosling, D. J. McComas, R. D. Zwickl, J. A. Slavin, E. J. Smith, and B. T. Tsurutani, Structure of the magnetotail at 220 R_E and its response to geomagnetic activity, *Geophys. Res. Letters*, *11*, 5-7, 1984.

Lennartsson, W., R. D. Sharp, E. G. Shelley, R. G. Johnson, and H. Balsiger, Ion composition and energy distribution during 10 magnetic storms, *J. Geophys. Res.*, *86*, 4628-4638, 1981.

Lockwood, M., J. H. Waite, Jr., T. E. Moore, J. F. E. Johnson, and C. R. Chappell, A new source of suprathermal O^+ ions near the dayside polar cap boundary, *J. Geophys. Res.*, *90*, 4099-4116, 1985.

Miyake, W., T. Mukai, and N. Kaya; On the evolution of Ion Conics along the field line from Exos-D observations, *J. Geophys. Res.*, *98*, 11127-11134, 1993.

Mukai, T., S. Machida, Y. Saito, M. Hirahara, T. Terasawa, N. Kaya, T. Obara, M. Ejiri, and A. Nishida, The Low Energy Particle (LEP) experiment onboard the GEOTAIL satellite, *J. Geomag. Geoelectr., in press*, 1993.

Orsini, S., E. Amata, M. Candidi, H. Balsiger, M. Stockholm, C. Huang, W. Lennartsson, and P. -A. Lindquist, Cold ion streams of ionospheric oxygen in the plasma sheet during the CDAW 6 event of March 22, 1979, *J. Geophys. Res.*, *90*, 4091-4098, 1985.

Orsini, S., M. Candidi, M. Stockholm, and H. Balsiger, Injection of ionospheric ions into the plasma sheet, *J. Geophys. Res.*, *95*, 7915-7928, 1990.

Peterson, W. K., R. D. Sharp, E. G. Shelley, R. G. Johnson, and H. Balsiger, Energetic ion composition of the plasma sheet, *J. Geophys. Res.*, *86*, 761-767, 1981.

Rostoker, G., and R. Bostrom, A mechanism for driving the gross Birkeland current configuration in the auroral oval, *J. Geophys. Res.*, *81*, 235-244, 1976.

Sharp, R. D., D. L. Carr, W. K. Peterson, and E. G. Shelley, Ion streams in the magnetotail, *J. Geophys. Res.*, *86*, 4639-4648, 1981.

Sharp, R. D., W. Lennartsson, W. K. Peterson, and E. G. Shelley, The origin of the plasma in the distant plasma sheet, *J. Geophys. Res.*, *87*, 10420-10424, 1982.

Singh, N., and R. W. Schunk, A possible mechanism for the observed streaming of O^+ and H^+ ions at nearly equal speeds in the distant magnetotail, *J. Geophys. Res.*, *90*, 6361-6369, 1985.

Slavin, J. A., E. J. Smith, B. T. Tsurutani, D. G. Sibeck, H. J. Singer, D. N. Baker, J. T. Gosling, E. W. Hones, Jr., and F. L. Scarf, Substorm associated traveling compression regions in the distant tail: ISEE-3 geotail observations, *Geophys. Res. Lett.*, *11*, 657-660, 1984.

Winglee, R. M., P. B. Dusenbery, H. L. Collin, C. S. Lin, and A. M. Persoon, Simulations and observations of heating of auroral ion beams, *J. Geophys. Res.*, *94*, 8943-8965, 1989.

T. Mukai, Y. Saito and A. Nishida, The Institute of Space and Astronautical Science, 3-1-1 Yoshinodai, Sagamihara, Kanagawa 229, Japan.

M Hirahara and T. Terasawa, Department of Earth and Planetary Physics, Faculty of Science, University of Tokyo, 7-3-1 Hongo, Bunkyo-ku, Tokyo 113, Japan.

S. Machida, Geophysical Institute, Faculty of Science, Kyoto University, Kitashirakawa-Oiwake, Sakyo-ku, Kyoto 606-01, Japan.

Temporal Evolution and Spatial Dispersion of Ion Conics: Evidence for a Polar Cusp Heating Wall

D. J. KNUDSEN, B. A. WHALEN, T. ABE[1], AND A. YAU

National Research Council of Canada, Ottawa, Ontario

Whalen et al. [1991] presented EXOS-D Suprathermal Ion Mass Spectrometer measurements of ion conic distributions which appear at a sharp equatorward boundary of the cusp/cleft region, exhibiting pitch angles which are initially perpendicular to the geomagnetic field, and which become more field aligned with increasing latitude. The folding of the distributions towards the (anti)field-aligned direction occurred more slowly for heavier ions. By solving the equation of motion for 150 eV ions subject to the gradient-B force, we model conic pitch angles as a function of latitude and ion mass for a single example and show that the measured pitch angle behavior for the observed suprathermal ions can be explained by a wall-like polar cusp heating region through which ions convect northward. We find that most of the heating took place within a region less than roughly 30 km wide, and which extended from 2,000 km altitude up to and probably beyond the satellite altitude of 8,000 km. While the heating wall plus northward drift scenario can account for the latitude and mass dependence of conic pitch angles, it does not explain the measured energy dependence, or rather lack thereof, to those angles. That is, lower energy ions should have smaller pitch angles than their higher energy counterparts for fixed time-of-flight, while the data show thermal and suprathermal conics with similar pitch angles. This discrepancy between our model and observations remains a topic for future study.

INTRODUCTION

The mid-altitude (~ 1 Re) polar cusp is populated by ions of both magnetosheath and of ionospheric origin. The often sharp equatorward boundary of this region coupled with the common occurrence of anti-sunward convection leads to distinct latitudinal profiles of precipitating magnetosheath ion energy, mass, and pitch angle. For example, the energy-dependent ion time-of-flight from the high altitude injection region produces latitude-energy dispersion [*Shelley et al.*, 1976; *Reiff et al.*, 1977] and pitch angle-energy dispersion [*Burch et al.*, 1982]. *Carlson and Torbert* [1980] measured the mass dependence of the energy dispersion effect from a sounding rocket. See *Peterson* [1985] for a review of these phenomena.

In this paper we study two other dispersion signatures in cusp region ions, namely the variation of ion pitch angle with latitude and with ion mass previously reported on EXOS-D (Akebono) with the Suprathermal Ion Mass Spectrometer (SMS) [*Whalen et al.*, 1991]. In contrast to properties of precipitating magnetosheath ions mentioned above, the signatures with which we are concerned appear in upflowing ion populations, at thermal (0-70 eV) and suprathermal (0.1-2 keV) energies. We will argue that these signatures, like those found in the populations of magnetosheath origin, are a consequence of a sharp equatorward boundary of the cusp, with an embedded ion energization source, and subsequent northward drift.

Ion heating and the formation of ion conics are fascinating phenomena in their own right, but there is added impetus to study them for the important role they play in transporting ionospheric ions to the magnetosphere [*Chappell et al.*, 1987]. While the polar wind alone, driven by ambipolar diffusion, supplies H^+, He^+ and O^+ to the magnetosphere [*Abe et al.*, 1993], additional energization mechanisms can intensify this flux and enhance heavy ion outflow [see the review by *Yau and Lockwood*, 1988]. The additional energization appears to take place in the direction perpendicular to the geomagnetic field, and hence is termed transverse ion energization (TIE).

Since the original identification of TIE [*Sharp et al.*, 1977], there have been substantial efforts towards understanding

---

[1] Now at College of Science and Engineering, Aoyama Gakuin University, Tokyo.

Solar System Plasmas in Space and Time
Geophysical Monograph 84
Published in 1994 by the American Geophysical Union.

the mechanisms responsible for the energization, and towards the related problem of identifying the geophysical conditions under which those mechanisms operate [see the review by *Yau and Whalen*, 1993]. TIE occurs most frequently in the nightside, low-altitude auroral zone and in the polar cusp and cleft regions. Our particular interest in this study is the polar cusp TIE region.

To explain DE-1 observations, *André et al.* [1990] invoked a narrow heating region at the equatorward boundary of the cusp/cleft through which plasma is heated as it convects northward. If it is narrow enough in latitude, we show that such a region leads to and can explain Akebono SMS observations of ion conic pitch angles which become more anti-field-aligned with increasing latitude and with decreasing ion mass.

INSTRUMENTATION

The SMS instrument on the EXOS-D (Akebono) spacecraft is a newly developed radio frequency ion mass spectrometer, and is fully described in *Whalen et al.* [1990]. Briefly, the instrument is composed of an energy bandpass and aperture selector section, r.f. velocity selector section, and total energy (mass) dispersing section. In this paper we discuss measurements made with the largest aperture, which is used mainly at mid to high altitudes. In the thermal mode, large aperture, the high energy end of the bandpass is approximately 70 eV and the low end is determined by the setting of the retarding potential analyzer (RPA) grid voltage. In the energetic mode, the instrument has an E/Q bandpass of 70 eV and a center energy E/Q range from 150 V to 4330 V. The mass selected by the instrument is determined by the r.f. frequency setting and has a range of $0.8 \leq m/Q \leq 70$ AMU/q.

The angular response and energy resolution in the thermal mode are complicated functions of aperture selected and incident ion energy distribution. For a typical ionospheric plasma, the large aperture has an angular half width ($\Delta\Theta$) in the spin-scanned plane of about 60° (i.e. ±30° about the central look direction).

EXOS-D was launched on February 21, 1989 from Kagoshima Space Center, Japan into a highly elliptical orbit (initial apogee 10,000 km and perigee 270 km). The spacecraft spin period is 8 s and the spin axis points to the sun within 1 degree.

DATA

Whalen et al. [1991] presented EXOS-D observations above 4000 km altitude showing a sudden onset of heating perpendicular to B_0 in H^+, He^+, and O^+ ions. As the satellite traveled northward, the observed conics "folded up", i.e. became more field-aligned, with H^+ folding more quickly than either He^+ or O^+.

Figure 1 illustrates these properties in a similar pass, on February 18, 1990. This day was moderately active magnetically, with $K_p = 4$ and the IMF B_z component weak and southward near the time of the pass. The top four panels show thermal ion counts (color coded) as a function of satellite spin angle and time. The geomagnetic field B_0 was oriented at an angle of 19° with respect to the satellite spin plane, thus a spin angle of 180° corresponds to ions flowing upwards with a pitch angle of 161°. The modulation of ion counts along the time axis is due to retarding potential analyzer (RPA) sweeps between 0 and 20 eV, shown by the red traces in the (small) fifth panel.

The bottom two panels show H^+ and O^+ ions, also as a function of spin angle, time, and energy, the latter of which is varied between 148.5 eV and 2 keV (white trace in the bottom panel) and has an energy bandpass of 70 eV. The H^+ panel shows an isotropic background population characteristic of magnetosheath precipitation. This observation along with intense soft electron precipitation and an energetic ion dispersion event detected by the EXOS-D Low Energy Particle (LEP) instrument (not shown) [T. Mukai, personal communication, 1993] indicate that Akebono was in the polar cusp during the event shown in Figure 1.

As the satellite traveled northward near local noon, it encountered an abrupt onset near 03:37 UT of fairly intense fluxes of warm (upwards of ~ 10 eV) H^+, He^+, and O^+. The peak flux was initially near 90° in spin angle (also 90° pitch angle), and migrated towards 180° spin angle with increasing time and latitude. The same feature is visible in the suprathermal energy panels (bottom), both in H^+ and (at lower intensities) O^+. Although not shown in this display, the suprathermal (148.5 eV) He^+ ions behave similarly. The suprathermal ions exhibit pitch angle distributions which are much narrower than their thermal counterparts, in large part due to the energy-dependent angular resolution of the instrument. The event continued for several minutes after onset, although the increase in pitch angles reverses after 120 s or so.

Figure 2 shows thermal ion counts for three ion masses versus retarding potential analyzer (RPA) setting (horizontal axes) and time (top to bottom). These RPA scans are taken in the direction of peak ion flux. The topmost plots in Figure 2 show data taken at or slightly before the onset of intense thermal ion fluxes, and demonstrate that thermal ion flux into the instrument is strongly attenuated when the RPA potential reaches a few volts, which is consistent with fairly cold, rammed plasma. Sixteen seconds later (two spin periods) significant ion counts persist above an RPA setting of +15 V, and in the following panels (16 and 32 s later) the RPA has an even smaller effect on ion count, which is consistent with ion temperatures of several tens of eV.

The measured RPA profiles may be used to derive ion drift velocity, spacecraft potential, and ion temperature. To a first approximation, the shallowing of the RPA profiles with time indicates a core ion heating rate of a few eV/s, with most of the heating taking place within 30 s or so as measured by the spacecraft. As we will argue later, the northward ion drift speed is roughly 60% faster than the northward component of the spacecraft velocity, which al-

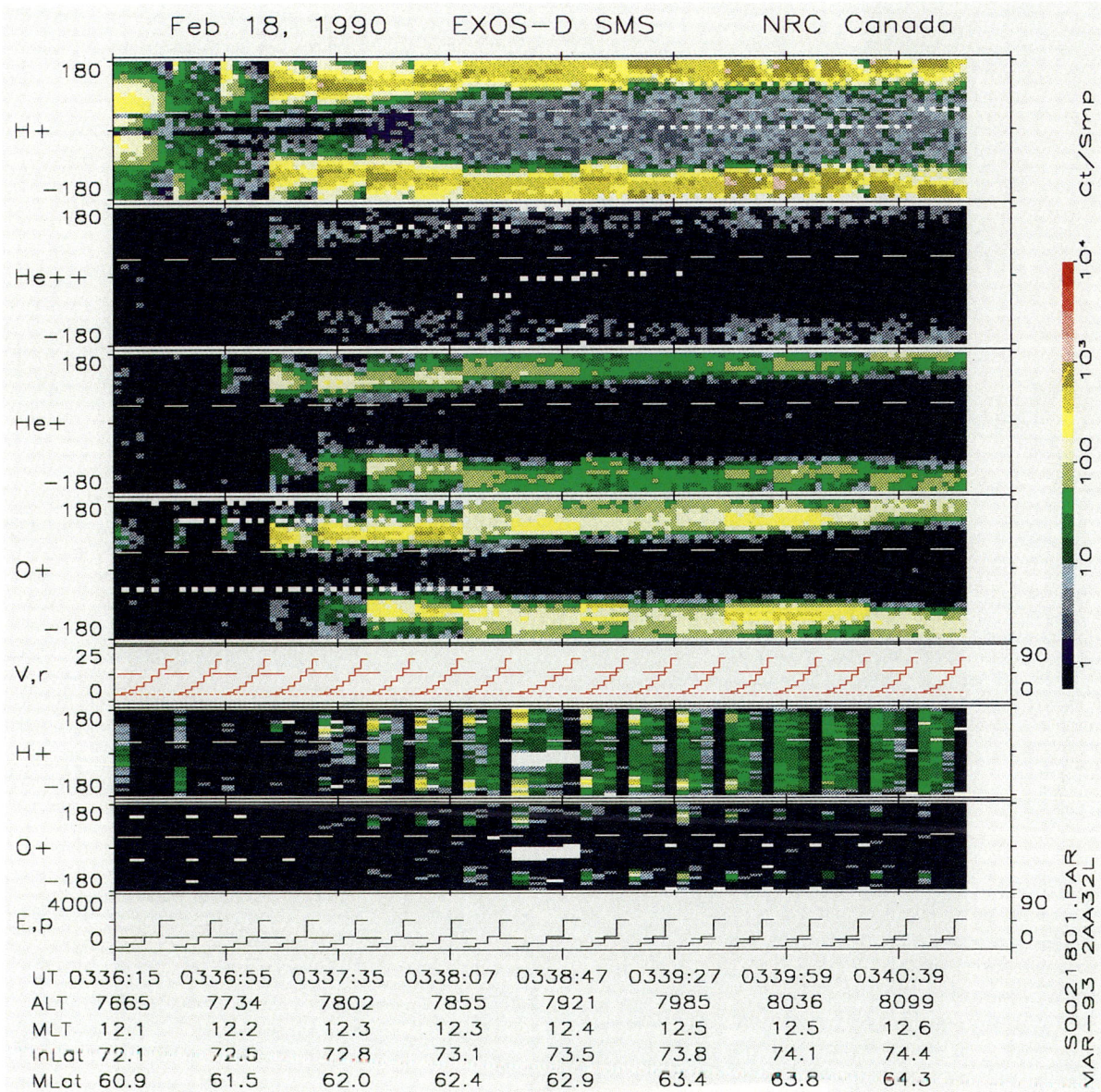

Fig. 1. Thermal (0-70 eV, top four panels) and suprathermal (0.1-2 keV, bottom two panels) ion counts versus spin angle (vertical axis) and time as Akebono traveled northward into the dayside cusp region. Upflowing conic distributions appear as double peaks in count rate at spin angles ϕ such that $|\phi| > 90°$. Apparent in this display are the increase of the spin (and pitch) angle of peak flux with increasing time (and latitude), and the variation of peak flux angle with ion mass. The RPA voltage for the thermal ions is shown in the (small) fifth panel.

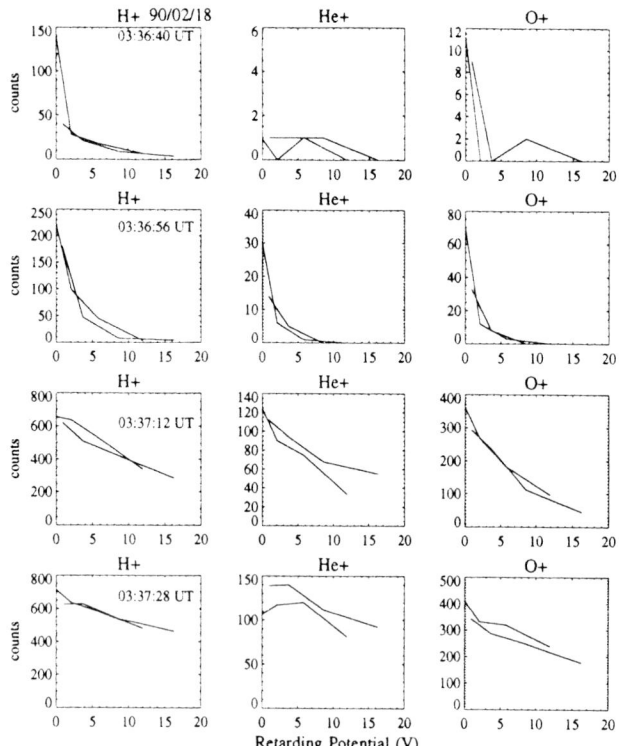

Fig. 2. Ion counts in the direction of peak ion flux for H$^+$, He$^+$, and O$^+$ versus retarding potential analyzer (RPA) voltage and time (top to bottom). The shallowing of the profiles over the 48 s covered by the plot is an indication of ion heating at a rate of a few eV/s for all species. The two lines in each frame result from the fact that the RPA makes two interleaving sweeps during the 64 ms required cover all eight RPA voltages, thus differences between the two curves indicate a change in ion flux between sweeps.

lows us to estimate the heating time in the plasma rest frame to be of the order of 20 s and the width of the heating region to be ~30 km in north-south extent.

As the core ion heating progresses, significant fluxes near 90° pitch angle appear in the 148.5 eV energy channels near 03:37:30 UT. As with the thermal ions, the pitch angles of the suprathermal ions increase with time. The average H$^+$ count rate is 5-10 times higher than that of the O$^+$ or He$^+$ (not shown). Some ions appear in the 318 eV energy channels, near the pitch angles of the 148.5 eV conics, but the count rate for these higher energy ions is an order of magnitude smaller than for the 148.5 eV ions.

PITCH ANGLE EVOLUTION

Measurements

Our intention is to exploit pitch angle information in order to model the morphology of the cusp ion heating region. For a given mass and energy, SMS made a measurement in the suprathermal mode every 11.25° in spin angle. Typically,

count rate peaks in 150 eV conics were ~20° wide in pitch angle, and we estimated the angle of peak flux by fitting a Gaussian angular profile to the measured counts.

The symbols in Figure 3 show fitted peak pitch angles for H$^+$, He$^+$, and O$^+$ in the 110-180 eV channel during the first two minutes of the heating event. The northward component of the satellite velocity was ~ 1 km/s, therefore the plot covers roughly 150 km in latitudinal extent. Pitch angles show a clear increase with latitude and decrease with ion mass, although this trend breaks down after 100 s or so. The solid lines are the modeled pitch angles described below.

Model

André et al. [1990] invoked a wall-like ion heating region near the equatorward boundary of the cusp/cleft to explain features of DE-1 conic observations. In an attempt to explain the pitch angle behavior measured by EXOS-D/SMS, we consider a similar structure with the following properties: (1) the heating region is a two-dimensional sheet extended in altitude and longitude, (2) cold ions of three ion species (H$^+$, He$^+$, and O$^+$) convect through this region with northward velocity $V_{d,N}$, and (3) some fraction of each cold ion population is energized instantaneously to 150 eV, after which these ions are subject only to the gradient-B force:

$$\ddot{r} = -\frac{\varepsilon_\perp}{m_i}\frac{\nabla B}{B} \quad (1)$$

where a dot denotes a time derivative, $\ddot{r}$ is the upward acceleration of a single ion, $\varepsilon_\perp$ is the perpendicular energy, m_i is ion mass, and B is the magnitude of the geomagnetic field, which we assume varies as r^{-3}.

Equation (1) is solved as a two-point boundary value problem which we integrate numerically, using the satellite

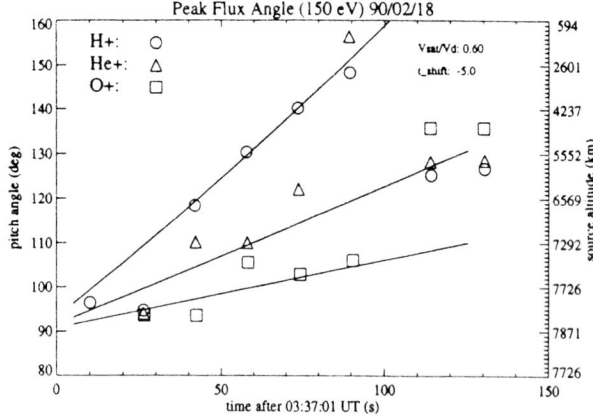

Fig. 3. The symbols are measured pitch angles of peak flux in the 110-180 eV energy channel for three different ion species versus time. The solid lines are modeled pitch angles assuming all ions were energized to 150 eV at 03:36:56 UT and subsequently drifted northward at 1.6 km/s while accelerating upward due to the gradient-B force.

altitude as the upper boundary of the integration, while the lower boundary follows from the assumption that the initial field-aligned velocity of ions is negligible, i.e.

$$\dot{r}(t=0) = 0. \quad (2)$$

A third constraint on the integration is that time-of-flight is fixed by the time it takes an ion to convect from the heating region to the invariant latitude at which it is measured by EXOS-D.

We solve (1) and (2) for the initial altitude $r(0)$, which allows us to predict the measured pitch angle α assuming a dipole geomagnetic field and conservation of the first adiabatic invariant:

$$\sin^2(180° - \alpha) = (r(0)/r_{satellite})^3 \quad (3)$$

The 180° shift in α is necessary to produce pitch angles greater than 90° (i.e. moving antiparallel to $\mathbf{B_0}$) for upward flowing ions in the northern hemisphere.

The fit to data produced by the above model is a function of two parameters, namely the time of energization and the ratio of the northward components of the satellite and plasma velocities, $V_{sat}/V_{d,N}$. One can get an estimate of the time of energization directly from the summary plot in Figure 1, and minor adjustments can be made to this value to obtain an optimum fit. In the model curves shown in Figure 3, we used 03:36:56 UT as the time of energization and a northward drift velocity of 1.6 km/s. The latter value was chosen for good fit (by eye) to measured pitch angles for all three ion species. The fit to the data is quite good for all three ion masses during the first 120 s, lending strong support to the "heating wall" model. After 120 s, the ion pitch angles do not seem to follow any obvious pattern.

In addition to the wall-like morphology of the heating region, the second major assumption we make in our model is that there is anti-sunward flow through the cusp during the event of interest. Data from the EXOS-D electric field instrument (EFD) (courtesy of H. Hayakawa, not shown here) indicate antisunward convection at 1-2 km/s just before the event onset, although inside the cusp the electric field becomes highly structured, precluding accurate measurement of the average flow component. Nevertheless, the EFD data are in general agreement with our assumption of antisunward drift. A more detailed comparison between drifts measured by EFD and fits from our model will be included in a follow-on study.

To this point we have modeled only ions in the 110-180 eV energy channel. In Figure 4 we have re-plotted the fitted pitch angles for suprathermal ions shown in Figure 3, but we have added fits to the thermal (0-70) eV ion pitch angles as well, shown as solid symbols. We have also plotted model curves for 150 eV and 50 eV H^+ obtained by integrating Equation (1) as before. Here we find that the model does not correctly predict thermal ion pitch angles. Instead, the

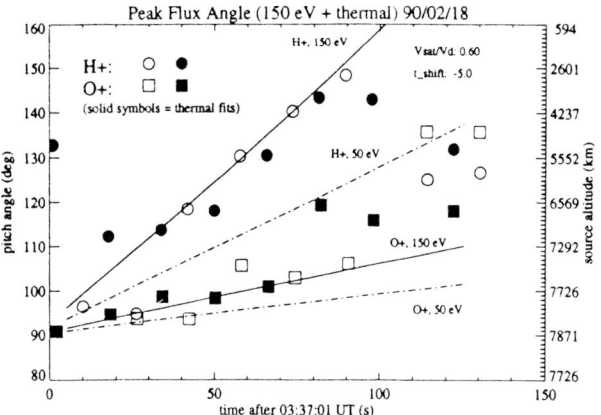

Fig. 4. Similar to Figure 3, with the addition of measured pitch angles of peak flux for thermal (0-70 eV) ions, shown by solid symbols. Also shown are modeled pitch angles for 50 eV ions, shown by dot-dashed lines. This plot illustrates that while the latitudinal (proportional to time) and mass dependencies of the pitch angles match the model curves well, the energy dependence does not. That is, the thermal ion pitch angles more closely follow the 150 eV model curves than the dot-dashed 50 eV model curves.

pitch angles of lower energy ions are very similar to the 150 eV pitch angles. We will discuss this point below.

CONCLUSIONS AND DISCUSSION

EXOS-D data shown here and by *Whalen et al.* [1991] contain examples near the mid-altitude polar cusp/cleft region in which several ion species were energized in the direction perpendicular to the geomagnetic field, with ion pitch angles becoming more field-aligned as the satellite traveled northward, and with light ion distributions folding upward more quickly than heavier ones. While we have treated only one example in this paper, these features are observed fairly regularly when EXOS-D is near the cusp/cleft region.

The mass and latitude dependence of 110-180 eV ion pitch angles in the example presented here fit well with a model comprised of (1) an L-shell-aligned sheet- or wall-like heating region near the equatorward boundary of the cusp, extended in altitude and longitude, (2) sudden energization to 150 eV of H^+, He^+, and O^+ ions, (3) northward drift out of the heating region at 1.6 km/s, and (4) upward adiabatic acceleration via the gradient-B force. These components and their effects on ion motion are summarized in Figure 5.

The agreement between model and measurements is good, but only within 100-200 km of the equatorward boundary of the cusp. Northward of this region, ion pitch angles do not depend in an obvious way on latitude or mass.

Even close to the ion heating region, an as yet unexplained disparity between our model and observations is that we measure similar pitch angles for both thermal and suprathermal ions, while acceleration by the gradient-B force would predict more field-aligned pitch angles for suprathermal ions given fixed time-of-flight. That is, higher energy ions should

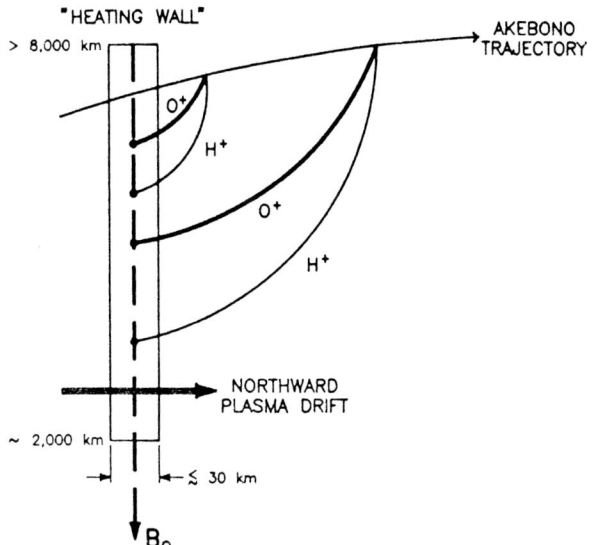

Fig. 5. Schematic drawing of the heating wall plus northward drift model and its effect on trajectories of ions with differing masses, leading to latitudinal and mass dispersion of ion conic pitch angles.

have larger pitch angles, while we measure no strong energy dependence. Future improvements to our model which may resolve this disparity could include macroscopic forces from, for example, (small) parallel electric fields, and/or microphysics such as velocity space and pitch angle diffusion which could in some way couple together pitch angles at different energies.

While we do not attempt to identify the actual ion heating source, our observations do provide some constraints on candidate mechanisms. Firstly, during the time between consecutive spins measuring suprathermal ions (16 s) we do not detect any substantial mass dependence in the heating rate, nor in the maximum energy attained by the ions. Some fraction of all three ion species is heated to between 110-180 eV at a rate of at least 10 eV/s. Few ions appear in the next higher energy channel (280-350 eV), implying an average heating energy not exceeding, say, 200 eV. While the average energy and ion heating rate are not strongly mass dependent, the flux of H^+ ions at 150 eV is 5-10 times more intense than the He^+ and O^+ fluxes. This could reflect the mass distribution of the cold plasma drifting into the heating region. Second, a consequence of our model is that ions with larger pitch angles were energized at lower altitudes. Since we measure similar suprathermal ion energies at all pitch angles, we infer that the ion energization rate is to first order independent of altitude between 2,000 and 8,000 km. A third constraint provided by the measurements is that the most intense energization occurs within a narrow latitudinal band of the order of or less than 30 km wide.

Finally, we briefly consider an alternative heating region morphology. *Whalen et al.* [1978] measured $\sim$ 100-400 eV ion conics in the topside, night-time ionosphere exhibiting $90°$ pitch angles at 400 km altitude and larger pitch angles above that altitude. Furthermore, the pitch angles were not strongly dependent on ion energy. Both of these attributes are consistent with a horizontal layer in which ions were heated. While such a layer could account for the lack of energy dependence in our dayside cusp measurements, it can explain neither the mass nor latitude variation of pitch angles, unless a) each species is heated in a different layer at a different altitude, and b) the altitude of each layer is a function of latitude. An alternative to a) would be that ions continue to be heated at a mass-dependent rate as they travel upwards. While it is unlikely that such an explanation is responsible for our measurements in the vicinity of the polar cusp boundary, horizontally stratified heating layers may be responsible for some of the features observed further north in the cusp.

Acknowledgments. The authors are grateful for supporting data supplied by H. Hayakawa and T. Mukai. Development of the SMS instrument was supported by the Canadian Space Agency.

REFERENCES

Abe, T., B. A. Whalen, A. W. Yau, R. E. Horita, S. Watanabe, and E. Sagawa, EXOS D (Akebono) suprathermal mass spectrometer observations of the polar wind, *J. Geophys. Res.*, *98*, 11,191, 1993.

André, M., G. B. Crew, W. K. Peterson, A. M. Persoon, C. J. Pollock, and M. J. Engebretson, Ion heating by broadband low-frequency waves in the cusp/cleft, *J. Geophys. Res.*, *95*, 20,809, 1990.

Burch, J. L., P. H. Reiff, R. A. Heelis, J. D. Winningham, W. B. Hanson, C. Gurgiolo, J. D. Menietti, R. A. Hoffman, and J. N. Barfield, Plasma injection and transport in the mid-altitude polar cusp, *Geophys. Res. Lett.*, *9*, 921, 1982.

Carlson, C. W., and R. B. Torbert, Solar wind ion injections in the morning auroral oval, *J. Geophys. Res.*, *85*, 2903, 1980.

Chappell, C. R., T. E. Moore, and J. H. Waite, The ionosphere as a fully adequate source of plasma for the Earth's magnetosphere, *J. Geophys. Res.*, *92*, 5896, 1987.

Peterson, W. K., Ion injection and acceleration in the polar cusp, in *The Polar Cusp*, J. A. Holtet and A. Egeland (eds.), pp.67-84, D. Reidel, Norwell, Mass., 1985.

Peterson, W. K., E. G. Shelley, S. A. Boardsen, D. A. Gurnett, B. G. Ledley, M. Sugiura, T. E. Moore, and J. H. Waite, Transverse ion energization and low-frequency plasma waves in the mid-altitude auroral zone: A case study, *J. Geophys. Res.*, *93*, 11,405, 1988.

Reiff, P. H., T. W. Hill, and J. L. Burch, Solar wind plasma injection at the dayside magnetospheric cusp, *J. Geophys. Res.*, *82*, 479, 1977.

Sharp, R. D., R. G. Johnson, and E. G. Shelley, Ovservation of an ionospheric acceleration mechanism producing energetic (keV) ions primarily normal to the geomagnetic direction, *J. Geophys. Res.*, *82*, 3324, 1977.

Shelley, E. G., R. D. Sharp, and R. G. Johnson, Satellite observations of an ionospheric acceleration mechanism, *Geophys. Res. Lett.*, *3*, 654, 1976.

Whalen, B. A., W. Bernstein, and P. W. Daly, Low altitude acceleration of ionospheric ions, *Geophys. Res. Lett.*, *5*, 55, 1978.

Whalen, B. A., S. Watanabe, and A. W. Yau, Observations in the transverse ion energization region, *Geophys. Res. Lett.*, *18*, 725-728, 1991.

Whalen, B. A., J. R. Burrows, A. W. Yau, E. E. Buczinski, A. M. Pilon, I. Iwamoto, K. Marubashi, S. Watanabe, H. Mori, and E. Sagawa, The suprathermal ion mass spectrometer (SMS) onboard the Akebono (EXOS-D) satellite, *J. Geomag. Geoelectr., 42,* 511, 1990.

Yau, A. W., and M. Lockwood, Vertical in flow in the polar ionosphere, in *Modeling Magnetospheric Plasma, Geophys. Monogr. Ser.,* vol. 44, edited by T. E. Moore and J. H. Waite, Jr., pp. 229-240, AGU, Washington, D.C., 1988.

Yau, A. W., and B. A. Whalen, Ion acceleration in the low- and mid-altitude auroral ionosphere, in *Auroral Plasma Dynmaics, Geophys. Monogr. Ser.,* edited by B. Lysak, AGU, Washington, D. C., 1993.

T. Abe, College of Science and Engineering, Aoyama Gakuin University, 6-16-1, Chitosedai, Setagaya-ku, Tokyo 157, JAPAN

D. J. Knudsen, B. A. Whalen, A. W. Yau, National Research Council of Canada, 100 Sussex Drive, Ottawa, Ontario, CANADA K1A 0R6

Temporal and Spatial Signatures in the Injection of Magnetosheath Plasma into the Cusp/Cleft

R. M. Winglee,[1] J. D. Menietti,[2] W. K. Peterson,[3] J. L. Burch and J. H. Waite, Jr.[4]

Within the cusp/cleft region, there is a mixing of magnetosheath and ionospheric plasma. It is shown through high time resolution observations from HAPI and EICS on board DE 1 that the magnetosheath injection is highly structured, and leads to the modulation of upwelling ionospheric ions. In particular, it is shown that there can be a direct transfer of momentum from the downflowing magnetosheath up to the upwelling ions, raising the energy of the ionospheric ions from about 40 eV to about 100 eV with a tail extending out to about 400 eV. Whether the observed modulations are temporal and/or spatial cannot be unambiguously determined from the observations. Further insight is attained through a comparison of the observations with meso-scale particle simulations for the injection of magnetosheath plasma into the cusp/cleft region and its interaction with the ionospheric plasma. Such simulations are able to directly isolate temporal and spatial effects. Important spatial effects are shown to arise from a limited source region for the magnetosheath injection and the convection of various plasmas across the cusp. Temporal effects arise from changes in the characteristics of the injected magnetosheath plasma. These combined temporal and spatial effects give rise to several of the unique features observed in the particle distributions.

1. INTRODUCTION

The cusp/cleft is a region of dayside high latitude magnetosphere, extending from less than 0900 to more than 1500 magnetic local time and with a latitudinal width of a few degrees. Energy from the solar wind flows into this region [*Rosenbauer et al.*, 1975] and can produce strong heating of ionospheric plasma. As a result of this heating, there is an enhanced outflow of ionospheric plasma out into the magnetosphere, making the cusp/cleft an important source of plasma to the magnetosphere [e.g., *Waite et al.*, 1985; *Peterson*, 1985; *Moore et al.*, 1986; *Chappell et al.*, 1987; *Roberts et al.*, 1987; *Pollock et al.*, 1990].

One of the outstanding issues at this time is identifying the actual processes coupling the energy from the shocked solar wind (magnetosheath) plasma and the ionospheric plasma. *Roth and Hudson* [1983, 1985] proposed heating by lower hybrid waves driven by a ring distribution (i.e., one which is peaked at $v_\parallel = 0$ and $v_\perp \neq 0$) associated with the mirroring of magnetosheath ions. While this mechanism can generate conic distributions particularly in the H^+ ions, observations indicate that conics can be present in the apparent absence of any magnetosheath injection. Thus, the relative importance of this mechanism in generating ionospheric conics has not been established.

An alternate mechanism proposed by *André et al.* [1988, 1990] is heating via broadband waves near the ion cyclotron frequency. This mechanism is an extension of recent models for the perpendicular heating of ions in the nightside auroral region [*Chang et al.*, 1986; *Crew et al.*, 1990]. While this mechanism can produce many of the features of the observed conics, the source of the postulated waves has not been established, although the presence of field-aligned currents could be an important source [cf., *Winglee et al.*, 1987, 1988].

In this paper, observations from DE 1 in conjunction with particle simulations are used to investigate the coupling between the magnetosheath and ionospheric plasmas. The results from the observations

---

[1]Geophysics Program, University of Washington, Seattle, Washington.
[2]Department of Physics and Astronomy, University of Iowa, Iowa City, Iowa.
[3]Space Science Laboratory, Lockheed, Palo Alto, California.
[4]Instrumentation and Space Research Division, Southwest Research Institute, San Antonio, Texas.

and simulations are summarized in the schematic in Figure 1. The injection of the magnetosheath plasma into the cusp/cleft region is associated with an alternating field-aligned current system driven by the slightly different convection patterns of the magnetosheath electrons and ions. Cold upwelling ions are seen in association with both upward and downward components of these field-aligned currents. However, the magnetosheath injection is seen to be modulated on short time scales of about 18–30 seconds in the spacecraft frame. This modulation is also seen in the flux of the upwelling ions which suggests a direct causal link between the upwelling and the magnetosheath plasma. In particular, it is shown that during periods of high magnetosheath ion intensity, the cold upwelling ions experience additional heating through an ion-ion streaming instability that develops between the ionospheric and magnetosheath ions. This additional interaction raises the average energy of the upwelling ions to about 100 eV with a high energy tail out to about 400 eV. During some of the injection events there is clear evidence that a downward conic, of presumably ionospheric origin, is produced by the interaction.

2. HIGH TIME RESOLUTION OBSERVATIONS FROM DE 1

The present study utilizes observations from DE 1 on Day 81295. This crossing has been previously discussed by *Burch et al.* [1985] in terms of global flows and currents. The encounter was at high altitudes at about 13,700 km at about 9am MLT (for more details see *Winglee et al.* [1993a]). Here we re-examine the crossing using high time resolution (6–8 s) data from High Altitude Plasma Instrument (HAPI) and Energetic Ion Composition Spectrometer (EICS) on DE 1. As an example of the changes of the distributions of both the ionospheric and magnetosheath ions seen during the encounter, Figure 2 shows the ion distributions as determined from HAPI for 6 consecutive 6 s intervals. In Figures 2a and 2b, the intensity of the cold upwelling ions is relatively strong as indicated by a local maximum in the distribution at about $v_\parallel \simeq 90$ km/s. In the first of these two intervals, the magnetosheath ion intensity is very weak but starts to rise by the second interval. The intensity of these magnetosheath ions reaches a local maximum in Figure 2c, then partially declines in Figure 2d, with a modest intensification in Figure 2e.

Note that during the period of high magnetosheath ion intensity (Figures 2c–2e), the cold upwelling feature disappears. In its stead, there is an enhancement in the ion flux at intermediate energies (100 - 400 eV) in the downward direction. The velocities of this downward intermediate component are not restricted to small pitch angles but extend out to about 45–60°. Because of this extension in pitch angle, this component is hereafter identified as a "downward conic". At the end of the magnetosheath injection, the downward conic feature dissipates and the cold upwelling ion feature reappears.

The downward conic is distinct in velocity space from the magnetosheath ions which are peaked at energies of about 1 - 1.4 keV and are therefore presumably not of magnetosheath origin. The more probable explanation is that they are heated ionospheric ions. This heating would also account for

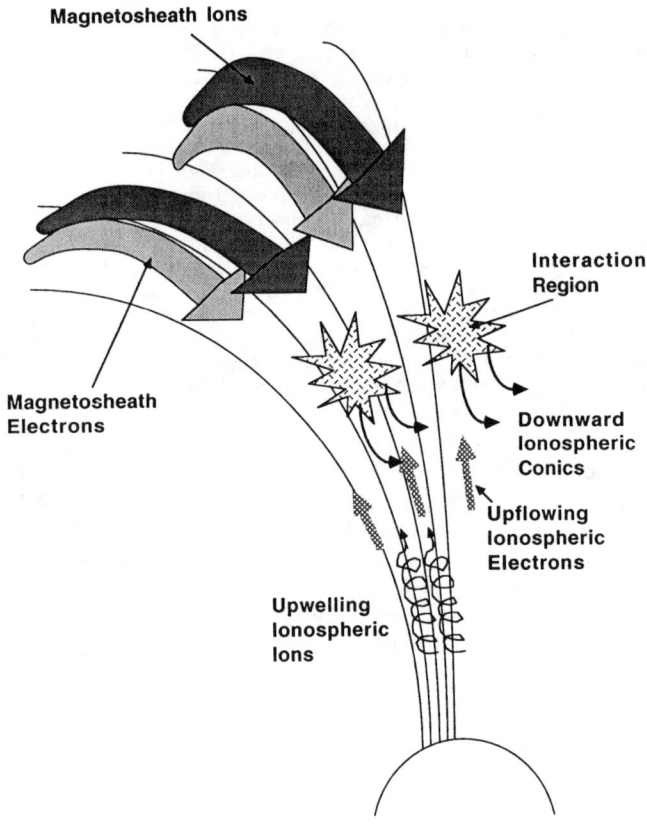

Fig. 1. Schematic diagram showing the interaction between the magnetosheath and ionospheric plasmas in the cusp/cleft region. The injection of the magnetosheath plasma can be modulated and, due to the different convection patterns of the electrons and ions, an alternating field-aligned current system can be generated. Cold (~ 40 eV) upwelling ionospheric ions are seen in association with these field-aligned currents. However, on the field lines where the magnetosheath ion intensity is large, there can be a direct momentum exchange between the magnetosheath and ionospheric ions, leading to the temporary production of downward ionospheric conics. However, due to the action of the mirror force, these conics are eventually forced back up the field lines, so that upwards conics are observed after the magnetosheath injection but with their energies raised from a few tens of eV to a few hundred eV.

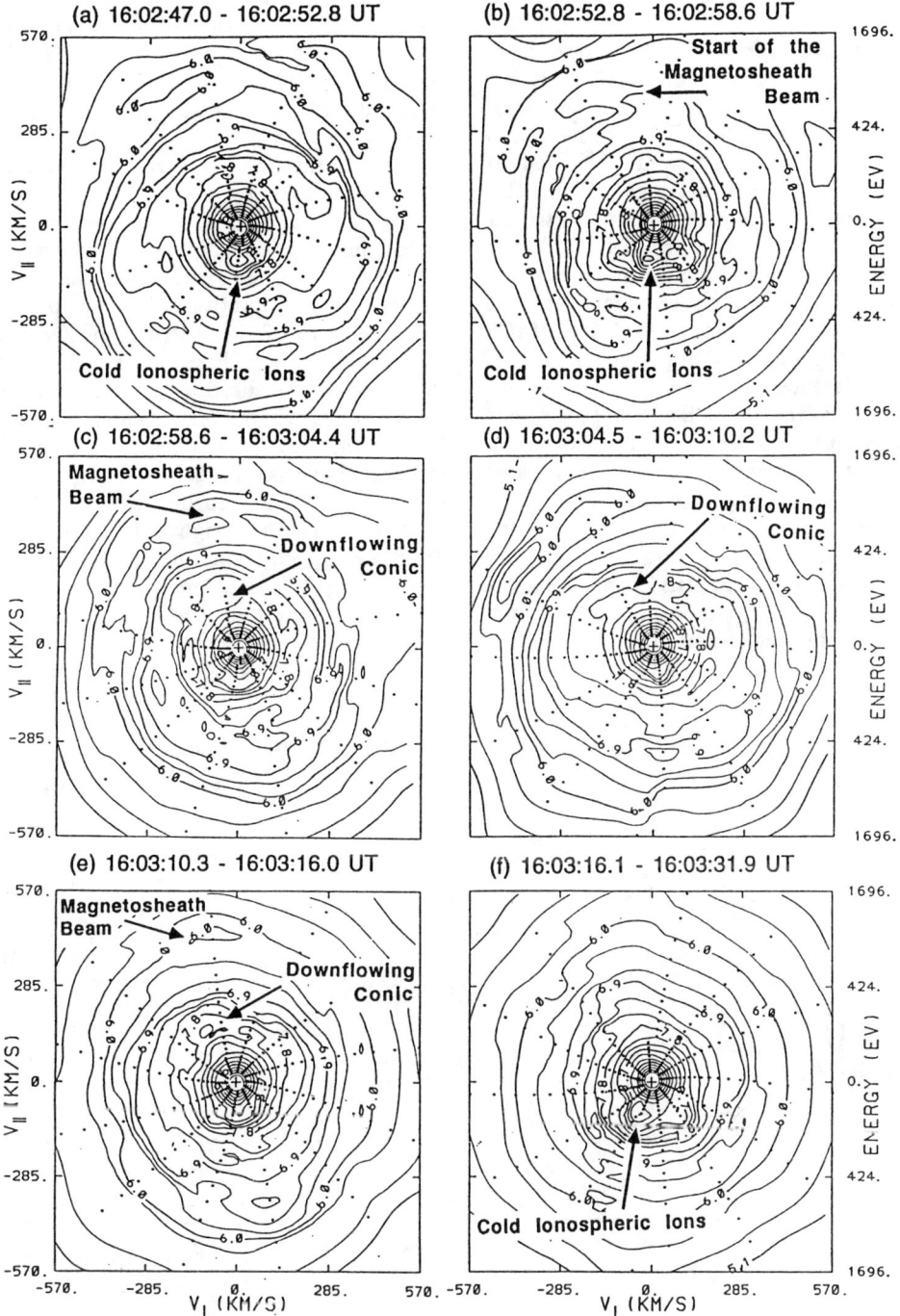

Fig. 2. Evolution of the ion distribution as observed by HAPI during part of the cusp/cleft encounter. The upwelling ions are seen as a highly localized (cold) feature at about $v_\parallel \simeq 100$ km/s. During the injection, the distribution has a downflowing component that is distinct from the magnetosheath ions. This feature presumably consists of ionospheric ions which have exchanged momentum with some of the downflowing magnetosheath ions and experienced additional heating. After the injection, the cold feature in the upwelling ions is again seen.

the simultaneous disappearance of the cold upwelling ion feature. This data is likely the first published evidence for the presence of downward flowing ionospheric conics.

Additional supporting evidence for the presence of these downward conics comes from EICS. During this particular encounter it was in a high temporal resolution mode in which the detector measured on successive 6 s spins the H^+ and O^+ ions at two different minimum energy settings. With this mode, 6 s distributions are attained for a particular ion species and energy range every 24 s. While this sampling badly aliases the pulsations described above, it provides an important complement to the HAPI data. Figure 3 shows the two H^+ EICS distributions corresponding to the period in Figure 2. An important difference between the distributions in Figures 2 and 3 is that the HAPI distributions are derived from about 15 energy channels over the relevant energy range whereas EICS has only three so that there is some smearing of features.

In the Figure 3b, a well defined upward flowing conic is seen. The observing period overlaps the last period in Figure 2 where HAPI also sees a well defined upward conic. On comparing Figure 3b with 3a, it can be seen that the distribution has been strongly modified, with a decrease in the upflowing component at high parallel speeds and an enhancement in the downward component. Indeed the downward component has the form of a conic with the distribution having a local minimum at field-aligned pitch angles. The observing period at this time overlaps the period when the HAPI distributions also indicate the presence of a "downward conic".

The alternating appearance between presence of the cold upwelling ions and the injection of the magnetosheath ions is not restricted to just the period indicated in Figure 2 but was seen throughout the encounter with the cusp/cleft. In order to show this quantitatively, we parameterized the characteristics of the distribution according to (i) the peak height of the distribution and (ii) the energy at the peak in the distribution for both the downgoing magnetosheath ions and upwelling ionospheric ions as determined from one-dimensional cuts through the distributions. Figure 4 shows these quantities as functions of time during the cusp/cleft encounter. In the figure the height of the distribution of the downflowing ions has been multiplied by 10^3 and their energy divided by a factor of 15 so that they could be overlaid on the same graphs.

The pulsations in the magnetosheath injection are clearly evident as orders of magnitude change in

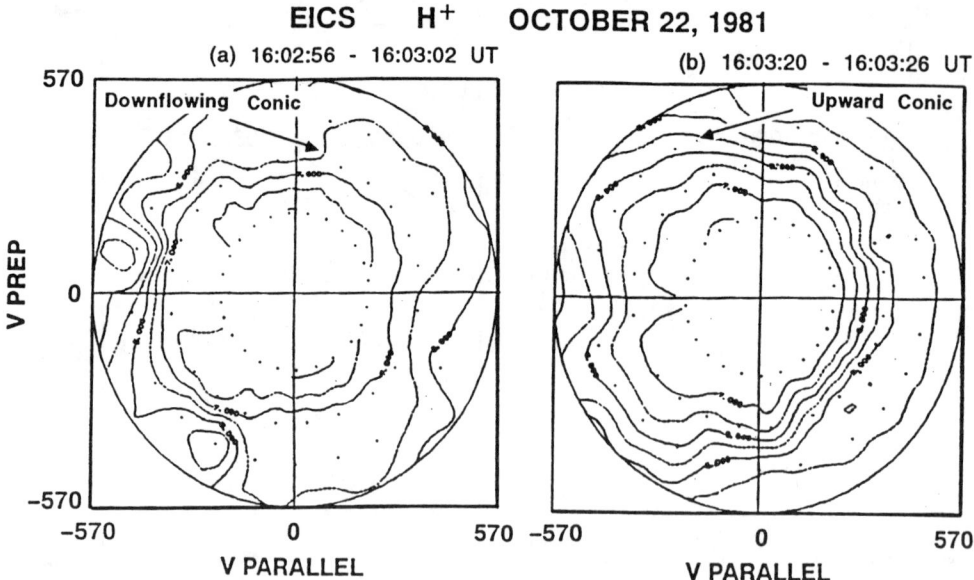

Fig. 3. The EICS H^+ distributions observed during the period in Fig. 2. These distributions are produced every 24 s, with a 6 s integration period. The distributions appear rotated by 90° relative to the HAPI format. The EICS energy resolution at these low energies is somewhat reduced. Nevertheless, a well defined upflowing conic is seen on the distribution in (b) and this period coincides with the period where HAPI also sees an upflowing conic. The distribution on the left hand side is markedly different with an enhancement in the intensity of the downflowing ions, with this component of the distribution resembling a conic, only it is flowing downwards. The time corresponds to the time when HAPI also indicates the presence of a downward conic.

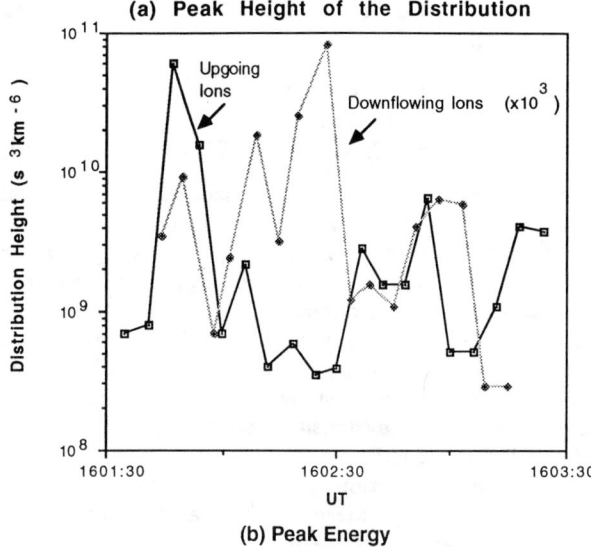

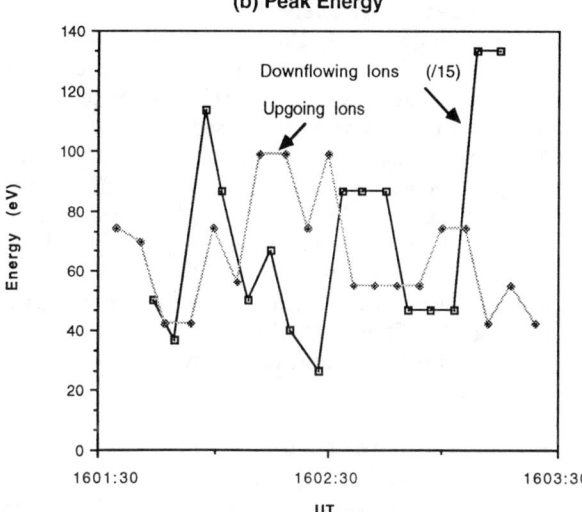

Fig. 4. Summary of the properties of the downflowing magnetosheath ions and the upflowing ionospheric ions. The top panel shows the values of the peak height in the distribution of these ions and the bottom panel shows the energy at the peak intensity of these features. In order to overlay these quantities on the same figures, the intensity of the magnetosheath ions has been multiplied by 10^3 and their energy by 1/15. The peak energy of the upflowing ions closely tracks the intensity of the magnetosheath ions.

the height of the peak in the distribution of the downgoing ions. Local maxima in the height appear at intervals of about 18 to 30 s across the full length of the encounter. The intensity of the ionospheric ions is very high as the region is first encountered at 1601:40. However, during the intense magnetosheath injection between 1602:00 and 1602:30 the ionospheric is at a local minimum. In the next 20-30 s period, the ionospheric intensity is elevated relative to the previous 30 s period and that of the magnetosheath ions is down by nearly an order of magnitude. Shortly after the next magnetosheath enhancement between 1602:45-1603:10, the ionospheric intensity again falls to a depressed value.

The energy of the peak in the distributions as illustrated in the lower panel of Figure 4 shows a similar pattern of alternating peaks in the energies of both the downflowing magnetosheath ions and the upflowing ionospheric ions. The most interesting feature is that the energy of the upflowing ionospheric ions closely tracks the intensity of the downflowing magnetosheath ions (dotted lines in both panels). This tie between the properties of the magnetosheath and ionospheric ions suggests that the upwelling ions experience additional heating when magnetosheath ion intensity becomes large.

While Figure 4 is indicative of the interaction between the magnetosheath and ionospheric ions, time lags between injection, interaction, and recovery mask some of the effects. In order to remove this masking, a two interval running average of the peak of the magnetosheath distribution and the peak energy of the magnetosheath ions was taken. The results are shown in Figure 5. It is seen that, indeed, the intensity of the magnetosheath ions is well correlated (with a correlation factor of about 0.8) with the peak energy of the ionospheric ions (which in turn is anticorrelated with their phase space density). This correlation further supports the suggestion that

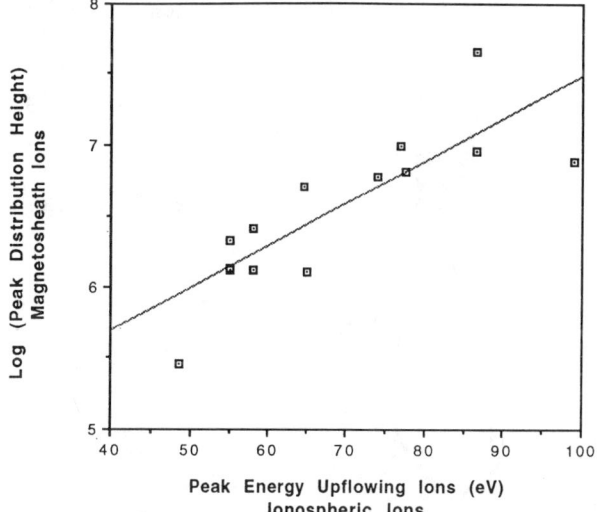

Fig. 5. The peak height of the magnetosheath ions versus the peak energy of the upwelling ions. In order to remove variations associated with time lags between injection and interaction between the two plasma, a two step running average was taken. The intensity of the magnetosheath ions is well correlated with the energy of the upwelling ions.

the upwelling ions are experiencing additional heating during the injection of magnetosheath ions into the cusp/cleft region.

3. PARTICLE SIMULATIONS

The observations show that the above magnetosheath injection is subject to rapid modulation in the spacecraft frame. This modulation is also seen in the properties of the upwelling ions, indicating a direct causal relationship between the upwelling ions and the magnetosheath injection. In order to establish the actual processes involved with this interaction, as well as to differentiate between temporal and spatial effects, the above observations are now compared with a meso-scale particle simulations model for the injection of magnetosheath plasma into the cusp/cleft region. These simulations are an important improvement over existing modelling because they are able to incorporate many of the global influences driving the system as well as the wave-particle interactions that are responsible for the coupling of energy between the magnetosheath and ionospheric plasmas.

3.1 Simulation Model

A schematic of the simulation model is shown in Figure 6. The simulations include gradients in both the ambient magnetic field and plasma density. Gradients in the magnetic field provide the mirror force that provides partial reflection of some of the magnetosheath plasma injected into the region as well as a pushing force on the upwelling ionospheric ions.

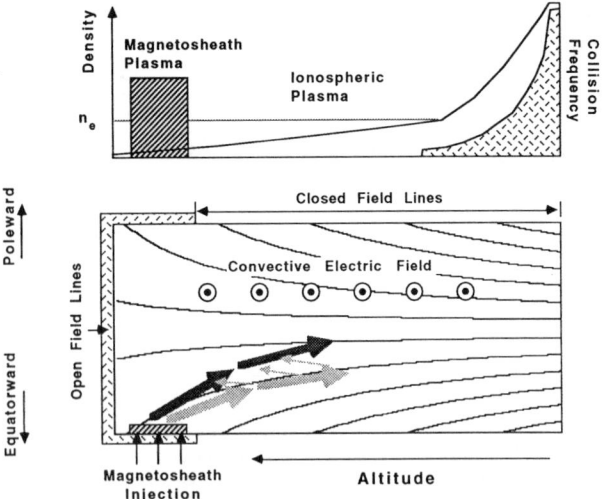

Fig. 6. Schematic of the simulation model showing (a) the assumed density profile and the variation of the collision frequency through the ionosphere and (b) the assumed magnetic field, the characteristics of the magnetosheath injection, its convection across the region, and its interaction with the ionospheric plasma.

The magnetic field is represented in the simulations by a field of the form

$$B_x = B_{xo}(1 + \beta x/L_x) \qquad (1)$$

$$B_y = -B_{xo}\beta(y - L_y/2)/L_x . \qquad (2)$$

where L_x and L_y are the horizontal and vertical dimensions of the simulations system (equal to $1024\Delta \times 128\Delta$ with Δ being a grid spacing), $\beta = 4$ so that $B_{min}/B_{max} = 0.20$, and B_{xo} is set so that electron cyclotron frequency at $x/\Delta = 0$ (i.e., high altitude) is equal to the ambient plasma frequency. The ion-to-electron mass ration is 50 and the heavy ionospheric ions are assumed to have a mass four times larger with a density comprising 20% of the initial plasma density.

Since the interaction between plasmas can vary according to the relative density and temperatures of the plasmas, a gradient in the ambient plasma density is included with the density increasing by a factor of 4 linearly with decreasing altitude until the lower 20% of the system. In this lower section of the system, the density is increased by an additional factor of 4 so that it can provide a ready source of cold plasma and whose properties do not change significantly during the injection. The relative ion and electron temperatures are assumed to be initially equal throughout the system but the absolute value of the temperatures is varied so that pressure balance is maintained. Thus, at the high altitude side of the simulations the ambient plasma temperature is about 100 eV and near the lower end it is about 25 eV.

Multi-charged particles are used in the high density regions to minimize computational requirements. A total of 350,000 superparticles is used in the simulations. In low density regions, the particle per cell is fractionally less than unity. However, in the region of interaction between the magnetosheath and ionospheric plasmas, the number of particles per cell is on the order of 4, which is sufficient to model most wave instabilities with wave lengths much greater than a Debye length. The stability of the system was verified through comparisons with and without injection of the magnetosheath plasma.

A collisional operator is applied to the plasma in the lower part of the system. The inclusion of collisions incorporates key properties of the ionosphere which provides (i) a sink for the energy of precipitating energetic particles and (ii) a means for providing Pederson conductivity which is needed to close any field-aligned currents. The algorithm for the collisional operator is described in *Winglee* [1990]. The collisional cross-section ramps up, starting from near the foot of the low-altitude density jump associated with the collisional ionosphere, to a maximum electron-neutral collision frequency of about 10% of the local electron cyclotron frequency at the end of the system.

Another important feature incorporated in the simulations is the inclusion of open and closed boundary conditions on different field lines as indicated in Figure 6b. Specifically, at the boundaries in the lower 80% of the system (i.e., $x/\Delta > 200$) standard reflection boundary conditions are used. On the remaining field lines, the boundary conditions are changed to model open field lines where any ion, irrespective of their origin, and the nearest electron are allowed to leave the simulation system as they approach the open field line boundary. Due to the introduction of these open boundary conditions, there is always a net upwelling of ionospheric plasma along these open field lines but their energy remains comparable to their initial energy.

The injection of the magnetosheath plasma occurs along a restricted portion of the open field lines, specifically just from the region at $y = 0$ and $100 < x/\Delta < 200$. This restricted source region is consistent with observations that indicated that the actual source of the injection region is spatially limited [e.g., *Menietti and Burch*, 1987]. The actual width of the injection region is not an important factor except that not all the injected magnetosheath plasma has access to all the open field lines. This limited access, as shown in the following, produces velocity filtering of the magnetosheath ions and leads to very distinct signatures in the evolution of the particle distributions of both the magnetosheath and ionospheric ions.

The simulations are first run for $\omega_{pe}t = 800$ without any magnetosheath injection to establish equilibrium. The injected magnetosheath density is then ramped up, reaching a peak density of 4 times the ambient density at high altitudes at $\omega_{pe}t = 2400$. After this time the injected magnetosheath density is ramped down to zero at $\omega_{pe}t = 4000$. This time scale is chosen so that there can be several electron transit times across the length of the system during the injection, but the bulk of the ions in general have time only for about one or two transits across the system.

The last important feature incorporated into the simulations is a convective electric field which drives convection across the cap similar to that observed when the interplanetary magnetic field is southward. It is also an important factor in the development of velocity filtering effects on the particle distributions. The applied convection electric field is sufficiently small that it does not produce any appreciable acceleration of plasma directly. The induced $\boldsymbol{E} \times \boldsymbol{B}$ drift is equal to $0.1\, v_{TH}$ at the highest altitudes and, because of the increasing magnetic field strength, decreases to $0.02\, v_{TH}$ at the lowest altitudes.

3.2 Evolution of the Ion Distribution

The influence of the magnetosheath injection on the upwelling ions is illustrated in Figure 7 which shows an arrow plot for the current density of the ionospheric ions (normalized to $en_e v_{Te}$) at four different times during the simulation. The length of the arrow indicates the magnitude of the current density, and regions of predominantly downward flow are drawn in black while those regions of net upflow are shown as gray arrows.

The flows just prior to the start of the magnetosheath injection (Figure 7a) are those of the quiescent cusp with a new upward flow along the open field lines and a weak drift to higher y, i.e., poleward, arising from the applied convection electric field. Note that the relatively high poleward convection seen in the ionosphere is due to its large density and not to a high speed.

With the magnetosheath injection, the quiescent outflow in the cusp current region reverses, becoming downward. These flows intensify and become peaked more sharply in the cusp current region (Figures 7b and 7c). After the injection ceases, this ionospheric downflow quickly dissipates (Figure 7d) due to the action of the mirror force and losses of low pitch angle particles to the collisional atmosphere. At this stage, the outflow of ionospheric ions is similar to that of the quiescent cusp in Figure 7a but the magnitude, particularly at mid-to-low altitudes, is enhanced by a factor of nearly 2. In other words, even though the magnetosheath-ionospheric interaction temporarily leads to a downward flux of H^+ ions, the net result after the injection is an enhanced outflow of heated ionospheric ions.

The changes in the H^+ distribution that are associated with these modified ionospheric flows are illustrated in Figure 8 which shows the total (magnetosheath and ionospheric) H^+ distributions in the ion interaction region at low altitudes (left hand side) and above the ion interaction region at high altitudes (right hand side). The exact positions of the sampling points are indicated by hatched rectangles in Figure 7c. Magnetosheath ions with low pitch angles can reach both regions, although due to the bending of the field lines, only about a third of the magnetosheath ions actually pass through the high altitude region compared with the low altitude region.

For the low altitude distribution, the magnetosheath ions start arriving after about $\omega_{pe}t = 2400$ (Figure 8b), giving the distributions a "banana" like feature similar to the observations in the companion paper. As the magnetosheath ion density continues to increase, the heating of the ambient ions is seen to increase steadily (Figures 8c and d) with a net transfer of downward momentum from the magne-

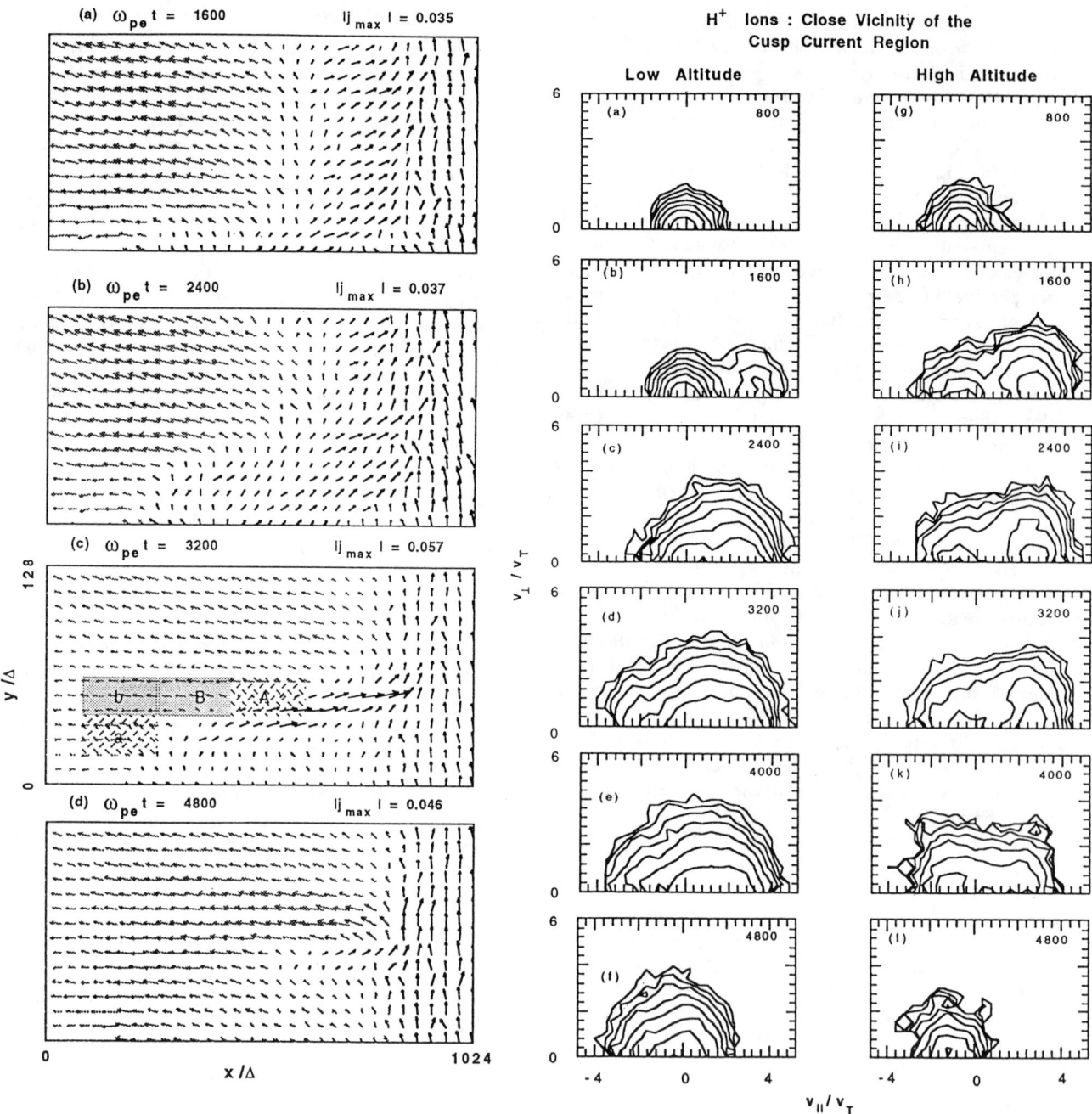

Fig. 7. The induced currents for the ambient H^+ ions only. The magnetic field is indicated by the dotted lines and the position of the assumed open field-lines are indicated in (a). Regions of predominant outflow are indicated by the gray arrows and that of downward flow by the black arrows. The currents are normalized to $e n_e v_{T_e}$ and their maximum values are indicated for each panel. Strong downward flows are seen to develop in the vicinity of the cusp current region during the magnetosheath injection. However, after the injection, the flow is primarily upward through most of the cusp. The hatched and dotted rectangles indicate the sampling regions for the ions distributions shown in Figures 8 and 9, respectively.

Fig. 8. The evolution of the total (magnetosheath and ionospheric) H^+ ions. The distributions on the left hand side are region "A" in Figure 7c corresponding to the cusp current region and show strong heating of the initial cold ionospheric component and the formation of a downward ionospheric flow. The distributions on the right are for region "a" in Figure 7c which fractionally above the cusp current region. The flux of magnetosheath flux is smaller through this region. As a result, there is less local heating of the ionospheric plasma and a sustained upflowing conic is seen during the injection. The observed heating of the distribution arises from the interactions at lower altitudes.

tosheath ions to the ambient H^+ ions. At the same time there is a reduction in the height of the ambient distribution, particularly the upflowing component. This process is analogous to the disappearance of cold upwelling ions during magnetosheath injection described in Section 2.

This momentum transfer and particle heating is responsible for the downward fluxes seen in the ion flows discussed in Figure 7. It occurs through the development of an ion-ion streaming instability downflowing magnetosheath ions and the upwelling ionospheric ions in much the same way as proposed for the nightside auroral region [e.g., Winglee et al., 1988 and references therein]. One important difference is that in the cusp/cleft the free energy is provided by the downflowing magnetosheath ions region while in the nightside auroral region light ionospheric ions that accelerated upwards by quasi-static fields provide the free energy.

As the magnetosheath injection declines, the downward component fades (Figure 8e) and eventually an upflowing component is again seen but with an increase of nearly a factor of three in their velocities or a factor of about 10 in their temperatures (Figure 8f). This heated distribution is a mixture of magnetosheath plasma that has mirrored and ionospheric plasma that has been heated.

The distributions at the higher altitude on the right hand side of Figure 8 show a different evolution due to the limited access of some of the injected magnetosheath plasma. Because of the higher altitude the magnetosheath intensity is seen to rise faster in Figure 8h than in Figure 8b. With the arrival of the magnetosheath ions there is again heating of the ionospheric plasma as evidenced by the filling of the region $0 \lesssim v_{\parallel}/v_{TH} \lesssim 2$ between the ionospheric and magnetosheath distributions (Figures 8i and 8j). However, there is always a distinct peak in upward direction of the distribution associated with the ionospheric ions, unlike the distributions on the left hand side. This component is seen to undergo enhanced perpendicular heating (Figure 8k) to produce a conic extended in energy. The continued presence of a cold upflowing component is due to the continued convection of ionospheric plasma into the region. After the injection, the bulk of the distribution is similar to that seen just prior to the injection (Figure 8l and 8a) except there is a high energy tail which gives the distribution a conic-like appearance.

While the distributions in Figure 8 show the presence of enhanced heating and net transfer of downward momentum, they do not show the presence of any downward conics. This lack of downward conics is actually a spatial effect because low pitch angle magnetosheath ions have access to both regions. However, these particles cannot reach regions further poleward of the cusp current region because their high field-aligned velocity carries them too far down the field lines before they have a chance to convect poleward. Thus, on the poleward edge of the cusp current region, there can be substantial changes in the evolution of the ion distribution as illustrated in Figure 9.

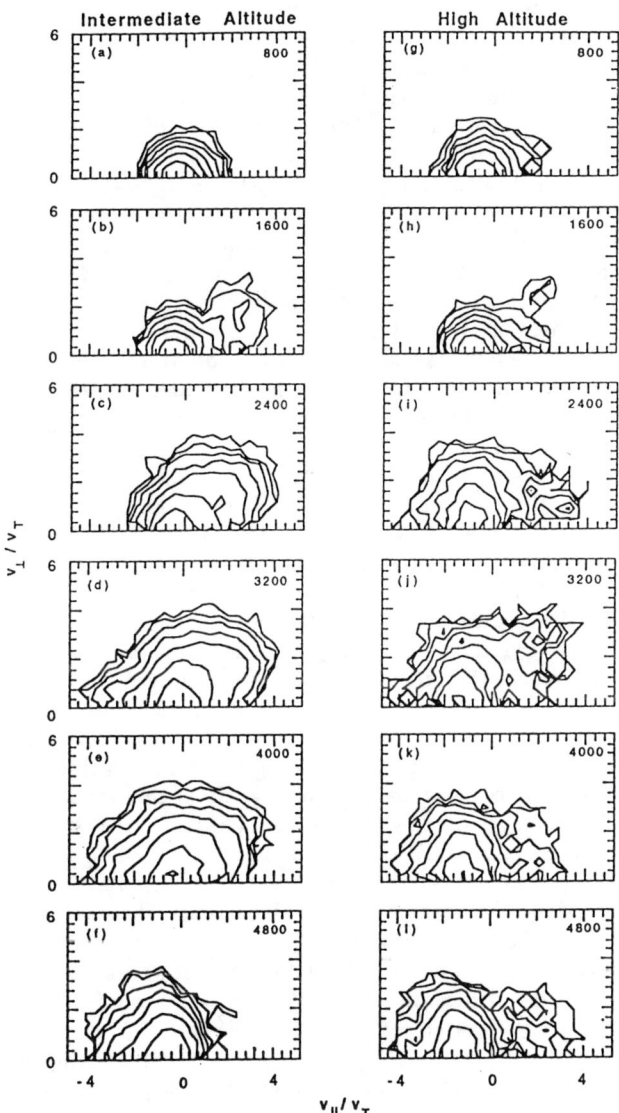

Fig. 9. As in Figure 8, but the distributions on the left hand side correspond to the region "B" in Figure 7c and the distributions on the right hand side to region "b". These regions are poleward of those in Figure 8 so that magnetosheath ions with low pitch angles cannot access these regions. As a result, the magnetosheath ions have a ring-beam distribution, as opposed to the banana-like distribution in Figure 8. The resultant heating of the ionospheric ions leads to the production of a downward ionospheric conic.

The sampling regions for these distributions are indicated by the dotted rectangles in Figure 7. It is seen in Figure 9b that, due to the limited access of magnetosheath ions with to the region, the distribution is peaked at pitch angles of about 30°, giving the magnetosheath ions a ring beam distribution. An ion-ion streaming instability can still develop but the ambient ions in this case are pulled out or heated along pitch angles of about 30° (Figure 9c). As a result of this heating, the distribution has the appearance of a downward conic (Figures 9c - 9d) as opposed to a downward stream as in Figure 8c and 8d. This downward conic has a similar form as described in section 2.

During late times (Figure 9e), some of the heated particles mirror and there is also an enhancement in the perpendicular energy of the ions flowing back up the field lines. After the injection ceases, the downward component rapidly diminishes (due to ions propagating further down the field lines and their mirroring). The resultant distribution has the form of a heated upward conic. The temperature of this conic is again raised by a factor of about 10 from that seen in the quiescent cusp. It is also fractionally hotter than the conic seen equatorward of the region (Figure 8l).

The distributions in the right hand side of Figure 9 are further poleward and at sufficiently high altitudes above the magnetosheath injection region, that only the very few magnetosheath ions with high energy and pitch angle can reach the region. As a result the development of local instabilities is suppressed. Ionospheric ions, excluding those initially present, can only reach this region if they first convect through the cusp current region. Thus, the particles observed in this region carry information about their past interactions and are not strongly influenced by the local development of instabilities. It is seen that the distributions in this region are characterized by an upward conic at all times and that the magnetosheath ions are not in strong evidence. However, during the injection (Figures 9i - 9l) these ions show strong perpendicular heating and have an enhanced upward bulk velocity due to interactions at lower altitudes. In other words, the magnetosheath interaction can produced elevated conics well poleward of the cusp current region so that a one-to-one correspondence between magnetosheath injection and these conics is not expected.

4. SUMMARY

In this paper, high time resolution (6 s) observations from DE 1 and two dimensional (three velocity) electrostatic particle simulations are used to investigate the particle dynamics of the cusp/cleft region. The observations show that the magnetosheath injection can pulsate in the spacecraft frame with periods as small as 18 - 30 s. Cold (~ 40 eV) upwelling ions are seen through the entire region, but their intensity is strongly modulated during the appearance of the magnetosheath ions. In particular, during the magnetosheath ion injection the intensity of the cold ionospheric is seen to rapidly decrease and a mildly energetic ($\gtrsim 100$ eV) component is seen to replace the cold component. During a couple of events this mildly energetic component is seen to be actually propagating downwards. This downward component is distinct from the injected magnetosheath ions and is presumably of ionospheric origin. They also exhibit some enhanced perpendicular heating and are referred to as downwards conics. The data presented here are probably the first detection of such downward ionospheric conics and provide direct evidence for transfer of momentum from the magnetosheath ions to the upwelling ionospheric ions.

Because the observations are limited to a single spacecraft, one cannot determine whether the observed variations in the particle properties are temporal or spatial or both. This uncertainty can be removed by the use of the meso-scale particle simulations presented here. These simulations are an improvement over existing models because they allow the self-consistent investigation of the interplay between (i) global influences, such as the convection of plasma across the cusp, the action of the mirror force, and the injection of magnetosheath plasma, and (ii) wave-particle interactions which produce the actual coupling between the magnetosheath and ionospheric plasmas.

These meso-scale simulations are able to show that the coupling between the downflowing magnetosheath ions and the cold upwelling ions is due to an ion-ion streaming instability. The characteristics of this instability change across the injection region due to velocity filtering effects imposed by the global influences described above. Due to these variations, the simulations are able to account for (i) the suppression of cold ionospheric flows during magnetosheath injection, (ii) the increase in average energy of the ionospheric ions during injection, (iii) the generation of downward conics, (iv) the rapid reappearance of cold ionospheric conics after magnetosheath injection, and (v) the appearance of elevated conics away from any magnetosheath injection. Further, the suppression of cold upwelling ions and the appearance of downward conics are tied to spatial effects, primarily located near the center and the poleward edge of the cusp current region, respectively, while the reappearance of the cold conics can be linked to both spatial and temporal effects.

Understanding the particle interactions in the cusp/cleft region is important because it represents a region of confluence between the ionosphere and magnetosheath plasma. The region is routinely monitored by polar orbiting spacecraft and as such provides a ideal opportunity to study the coupling between the magnetosheath and ionospheric plasma. In addition, it provides an important opportunity to remotely sense the properties of the magnetopause by the properties of the magnetosheath particles that are able to enter the magnetosphere. However, in order to achieve this type of remote sensing, further observational and computational work needs to be undertaken to link the conditions observed in the cusp to the prevailing solar wind conditions and, in particular, establishing exactly where the observed magnetosheath plasma crosses the magnetopause.

Acknowledgments. This research was supported by NASA grants NAGW-2471, NAGW-2412, and NAGW-1936, and NSF grant ATM-9296075. The simulations were performed on the CRAY Y-MP at the San Diego Supercomputing Center which is funded by the National Science Foundation.

REFERENCES

André, M., M. Temerin, and D. Gorney, Resonant generation of ion waves on auroral field lines by positive slopes in ion velocity space, *J. Geophys. Res.*, *91*, 3145, 1986.

André, M., G. B. Crew, W. K. Peterson, A. M. Persoon, C. J. Pollock and M. J. Engebretson, Ion heating by broadband low-frequency waves in the cusp/cleft, *J. Geophys. Res.*, *95*, 20809, 1990.

Burch, J. L., P. H. Reiff, J. D. Menietti, R. A. Heelis, W. B. Hanson, S. D Shawhan, E. G. Shelley, M. Sugiura and J. D. Winningham, IMF B_y-dependent plasma flow and Birkeland currents in the dayside magnetosphere: 1. Dynamics Explorer observations, *J. Geophys. Res.*, *90*, 1577, 1985.

Chang, T., G. B. Crew, N. Hershkowitz, J. R. Jasperse, J. M. Retterer, and J. D. Winningham, Transverse acceleration of oxygen ions by electromagnetic ion cyclotron resonance with broadband left-hand-polarized waves, *Geophys. Res. Lett.*, *13*, 636, 1986.

Chappell, C. R., T. E. Moore and J. H. Waite, Jr., The ionosphere as a fully adequate source of the earth's magnetosphere, *J. Geophys. Res.*, *92*, 5896, 1987.

Crew, G. B., T. Chang, J. M. Retterer, W. K. Peterson, D. A. Gurnett and R. L. Huff, Ion cyclotron resonance heated conics: Theory and observations, *J. Geophys. Res.*, *95*, 3959, 1990.

Moore, T. E., M. Lockwood, J. H. Waite, Jr., M. O. Chandler, W. K. Peterson, D. Wiemer and M. Sugiura, Upwelling O^+ source characteristics, *J. Geophys. Res.*, *91*, 7019, 1986.

Peterson, W.K., Ion injection and acceleration in the polar cusp, in *The Polar Cusp*, J.A. Holtet and A. Egeland (eds.), D. Reidel Publishing Company, p.67, 1985.

Pollock, C. J., M. O. Chandler, T. E. Moore, J. H. Waite, Jr., C. R. Chappell, and D. A. Gurnett, A survey of upwelling ion event characteristics, *J. Geophys. Res.*, *95*, 18969, 1990.

Roberts, W. T., J. L. Horwitz, R. H. Comfort, C. R. Chappell, J. H. Waite, Jr. and J. L. Green, Heavy ion density enhancements in the outer plasmasphere, *J. Geophys. Res.*, *92*, 13499, 1987.

Rosenbauer, H., H. Grunwaldt, M. D. Montegomery, G. Paschmann, and N. Sckope, Heos 2 plasma observations in the distant polar magnetosphere: The plasma mantle, *J. Geophys. Res.*, *80*, 2723, 1975.

Roth, I., and M. K. Hudson, Particle simulations of electrostatic emissions near the lower hybrid frequency, *J. Geophys. Res.*, *88*, 483, 1983.

Roth, I., and M. K. Hudson, Lower hybrid heating of ionospheric ions due to ion ring distributions in the cusp, *J. Geophys. Res.*, *90*, 4191, 1985.

Waite, J. H., Jr., T. Nagai, J. F. E. Johson, C. R. Chappell, J. L. Burch, T. L. Killeen, P. B. Hays, G. R. Carignan, W. K. Peterson, and E. G. Shelley, Escape of suprathermal O^+ ions in the polar cap, *J. Geophys. Res.*, *90*, 1619, 1985.

Winglee, R. M., Electron beam injection during active experiments 2. Collisional effects, *J. Geophys. Res.*, *95*, 6191, 1990.

Winglee, R. M., M. Ashour-Abdalla, and R. D. Sydora, Heating of ionospheric O^+ ions by shear Alfvén waves, *J. Geophys. Res.*, *92*, 5911, 1987.

Winglee, R. M., R. D. Sydora, and M. Ashour-Abdalla, Particle simulations of the heating of ionospheric O^+ ions by current driven shear Alfvén waves, in *Modelling Magnetosphere Plasmas*, Geophysical Monograph 44, edited by T. E. Moore and J. H. Waite, Jr., p. 205, 1988.

Winglee, R. M., J. D. Menietti, W. K. Peterson, J. L. Burch, and J. H. Waite, Jr., Magnetosheath-ionospheric plasma interactions in the cusp: 1 Observations of modulated injections and upwelling ion fluxes, *J. Geophys. Res.*, in press, 1993a.

Winglee, R. M., J. D. Menietti, and C. S. Lin, Magnetosheath-ionospheric plasma interactions in the cusp: 2 Meso-scale particle simulations, *J. Geophys. Res.*, in press, 1993b.

J. L. Burch and J. H. Waite, Jr., Instrumentation and Space Research Division, P.O. Drawer 28510, Southwest Research Institute, San Antonio, TX 78228-0510.

J. D. Menietti, Department of Physics and Astronomy, University of Iowa, Iowa City, IA 52242.

W. K. Peterson, Space Science Laboratory, Department 91/20, Locheed, Building 255, 3251 Hanover Street, Palo Alto, CA 94304.

R. M. Winglee, Geophysics Program AK-50, University of Washington, Seattle, WA 98195.

The Location of Magnetopause Reconnection for Northward and Southward Interplanetary Magnetic Field

T. G. Onsager

Institute for the Study of Earth, Oceans, and Space and Department of Physics, University of New Hampshire, Durham, New Hampshire

S. A. Fuselier

Lockheed Palo Alto Research Laboratory, Palo Alto, California

It has been well established, both experimentally and theoretically, that magnetic reconnection occurs at the Earth's magnetopause. Although the occurrence of reconnection has been demonstrated, its location on the magnetopause surface is not well known. We have used He^{++} distributions measured by the AMPTE CCE spacecraft in the dayside low latitude boundary layer to estimate the location of magnetic reconnection on the magnetopause. We present two cases, one for which the interplanetary magnetic field (IMF) had a southward component and one for which the IMF was strongly northward. In both cases, the He^{++} distributions exhibited characteristic velocity-space features that we interpret as resulting from dayside reconnection. From these characteristic distribution functions we are able to estimate the distance from the spacecraft to the reconnection site. We find the reconnection site to be within about 6 Earth radii of the subsolar magnetopause. Independent of this determination, the velocity-space features in the He^{++} distributions indicate that even in the northward IMF case, reconnection occurred equatorward of the cusp, i.e., on magnetospheric field lines that prior to reconnection intersected the dayside equatorial plane. In addition, these observations suggest that the perpendicular convection of the reconnected flux tubes was very slow, such that the spacecraft near the subsolar magnetopause detected the flux tubes as much as 10 min after the onset of reconnection.

INTRODUCTION

Magnetic reconnection at the dayside magnetopause [*Dungey*, 1961] is thought to be one of the more important processes contributing to the coupling of solar wind particles and energy into planetary magnetospheres. *In situ* measurements near the subsolar magnetopause [*Paschmann et al.*, 1979, 1989; *Sonnerup et al.*, 1981; *Gosling et al.*, 1982; *Smith and Rodgers*, 1991; *Fuselier et al.*, 1991] and in the cusp [*Hill and Reiff*, 1977; *Reiff et al.*, 1980; *Newell et al.*, 1989; *Lockwood and Smith*, 1992; *Onsager et al.*, 1993] have been shown in many cases to be consistent with the reconnection picture. Particularly under conditions when the interplanetary magnetic field (IMF) has a strong southward component, reconnection has been invoked as the dominant process responsible for driving magnetospheric convection.

Although there is considerable observational evidence that reconnection does occur at the Earth's magnetopause, the location of reconnection and its spatial and temporal extents are not well known. Early analysis of Heos 2 data [*Haerendel et al.*, 1978] indicated that reconnection was occurring near the cusp regions of the magnetopause. The fact that accelerated plasma flows were not detected initially in the low latitude boundary layer led these researchers to conclude that reconnection was not occurring at low latitudes. A theoretical analysis [*Crooker*, 1979] identified the expected location of magnetic reconnection under the assumption that reconnection would occur where the magnetosheath magnetic field was antiparallel to the Earth's magnetic field.

Solar System Plasmas in Space and Time
Geophysical Monograph 84
Copyright 1994 by the American Geophysical Union.

This study concluded that reconnection would occur along lines that extend out from the cusps, and that only under the singular case of a purely southward IMF would reconnection occur at the subsolar point.

Further experimental evidence, however, did show that accelerated flows are present in the low latitude boundary layer [*Paschmann et al.*, 1979, 1985, 1986, 1989; *Sonnerup et al.*, 1981; *Gosling et al.*, 1982, 1986]. In these events, the proton bulk flow speed in the boundary layer exceeded the flow speed in the adjacent magnetosheath. These accelerated flows were observed predominantly when the IMF had a southward component.

Another characteristic signature of reconnection is the reversal of the proton bulk flow direction in the low latitude boundary layer relative to that in the magnetosheath [*Gosling et al.*, 1990a]. The boundary layer flow reversals result when the curvature force on the reconnected fluid element is directed opposite to the magnetosheath flow and is sufficiently large so as to overcome the magnetosheath flow. Dawnward boundary layer flows were observed in the northern dusk hemisphere and duskward flows were observed in the southern dawn hemisphere when B_y was positive, and duskward flows were observed in the northern dawn hemisphere and dawnward flows were observed in the southern dusk hemisphere when B_y was negative. In 15 of the cases shown by *Gosling et al.* [1990a] for which the flow reversal occurred in the $\hat{y}$ component, the $\hat{z}$ component of the flow was directed away from the subsolar region, implying that reconnection was occurring near the equatorial plane. In the remaining two cases, a reversal was observed in the $\hat{z}$ component of the flow. Additional examples of reversals in the $\hat{z}$ component of the flow have recently been presented [*Scurry et al.*, 1993].

When the IMF is directed northward, it is generally assumed that if reconnection occurs, it will take place at high latitudes on the magnetopause [*Russell*, 1972; *Maezawa*, 1976; *Crooker*, 1979, 1992]. Reconnection at high latitudes may result in sunward-directed plasma flows in the polar cap regions, as has been inferred from ground-based observations [*Maezawa*, 1976] and low-altitude spacecraft measurements (see review by *Reiff* [1984]). Measurements obtained in the high-latitude magnetopause boundary layer also indicate the presence of accelerated plasma flows, consistent with reconnection occurring in that region [*Gosling et al.*, 1991]. In these events, however, the main shear in the magnetic field was in the $\hat{x}$ and $\hat{y}$ components, indicating that at high latitudes the $\hat{z}$ component of the IMF may play a secondary role. Reconnection in the high-latitude cusp has been discussed as a possible means of forming the dayside low latitude boundary layer when the IMF is northward [*Song and Russell*, 1992].

Evidence that reconnection is occuring at the dayside magnetopause is also found in the detailed distribution functions measured in the low and high latitude boundary layers. One of the characteristic features seen in these distribution functions is the restricted regions of velocity space where particles arriving at a spacecraft from different locations, such as the magnetosheath or the magnetosphere, are found. For example, "D" shaped proton distributions [*Smith and Rogers*, 1991] and a mixture of various reflected and transmitted (at the magnetopause) species have been measured [*Fuselier et al.*, 1991] and have been shown to be consistent with the expectations based on magnetic reconnection [*Cowley*, 1980, 1982].

In this paper we use electron, proton, He^{++}, and O^+ measurements made by the AMPTE Charge Composition Explorer (CCE) spacecraft at the low-latitude magnetopause and in its boundary layers to estimate the location of magnetic reconnection on the magnetopause surface and to investigate the convection velocity of the boundary layer plasma following reconnection. Because He^{++} is abundant in the magnetosheath and absent in the magnetosphere (relative to the instrument detection threshold and energy range), it provides a tracer for magnetosheath particles detected inside the magnetopause on the reconnected field lines in the low latitude boundary layer. Similar to previous observations, the He^{++} distributions have the characteristic velocity-space features expected to exist on recently reconnected field lines in the low latitude boundary layer [*Fuselier et al.*, 1991, 1992].

The low latitude boundary layer He^{++} distributions we show here contain two distinct populations in velocity space. One component consists of the ions that are observed at the spacecraft coming directly from the magnetopause. The second population is interpreted as magnetosheath ions that have crossed the magnetopause, travelled along the magnetospheric field to low altitudes where they have mirrored, and are subsequently detected on their way back toward the magnetopause. Both components exhibit the characteristic "D" shape and have bulk velocities directed parallel and anti-parallel to the magnetic field. The interpretation we adopt here has been used to explain counterstreaming ion distributions in the Earth's magnetotail [*Forbes et al.*, 1981; *Andrews et al.*, 1981], in the high latitude boundary layer [*Gosling et al.*, 1991] and in the low latitude boundary layer [*Fuselier et al.*, 1992].

When both the directly entering and the mirrored components are simultaneously observed, it is possible to estimate the field-aligned distance from the spacecraft to the reconnection site [*Onsager et al.*, 1990; *Gosling et al.*, 1990b; *Phillips et al.*, 1993]. This determination of the reconnection site location assumes that the low-speed cutoffs of both the directly entering and the mirrored ions crossed the magnetopause near the reconnection site. Since the ions at these two cutoff speeds have travelled different distances to the spacecraft in approximately the same amount of time, the speeds at which these cutoffs are observed can be used to calculate the difference in the path lengths the ions have travelled. With an estimate of the distance from the spacecraft to the near-Earth mirror point (in this case, obtained from a magnetic field model of the magnetosphere), one can then estimate the distance from the spacecraft to the reconnection site.

We report two cases when counterstreaming He^{++} were observed in the low latitude boundary layer, one when the IMF had a southward component and one when the IMF was northward. In the case where the IMF was directed southward, reconnection is estimated to have occurred roughly 6 Earth radii (R_E) above the equatorial plane on the dayside magnetopause. In this event, the boundary layer ions that mirrored at low altitudes and returned to the spacecraft required approximately a 10 min travel time from when the flux tube reconnected to when the ions were detected. This long travel time requires that the plasma convection following reconnection had to have been extremely slow, in order for the spacecraft near the subsolar magnetopause to detect these ions [see also *Fuselier et al.*, 1992]. The $\hat{y}$ component of the IMF during this event was directed such that the field-line tension following reconnection would initially be opposed to the magnetosheath flow velocity, consistent with the slow convection speed inferred from the He^{++} measurements. This reconnection geometry would result initially in a deceleration of the boundary layer plasma relative to the adjacent magnetosheath [*Crooker*, 1979].

In the event we describe for which the IMF was directed northward, the reconnection site was estimated to be within a few Earth radii of the subsolar point. The conclusion from these observations is that even with a northward IMF, reconnection occurred on the dayside magnetopause equatorward of the cusp, i.e., on magnetospheric field lines that cross the dayside equatorial plane, rather than on the high-latitude field lines that extend back into the magnetotail. In this case the reconnection appeared to be due to a shear in the $\hat{y}$ component of the IMF, even though the $\hat{z}$ component of the field was considerably larger.

INSTRUMENTATION

Plasma observations reported here are from the Hot Plasma Composition Experiment (HPCE) onboard the AMPTE/Charge Composition Explorer (AMPTE/CCE) [*Shelley et al.*, 1985]. The HPCE instrument package consisted of an energetic ion mass spectrometer with an energy per charge range from the spacecraft potential to 17 keV/e and a separate electron spectrometer that measured the electron distribution in 8 broad energy bands from 50 eV to 25 keV. For the time period discussed in this paper, the ion mass spectrometer was operated in a mode to detect only the major ion species (H^+, He^+, He^{2+}, and O^+) sequentially every 38 ms while the energy was fixed for approximately a spacecraft spin ($\approx$ 6 s). After 15 energy steps (96 s), a background measurement, and a short period of very low energy ion measurements and O^{2+} measurements, the cycle was repeated. Thus, in approximately 2 minutes the phase space distributions of four major ion species were measured at 15 energies and with an azimuthal (spin plane) angular resolution of about 10 degrees.

The electron spectrometer data will be used to identify different plasma regimes with a time resolution much higher than that for the ion measurements. Both the ion and electron instruments view perpendicular to the spacecraft spin axis. Since the spacecraft spin axis is approximately parallel to the Earth-Sun line, all HPCE measurements are made in a plane approximately perpendicular to the undeflected solar wind flow and approximately tangent to the subsolar magnetopause surface.

The magnetic field measurements were obtained from a tri-axial, fluxgate magnetometer [*Potemra et al.*, 1985]. This instrument obtained approximately 8.6 vector measurements each second.

OCTOBER 19, 1984, SOUTHWARD IMF

The first observations we describe were obtained on October 19, 1984 (day 293) when CCE was located at approximately (8.3, 0.3, 1.3) R_E in GSE coordinates. Magnetic field data from CCE during a 30 min time interval containing multiple crossings of the magnetopause are shown in Figure 1. At the beginning of this time period, the spacecraft crossed the magnetopause into the magnetosphere. Subsequent magnetopause crossings occurred at approximately 0805:30 UT, 0811 UT, and 0815:30 UT. The spacecraft was then located in the magnetosheath between about 0815:30 UT and 0830 UT. Note that these magnetopause crossings have also been described by *Paschmann et al.* [1989].

The main rotation in the magnetic field across the magnetopause is seen in the B_y and B_z components. The B_z component of the magnetic field was predominately negative (southward) in the magnetosheath and positive (northward) in the magnetosphere. The magnetic field in the magnetosphere had the expected direction given the location of the spacecraft, i.e., small, negative B_y and positive B_z in the northern, dusk sector.

Electron and proton measurements from this same 30 min time interval are shown in Figure 2. From top to bottom the panels include electron number flux at ener-

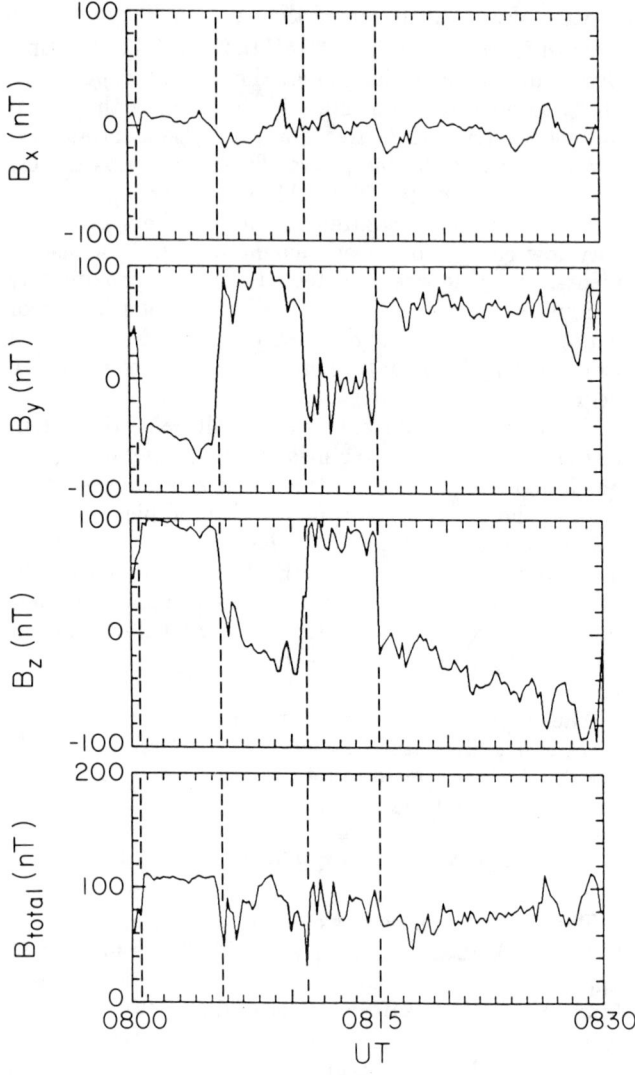

Fig. 1. Three components of the magnetic field and the magnitude of the field during multiple crossings of the magnetopause made by the AMPTE CCE spacecraft on October 19, 1984. The $\hat{z}$ component of the magnetic field was southward (negative) in the magnetosheath and northward in the magnetosphere. The approximate magnetopause crossings are indicated with vertical dashed lines.

magnetopause, we will use the common terminology, low latitude boundary layer.

The boundary layers generally contain levels of electron flux that are intermediate to those on either side of the magnetopause. For example, the 151 eV electron flux (top trace) in the boundary layer is comparable to

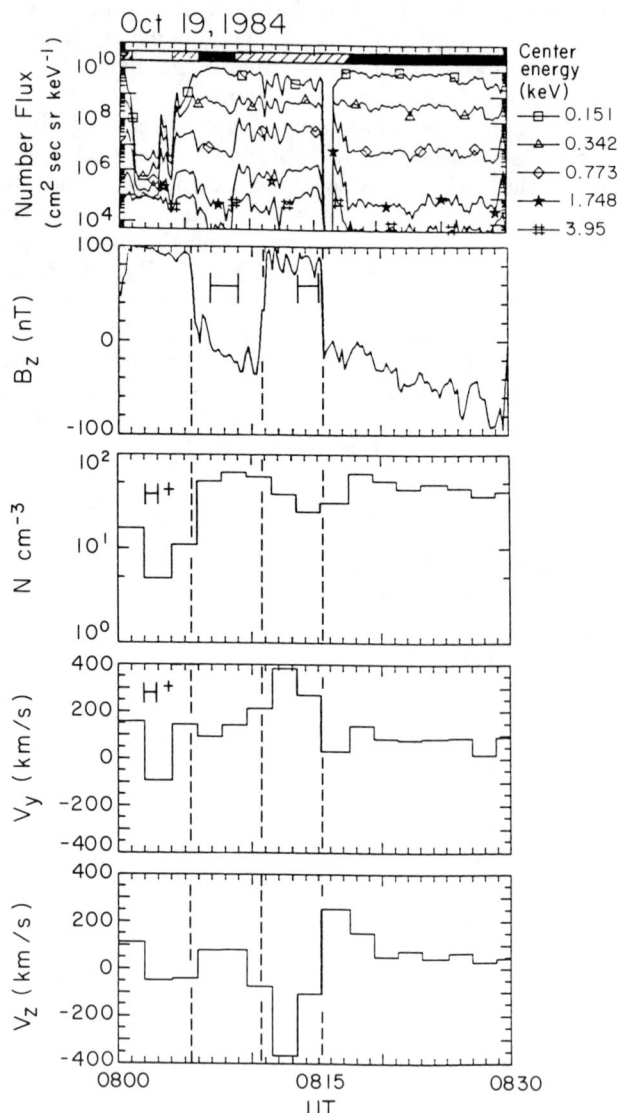

Fig. 2. Plasma and magnetic field data from the same 30 min period shown in Figure 1. From top to bottom the panels contain the electron number flux at energies between 151 eV and 3.95 keV, the $\hat{z}$ component of the magnetic field, the proton density, the proton $\hat{y}$ velocity, and the proton $\hat{z}$ velocity. The horizontal bar drawn above the electron flux indicates the times when the spacecraft was in the (white) magnetosphere, (stripped) low latitude boundary layer, and (black) magnetosheath. The horizontal bars drawn in the panel containing B_z indicate the times from which the He^{++} distributions shown in Figure 3 were obtained.

gies ranging from 151 eV to 3.95 keV, the $\hat{z}$ component of the magnetic field, the proton number density, the $\hat{y}$ component of the proton bulk flow, and the $\hat{z}$ component of the proton bulk flow. At the top of the upper panel, three regions are indicated by the horizontal bar: magnetosheath (black), boundary layer (stripped), and magnetosphere (white). The boundary layer regions are present on both sides of the magnetopause, in both the magnetosheath and the magnetosphere. When discussing the boundary layer that lies earthward of the

the flux at this energy in the magnetosheath and about three orders of magnitude higher than in the magnetosphere. On the other hand, the 3.95 keV electron flux in the boundary layer is comparable to the flux at this energy in the magnetosphere yet about an order of magnitude higher than in the magnetosheath. The proton density (middle panel) also indicates that the density in the boundary layers is intermediate between that in the magnetosheath and in the magnetosphere.

The bottom two panels of Figure 2 contain the v_y and v_z components of the proton velocity. In the magnetosheath, the proton velocity is in the positive $\hat{y}$ and positive $\hat{z}$ direction, as expected for flow around the magnetosphere at the spacecraft location. Between 0811 UT and 0815 UT, CCE was located inside the magnetopause in the low latitude boundary layer. In this region, CCE detected strong proton flows in the positive $\hat{y}$ direction and in the negative $\hat{z}$ direction (see also *Paschmann et al.* [1989]). This enhancement of v_y and reversal of v_z is similar to previously reported flow reversals associated with reconnection at the magnetopause [*Gosling et al.*, 1990a] and is consistent with the location of reconnection inferred from the simultaneous He^{++} measurements. The reversal in the $\hat{z}$ component described here places the reconnection site at least as far above the equatorial plane as the spacecraft, and as estimated below, about 6 R_E above the equatorial plane.

Two-dimensional contour plots and one-dimensional cuts of the He^{++} distributions at two locations, one in the magnetosheath and one in the low latitude boundary layer, are shown in Figure 3. The contour plots that we present in this paper were constructed using a threshold count rate of at least 3 counts per sample above the instrument background count level. The contours represent constant phase space density levels in the $v_y - v_z$ plane, i.e., in the plane that is approximately tangent to the magnetopause surface. The center of the contour plots corresponds to zero velocity in the spacecraft reference frame. Positive $\hat{y}$ velocities are duskward and positive $\hat{z}$ velocities are northward. The 1-D cuts are taken along the background magnetic field direction projected into the y - z plane. The projection of the magnetic field into the y - z plane is indicated in each of the contour plots. For the events discussed in this paper, the magnetic field was nearly in the plane of the plasma measurements.

The panels on the left in Figure 3 illustrate the He^{++} distribution in the magnetosheath, sunward of the boundary layer. This distribution has a fairly large temperature anisotropy, with $T_\perp > T_\parallel$. There is also a small bulk flow in the positive $\hat{y}$ and positive $\hat{z}$ direction, indicated by the dot drawn on the magnetic field vector. The right hand panels contain the He^{++} distribution in the low latitude boundary layer, just earthward of the magnetopause. This distribution is composed of two distinct populations, a population flowing predominantly up along the magnetic field ($+\hat{z}$) and a population flowing down along the field ($-\hat{z}$). There is a well-resolved absence of He^{++} at low parallel speeds (parallel to the background magnetic field), even though particles were present at these speeds in the magnetosheath.

These two counterstreaming ion populations also have a substantial velocity in the $\hat{y}$ direction, nearly perpendicular to the magnetic field. The magnetic field vector has been drawn on the contour plot with an offset in the perpendicular direction of roughly 300 km/sec. This is the value estimated for the ExB drift velocity at this time. The high ExB drift velocity can also be seen in the O^+ measurements shown in Figure 4.

In Figure 4, we show two-dimensional contour plots of the O^+ distribution in the magnetosphere proper, i.e., earthward of the low latitude boundary layer, (left) and in the low latitude boundary layer (right). The O^+ ions in the magnetosphere proper are present predominantly at low velocites, with a drift in the $-\hat{y}$ direction (dawnward). The low latitude boundary layer O^+ distribution was measured simultaneously with the He^{++} distribution shown in the right panels of Figure 3. In the low latitude boundary layer, the O^+ ions have a higher temperature than in the magnetosphere proper and also have a large bulk velocity perpendicular to the magnetic field and in the $+\hat{y}$ direction. The O^+ ions have nearly the same perpendicular velocity as the low latitude boundary layer He^{++} ions but have a much lower parallel speed (see also *Fuselier et al.* [1993]).

As is sometimes the case in the low latitude boundary layer, the magnetosheath particles are detected flowing both directions along the magnetic field. This is typically the case with electrons because of their high speeds [*Gosling et al.*, 1990b; *Fuselier et al.*, 1992], but often not the case for protons and He^{++}. We have interpreted the two counterstreaming populations as those particles that are directly entering across the magnetopause and those that have crossed the magnetopause, travelled to low altitudes and mirrored, and are then detected while returning to the magnetopause.

Given this interpretation, the low-speed cutoffs seen in the He^{++} distributions arise from the velocity filter effect. As the He^{++} ions flow across the magnetopause along the magnetic field, they also ExB drift perpendicular to the field. At a given spacecraft location, only particles with sufficiently high parallel velocity will be able to reach the spacecraft from the magnetopause prior to ExB drifting past the magnetic field line that connects the spacecraft to the magnetopause. The low-speed cutoff represents the particles that have crossed the magnetopause nearest to the reconnection site and farthest from the spacecraft [*Onsager et al.*, 1990; *Gosling et al.*, 1990b; *Lockwood and Smith*, 1992], whereas higher speed particles have crossed the magnetopause on the open field line some time after recon-

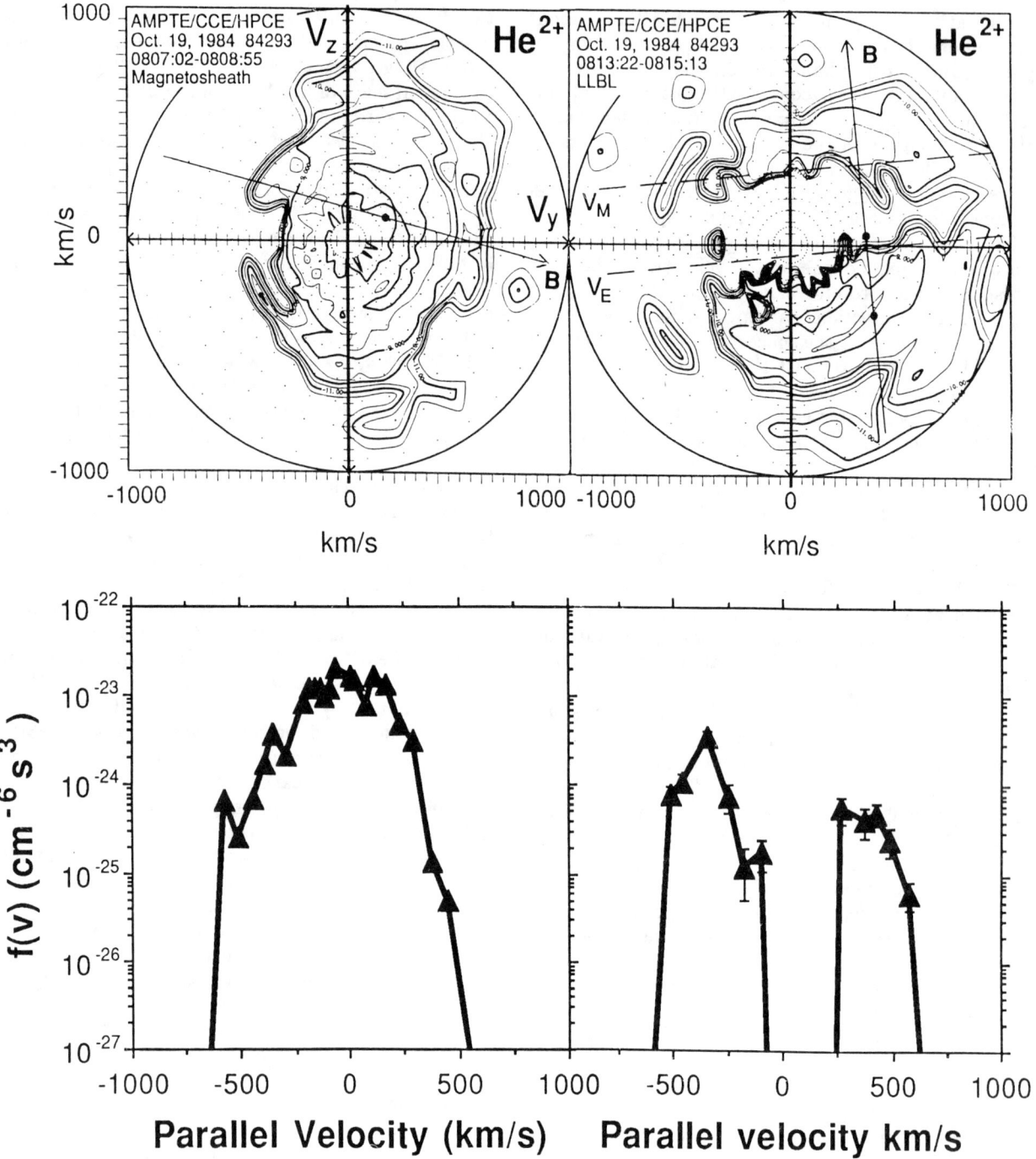

Fig. 3. (Top) He^{++} two-dimensional contour plots and (bottom) one-dimensional cuts along the background magnetic field for distribution functions measured in the (left) magnetosheath and (right) low latitude boundary layer (earthward of the magnetopause). The projection of the magnetic field onto the spacecraft spin plane is indicated by the arrow in each contour plot. The distribution function in the magnetosheath is anisotropic ($T_\parallel < T_\perp$) with a small bulk velocity in the $+\hat{y}$ and $+\hat{z}$ direction. In the low latitude boundary layer, the He^{++} ions consist of two distinct populations counterstreaming along the magnetic field. The approximate low-speed cutoffs of these two populations are indicated with the dashed lines drawn perpendicular to the magnetic field, v_E for the earthward-directed ions and v_M for the mirrored ions.

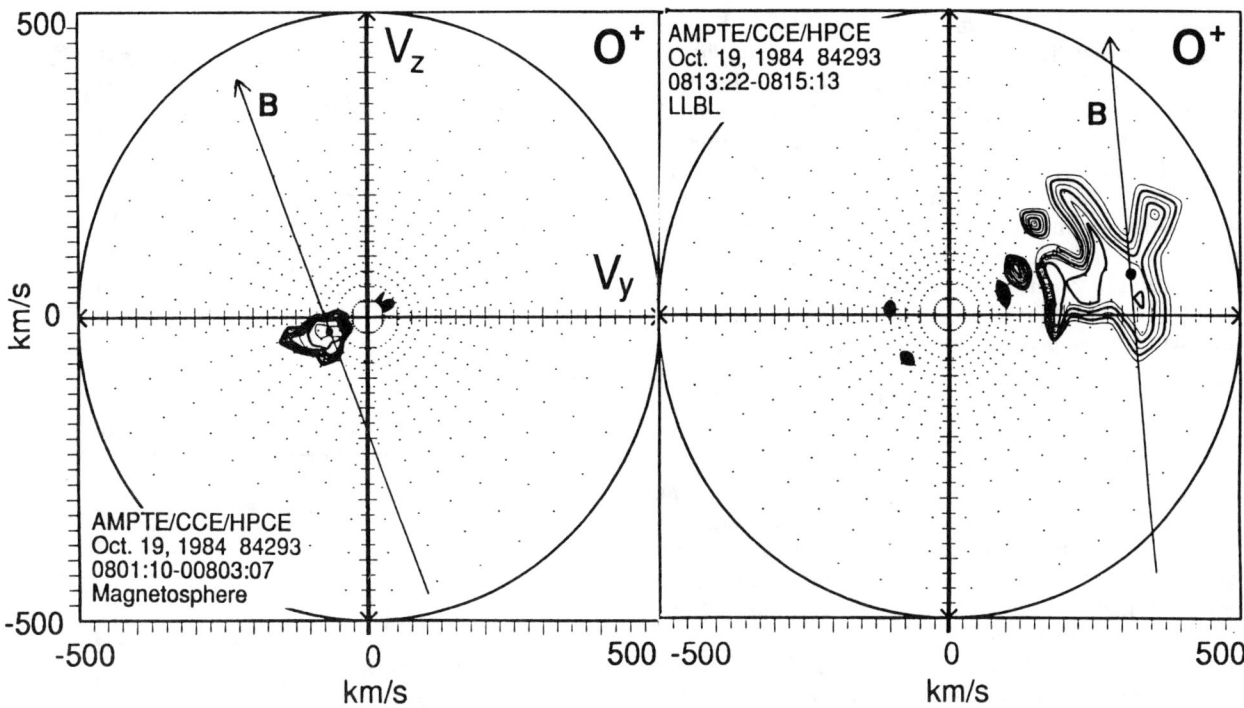

Fig. 4. Two-dimensional contour plots of O$^+$ measured in the (left) magnetosphere and (right) low latitude boundary layer. The magnetospheric oxygen has a small bulk velocity in the -$\hat{y}$ direction. The O$^+$ ions in the low latitude boundary layer have a large bulk velocity (about 300 km/sec) in the +$\hat{y}$ direction.

nection occurred and closer to the spacecraft. The constraints imposed by the velocity filter effect also apply to the mirrored particles, and, therefore, there should be a low-speed cutoff in this population as well.

Since the low-speed cutoffs of both the directly entering and the mirrored particles represent the speeds of particles arriving from near the reconnection site, these two cutoff speeds can be used to estimate the field-aligned distance from the spacecraft to the reconnection site [*Onsager et al.*, 1990; *Gosling et al.*, 1990b; *Phillips et al.*, 1993]. We have indicated the approximate low-speed cutoffs on the earthward directed ions, v_E, and on the mirrored ions, v_M, on the low latitude boundary layer distribution shown in Figure 3. We have also indicated zero parallel speed with the dot drawn on the magnetic field vector in the figure for reference. Note that this same velocity is also indicated on the magnetic field vector drawn on the low latitude boundary layer O$^+$ distribution in Figure 4. Because the mirrored ions have a larger distance to travel to the spacecraft than the directly entering ions, they will have a larger low-speed cutoff. We can immediately conclude, therefore, that the reconnection is above (at higher latitudes than) the spacecraft, since the directly entering ions (with the lower low-speed cutoff) are travelling in the -$\hat{z}$ direction. This same result can be inferred by the direction of the flow reversal seen in Figure 2 and was also obtained by *Paschmann et al.* [1989] for this event based on fluid momentum balance.

Using the low-speed cutoffs identified in the He^{++} distribution, we can estimate the distance from the spacecraft to the reconnection site, x_r, using

$$x_r = x_m \frac{2\,v_E}{v_M - v_E} \quad (1)$$

where x_m is the field-aligned distance from the spacecraft to the mirror point of the particles [*Onsager et al.*, 1990]. Estimating $v_E \approx 50 \pm 30$ km/sec, $v_M \approx 310 \pm 30$ km/sec, and $x_m \approx 10.5 \pm 1$ R$_E$, we obtain the approximate distance from the spacecraft to the reconnection site, $x_r \approx 4.7 \pm 1$ R$_E$. The field aligned distance from the spacecraft to the low-altitude mirror point was estimated using the *Stern* [1985] magnetic field model.

From the approximate distance to the reconnection site and the speed of the particles that have arrived from that location, we estimate that the He^{++} ions have taken roughly 10 min to travel from the magnetopause to the spacecraft. It is interesting, then, to compare the average ExB drift over the 10 min prior to the detection of the He^{++} ions with the ExB drift determined from the O$^+$ distribution. We do not know the perpendicular distance from the spacecraft to the separatrix between open and closed field lines. However, if we roughly estimate this distance to be on the order of 1 R_E, then the average ExB drift speed of the He^{++} ions over the 10 min that they take to travel from the reconnection site to the spacecraft would be about 10 km/sec. Since the ExB drift speed at the time of the He^{++} measurements is approximately 300 km/sec, we conclude that there was a substantial acceleration in the perpendicular direction between the time when the flux tube reconnected and when these observations were made. This point will be discussed in more detail below.

OCTOBER 6, 1984, NORTHWARD IMF

The second low latitude boundary layer observations we describe were obtained on October 6, 1984, when CCE was located at approximately (8.07, -1.18, -0.17) R_E GSE. In Figure 5, we show the electron flux from 151 eV to 3.95 keV (top panel) and the magnetic field (bottom panels) over a 30 min time period that included multiple magnetopause crossings. The magnetopause locations are identified by the abrupt changes in the electron fluxes. In particular, electrons with energies at and above about 1.75 keV are not present at levels above background in the magnetosheath. The fluxes at these energies are enhanced between about 0237 UT and 0245 UT and again between about 0250 UT and 0253 UT, indicating that the spacecraft was earthward of or near the magnetopause at these times.

The magnetopause crossing that we will concentrate on in this example occurred at roughly 0245 UT. As seen in the previous example, the high-energy electron flux can also be enhanced in the boundary layer outside the magnetopause, so that the presence of these electrons does not necessarily indicate that the spacecraft is earthward of the magnetopause. High time resolution electron measurements across this boundary (not shown) indicate isotropic electron fluxes just prior to 0245 UT (inside the magnetopause), and unidirectional, field-aligned streaming electrons in only the outermost edge of the region with enhanced electron flux. The field-aligned electrons are directed sunward, i.e., outward from the magnetopause, and thus indicate that a thin boundary layer was present just on the outer edge of the magnetopause.

The magnetopause crossing at 0245 UT occurred under conditions of strongly northward interplanetary magnetic field. The $\hat{z}$ component of the field changed only slightly across the magnetopause. The predominant shear in the magnetic field was seen in the $\hat{y}$ component, which went from about -40 nT in the low latitude boundary layer to nearly zero and occasionally positive in the magnetosheath.

He^{++} distributions measured in the low latitude boundary layer and in the magnetosheath just outside

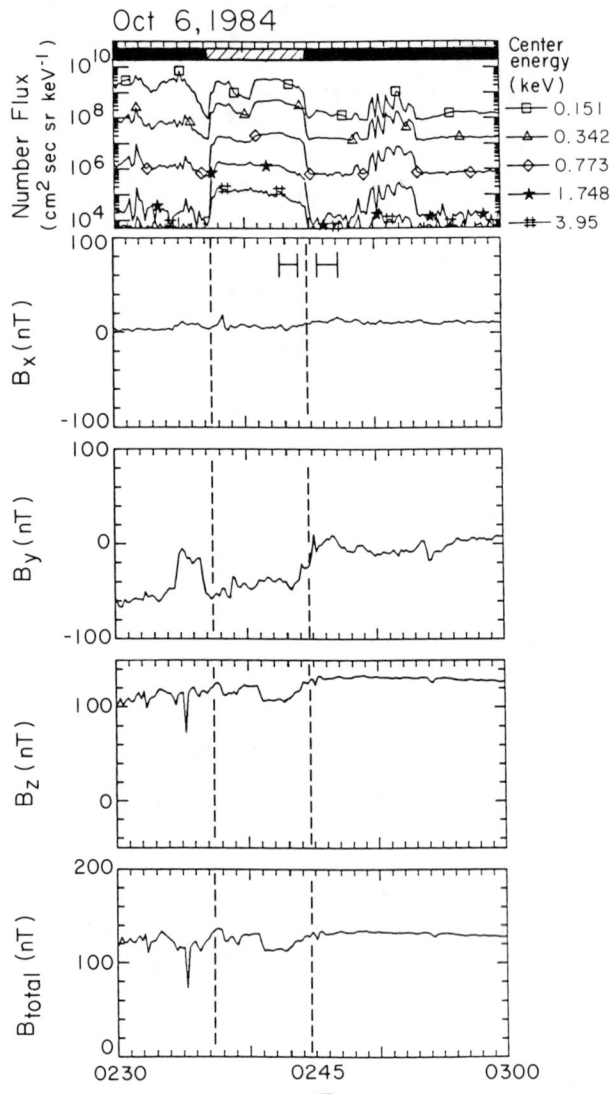

Fig. 5. Electron number flux and magnetic field obtained during two crossings of the dayside magnetopause by the AMPTE CCE spacecraft on October 6, 1984. The horizontal bar drawn above the electron flux indicates the times when the spacecraft was in the (black) magnetosheath and (stripped) low latitude boundary layer (inside the magnetopause). During this time period the IMF was strongly northward. The two magnetopause crossing are identified by the changes in the electron number flux and indicated with vertical dashed lines. The main rotation in the magnetic field at the magnetopause is seen in the $\hat{y}$ component of the magnetic field. The horizontal bars in the panel containing B_x indicate the times from which the He^{++} measurements shown in Figure 6 were obtained.

the magnetopause are shown in Figure 6. As with the previous example, the magnetosheath distribution (left) indicates that the He^{++} just upstream from the magnetopause was quite anisotropic, with $T_\perp > T_\parallel$. This

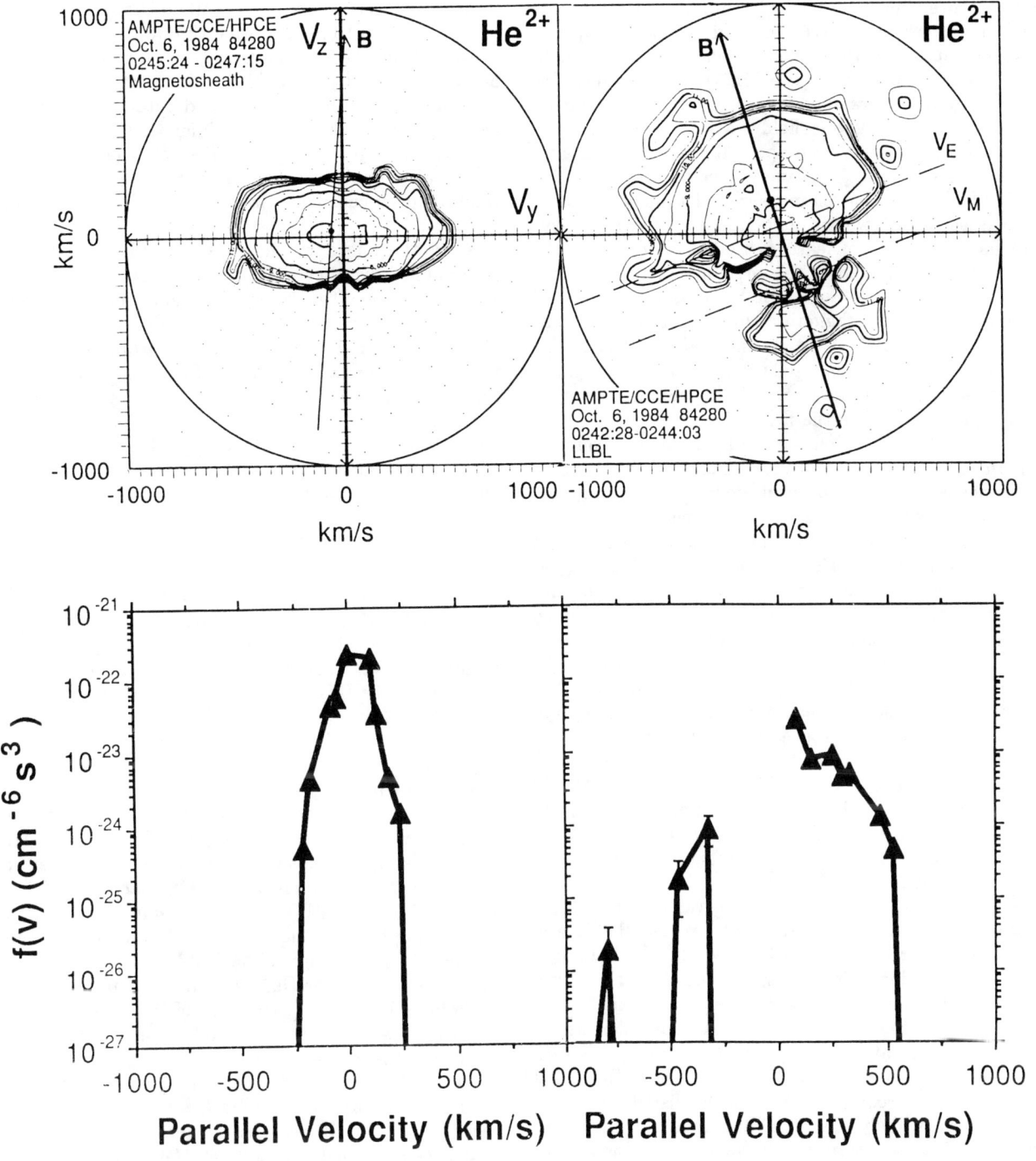

Fig. 6. (Top) He^{++} two-dimensional contour plots and (bottom) one-dimensional cuts along the background magnetic field for distribution functions measured in the (left) magnetosheath and (right) low latitude boundary layer (earthward of the magnetopause). The projection of the magnetic field onto the spacecraft spin plane is indicated by the arrow in each contour plot. The distribution function in the magnetosheath is anisotropic ($T_\parallel < T_\perp$) with a small bulk velocity in the $-\hat{y}$ and $+\hat{z}$ direction. In the low latitude boundary layer, the He^{++} ions consist of two distinct populations counterstreaming along the magnetic field. The approximate low-speed cutoffs of these two populations are indicated with the dashed lines drawn perpendicular to the magnetic field, v_E for the earthward-directed ions and v_M for the mirrored ions.

temperature anisotropy is a common feature of the depletion layer just upstream from the magnetopause [*Crooker et al.*, 1979; *Anderson and Fuselier*, 1993]. The magnetic field projection in the ŷ-ẑ plane is indicated by the arrow superimposed on the distribution function. The magnetic field was mostly northward, with a small positive y component. Also, the magnetosheath flow in the ŷ-ẑ plane was small.

The He^{++} distribution measured just inside the magnetopause in the low latitude boundary layer (right panels) has a number of features that are similar to the distribution described in the previous example. First, there is an absence of ions with low speeds parallel and antiparallel to the magnetic field, even though ions at these speeds were present in the magnetosheath. In addition, there are two populations of ions, one component flowing parallel to the magnetic field and the other flowing antiparallel to the field.

A natural explanation for the field-aligned, counter-streaming populations and the lack of ions with low parallel speeds is that the low latitude boundary layer measurement was made on recently reconnected field lines. In that case, as with the previous example, we interpret the parallel streaming He^{++} population as those magnetosheath ions that have crossed the magnetopause into the low latitude boundary layer, and the anti-parallel streaming population as those ions that have also come from the magnetosheath but have mirrored at low altitudes and are now returning to the magnetopause. Because of the velocity filter effect, each of these components will have a low-speed cutoff, thus giving rise to the absence of particles at low parallel speeds. Also, the low-speed cutoff of the directly entering ions will be at a lower speed than that of the simultaneously observed mirrored ions. In this case the low-speed cutoff of the directly entering ions is difficult to distinguish from zero parallel speed.

We have indicated approximate values for v_E, the low-speed cutoff of the earthward directed He^{++}, and v_M, the mirrored He^{++}, on the low latitude boundary layer distribution. The low-speed cutoff on the earthward directed ions was based largely on the higher energy contours that have a sharp break from their nearly circular shapes at about v_E. The value of v_E is estimated to be about $v_E \approx 30 \pm 20$ km/sec. Note that about 95 s was required to obtain the distribution, and therefore it is likely that some broadening of the distribution function features may have occurred due to time variability in the particle fluxes. A negative low-speed cutoff, i.e., antiparallel to the magnetic field in this example, would be inconsistent with the model we have proposed and would require a different interpretation.

The value of v_M was taken to be about 220 ± 50 km/sec, corresponding roughly to the low-speed edge of the anti-parallel streaming ion population. Using these values for v_E and v_M and assuming the distance from the spacecraft to the low-altitude mirror point to be about $x_m \approx 10 \pm 1 \, R_E$, we estimate the field-aligned distance from the spacecraft to the reconnection site to be $x_r \approx 3.2 \pm 2 \, R_E$. Since the directly entering ion population (with the lowest low-speed cutoff) was directed northward along the field, we infer that the reconnection site was approximately 3 R_E below (south of) the spacecraft. The large error in x_r reflects the fact that the low speed cutoff, v_E, is difficult to distinguish from zero.

From the speed of the ions and the approximate field-aligned distance from the reconnection site to the spacecraft, we estimate that reconnection of the flux tube occurred about 10 minutes prior to these measurements. Similar to the previous example, this interpretation demands that the ExB convection of the flux tube was quite slow between the time when reconnection occurred and when the flux tube arrived at the spacecraft. However, in this case, the ExB drift speed in the ŷ-ẑ plane at the spacecraft is nearly zero, consistent with this interpretations.

DISCUSSION

In this paper we have presented measurements from the AMPTE CCE spacecraft during two encounters of the subsolar low latitude boundary layer. One of these events occurred when the IMF was southward, and the other when the IMF was strongly northward. We have concentrated primarily on the He^{++} distributions measured in the low latitude boundary layer and in the adjacent magnetosheath. The He^{++} ions are a unique identifier of plasma originating in the magnetosheath, and therefore, provide valuable clues regarding the interconnection of magnetospheric field lines to the IMF.

The events that we have reported here are somewhat unusual in two ways. First, the CCE spacecraft with an apogee of 8.8 R_E will cross the dayside magnetopause only during times of high solar wind dynamic pressure. A more unique feature of these observations is that for both these events, the He^{++} distributions in the low latitude boundary layer (earthward of the magnetopause) contained two distinct populations, one with velocity parallel to the magnetic field and the other anti-parallel to the field. This feature in the He^{++} distributions has been reported previously [*Fuselier et al.*, 1992] and, of the 27 low latitude boundary layer crossings by CCE that have been analyzed in detail [*Fuselier et al.*, 1993], it has been found in four of them.

Aside from these unusual features, the He^{++} distributions in the low latitude boundary layer have many features that are similar to previously reported proton, He^{++}, and electron distributions in this region. Proton and He^{++} distributions have been shown at times to exhibit a characteristic "D" shape [*Smith and Rogers*, 1991; *Fuselier et al.*, 1991], as predicted by *Cowley*

[1980]. In these observations, the magnetosheath protons and He^{++} were only flowing in one direction along the magnetic field, i.e., directed earthward from the magnetopause. The oppositely directed protons and He^{++} that mirrored at low altitudes were not observed, presumably because the convection of the reconnected flux tube across the magnetopause was sufficiently rapid that the ions were not able to return to the spacecraft near the subsolar region between the time when reconnection occurred and when the flux tube convected past the spacecraft.

Electrons, on the other hand, with their much higher speeds typically will return to the subsolar region before the reconnected flux tube convects past the spacecraft. In fact, with electrons the opposite situation is usually the case. Namely, the electron distributions nearly always appear as two field-aligned, counterstreaming components. Only in the portion of the low latitude boundary layer closest to the separatrix will uni-directional magnetosheath electrons be observed [*Gosling et al.*, 1990b].

An important feature of the He^{++} ions shown here is the absence of ions at low parallel speeds in the low latitude boundary layer. The absence of the ions below some cutoff speed is what gives the distributions the characteristic "D" shape. A natural explanation for this feature is that at the time of the measurement the spacecraft was located inside the magnetopause on recently reconnected field lines. Given this interpretation, these observations constrain the possible locations of the magnetic reconnection site.

The He^{++} ions measured in the low latitude boundary layer can be used to distinguish between two general locations of reconnection, either equatorward or poleward of the cusp. We refer to reconnection as being equatorward of the cusp if it occurs on magnetospheric field lines that cross the dayside equatorial plane, and poleward of the cusp if it occurs on magnetospheric field lines that extend back into the tail. For southward IMF, it is common to assume that reconnection occurs equatorward of the cusp. However, with northward IMF, it is often assumed that reconnection will occur poleward of the cusp.

A sketch of the field line configuration for reconnection poleward of the cusp is shown in Figure 7 (after *Gosling et al.* [1991]). This diagram is a view of the noon-midnight meridian as seen from the dusk side of the Earth. We have indicated the magnetosheath magnetic field with thick lines and the magnetospheric field with thin lines. Prior to reconnection, the magnetosheath flux tube will contain only magnetosheath plasma. Following reconnection, the plasma at the equatorial region of the flux tube will still contain the magnetosheath plasma, with perhaps the addition of high-speed, field-aligned particles streaming out of the magnetosphere. The important point is that the mag-

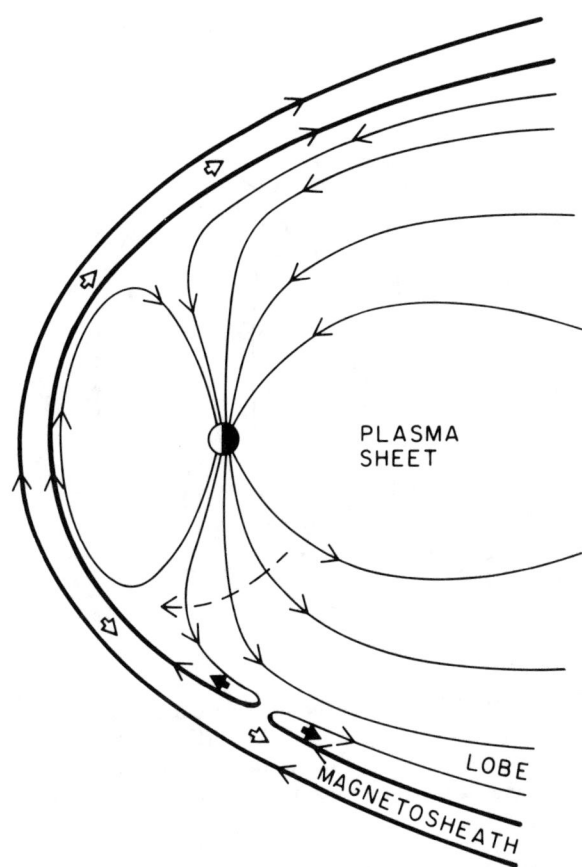

Fig. 7. Sketch of the noon-midnight meridianal plane illustrating reconnection occurring poleward of the cusp when the IMF is northward (after *Gosling et al.* [1991]). The thick lines indicate the IMF and the thin lines indicate the magnetospheric magnetic field.

netosheath particles with low parallel speeds will continue to be present at the equatorial region of the flux tube, even following reconnection. The closed, dayside magnetospheric field line will contain little or no He^{++} and, therefore, will not be identified as part of the low latitude boundary layer. If reconnection were occurring poleward of the cusp, some additional process would need to be invoked to account for the absence of He^{++} ions at low parallel velocities, as seen in Figures 3 and 6. However, this feature in the distribution functions can be readily understood if reconnection occurs equatorward of the cusp.

Sketches of the magnetic field geometry inferred from the observations presented here for the case with northward IMF (left) and southward IMF (right) are shown in Figure 8. These sketches are views of the dayside magnetopause as seen from the sun (after *Gosling et al.* [1990a]). The straight lines represent the IMF and the curved, dipole-like lines represent the magnetospheric

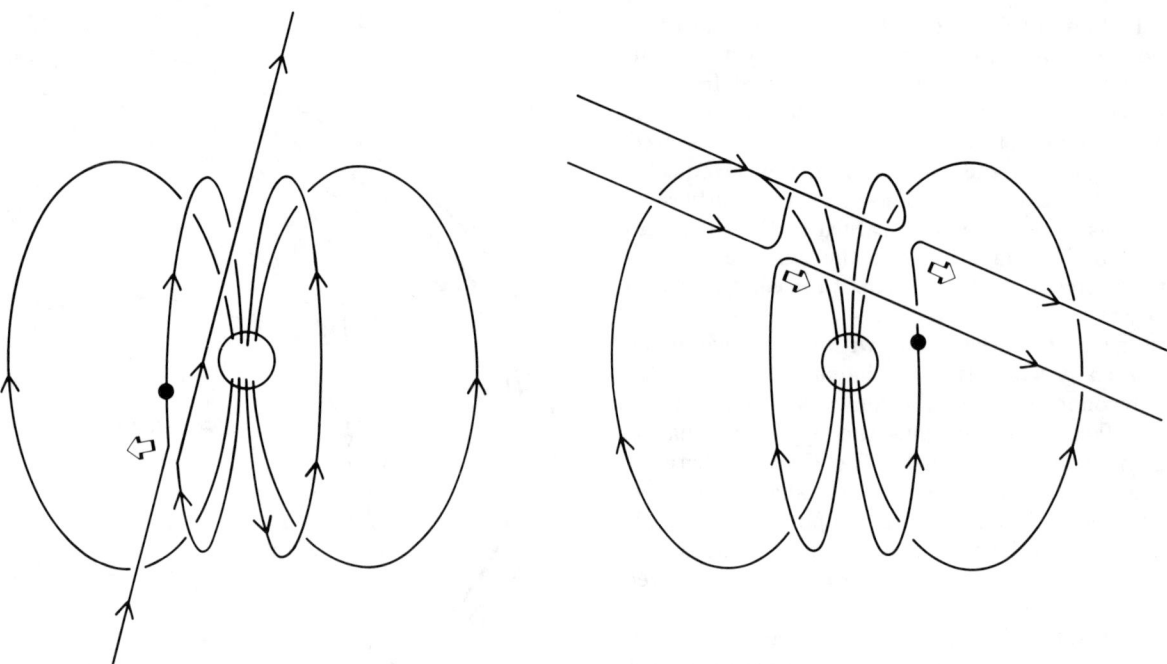

Fig. 8. Sketches of the dayside magnetopause as viewed from the sun with reconnection occurring equatorward of the cusp for (left) northward IMF and (right) southward IMF. The approximate location of the spacecraft for the two events described in this paper are indicated with the black dots. The reconnection site locations are based on the analysis of He^{++} distributions measured in the low latitude boundary layer.

field. The black dots indicate the approximate locations of the spacecraft during the low latitude boundary layer measurements. From the He^{++} measurements described above, we have estimated the reconnection site location to be about 3 R_E below the spacecraft during the northward IMF case and about 5 R_E above the spacecraft during the southward IMF case. Note also that since the reconnected flux tubes were detected some 10 min after reconnection occurred, the magnetospheric flux tube was further duskward (dawnward) of the spacecraft when reconnection occurred in the northward (southward) IMF case.

In the sketch showing northward IMF, we have indicated a possible reconnection location equatorward of the cusp. As illustrated, reconnection would be due to a shear in B_y, rather than in B_z (see Figure 5). In this case, the magnetospheric field line that reconnects is one that crosses the dayside equatorial plane. A spacecraft located inside the magnetosphere near the subsolar point on a recently reconnected field line will only detect entering magnetosheath particles with velocities above some cutoff. The low-speed cutoff is given roughly by the parallel distance from the reconnection site to the spacecraft divided by the time it takes the flux tube to ExB drift to the spacecraft location following reconnection. If the perpendicular convection time from the reconnection site to the spacecraft is sufficiently long, the entering magnetosheath particles that have flowed along the magnetospheric magnetic field to low altitudes, mirrored, and returned to the magnetopause will also be observed.

The location of the reconnection site relative to the cusp, i.e., either on magnetospheric field lines that cross the dayside equatorial plane or those that extend into the tail, should lead to distinctly different plasma distributions on the recently reconnected magnetic field lines. The low latitude boundary layer He^{++} distribution we have shown during strongly northward IMF indicates the presence of two populations counterstreaming along the magnetic field (Figure 6). We interpret these observations as evidence that the spacecraft was located on recently reconnected field lines, with the reconnection site equatorward of the cusp.

Another interesting aspect of these observations for both northward and southward IMF is the long convection time required following reconnection in order for the spacecraft to have detected the magnetosheath He^{++} ions that mirrored at low altitudes. As stated above, it is usually the case that in the low latitude boundary layer only the directly entering magnetosheath ions are detected. Therefore, for both of the cases described here, the reconnection geometry must have been such that the perpendicular plasma convection following reconnection was very slow.

The approximate magnetic field geometry for the event we have described for a southward IMF is illustrated in the right-hand panel of Figure 8. From the observations described above, we conclude that the low latitude boundary layer observations were made approximately 10 min after the onset of reconnection for the flux tube that the spacecraft detected. When the low latitude boundary layer He^{++} ions were detected, the plasma was seen to have a large ExB drift (about 300 km/sec) in the $+\hat{y}$ direction (duskward), as shown in Figure 2. This is the expected direction given the magnetosheath flow direction and the direction of the post-reconnection field-line tension. However, the ten minute convection time implies a very slow average convection for the flux tube on the dayside magnetopause (on the order of 10 km/sec).

A possible scenerio to understand the approximate 10 min duration between reconnection and the arrival of the flux tube at the spacecraft location would be if reconnection occurred slightly dawnward of local noon. In this case, the field-line tension (directed duskward) would be opposed by the magnetosheath flow (directed dawnward). The competition between these two effects could reduce the initial convection velocity of the newly reconnected flux tube [*Crooker*, 1979; *Cowley and Owen*, 1989]. As the intersection point of the flux tube with the magnetopause moved duskward of local noon, the flux tube would accelerate and eventually reach high velocities.

The slow convection required for the northward IMF case may have arisen through somewhat different circumstances. Since the IMF is strongly northward, the tension in the reconnected field line is likely to be weak. Also, with reconnection occurring near the noon-midnight meridian where the $\hat{y}$ component of the magnetosheath flow stagnates, we would expect this contribution to the convection to be small as well. In this case the slow convection does not necessarily require a competition between field-line tension and magnetosheath convection. If reconnection had occurred slightly duskward of local noon, the field-line tension would have opposed the magnetosheath flow, further tending to slow the convection.

Another interesting observation is that when the magnetopause is crossed by the spacecraft in the events described, the electron flux measurements (not shown) indicate that reconnection is occurring at these times as well. The combination of the He^{++} measurements in the low latitude boundary layer and the electron measurements near the magnetopause crossing suggest that reconnection had been occurring on a quasi-steady basis for at least 10 minutes.

In summary, the He^{++} ions detected during two crossings of the low latitude boundary layer by the AMPTE/CCE spacecraft exhibit features in velocity space that are characteristic of the distribution functions expected to result from magnetic reconnection. Given our interpretation of the observed features as resulting from reconnection, the measurements have been used to determine the general location of magnetic reconnection, either equatorward of poleward of the magnetospheric cusps. In addition, the low-speed cutoffs observed on the He^{++} distributions have been used to estimate the magnetic-field-aligned distance from the spacecraft to the reconnection site. In both cases presented, we estimate that reconnection occurred within about 6 R_E of the subsolar magnetopause, equatorward of the cusp.

For these events where both the directly entering and the mirrored magnetosheath ions are observed, we have concluded that the flux tube convection on the dayside magnetopause must have been very slow. This slow convection can result in one of two ways. One possibility, as discussed regarding the southward IMF case, is that the curvature force (field-line tension) on the newly reconnected flux tube is opposed to the magnetosheath flow. The competition between these two forces could lead to slow convection. Another possibility, discussed regarding the northward IMF case, is for both the curvature force and the magnetosheath flow to be small. The curvature force will be small when there is a small shear in the reconnecting fields, and the magnetosheath flow will be small near the subsolar magnetopause and near the cusps. A topic of future research will be to identify additional cases of counterstreaming magnetosheath ions in the low latitude boundary layer to further investigate the reconnection site location and the conditions leading to slow convection of the reconnected flux.

Acknowledgments. We would like to acknowledge valuable discussions with M. F. Thomsen, J. T. Gosling, N. U. Crooker, and J. Fedder. The work at the University of New Hampshire was supported through NASA Supporting Research and Technology grant NAGW-2505 and NSF grant ATM-9111754. Research at Lockheed was supported by NASA through contract NAS5-30565.

REFERENCES

Anderson, B. J., and S. A. Fuselier, Magnetic pulsations from 0.1 to 4.0 Hz and associated plasma properties in the Earth's subsolar magnetosheath and plasma depletion layer, *J. Geophys. Res., 98*, 1461, 1993.

Andrews, M. K., P. W. Daly, and E. Keppler, Ion jetting at the plasma sheet boundary layer: Simultaneous observations of incident and reflected particles, *Geophys. Res. Lett., 8*, 987, 1981.

Cowley, S. W. H., Plasma populations in a simple open model magnetosphere, *Space Sci. Rev., 26*, 217, 1980.

Cowley, S. W. H., The causes of convection in the Earth's magnetosphere: A review of developments during the IMS, *Rev. Geophys., 20*, 531, 1982.

Cowley, S. W. H., and C. J. Owen, A simple illustrative model of open flux tube motion over the dayside magnetopause, *Planet. Space Sci., 37,* 1461, 1989.

Crooker, N. U., Dayside merging and cusp geometry, *J. Geophys. Res., 84,* 951, 1979.

Crooker, N. U., Reverse Convection, *J. Geophys. Res., 97,* 19,363, 1992.

Crooker, N. U., T. E. Eastman, and G. S. Stiles, Observations of plasma depletion in the magnetosheath at the dayside magnetopause, *J. Geophys. Res., 84,* 869, 1979.

Dungey, J. W., Interplanetary field and the auroral zones, *Phys. Rev. Lett., 6,* 47, 1961.

Forbes, T. G., E. W. Hones, Jr., S. J. Bame, J. R. Asbridge, G. Paschmann, N. Sckopke, and C. T. Russell, Evidence for the tailward retreat of a magnetic neutral line in the magnetotail during substorm recovery, *Geophys. Res. Lett., 8,* 261, 1981.

Fuselier, S. A., D. M. Klumpar, and E. G. Shelley, Ion reflection and transmission during reconnection at the Earth's subsolar magnetopause, *Geophys. Res. Lett., 18,* 139, 1991.

Fuselier, S. A., D. M. Klumpar, and E. G. Shelley, Counterstreaming magnetosheath ions in the dayside low latitude boundary layer, *Geophys. Res. Lett., 19,* 425, 1992.

Fuselier, S. A., E. G. Shelley and D. M. Klumpar, Mass density and pressure changes across the subsolar magnetopause, *J. Geophys. Res., 98,* 3935, 1993.

Gosling, J. T., J. R. Asbridge, S. J. Bame, W. C. Feldman, G. Paschmann, N. Sckopke, and C. T. Russell, Evidence for quasi-stationary reconnection at the dayside magnetopause, *J. Geophys. Res., 87,* 2147, 1982.

Gosling, J. T., M. F. Thomsen, S. J. Bame, and C. T. Russell, Accelerated plasma flows at the near-tail magnetopause, *J. Geophys. Res., 91,* 3029, 1986.

Gosling, J. T., M. F. Thomsen, S. J. Bame, R. C. Elphic, and C. T. Russell, Plasma flow reversals at the dayside magnetopause and the origin of asymmetric polar cap convection, *J. Geophys. Res., 95,* 8073, 1990a.

Gosling, J. T., M. F. Thomsen, S. J. Bame, T. G. Onsager, C. T. Russell, The electron edge of the low latitude boundary layer during accelerated flow events, *Geophys. Res. Lett., 17,* 1833, 1990b.

Gosling, J. T., M. F. Thomsen, S. J. Bame, R. C. Elphic, and C. T. Russell, Observations of reconnection of interplanetary and lobe magnetic field lines at the high-latitude magnetopause, *J. Geophys. Res., 96,* 14,097, 1991.

Haerendel, G., G. Paschmann, N. Sckopke, H. Rosenbauer, and P. C. Hedgecock, The frontside boundary layer of the magnetosphere and the problem of reconnection, *J. Geophys. Res., 83,* 3195, 1978.

Hill, T. W., and P. H. Reiff, Evidence of magnetospheric cusp proton acceleration by magnetic merging at the dayside magnetopause, *J. Geophys. Res., 82,* 3623, 1977.

Lockwood, M., and M. F. Smith, The variation of reconnection rate at the dayside magnetopause and cusp ion precipitation, *J. Geophys. Res., 97,* 14,841, 1992.

Maezawa, K., Magnetospheric convection induced by the positive and negative z components of the interplanetary magnetic field: Quantitative analysis using polar cap magnetic records, *J. Geophys. Res., 81,* 2289, 1976.

Newell, P. T., C.-I. Meng, D. G. Sibeck, and R. Lepping, Some low-altitude cusp dependencies on the interplanetary magnetic field, *J. Geophys. Res., 94,* 8921, 1989.

Onsager, T. G., M. F. Thomsen, J. T. Gosling, and S. J. Bame, Electron distributions in the plasma sheet boundary layer: Time-of-flight effects, *Geophys. Res. Lett., 17,* 1837, 1990.

Onsager, T. G., C. A. Kletzing, J. B. Austin, and H. MacKiernan, Model of magnetosheath plasma in the magnetosphere: Cusp and mantle particles at low altitudes, *Geophys. Res. Lett., 20,* 479, 1993.

Paschmann, G., B. U. Ö. Sonnerup, I. Papamastorakis, N. Sckopke, G. Haerendel, S. J. Bame, J. R. Asbridge, J. T. Gosling, C. T. Russell, and R. C. Elphic, Plasma acceleration at the Earth's magnetopause: evidence for reconnection, *Nature, 282,* 243, 1979.

Paschmann, G., I. Papamastorakis, N. Sckopke, B. U. Ö. Sonnerup, S. J. Bame, and C. T. Russell, ISEE observations of the magnetopause: Reconnection and the energy balance, *J. Geophys. Res., 90,* 12,111, 1985.

Paschmann, G., I. Papamastorakis, W. Baumjohann, N. Sckopke, C. W. Carlson, B. U. Ö. Sonnerup, and H. Lühr, The magnetopause for large magnetic shear: AMPTE/IRM observations, *J. Geophys. Res., 91,* 11,099, 1986.

Paschmann, G., S. A. Fuselier, and D. M. Klumpar, High speed flows of H^+ and He^{++} ions at the magnetopause, *Geophys. Res. Lett., 16,* 567, 1989.

Phillips, J. L., S. J. Bame, R. C. Elphic, J. T. Gosling, M. F. Thomsen, and T. G. Onsager, Well-resolved observations by ISEE-2 of ion dispersion in the magnetospheric cusp, *J. Geophys. Res.,* in press, 1993.

Potemra, T. A., L. J. Zanetti, and M. H. Acuna, The AMPTE CCE magnetic field experiment, *IEEE Trans. Geosci. Remote Sens., GE-23,* 246, 1985.

Reiff, P. H., J. L. Burch, and R. W. Spiro, Cusp proton signatures and the interplanetary magnetic field, *J. Geophys. Res., 85,* 5997, 1980.

Reiff, P. H., Evidence of magnetic merging from low-altitude spacecraft and ground-based experiments, in *Magnetic Reconnection in Space and Laboratory Plasmas, Geophys. Monogr. Ser.,* vol. 30, edited by E. W. Hones, pp. 104-113, AGU, Washington, D. C., 1984.

Russell, C. T., The configuration of the magnetosphere, in *Critical Problems of Magnetospheric Physics,* edited by E. R. Dyer, pp. 1-16, Inter-Union Commission on Solar-Terrestrial Physics, Secretariat, National Academy of Sciences, Washington, D. C., 1972.

Scurry, L., C. T. Russell, and J. T. Gosling, A statistical study of accelerated flow events at the dayside magnetopause, *J. Geophys. Res.,* in press, 1993.

Shelley, E. G., A. Ghielmetti, E. Hertzberg, S. J. Battel, K. Altwegg-Von Burg, and H. Balsiger, The AMPTE CCE Hot-Plasma Composition Experiment (HPCE), *IEEE Trans. Geosci. Remote Sens., GE-23,* 241, 1985.

Smith, M. F., and D. J. Rodgers, Ion distributions at the dayside magnetopause, *J. Geophys. Res., 96,* 11,617, 1991.

Song, P., and C. T. Russell, Model of the formation of the low latitude boundary layer for strongly northward interplanetary magnetic field, *J. Geophys. Res., 97,* 1411, 1992.

Sonnerup, B. U. Ö., G. Paschmann, I Papamastorakis, N. Sckopke, G. Haerendel, S. J. Bame, J. R. Asbridge, J. T. Gosling, and C. T. Russell, Evidence for magnetic field reconnection at the Earth's magnetopause, *J. Geophys. Res., 86,* 10,049, 1981.

Stern, D. P., Parabolic harmonics in magnetospheric modelling: The main dipole and the ring current, *J. Geophys. Res., 90,* 10,851, 1985.

S. A. Fuselier, Dept. 91-20 Bldg. 255, Lockheed Palo Alto Research Laboratory, 3251 Hanover Street, Palo Alto, CA 94304.

T. G. Onsager, Institute for the Study of Earth, Oceans, and Space, Morse Hall, University of New Hampshire, Durham, NH 03824.

The Shape and Size of Convection Cells in the Jovian Magnetosphere

T. W. Hill

Space Physics and Astronomy Department, Rice University, Houston, Texas

Plasma convection in Jupiter's magnetosphere is driven, at least in part, by the centrifugal instability of the plasma torus produced by the satellite Io. Very little is known, however, about the shape and size of the resulting convection cells; in the absence of definitive observations, we have a sufficient variety of theoretical models to accommodate any desired shape and size. There is indirect evidence for a global-scale corotating convection pattern, as well as for significant superimposed "mesoscale" structure, having scale sizes much larger than the ion gyroradius but much smaller than the magnetosphere. If the characteristic shape of mesoscale convection cells is assumed, then their characteristic scale size can be estimated from observed parameters with the help of theoretical considerations. If the cells are assumed to be basically circular, as in the interchange diffusion model, then their scale size is estimated to be ~0.2 R_J. If they are assumed to be basically linear (much longer in radial than azimuthal extent), then their azimuthal scale size is estimated to be ~2 R_J. Both estimates scale with the square root of the ratio of mass-loading rate to ionospheric conductance.

INTRODUCTION

The term "magnetospheric convection" applies to the transport of plasma within or through a planetary magnetosphere by the motion of magnetic flux tubes which, by definition, have transverse ($\perp \mathbf{B}$) dimensions exceeding the largest plasma ion gyroradius, and transverse velocities $\mathbf{v}$ satisfying the MHD approximation

$$\mathbf{E} + \mathbf{v} \times \mathbf{B} = 0 \qquad (1)$$

In a steady state, or in a sufficiently long-term average of a time dependent convection process, the electric field $\mathbf{E}$ must also satisfy the steady-state version of Faraday's law

$$\nabla \times \mathbf{E} = -\partial \mathbf{B}/\partial t = 0 \qquad (2)$$

so that the transverse velocity field given by (1) tends to form closed circulation cells or "convection cells."

The shape and size (and hence multiplicity) of these convection cells can be influenced by a number of factors, the most obvious being any spatial structure that is intrinsic to the source(s) of plasma and energy that drive the convection system. For example, convection in Earth's magnetosphere is driven primarily by the solar wind, which operates on a global scale, thus imposing a global-scale pattern on the flow (antisunward convection on open polar-cap field lines that connect to the solar wind, and sunward return flow on closed magnetospheric field lines). Superimposed on this global pattern, however, there is evidence of significant mesoscale structure, that is, structure with scale sizes much larger than the ion gyroradius but much smaller than the magnetosphere [e.g., *Angelopoulos et al.*, 1992, and references therein]. Although the cause of this mesoscale structure is not well understood, it can have global consequences on the structure and dynamics of the terrestrial magnetotail [*Pontius and Wolf*, 1990].

It is likely that convection is the primary mechanism of plasma transport in planetary magnetospheres generally, although Earth's is the only magnetosphere for which we have empirical knowledge of the spatial and temporal structure of that convection. Jupiter's magnetosphere provides a particularly useful testbed for theories of magnetospheric convection, despite the relative paucity of *in situ* observations compared to Earth's, for a number of reasons:

Solar System Plasmas in Space and Time
Geophysical Monograph 84
Copyright 1994 by the American Geophysical Union.

(1) the driving force (centrifugal force) can be specified in a simple, model-independent, algebraic form, (2) the location and magnitude of the plasma source are relatively well known, and (3) the source region and much of the surrounding convection region are characterized by relatively small values of β (= plasma pressure/magnetic-field pressure) and of M_A (= flow speed/Alfven speed), so that the magnetic field configuration may be taken as given (indeed, may be taken as that of a spin-aligned centered dipole without much loss of physics). Thus we turn to Jupiter's magnetosphere to investigate the question, what determines the shape and size of magnetospheric convection cells produced by given global-scale sources of plasma and energy?

REVIEW OF JOVIAN CONVECTION THEORIES

In Jupiter's magnetosphere, convection is most likely driven, not by the solar wind, but by Jupiter's rotation. Rotational energy is extracted by the ionization and subsequent outward transport of plasma from the gas torus derived from, and approximately co-orbital with, the satellite Io [e.g., *Dessler*, 1980, *Eviatar and Siscoe*, 1980]. Convection cells, irrespective of their shape or size, are energized by the centrifugal instability of the plasma distribution injected by the Io-torus source, which is located well within the corotation-dominated plasmasphere yet well outside the spin-synchronous orbital distance, so that the centrifugal force dominates both external forces and gravity. (Note that no analogous region exists in Earth's magnetosphere.) The centrifugal force ($\Omega_J^2 r$ per unit mass) is approximately balanced by the Lorentz force $\mathbf{J}_c \times \mathbf{B}$ (where $\mathbf{J}_c$ is the centrifugal drift current density [*Hill*, 1983, and references therein]). Any radial perturbation of the outer torus boundary (Figure 1) produces a divergence of $\mathbf{J}_c$, which produces a polarization of charge, which produces an electric field, which produces a velocity perturbation (by (1), which causes the perturbation to grow with time. The centrifugal drift current J_c is coupled by Birkeland (magnetic-field-aligned) currents $J_{\|}$ to Pedersen currents J_P in the Jovian ionosphere, whereby the ionospheric conductivity determines the magnitude, but not the sign, of the charge polarization, and hence the speed, but not the direction, of the resulting magnetospheric flow. This basic instability mechanism underlies virtually all theories of Jovian plasma transport, although the theories differ widely in their assumptions about the shape and size of the resulting convection cells.

The rotational energy source is available on a global scale and imposes no obvious mesoscale structure on the resulting convection. The plasma source, however, introduces intrinsic mesoscale structure both in its radial confinement,

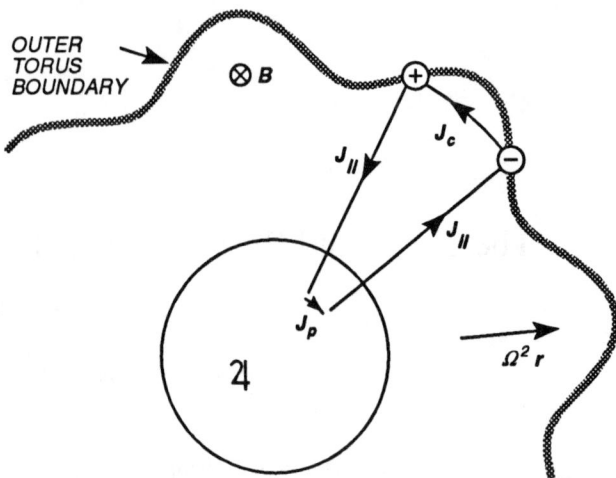

Fig. 1. Cartoon of the outer boundary of the Io plasma torus, illustrating the mechanism of centrifugal instability regulated by Jovian ionospheric conductivity. The centrifugal drift current J_c is connected by Birkeland currents $J_{\|}$ to the Pedersen current J_p in Jupiter's ionosphere.

which is well established, and probably also in its azimuthal variations, which are less well documented observationally and hence more open to theoretical debate, although they are undoubtedly present. For example, a variety of indirect evidence suggests that the plasma source has a persistent global-scale longitudinal asymmetry in the corotating (System III) coordinate system [e.g., *Hill et al.*, 1983, and references therein], which would impose a global-scale two-cell convection pattern in the corotating frame of reference [*Vasyliunas*, 1978; *Hill et al.*, 1981]. A simple model of such a corotating convection system [*Liu and Hill*, 1990] is illustrated in Figure 2.

Although there is considerable indirect evidence in favor of such a global corotating convection pattern [e.g., *Hill et al.*, 1983, and references therein], and no direct evidence against it, there is no reason to doubt the existence of smaller scale (mesoscale) convection structures superimposed on this global-scale pattern. There is, indeed, considerable evidence of such mesoscale structure in Voyager 1 plasma wave data [*Ansher et al.*, 1992], and in Ulysses radio wave observations during Jupiter encounter [*Stone et al.*, 1992]. Analytic theories of Jovian convection have primarily emphasized mesoscale structures, probably because they are easier to treat mathematically than global-scale structures. Such analytic theories are of two types.

The "interchange diffusion" model [e.g., *Siscoe and Summers*, 1981; *Siscoe et al.*, 1981; *Vasyliunas*, 1989] is based on the assumption that convection is dominated by multiple circulation cells or eddies (Figure 3) that are

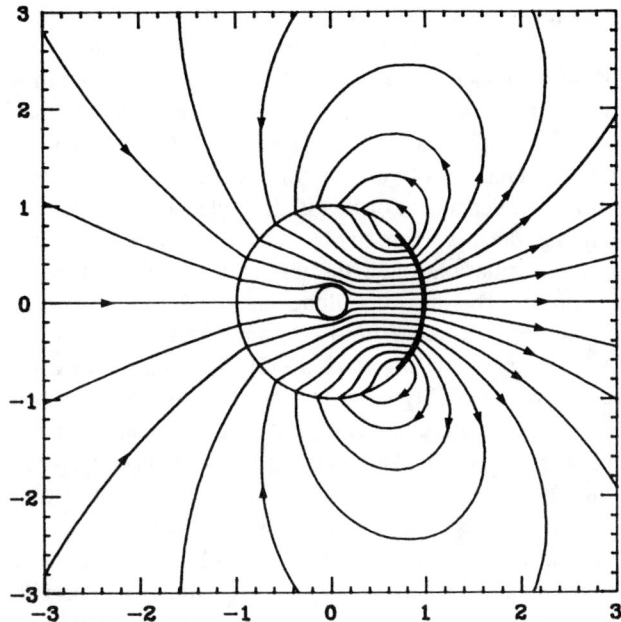

Fig. 2. Plasma streamlines (and equipotential contours) for the corotating convection model of *Liu and Hill* [1990]. Plasma flows outward from the overdense active sector (heavy line) of the Io torus, and inward in the remaining longitude sector. The entire pattern corotates with Jupiter. The unit of distance is $6\,R_J$.

intermittent in time and much smaller in their transverse dimension than any macroscopic lengthscale. Azimuthal symmetry on the global scale is always taken as a simplifying assumption in models of this type (although it does not appear to be required by the nature of the process), and the radial transport of plasma is formulated in terms of a diffusion equation that describes a random walk of flux tubes in the equatorial radial coordinate. As in any purely diffusive process, the time-averaged flow velocity vanishes everywhere but the time-averaged mass flux remains finite (and is directed from source to sink, in this case radially outward) because outward-moving flux tubes tend to carry more mass (per unit magnetic flux) than inward-moving ones. The circulation of these eddies is driven by the centrifugal instability of the mass-density distribution produced by the Io torus source:

$$\partial \eta / \partial L < 0 \qquad (3)$$

where $L \equiv r/R_J$ and

$$\eta \equiv \int \rho ds/B \qquad (4)$$

is the mass contained per unit magnetic flux. (In (4), $\rho =$ volume mass density and the integral is along a field line.)

The circulation rate is governed by the Jovian height-integrated ionospheric Pedersen conductivity Σ_P, which determines the magnitude of the electric field (hence velocity, by (1)) that is needed to provide ionospheric closure of a given diverging magnetospheric centrifugal drift current as illustrated in Figure 1. These considerations imply [*Siscoe and Summers*, 1981] that the eddy circulation timescale is

$$\tau_c \sim \frac{\Sigma_p B_J}{\Omega_J^2 L^4 \left|\frac{\partial \eta}{\partial L}\right|} \qquad (5)$$

independent of the size of the eddy (provided that it is small in the sense defined above). In (5), B_J is Jupiter's surface equatorial dipole field strength (a spin-aligned dipole is assumed throughout this paper). If $R_J \Delta L$ is the diameter of the circulation cell, then the speed associated with its circulation is

$$v \sim \frac{\Omega_J \Delta \eta}{\Sigma_P B_J} \Omega_J R_J L^4 \qquad (6)$$

where $\Delta \eta = \Delta L \partial \eta / \partial L$ is the change in the ambient value of η across the diameter of the cell.

The other type of mesoscale convection model is the "transient convection" model [*Pontius et al.*, 1986, *Pontius and Hill*, 1989] illustrated in Figure 4. Here it is assumed that discrete mass-loaded flux tubes are shed from the outer edge of the Io torus source structure and are subsequently accelerated outward through a relatively empty background under the influence of the centrifugal buoyancy force. With

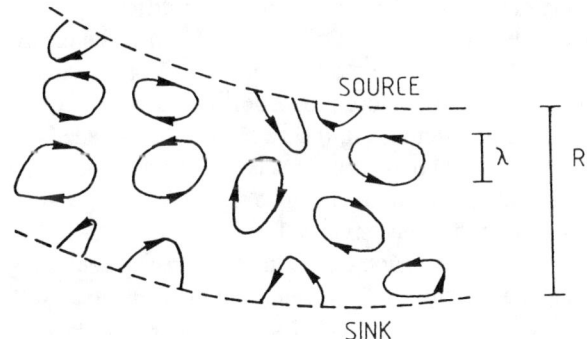

Fig. 3. Cartoon illustration, from *Vasyliunas* [1989], of the interchange diffusion process projected on the equatorial plane. The source is located in the Io torus at $L \approx 6$, and the sink is at an unspecified distance ($L > 6$) in the magnetosphere. The characteristic eddy scale size, labeled λ in the figure, is $R_J \Delta L$ in the notation of the text.

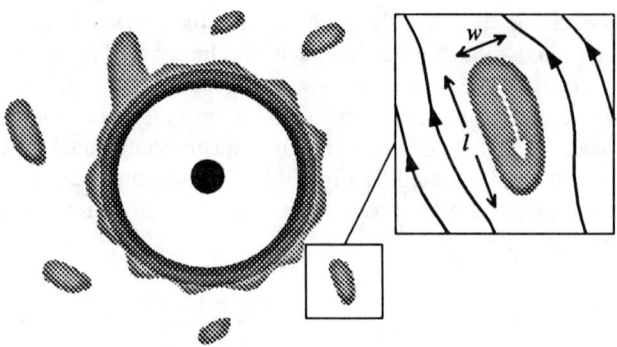

Fig. 4. Cartoon illustration of the transient convection model of *Pontius and Hill* [1989], in which discrete mass-loaded flux tubes break away from the Io torus and accelerate outward under the centrifugal force. The flux tubes are assumed to have elliptical cross section in the equatorial plane, with radial length l and azimuthal width w.

the simplifying assumptions that these flux tubes have mesoscale dimensions and elliptical equatorial cross section, and are electrodynamically isolated one from another, their radial velocity can be estimated as

$$v_r = \left(\frac{l}{l+2w}\right)\frac{\Omega_J \Delta \eta}{\Sigma_P B_J}\Omega_J R_J L^4 \qquad (7)$$

where l and w are the radial and azimuthal dimensions of the flux tube cross section, respectively, and $\Delta \eta$ is now the difference in η between the mass-loaded flux tube and its surroundings. The formal similarity of the results for interchange diffusion (6) and transient convection (7) reflects the fact that both models solve the same basic magnetosphere-ionosphere coupling problem implied by the current system of Figure 1. The geometric factor in parentheses in (7) is expected to be of order unity. The results differ essentially in the different definitions of $\Delta \eta$; in (6), $\Delta \eta$ is the change in the background η across the diameter ΔL of the circulation cell (and is therefore proportional to ΔL), whereas in (7), $\Delta \eta$ is the difference in η between the mass-loaded flux tube and its surroundings, and is taken as a parameter in the model, independent of the flux-tube size.

The size of the convection cell is a free parameter in both the interchange diffusion model and the transient convection model, subject only to the assumption that it be mesoscale as defined above. The effect of this free parameter is, however, different in the two models. In the interchange diffusion model, the cell size determines $\Delta \eta$ as described above, and hence influences the radial mass flux or transport rate. For nominal values of the transport rate $\dot{M} \sim 10^3$ kg/s) and the ionospheric conductivity ($\Sigma_P \sim 1$ mho), the interchange diffusion model requires a cell size $\Delta L \sim 0.2$ for self-consistency [*Siscoe and Summers*, 1981; *Huang and Hill*, 1991]. In the transient convection model, the parameter $\Delta \eta$ has no obvious dependence on the flux-tube size, and thus there is no similarly simple relationship between flux-tube size and transport rate (see, however, the following section).

If the radial density distribution $\eta(L)$ is taken as a given function (independent of longitude angle ϕ), one can solve an initial-value problem to obtain the time evolution of this initial distribution under the influence of the centrifugal instability. This has been done analytically for the linear regime [*Huang and Hill*, 1991] and numerically for the nonlinear regime [*Yang et al.*, 1992]. The nonlinear evolution of the instability takes the form of radially elongated fingers of torus plasma flowing outward through the magnetosphere (Figure 5). The transverse (azimuthal) dimension of these fingers is comparable to the radial lengthscale of the initially assumed radial density gradient ($L\Delta\phi \sim \eta/|\partial\eta/\partial L|$), consistent with expectations based on the linear analysis. This initial-value problem, although internally self-consistent, does not provide a consistent description of the Jovian magnetosphere because the initial plasma distri-

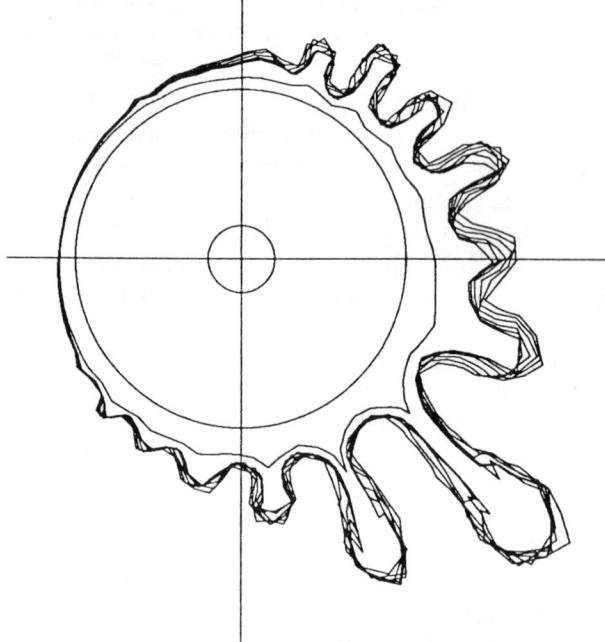

Fig. 5. Simulation result of *Yang et al.* [1992] showing contours of flux-tube content in the equatorial plane. The initially assumed torus distribution has evolved into radially elongated "fingers" of outward-moving plasma. In this particular case, the fingers occur preferentially on one side of Jupiter because the initially assumed torus distribution was slightly heavier on that side, as assumed in the corotating convection model. The innermost (circular) contour is at $L = 6$.

bution, chosen to match observations in some sense, quickly evolves into a very different distribution. If the initial distribution matches observations (a debatable point), the evolved distribution cannot.

In principle, it is the distribution of the plasma source rate, not the resulting distribution of plasma density, that is "given," that is, determined extrinsically, at least in part, by non-magnetospheric processes. A given source distribution determines, in principle, a resultant convection pattern that combines with the source distribution to produce a resultant density distribution. The numerical simulation scheme of Yang et al. is presently being revised [*R. A. Wolf and Y. S. Yang*, private communication, 1993] to accommodate a continuous plasma source. A recent analytic attempt along the same lines [*T. W. Hill*, submitted to *JGR*, 1993; hereinafter "H93"] is summarized in the following section.

A SIMPLE CONVECTION MODEL WITH PLASMA SOURCE

We adopt a simple model of convection driven by the Io torus plasma source, with the explicit purpose of estimating the effect of the source geometry on the scale size of the resulting convection cells. The assumed form of the convection cells is illustrated in Figure 6. The source rate is assumed to be uniform for $L_s - \Delta L_s < L < L_s$ (the shaded region) and zero elsewhere. The convection just outside the source region is assumed to consist of alternating azimuthal sectors of purely radial inflow and outflow. We do not solve for the flow pattern within the source region, but we assume that its average azimuthal component is given by the 3-5% corotation lag [*Brown*, 1983] that is needed to close the pick-up current associated with mass loading [*Pontius and Hill*, 1982].

The radial speed just outside the source region is

$$v_r = \frac{\Omega_J(\eta - \langle \eta \rangle)}{\Sigma_P B_J} \Omega_J R_J L_s^4 \qquad (8)$$

[*Hill et al.*, 1981], where $\langle \eta \rangle$ is the azimuthally averaged value of η at $L = L_s$ (≈ 6). Note that if a single, narrow outflow sector is embedded in an empty background, then $\langle \eta \rangle \approx 0$ and $\eta - \langle \eta \rangle \approx \Delta \eta$, so that (8) reduces to (7) in the limit of a long narrow transient ($l >> w$). If, however, the outflow and inflow sectors (subscript "o" and "i" respectively) are equally spaced, as illustrated in Figure 6, then $\langle \eta \rangle \approx (\eta_o + \eta_i)/2$ and $\eta_{o,i} - \langle \eta \rangle \approx \pm \Delta \eta / 2$ where $\Delta \eta \equiv \eta_o - \eta_i$, and

$$(v_r)_{o,i} = \pm \frac{\Omega_J^2 R_J \Delta \eta}{2 \Sigma_P B_J} L_s^4 \qquad (9)$$

The net rate of radial mass transport is then

$$\dot{M}_t = \sum_{i,o} \pi r B \eta v_r = \frac{\pi \Omega_J^2 R_J^2 (\Delta \eta)^2 L_s^2}{2 \Sigma_P} \qquad (10)$$

whereas the mass loading rate in the source region is

$$\dot{M}_s = \dot{\eta} \left(\frac{B_J}{L_s^3}\right)\left(2\pi R_J^2 L_s \Delta L_s\right) \qquad (11)$$

For steady state, we have $\dot{M}_t = \dot{M}_s$, which implies, from (10) and (11)

$$(\Delta \eta)^2 = \frac{4 \Sigma_P B_J \dot{\eta} \Delta L_s}{\Omega_J^2 L_s^4} \qquad (12)$$

Another relation between $\Delta \eta$ and $\dot{\eta}$ comes from the continuity equation

$$\Delta \eta = \dot{\eta} \tau \sim \dot{\eta} \frac{R_J L_s \Delta \phi}{2 \langle v_\phi \rangle} \qquad (13)$$

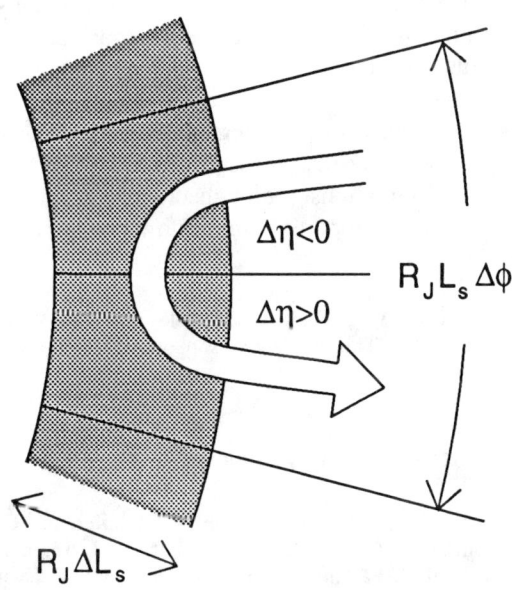

Fig. 6. Illustration, in the equatorial plane, of the simple convection model assumed here in order to investigate the dependence, if any, of the azimuthal size $\Delta \phi$ of the convection cells on the radial width ΔL_s of the plasma source region (shaded).

where τ is the residence time of a flux tube in the source region and $\langle v_\phi \rangle$ is its average azimuthal velocity therein. Equations (11), (12), and (13) can be combined to give

$$\Delta\phi = \frac{4\langle v_\phi\rangle B_J \Delta L_s}{\Omega_J L_s^4}\left(\frac{2\pi\Sigma_p}{\dot{M}}\right)^{1/2} \quad (14)$$

The mean azimuthal speed $\langle v_\phi \rangle$ in the source region corresponds to the corotation lag due to mass loading

$$\langle v_\phi \rangle = \frac{\xi L_s^6 \dot{M}(\Omega_J - \Omega_{Io})}{8\pi\Sigma_p R_J B_J^2 \Delta L_s} \quad (15)$$

[*Pontius and Hill*, 1982], where Ω_{Io} ($\approx \Omega_J/4$) is the angular frequency of a circular Keppler orbit at $L=L_s$ and

$$\xi \equiv \frac{\text{Ionization Rate} + \text{Charge-Exchange Rate}}{\text{Ionization Rate}} \sim 4 \quad (16)$$

The factor ξ is needed because charge-exchange collisions do not contribute to the net mass-loading rate $\dot{M}$, but they do contribute to the requirement for spin-up torque, and hence to $\langle v_\phi \rangle$ [*Pontius and Hill*, 1982]. (In principle, a correction to (15), of first order in the ratio $\langle v_\phi\rangle/\Omega_J R_J L_s$, is needed to prevent this ratio from exceeding unity for large values of $\dot{M}/\Sigma_p \Delta L_s$ [H93], but this correction is not needed for Jupiter because the ratio $\langle v_\phi\rangle/\Omega_J R_J L_s$ is known empirically to be ~0.03 [*Brown*, 1983].)

Finally, combining (14) and (15) we obtain

$$\Delta\phi = \xi\left(1 - \frac{\Omega_{Io}}{\Omega_J}\right)\left(\frac{L_s}{L_c}\right)^2 \quad (17)$$

where

$$L_c \equiv \left(\frac{2\pi\Sigma_p R_J^2 B_J^2}{\dot{M}}\right)^{1/4} \quad (18)$$

is the characteristic distance for corotation breakdown under the Coriolis acceleration of outward-moving plasma [*Hill*, 1979]. Empirically, $L_c \sim 20$ [*Hill*, 1980], which, with $L_s=6$, $\xi=4$ and $\Omega_{Io}\approx\Omega_J/4$, gives $\Delta\phi \sim 0.3$ rad ≈ 20 deg, independent of ΔL_s. If we had retained the first-order correction in (15) mentioned earlier, a similar correction would appear in (18) [H93] which would make $\Delta\phi$ proportional to ΔL_s for sufficiently small ΔL_s, but for nominal Jovian parameters this correction is not important and we have $\Delta\phi$ independent of ΔL_s as in (18).

CONCLUSION

For the simple convection picture illustrated in Figure 6, motivated by the numerical simulation results of Figure 5, and for nominal Jovian parameters ($\dot{M} \sim 1000$ kg/s, $\Sigma_p \sim 1$ mho), we find that mass conservation and other considerations imply an azimuthal cell size $\Delta\phi \sim 20$ deg ($R_J L_s \Delta\phi \sim 2 R_J$), independent of the radial width ΔL_s of the source region. (The radial extent of the convection cells is much larger than this, by assumption.) This is in contrast to the interchange diffusion picture illustrated in Figure 3 and discussed earlier, wherein the same parameters require a radial width (and hence, by assumption, also an azimuthal width) $R_J \Delta L \sim 0.2 R_J$. Thus the two models differ widely, not only in the (assumed) shape of the convection cells, but also in the (derived) size of the cells implied by a given mass loading rate.

Existing observations that bear on this question are scarce and sometimes in apparent conflict. For example, *in situ* plasma observations made by the Voyager 1 spacecraft during its Jupiter encounter [*Richardson and McNutt*, 1987] appear to rule out any mesoscale structure having $\Delta\eta/\eta \gtrsim 0.1$ within the radial range $6 < L < 11$, whereas remote radio observations made by the Ulysses spacecraft during its Jupiter encounter [*Stone et al.*, 1992] appear to require such structure in the same radial range. (Remote observations from Earth or Earth orbit, while providing a wealth of information about the Io torus itself [e.g., *Brown et al.*, 1983], cannot detect the more tenuous plasma in the circum-torus convection region considered here.) It is, of course, possible in principle that the mode of plasma transport was fundamentally different during the encounters of Voyager 1 and Ulysses. It seems more likely, however, that further observations and/or theoretical insights will enable us to fit these partial clues into a single coherent picture.

Acknowledgments. I thank Alex Dessler and Dick Wolf for helpful comments. This work was supported in part by NSF grant ATM89-11031.

REFERENCES

Angelopoulos, V., W. Baumjohann, C. F. Kennel, F. V. Coroniti, M. G. Kivelson, R. Pellat, R. J. Walker, H. Lühr, and G. Paschmann, Bursty bulk flows in the inner central plasma sheet, *J. Geophys. Res.*, 97, 4027, 1992.

Ansher, J. A., W. S. Kurth, D. A. Gurnett, and C. K. Goertz, High resolution measurements of density structures in the Jovian plasma sheet, *Geophys. Res. Lett.*, 19, 2281, 1992.

Brown, R. A., Observed departure of the Io plasma torus from rigid corotation with Jupiter, *Astrophys. J.*, *268*, L47, 1983.

Brown, R. A., C. B. Pilcher, and D. F. Strobel, Spectrophotometric studies of the Io torus, in *Physics of the Jovian Magnetosphere*, edited by A. J. Dessler, chap. 6, Cambridge University Press, New York, 1983.

Dessler, A. J., Mass-injection rate from Io into the Io plasma torus, *Icarus*, *44*, 291, 1980.

Eviatar, A., and G. L. Siscoe, Limit on rotational energy available to excite Jovian aurora, *Geophys. Res. Lett.*, *7*, 1085, 1980.

Hill, T. W., Rotationally-induced Birkeland current systems, in *Magnetospheric Currents*, (T. A. Potemra, ed.) Geophysical Monograph 28, pp. 340-349, American Geophysical Union, Washington, D.C., 1983.

Hill, T. W., Inertial limit on corotation, *J. Geophys. Res.*, *84*, 6554, 1979.

Hill, T. W., Corotation lag in Jupiter's magnetosphere: A comparison of observation and theory, *Science*, *207*, 301, 1980.

Hill, T. W., A. J. Dessler, and L. J. Maher, Corotating magnetospheric convection, *J. Geophys. Res.*, *86*, 9020, 1981.

Hill, T. W., A. J. Dessler, and C. K. Goertz, Jovian magnetospheric models, in *Physics of the Jovian Magnetosphere*, edited by A. J. Dessler, chap. 10, Cambridge University Press, New York, 1983.

Huang, T. S., and T. W. Hill, Drift-wave instability in the Io plasma torus, *J. Geophys. Res.*, *96*, 14075, 1991.

Liu, W. W. and T. W. Hill, Convective transport of plasma in the inner Jovian magnetosphere, *J. Geophys. Res.*, *95*, 4017, 1990.

Pontius, D. H., Jr., and T. W. Hill, Departure from corotation of the Io plasma torus: Local plasma production, *Geophys. Res. Lett.*, *9*, 1321, 1982.

Pontius, D. H., Jr., T. W. Hill, and M. E. Rassbach, Steady state plasma transport in a corotation-dominated magnetosphere, *Geophys. Res. Lett.*, *13*, 1097, 1986.

Pontius, D. H., Jr., and T. W. Hill, Rotation driven plasma transport: the coupling of macroscopic motion and microdiffusion, *J. Geophys. Res.*, *94*, 15041, 1989.

Pontius, D. H., Jr, and R. A. Wolf, Transient flux tubes in the terrestrial magnetosphere, *Geophys. Res. Lett.*, *17*, 49, 1990.

Richardson, J. D., and R. L. McNutt, Jr., Observational constraints on interchange models at Jupiter, *Geophys. Res. Lett.*, *14*, 64, 1987.

Siscoe, G. L., and D. Summers, Centrifugally driven diffusion of Iogenic plasma, *J. Geophys. Res.*, *86*, 8471, 1981.

Siscoe, G. L., et al., Ring current impoundment of the Io plasma torus, *J. Geophys. Res.*, *86*, 8480, 1981.

Stone, R. G., et al., Ulysses radio and plasma wave observations in the Jupiter environment, *Science*, *257*, 1524, 1992.

Vasyliunas, V. M., A mechanism for plasma convection in the inner Jovian magnetosphere, COSPAR Abstracts, p. 66, Innsbruck, Austria, 29 May - 10 June 1978.

Vasyliunas, V. M., Maximum scales for preserving flux tube content in radial diffusion driven by interchange motions, *Geophys. Res. Lett.*, *16*, 1465, 1989.

Yang, Y. S., R. A. Wolf, R. W. Spiro, and A. J. Dessler, Numerical simulation of plasma transport driven by the Io torus, *Geophys. Res. Lett.*, *19*, 957, 1992.

T. W. Hill, Space Physics and Astronomy Department, Rice University, Houston, TX 77251

Structure of the Venus Tail

O. Vaisberg, V. Smirnov, A. O. Fedorov, L. Avanov, and F. Dunjushkin

Space Research Institute, Moscow, Russia

J. G. Luhmann and C. T. Russell

Institute of Geophysics and Planetary Physics, University of California, Los Angeles, USA

The steady-state tail of Venus at relatively close distances to the planet consists of two magnetic field lobes separated by a current sheet. It is believed to be formed by the mass loading of passing solar wind magnetic flux tubes by planetary ions. The lobes are separated by a current sheet presumably populated by hot plasma. The induced magnetic tail together with the cross-tail current sheet rotates around its axis due to rotation of the transverse component of the IMF. The magnetic field structure in the tail at 10-12 R_v downstream is always very complicated and shows a variety of magnetic structures and current layers. This appearance is usually interpreted in terms of motions of the tail.

Within the Venusian tail at close and intermediate downstream distances (from 0.5_v to ~ 5 R_v) Venera 9 and 10 measured relatively low-energy ion fluxes that appear to be nearly permanent. At least two plasma populations were identified: one within the tail lobes, and another more energetic one at the current layer. Pioneer Venus data also indicated the presence of two ion populations in the tail at ~ 10-12 R_v, with the higher energy population being interpreted as accelerated planetary oxygen ions. In this report we analyze different plasma regimes in the context of the observed magnetic field configurations in the tail.

INTRODUCTION

Venus is the most extensively studied case of the solar wind interaction with a non-magnetized planet. The ionosphere and upper atmosphere of the planet form an almost impenetrable obstacle to the solar wind flow. As a result, a bow shock forms that heats and deflects the supersonic solar wind plasma around the planet. The shocked solar wind plasma, while flowing around the planet, interacts with newly born ions from the upper atmosphere. The pick-up of newly born ions leads to significant modification of the flow close to the planet, and determines the configuration and properties of the Venus tail.

Solar System Plasmas in Space and Time
Geophysical Monograph 84
Copyright 1994 by the American Geophysical Union.

The data on the solar wind-Venus interaction were collected mostly by Venera 9 and 10 in 1975-1976 and by Pioneer Venus Orbiter (PVO) from 1978-1992 [see *Russell and Vaisberg*, 1983; *Phillips and McComas*, 1991, for reviews]. It was shown that Venus has an induced magnetic tail with two magnetic lobes formed by draping of interplanetary magnetic field (IMF) around the planet [*Yeroshenko*, 1979]. Distinct changes of the plasma regime from ionosheath flow to antisunward low-energy plasma flow in the tail were found at close downstream distances [*Vaisberg et al.*, 1976].

The purpose of this paper is to summarize existing plasma observations within the Venusian tail. We will consider plasma domains observed at different distances and within different magnetic field configurations, dynamics of the tail plasma, the possible connections between the plasma properties and magnetic field structures in the tail, and will speculate on the origin of the tail plasma. As several excellent review papers concerning mostly the magnetic tail are available, the emphasis here is on the plasma observations.

MAGNETIC TAIL

Magnetic Field at Close Distances

As shown in Figure 1 *Yeroshenko* [1979] found from Venera 9 and 10 measurements that the tail magnetic field at distances close to Venus has a two-lobe structure with the polarity of the radial component controlled by the transverse component of the interplanetary magnetic field. Venera 9 and 10 magnetic field measurements at distances of ~ 0.2 to 1.0 R_v showed a diverging magnetic field [*Dolginov et al.*, 1981], while Pioneer Venus measurements at ionospheric heights reveal converging fields [*Russell et al.*, 1980]. The dimension of the Venus tail was determined from plasma observations of the radius of the cavity formed by Venus in the solar wind flow, about 1.3 R_v [*Vaisberg et al.*, 1976].

Another observed feature of the nightside environment of Venus are ionospheric holes with low number density and strong quasivertical magnetic fields [*Brace et al.*, 1980, 1982, 1987]. The magnetic field in the surrounding ionosphere is usually weak and nearly horizontal [*Luhmann et al.*, 1982]. *Marubashi et al.* [1985] showed that the magnetic field polarities in the holes reverse across the magnetic meridian plane in accordance with the magnetic field draping around the Venus ionosphere. The magnetic field in ionospheric holes is generally dominated by the B_x component rather than a truly radial component.

Distant Magnetic Tail

Slavin et al. [1984] identified three regions in the tail of Venus: the lobes with strong magnetic fields oriented toward or away from the planet, the plasma sheet with weaker magnetic fields with more variable directions, and the boundary layer where the magnetic field becomes weaker and more variable, changing from the lobe orientation, and the ions assume a typical ionosheath distribution. They interpret the boundary layer as the result of leakage of solar wind protons into the outer portions of the tail lobe. *Slavin et al.* [1984] found from their statistical study that the main difference between the distant magnetotail structure of Venus and Earth lies in the higher mean plasmasheet β value for the Venus (8.5) than for the Earth (3.8).

Saunders and Russell [1986] carried out a detailed analysis of 4 tail crossing seasons and a statistical study of the magnetic tail properties. Determining the tail by the

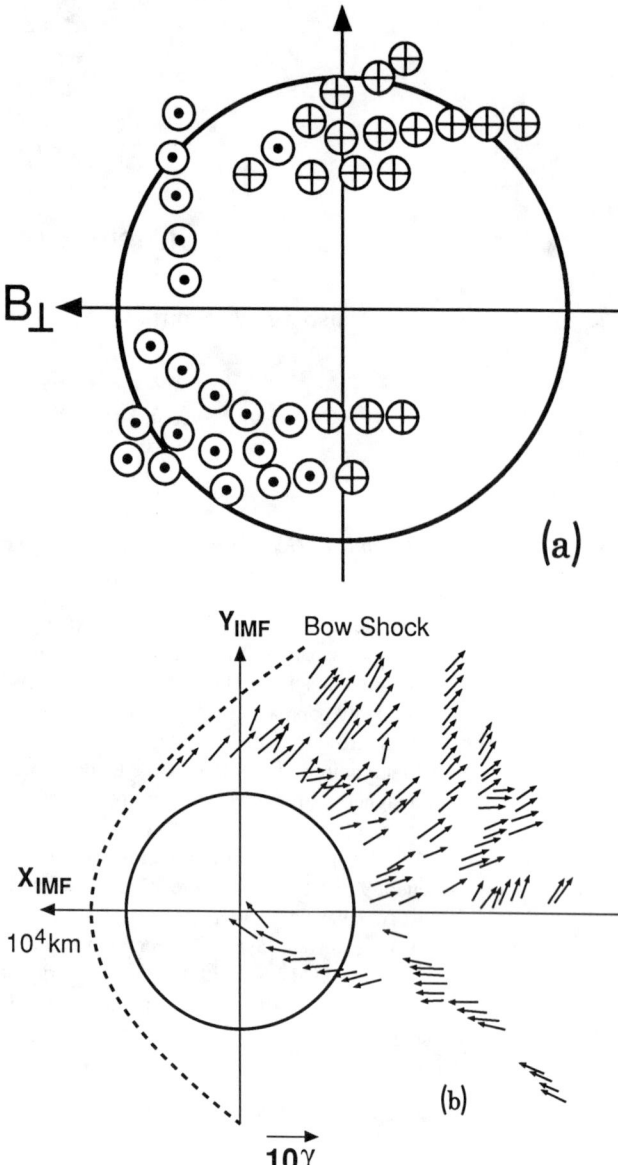

Fig. 1. (a) Polarity of the B_x - component in the Venus tail in the VxB coordinate system (solar wind velocity V is perpendicular to the picture plane, the IMF transverse component B is rotated into the Y-direction). This picture demonstrates that the two-lobe structure of the tail as controlled by the IMF [*Yeroshenko*, 1979]. (b) Another projection of the magnetic field in the ionosheath and in the tail of Venus (IMF transverse component B is rotated to be perpendicular to the picture plane) showing the draping pattern of the field [*Dolginov et al.*, 1981].

increased magnetic field, and the plasma sheet by magnetic field depression, they found that the tail boundary resembles a rotational discontinuity at the edges of the tail lobes and a tangential discontinuity at the edges of the plasma sheet.

Saunders and Russell [1986] also found that the cross-tail magnetic field is, on average, about 2 nT and is always approximately parallel to the IMF determined by the magnetic field direction in the solar wind or ionosheath observations during the same orbit. The radius of the tail at 12 R_v downstream is, on average, 2.3 R_v, and the tail is still flaring at these distances. The tail does not have a circular cross-section; the major axis to minor axis ratio is 1.2 with the major axis directed along the transverse IMF. The tail boundary normal component of the magnetic field B_n is about 1 nT, and its direction indicates that flux is added to the tail at the magnetopause.

Saunders and Russell [1986] discussed implications of their findings. The value of a normal component $B_n \sim 1$ nT for a tail radius of 2.3 R_v and an average transverse IMF component of 6 nT implies that the region of incident IMF which supplies the tail is 0.43 R_v (2600 km) thick. The associated cross-tail potential drop is 10 kV. The flux content of the distant tail is about 3 MWb, so that (1) it cannot close above the dayside ionosphere, or, as *Saunders and Russell* [1986] conclude it would fill the entire ionosheath directly in front of the planet out to the bow shock with 100 nT fields, and (2) it exceeds, by at least an order of magnitude, estimates for magnetic flux either in the dayside ionosphere or in the nightside ionosphere radial fields. Part of this flux (~ 1 MWb) may close in the plasma mantle at the dawn-dusk meridian (a 29 nT field in a layer of 1500 km thickness would accomplish this) but most of the flux must close across the tail center plane. With a cross-tail magnetic field of 2 nT, existing over an average tail width of 4 R_v between the planet and 11 R_v distance, all of the tail flux at 11 R_v could close behind the planet. The tail flux of 3 MWb at 11 R_v would exit the tail with the observed B_n magnitude of 1 nT within a distance of 13 R_v further downstream. Since the tail ionosheath plasma flows at about 400 km/sec, the tail flux can be resupplied in only a few minutes.

Using the cross-tail magnetic field direction as the direction of the IMF, *McComas et al.* [1986] studied the statistical properties of the magnetotail at large distances. They explain the observed variability of the magnetic field by the motion and reconfiguration of the tail on a time scale faster than the Pioneer Venus orbital motion. *McComas et al.* [1986] assumed that the tail has one current sheet, and that the relative thickness of the tail lobe (determined by an inclination of the B-vector to the average solar wind flow direction of 73-78°) and the current layer are determined from fractions of the time each configuration is observed. From this they concluded that an average diameter of the tail is 5.1 R_v (where the spacecraft observed the tail configuration > 50% of time) and that the thickness of the current layer is 28% of it, i.e., 1.4 R_v. They also determined

an average value of the cross-tail magnetic field of 4 nT. Using their derived thickness of the current sheet, *McComas et al.*, [1986] calculated a current density ~ 1.5 n A/m^2, and a JxB force, applied to the plasma in the current sheet, ~ 8×10^{-18} N/m^2.

PLASMA WITHIN THE TAIL

Several plasma detectors have flown around Venus and collected data within the tail. The Venera 9 and 10 satellites had two plasma spectrometers: a 2-D ion spectrometer consisting of 6 narrow-angle cylindrical electrostatic analyzers with channel electron multipliers (CEMs) RIEP [*Ainbund et al.*, 1972] and a combination sunward-looking differential ion Faraday cup and antisunward oriented integral electron Faraday cup D-127 [*Gringauz et al.*, 1976a]. The energy ranges were: 50 eV/Q to 20 keV (RIEP) and 0-4 keV/Q for ions and 0-400 eV for electrons (D-127). The temporal resolution of both spectrometers was 160 sec. Measurements of plasma and magnetic field on Venera 9 and 10 were performed on selected orbits within the October 1975-March 1976 time interval. The orbit of the Venera spacecraft allowed measurements within the tail from approximately 0.5 R_v behind the terminator to about 5 R_v downstream.

The 3-D plasma analyzer flown on PVO [*Intriligator et al.*, 1980] was of the quadrispheric electrostatic type with five collectors. The energy per unit charge (E/Q) range from 50 to 800 eV is scanned in 9 minutes. PVO operated in orbit around Venus from December 1978 until October 1992. Plasma and magnetic field measurements were made on essentially every 24 hour orbit. The orbit of PVO crossed the tail in two regions: near periapsis, just behind the planet, and near apoapsis at 8-12 R_v downstream.

Plasma Regions in the Close Downstream Tail

Figure 2 shows the directions of ion flow and ion temperature and velocity as determined with RIEP data on a Venera 10 pass through the interaction region on October 29, 1975 [*Vaisberg et al.*, 1976]. These data are superimposed on a hydrodynamic model of *Spreiter* [1976], showing reasonable agreement with it in terms of shock geometry and ionosheath flow directions. Below the "ionopause" of *Spreiter et al.* [1968a] a lower energy and lower temperature ion flux is detected, converging to the tail axis. These two ion components are seen together in the layer adjacent to the boundary. The velocity of the low energy component was calculated under the assumption that these ions are protons.

Figure 3a shows the plasma boundaries behind the terminator as deduced from observations of ion properties on

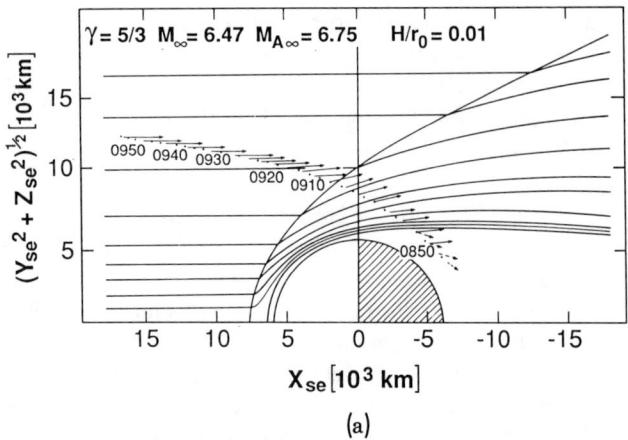

(a)

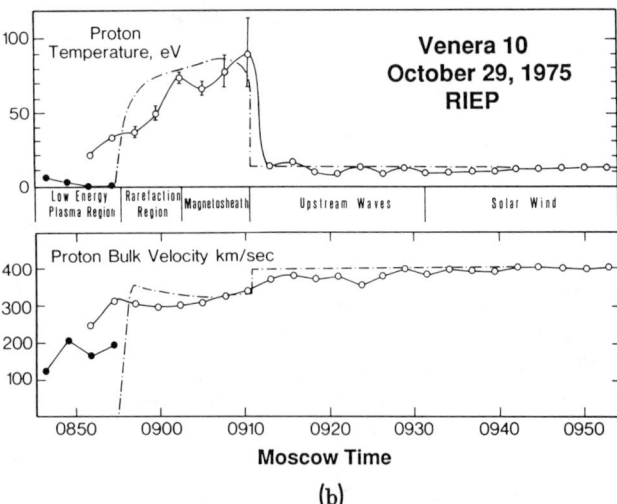

(b)

Fig. 2. The direction of ion flow (a) and ion temperature and velocity (b) in the solar wind, in the ionosheath, and in the tail as calculated from RIEP spectrometer data on a Venera 10 pass through the interaction region on October 29, 1975 [*Vaisberg et al.*, 1976]. Hydrodynamic model flow lines and flow velocity and temperature from *Spreiter* [1976] are shown for comparison. Two ion components are observed in the layer adjacent to the boundary.

Venera 9 and 10 with the RIEP plasma analyzer. In addition to the shock, three more boundaries were observed [*Vaisberg et al.*, 1976]: (1) The external boundary of the boundary layer, or rarefaction region where strong modifications of the external flow were observed, apparently associated with the mass-loading of this part of the external flow. Mariner 5 crossed this region during its fly-by in 1967 [*Bridge et al.*, 1967]. (2) A boundary separating the ionosheath from a low energy plasma domain that was initially called the ionopause. This term was adopted by *Vaisberg et al.* [1976] from the HD model of *Spreiter* [1976]. The boundary was assumed to be a continuation of the dayside ionopause that was assumed to be the solar wind-obstacle interface before

discovery of the magnetic barrier by *Elphic et al.* [1980]. This is evidently the tail boundary where the solar wind plasma flow terminates. (3) The innermost boundary is the one below which no measurable fluxes of ions with energies > 50 eV (the lower energy threshold of the RIEP ion spectrometer) have been observed. The region behind the planet that is within this boundary was called a cavity. *Gringauz et al.* [1976] reported observations of sporadic ion fluxes in a wide energy range in this region that was called the particulate shadow by the authors (see fifth spectrum from the left below in Figure 7a).

Two sets of ion spectra are shown in Figure 4 [*Vaisberg and Zelenyi*, 1984]. Panel A clearly demonstrates the difference between planetary domains including the low energy flow in the tail, and the boundary layer with the secondary spectral peak at low energies, whose energy, shape, and intensity indicate that it is the continuation of the

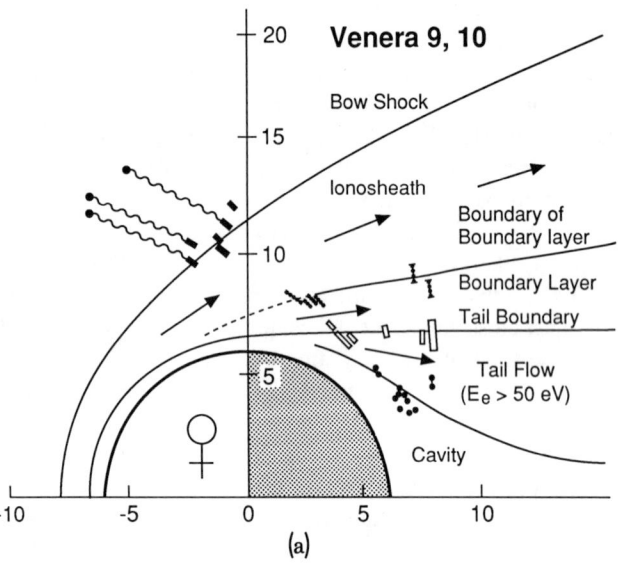

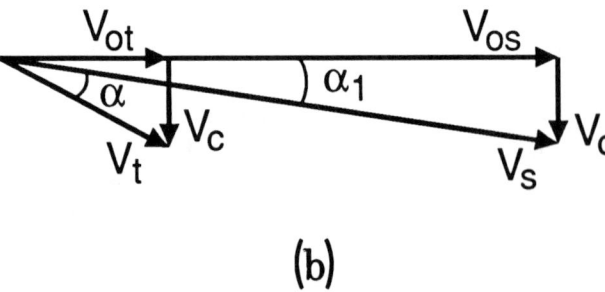

Fig. 3. (a) The plasma boundaries behind the terminator observed on Venera 9 and 10 [*Vaisberg et al.*, 1976]. (b) An estimate of the mass of ions in the tail under the assumption that the transverse velocity components of heavy ions and of the solar wind ions near the edge of the tail are equal [*Vaisberg and Zelenyi*, 1984].

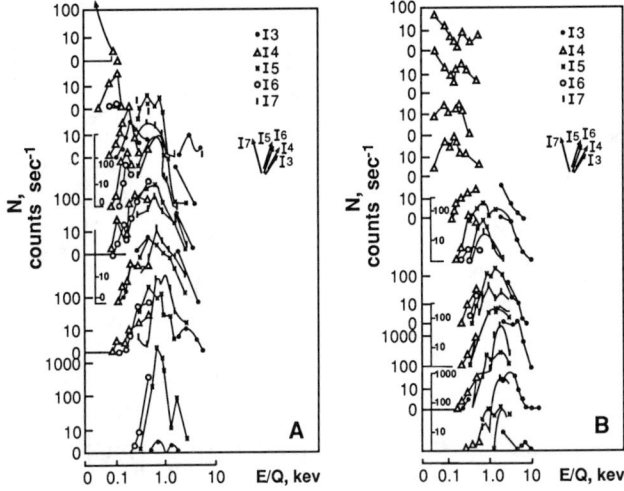

Fig. 4. Two sets of ion spectra taken on parts of orbits passing through outer part of the tail and ionosheath at low magnetic latitudes (A) and at high magnetic latitudes (B) in the VxB coordinate system [*Vaisberg and Zelenyi, 1984*]. Two upper spectra in (A) and four upper spectra in (B) show the difference in tail ion populations at different magnetic latitudes.

low energy flow in the tail. Similar plasma domains are seen on another Venera 10 pass in panel B except that the thickness of the layer of low-energy plasma in the tail is thicker, and the energy spectrum is wider. Again the traces of the low energy component are seen in the surrounding shocked solar wind flow.

Figure 5 [*Vaisberg, 1980*] shows the parts of the Venera 10 orbits projected on the YZ plane where tail plasma flow was observed. (Throughout this paper X will be taken to be opposite the direction of the solar wind flow vector, i.e., roughly sunward.) The orbits were rotated in order to coalign the Y-axis with the transverse IMF component, as determined in the solar wind on part of the trajectory closest to the observation period. It is seen that all four orbits passing through high magnetic latitude, i.e. far from the VxB plane, crossed the thick layer of the tail plasma flow, while the two low magnetic latitude orbits, close to the VxB plane, crossed a thin layer of low energy plasma in the tail. Though no measurements in the negative Z-hemisphere have been made with the RIEP plasma spectrometer in the near tail, Figure 5 strongly suggests that the inflow of low-energy plasma in the tail is controlled by the IMF orientation.

Plasma observed within the cavity formed by Venus within the solar wind flow has a significant converging velocity component, as if this plasma flow is filling the tail. Near the boundary of the plasma tail determined by the change of plasma characteristics, two plasmas are observed simultaneously within the same layer. The difference of directions of the two plasma components, one of shocked solar wind and the other of tail flow (see Figure 3b), allowed *Vaisberg* [1984] and *Vaisberg and Zelenyi* [1984], to make an estimate of the mass of the ions in the tail under the assumption that the converging velocities of these two plasma components are equal (see Figure 2b). Then the ratio of the two masses is given by

$$\frac{M_t}{M_s} = \frac{W_t}{W_s} \times \frac{\sin^2 \alpha_t}{\sin^2 \alpha_s} \quad (1)$$

where W is the energy per charge and α is the deflection angle of the flow due to the convergence. Subscripts t and s are for tail and sheath components, respectively. The values for the November 29, 1975 Venera 10 pass are W_s = 470 eV, W_t = 170 eV, α_t = 25°, and α_s is not known, but we assume that it is between 2° and 6°. The resulting M_t/M_s is between 61 and 9. As the bulk of the ionosheath ions are protons we may infer that the tail flow most probably consists of oxygen ions. The direction of the tail flow observed at relatively small distance behind the planet suggests that the source region of this plasma is located near the terminator.

Identification of the region where the tail flow was observed on the Venera 9 and 10 spacecraft as a distinct

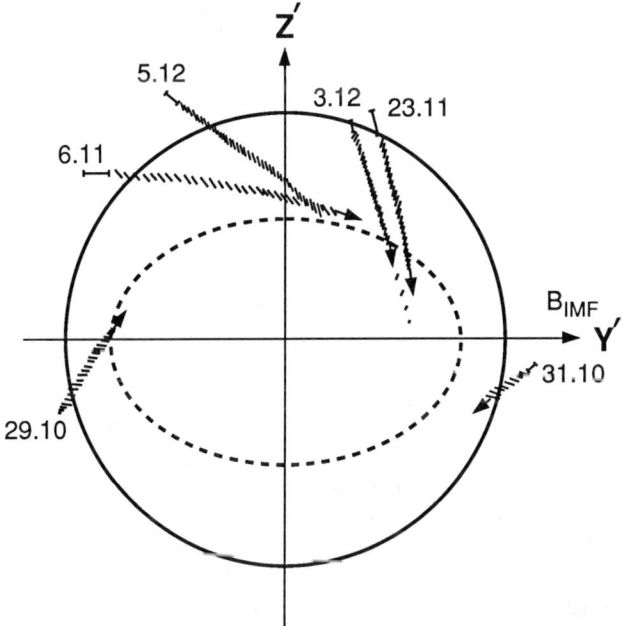

Fig. 5. The parts of the Venera 10 orbits in the Y'Z' plane of the VxB coordinate system where tail plasma flow was observed [*Vaisberg, 1980*]. This panel shows that the inflow of low-energy plasma in the tail is controlled by the IMF.

plasma region (the Venus mantle) was also made with the Pioneer Venus retarding potential analyzer data [*Spenner et al.*, 1980]. *Spenner et al.* [1980] observed that within this region the electron energy spectrum is intermediate between the energy spectrum of the ionosheath and that of the ionosphere, and that it exhibits significantly lower intensities than in the adjacent regions.

Figure 6a shows the structure of the tail boundary at close distances [*Romanov et al.*, 1978]. This tail boundary

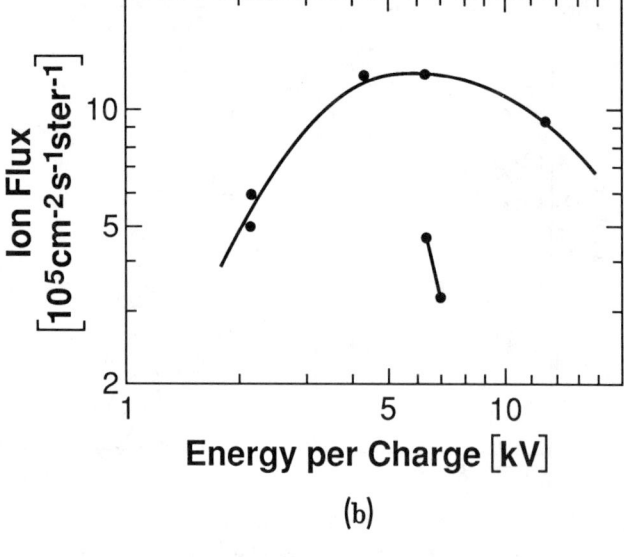

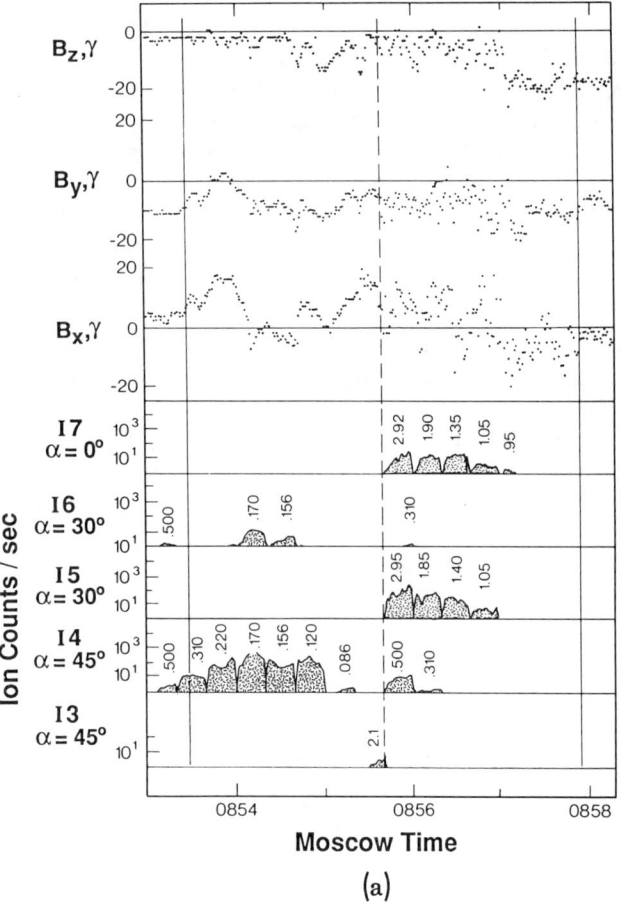

Fig. 6. Structure of the tail boundary (a) in the magnetic field (three upper panels) and according to the ion analyzers of RIEP (five lower panels). Electrostatic analyzers I4 and I6 cover the energy (E/Q) range 0.048-0.500 keV, analyzers I5 and I7 cover the energy range 0.32-2.9 keV, analyzer I3 covers the energy range 2.1-19.8 keV. Their orientation relative to the solar direction is shown on the left of each panel. The tail is on the left, the ionosheath is on the right side. The sharp tail boundary is seen in the change of ion energy, in the frequency of magnetic field fluctuations, and in the appearance of accelerated ions in the I3 data [*Romanov et al.*, 1978]. No solar wind ions penetrate the tail. (b) The composite spectrum of accelerated particles observed on different tail boundary crossings.

crossing occurred at high magnetic latitude in the VxB coordinate system. It is observed that the transition from the tail flow to ionosheath flow occurs within a time interval of 20 sec (the time when the RIEP plasma spectrometer was performing measurements at a fixed energy step). This fast transition is confirmed by a change in appearance of the magnetic field fluctuations, with higher frequency fluctuations observed in the external flow. The sudden transition from one plasma region to another at the tail boundary was also observed during other tail crossings close to Venus. It is not known what the velocity of the tail boundary is, but if it was stationary during the crossing the thickness of the boundary is estimated at about 100 km. Another feature of the boundary crossing is the observation of 2.1 keV ions by one of the RIEP analyzers exactly at the boundary. These accelerated particles were observed at every crossing of the tail boundary at different energies within a time scale much shorter than the energy scan time of the instrument. Figure 6b shows a composite spectrum of these particles that suggests that a wide energy spectrum of these high-energy ions is associated with the tail boundary. These ions may be locally accelerated due to the very high level of magnetic field fluctuations ($\Delta B/B \sim 1$) at the tail boundary. No analysis of this possible acceleration process has yet been made.

In summary, analyses of the plasma measurements and comparisons with magnetic field measurements give evidence that no measurable solar wind plasma is penetrating into the tail at close downstream distances, and that a two-component flow is observed in the external flow, outside the tail.

Plasma at Distances 4 - 5R_v

There were only two passes of Venera 10 through the intermediate (2.5 to 5 R_v downstream) tail of Venus where plasma and magnetic field measurements were made. Figure 7a is a composite of a distant Venera 10 pass on April 18/19, 1976 and a Venera 9 pass on November 1, 1975 at closer distances [*Gringauz et al.*, 1976; *Verigin et al.*, 1978]. This figure illustrates how the ion spectra change in the downstream region. The wide angle Faraday cup analyzer measured ion fluxes in the tail in a broad energy range, and ion fluxes were recorded regularly. With these measurements, *Verigin et al.*, [1978] separated the corpuscular umbra (the region of separate ion spikes in a wide energy range that have been observed in 30% of the telemetry samples), and the penumbra, which is characterized by lower (compared to ionosheath) ion flux densities and velocities. Because the major maximum of the ion spectrum systematically decreased as the satellite moved deeper into the ionosheath and penumbra regions, *Verigin et al.* [1978] concluded that the major maximum is determined by solar wind protons. Figure 7b shows the electron isodensity contours (normalized to solar wind density) in the downstream region near Venus [*Verigin et al.*, 1978] for electrons with energies above 10 eV. *Verigin et al.* [1978] suggested that the convergence of the electron isodensity surfaces indicated the convergence of the flow from the penumbra toward the tail axis. This conclusion was reached without consideration of the acceleration of ions along the tail (see below), which may eliminate the reason for it.

Ion velocities (under the assumption that the plasma everywhere consists of protons) and temperatures were calculated from RIEP data for the same April 18/19, 1976 orbit of Venera 10 (Figure 8a). This is the only reported tail boundary crossing of Venus at these distances. The boundary layer is identified by increasing temperature and decreasing velocity, both apparently due to mass loading of the external (solar wind) flow. The transition to the tail is easily identified by the sudden change of ion temperature, and this tail boundary crossing suggests the existence of a sharp tail boundary at this distance. No traces of solar wind penetration into the tail are observed. The calculated ion velocity (the energy of convective motion, to be more precise) does not show a jump at the tail boundary. This effect was observed near Mars at comparable downstream distances [*Vaisberg*, 1976; *Lundin et al.*, 1991], and was explained by *Lundin et al.* [1991] in terms of momentum balance in the mass-loading region.

Figure 8b shows the observation by *Verigin et al.* [1978] of the high energy ions at the magnetic field reversal within the tail. They calculated an ion temperature of about 1000 keV near the tail reversal and interpreted this as an isotropic plasma layer of the Venus tail. The same feature is apparently seen in Figure 8a as a spike in the ion temperature accompanied by a slight velocity decrease. It should be noted that, since neither instrument was measuring the 3-D ion distributions, definite conclusions regarding this complicated environment cannot be drawn. We suggest that the shape of the high-energy peak in the third panel of Figure 8b indicates the presence of a convected rather than an isotropic component. *Verigin et al.* [1978] reported two more observations of the high-energy ions near the B_x sign reversal at lower distances downstream.

On another pass of Venera 10 on March 26, 1976 measurements were carried out only within the tail, so that no boundary crossing was observed. The satellite was in a

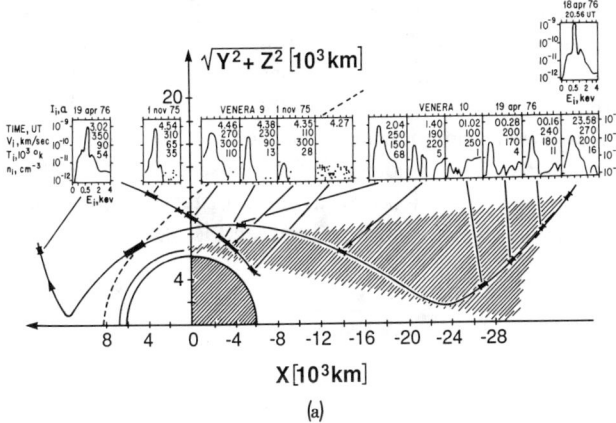

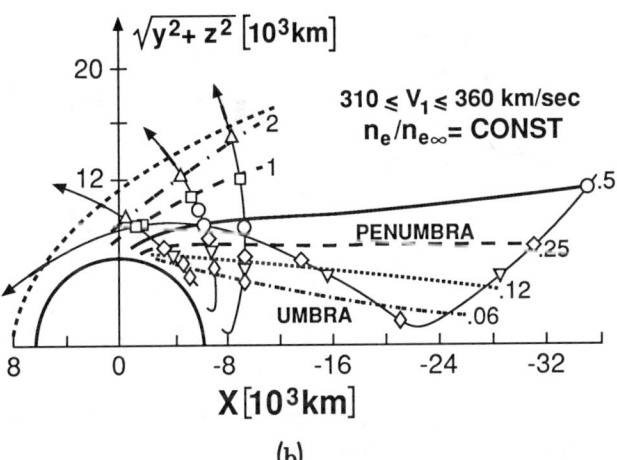

Fig. 7. (a) The change in ion spectra in the downstream region from the wide angle Faraday cup analyzer on the Venera 10 pass on April 18/19, 1976 and the Venera 9 pass on November 1, 1975 [*Gringauz et al.*, 1976; *Verigin et al.*, 1978]. The particulate shadow behind the planet and penumbra (shaded) are indicated. (b) The electron isodensities (normalized to solar wind density) in the downstream region near Venus [*Verigin et al.*, 1978].

field component. This pass also suggests that the Venusian tail is nearly permanently populated with plasma although there are short time intervals when no ions are observed. *Spenner et al.*, [1980] also reported a case of electron flux dropout at PVO). The mean energy of the ions is larger than at closer distances to Venus. Ions with significantly higher energy and lower temperature are observed in one-to-one correspondence with currents related to variations in the B_x component. In this case no hot ions were observed at the B_x reversal that appears to be the current sheet separating the two tail magnetic field lobes, although two small bursts of 2 keV/Q ions occurred at the edges of the current sheet. After the central current sheet crossing, the mean ion energy started to increase, probably due to the approach to the tail boundary. The average parameters of the ion flux are given in Table 1.

Figure 10 shows the variation of the calculated ion transport velocity within the tail with the distance as observed by the RIEP plasma spectrometer on Venera 10 [*Vaisberg et al.*, 1989]. Again the velocity was calculated under the assumption that the ions were protons, and normalized to the velocity of the ionosheath flow. Due to convergence of the ion flow at closer distances it is difficult to establish the relationships between the plasma regions at

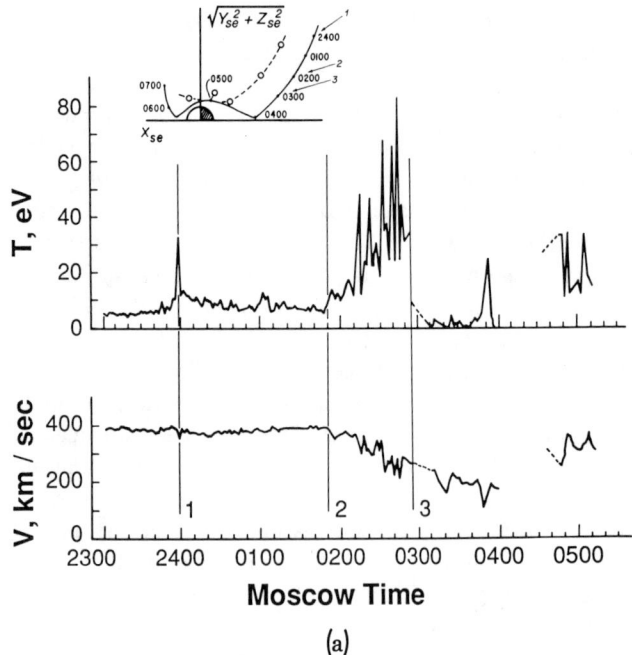

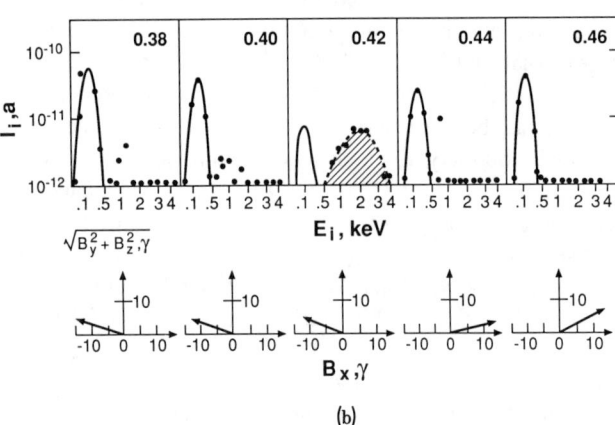

Fig. 8. (a) Ion velocity and temperature profiles along the Venera 10 orbit (shown above) on April 18/19, 1976 according RIEP spectrometer data (the ion velocity is calculated assuming that the plasma everywhere consists of protons). Crossings of the bow shock (1), external boundary of the boundary layer (2), and the tail boundary (3) are indicated [*Romanov et al.*, 1978]. (b) Upper panel - ion spectra measured by wide angle Faraday cup analyzer on the same Venera 10 orbit [*Verigin et al.*, 1978]. Lower panel - vector of magnetic field in components B_x, B_{yz}. Observation of the high energy ions at the magnetic field reversal within the tail is indicated by shading.

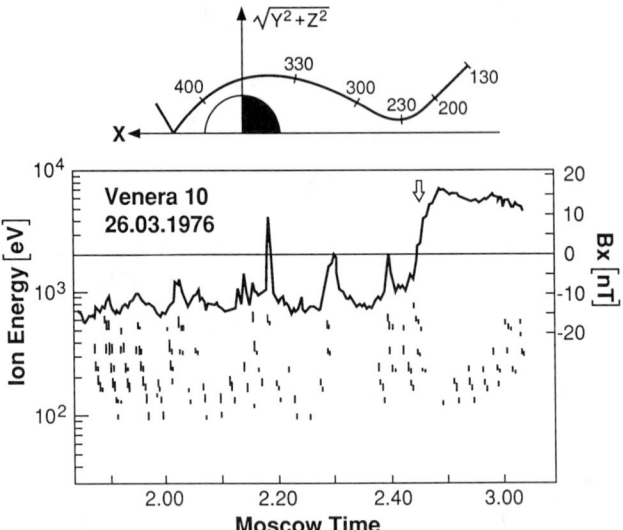

Fig. 9. B_x magnetic field component and dynamic spectrum of ions as measured by the RIEP spectrometer on a Venera 10 pass within the tail on March 26, 1976. The Venera 10 orbit is shown above in the cylindrical coordinate system. The length of the bar shows the logarithm of the ion counting rate at specific energy E/Q (scale is on the left). Note that whenever a B_x disturbance is observed, the energy of ions is increasing. The B_x reversal that seems to be the main current sheet is indicated by an arrow. [*Vaisberg et al.*, 1989]

spin-stabilized mode around its axis which was approximately directed toward the Sun, and the clock angle was unknown. Figure 9 shows the dynamic spectrum of ions as measured by the RIEP, along with the B_x magnetic

Table 1. Plasma Parameters within the Venusian Tail

Region	Downstream Distance(R_v)	Velocity (km/sec)	Temperature (eV)	Number Density (cm^{-3})	Number Flux (cm^{-2} sec^{-1})	Sources
Lobe	1	180 (p)	3-10	0.12 (p)	2×10^6	V & Z
		or 45 (O$^+$)		0.5 (O$^+$)	2×10^6	V & Z
		100 (p)		0.8 (p)	9×10^6	Vg
Lobe	4-5	200 (p)	4-35	0.3 (p)	5×10^6	Vs
		or 50 (O$^+$)		1.2 (O$^+$)		Vs
		100-200 (p)	100-200	1-4 (p)	4×10^7	Vg
Current sheet	4-5	300 (p)	10-75	0.2 (p)	6×10^6	Vs
		or 75 (O$^+$)		0.8 (O$^+$)		Vs
			1000			
Lobe	8-12	~ 400		0.07 (p)	3×10^7 (p)	McC
				0.005 (O$^+$)		McC
Current sheet	8-12	~ 400	500 (p)	0.9 (p)	3.5×10^7 (p)	McC
		or 8000 (O$^+$)		0.06 (O$^+$)	2.5×10^6 (O$^+$)	McC
		n × 10			$10^6 - 10^7$	M&B
					1.6×10^7	I

Note: No measurements of ion composition were made, so numbers indicated depend on assumed major ion.

Sources:
V&Z - Vaisberg and Zelenyi [1984]
Vg - Verigin et al. [1978]
Vs - Calculated by author from ion spectra shown in Figure 4 of Romanov et al. [1978]
McC - Derived from self-consistent analysis of magnetic structure of the tail by McComas et al. [1986]
M&B - Maximum flux provided by Mihalov and Barnes [1982]
I - Maximum flux provided Intriligator [1982]

different planetocentric distances. At the present stage we may argue that an acceleration of ions occurs between approximately $X = -1 R_v$ and $X = -5 R_v$ by using average observed velocities at the two distances. This gives an average acceleration between $X = -1 R_v$ and $X = -5 R_v$ of ~ 4×10^4 cm/sec^2 if the plasma consists of protons, and ~ 2.5×10^3 cm/sec^2 if the tail flow consists primarily of singly ionized oxygen ions [*Vaisberg et al.* 1989].

Plasma at Large Downstream Distances (8 - 12 R_v)

According to Pioneer Venus measurements [*Intriligator*, 1982; *Mihalov and Barnes*, 1982] ion fluxes in the tail are not regularly recorded at these distances. The tail is thus considered devoid of plasma. However, there are significant numbers of what are called O$^+$ events, namely high-energy ions that are usually, but not always, located at about a factor of 16 in E/Q value above any low-energy component.

Figure 11a shows some examples of these double-peaked ion spectra [*Mihalov and Barnes*, 1982]. The relative positions of the two spectral peaks allowed *Intriligator* [1982] and *Mihalov and Barnes* [1982] to tentatively identify the higher energy peak as O$^+$ ions of planetary origin. The O$^+$ events normally occur within the magnetotail. Their calculated flow speeds for the O$^+$ fairly uniformly cover the range from about 150 km/sec to the end of the analyzer scale (about 300 km/sec). The flux of these ions lies in the range of $10^6 - 10^7$ ions/cm^2 sec for 1/3 of their cases, with 2/3 of the observations below this level. The estimated temperature is typically several tens of eV. *Mihalov and Barnes* [1982] suggested that the pick-up of exospheric photoions was the source of this plasma component in the tail.

Intriligator [1989] performed a statistical study of these events in the distant tail and was able to show the asymmetric location of what she calls pick-up ionospheric ions in the magnetic coordinate system. The vast majority

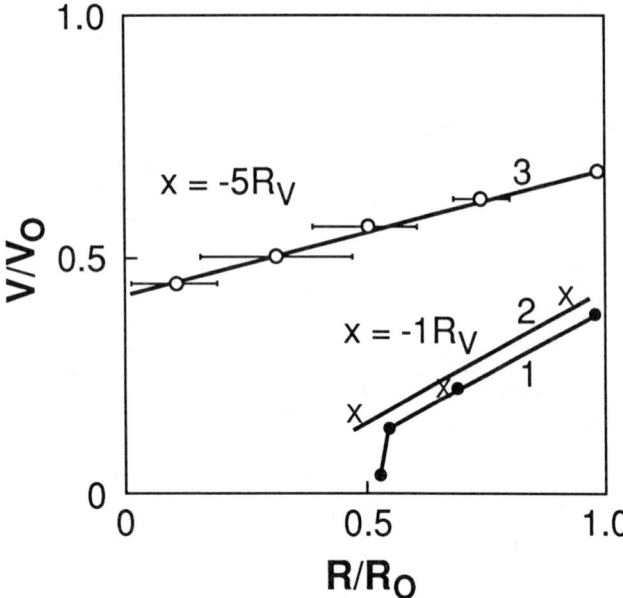

Fig. 10. The variation of calculated ion transport velocity in the tail (protons are assumed) with the distance from the tail axis for two downstream distances as observed by the RIEP plasma spectrometer on Venera 10. Values are normalized to the velocity of the ionosheath flow. [*Vaisberg et al.*, 1989]

of the O^+ detections are located in the hemisphere where the solar wind induced electric field points outward from the planet. Asymmetric pick up of planetary ions was initially proposed by *Cloutier et al.* [1974]. The signatures of this asymmetry were noticed in several ways including the magnetic field strengths in the ionosheath near the terminator and in the downstream tail, and in the asymmetry of the ionopause altitude near the terminator [*Phillips et al.* 1987, 1988]. Subsequently *Slavin et al.* [1989] carried out a detailed analysis of PVO plasma observations within the distant magnetotail. They found that increased fluxes of two plasma components identified as H^+ an O^+ are observed in the plasma sheet and adjacent tail lobes and confirmed the earlier finding that these fluxes are mostly concentrated downstream of the Venus hemisphere over which the solar wind motional electric field is directed away from the planet.

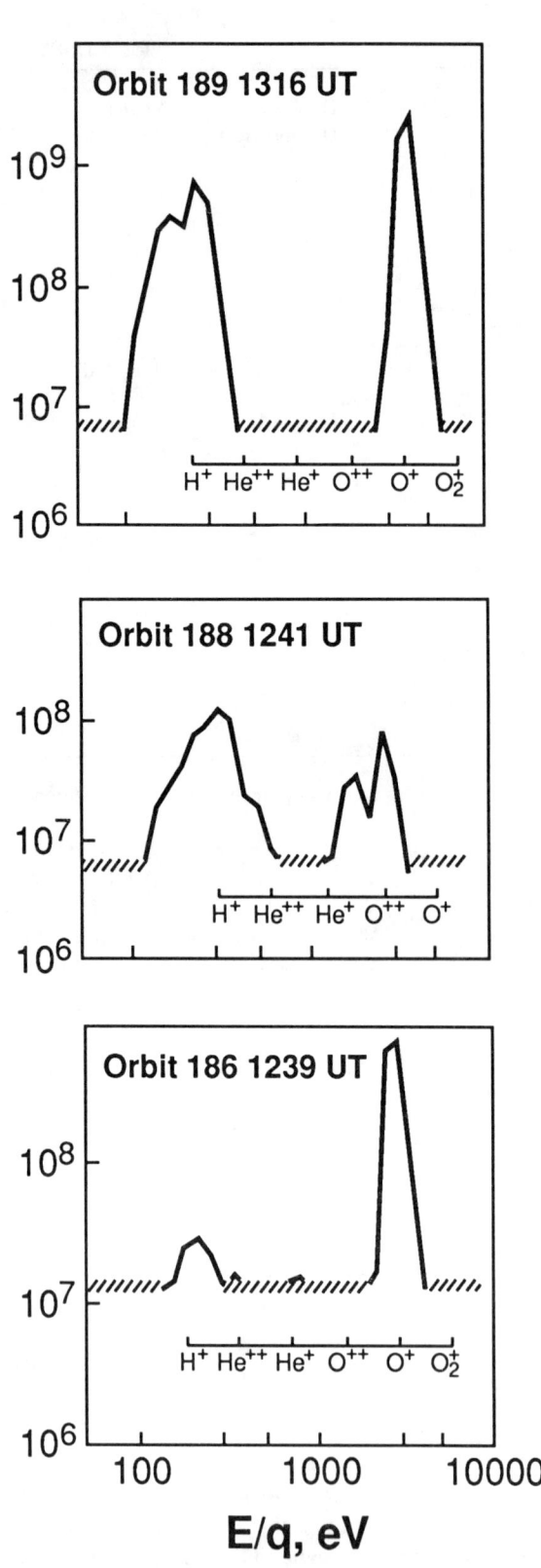

Fig. 11(a). Examples of double peaked ion spectra observed with the Pioneer Venus plasma analyzer at large downstream distances (8-12 R_V) [*Mihalov and Barnes*,1982]. These are called O^+ events. (b) Statistical study of these events in the distant tail by *Intriligator* [1989] and *Slavin et al.* [1989] showed that the vast majority of O^+ events are observed in the plasma sheet and adjacent tail lobes in the hemisphere where solar wind induced electric field points outward from the planet.

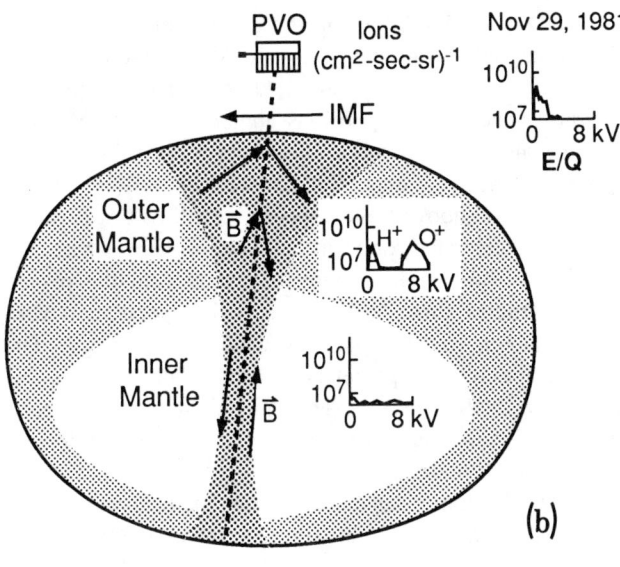

Fig. 11(b.)

The greatest concentration of those O^+ ions was found in the vicinity of the cross-tail current layer. Their interpretation of the plasma observations is given in Figure 11b. *Slavin et al.* [1989] also found that at the outer boundary of the Venus magnetotail where significant fluxes of E/Q = 0-8 keV ions are observed, the magnetotail-ionosheath interface is very broad and resembles a slow mode expansion fan with field strength slowly decreasing and plasma density gradually increasing with increasing distance from the tail axis. They suggest that downstream of the Venus hemisphere over which the solar wind motional electric field is directed toward the planet, no significant plasma fluxes are observed, except sometimes directly adjacent to the outer boundary of the tail. The outer boundary of the magnetotail in this hemisphere is typically a well-defined magnetopause-like current layer. *Moore et al.* [1990] most recently carried out a comprehensive statistical study of the locations of the O^+ ions relative to the magnetic field structure in this region of the distance tail. They show how the O^+ detections are located with respect to the average tail lobes and find a solar cycle dependence in the observed O^+ flux.

Plasma within the Current Sheets

There are a limited number of plasma observations within the cross-tail current sheets. *Verigin et al.* [1978] reported the observation of a hot ($T_i \sim 1000$ eV) ion population in one case at intermediate downstream distances of 4-5 R_V (see Figure 7b), and two more cases at unspecified distances.

The narrow-angle analyzer measurements suggest that the ion distribution function in the tail current sheet may be quite complicated.

At large distances (10-12 R_V) only the high-energy ion component (O^+) has been analyzed, and *Slavin et al.* [1989] have found that the greatest concentration of these O^+ ions is observed in the vicinity of the cross-tail current sheet. There have also been indirect assessments of the plasma properties in the current sheet. *Saunders and Russell* [1986] inferred that to maintain the pressure balance with the observed tail lobe field, the plasma sheet must have an energy density ~ 1 keV/cm^3. Using the cross-tail field variation with the X-coordinate, and assuming the continuity of the electric field across the ionosheath and the tail, *McComas et al.* [1986] calculated the inferred plasma velocity at different distances. They obtained estimated velocities of 250 km/sec at 8 R_V, 320 km/sec at 10 R_V and 420 km/sec at 12 R_V. With these values we can calculate the inferred plasma acceleration, which gives 4×10^4 cm/sec^2 for distances between 8 and 10 R_V and 1.6×10^5 cm/sec^2 for distances 10-12 R_V. From these calculated force and velocity variations *McComas et al.* [1986] estimated plasma number densities of 0.07 cm^{-3} in the tail lobes and 0.9 cm^{-3} in the current sheet, if the plasma consists of protons, and 0.005 cm^{-3} and 0.06 cm^{-3}, respectively, if singly ionized oxygen is the main constituent. From pressure balance between the tail lobes and the plasma sheet they determined an ion temperature of 6×10^6 K for protons and 9×10^7 for O^+.

SOURCES OF THE TAIL PLASMA

Plasma Mantle

The plasma mantle [*Vaisberg et al.*, 1976; *Spenner et al.*, 1980] is one of the important sources of the tail plasma. The ion flux entering the tail from the terminator region is of order 5×10^6 cm^{-2} sec^{-1} [*Vaisberg and Zelenyi*, 1984]. *Vaisberg and Zelenyi* [1984] proposed a model of the plasma mantle based on the expected acceleration of pick-up photoions within the magnetic barrier. The magnetic barrier is formed above the dayside ionosphere due to MHD deceleration of the flow at the obstacle and due to depletion of the piled-up field lines through the escape of the solar wind plasma along the field lines (e.g., the *Zwan and Wolfe* [1974] effect). Depleted field lines convect around the planet and are filled with the ions that are formed by photoionization of the upper atmosphere neutrals. The loaded field lines are accelerated from the subsolar region to the terminator by the magnetic pressure gradient within the magnetic barrier. The model gives reasonable values of the temperature of ions and the total ion influx to the tail (~5×10^{24} sec^{-1}), but higher transport energies than observed.

Ionospheric Clouds

Detached ionospheric clouds are frequently observed above the ionosphere [Brace et al., 1982a]. *Russell et al.* [1982] discussed a clear example of one such cloud and the forces applied to it by the surrounding draped magnetic field. They calculated the Maxwell stress exerted by the magnetic field at 4×10^{-8} dynes/cm^2. For an O$^+$ slab of 300 km thickness the acceleration would be 0.7 km/sec^2. If accelerated from the subsolar point, the plasma velocity will reach 90 km/sec. If the cloud is of 500 km height, it will supply to the tail ~ 2×10^{25} ions/sec.

Plasma in the nightside ionosphere

The nightside ionosphere can also provide a major source of tail plasma. It is very structured, and consists of dense rays separated by holes [Brace et al., 1980, 1982b; Taylor et al., 1980; Luhmann et al., 1982]. The boundaries of the depletion regions are sometimes marked by a superthermal ion signature [Taylor et al., 1980]. Pressure balance can occur throughout the regions of holes. Plasma β is > 1 outside the holes, and it is < 1 inside the holes.

Brace et al. [1987] found that the tail rays above 1000 km have dimensions of the order $1-3\times10^3$ km, decreasing in width at higher altitudes. The largest currents flow on the flanks of the rays. The ions in both the rays and troughs are almost exclusively superthermal oxygen with energies 9-16 eV and a flat angular distribution. Fast ions (> 40 eV) are a minor constituent and typically have a tailward component of velocity. Significant densities of fast ions are commonly associated with tail rays. The central ray is a common feature, with one or more rays often arranged symmetrically on either side. The average electron density N_e over the entire umbra at altitudes ~ 2000 km is 39 cm^{-3}. Adopting an average energy of 13 eV for the superthermal O$^+$, corresponding to a velocity of 12.5 km/s (about 4 km/sec greater than escape velocity), the resulting global mean flux is calculated as ~ 1×10^7 cm^{-2} sec^{-1}. The outer layers of tail rays appear to be accelerated and heated by some process.

Kasprzak et al. [1987] studied the energetic ions (E>40 eV), most frequently O$^+$ ions, that occur only on the nightside at solar zenith angles > 120°. They are observed at all altitudes sampled at PVO periapsis. The maximum energetic ion flux observed by the Orbiter neutral mass spectrometer between 1900 and 2500 km is < 4×10^6 ions/cm^2 sec. The average O$^+$ flux for ions > 40 eV is about 4×10^4 ions/cm^2 sec. Tailward flow is typical, but sunward flows are also observed.

Gasdynamic Model with Mass Loading

Moore et al. [1991] developed a gasdynamic model of the Venus magnetotail assuming a tapered ionopause obstacle and mass addition by exospheric O$^+$. Their model predicts central magnetotail oxygen ions number densities of about 0.2 cm^{-3} and temperatures of the order of 100 eV, flowing tailward at speeds of about 200 km/sec at ~ 10 R$_V$. The ions picked up well above the ionopause behave as test particles and are accelerated only in one hemisphere where the motional electric field is directed away from the planet. These particles form an asymmetric ion population with energies about 16 times the solar wind proton energy. The total escape rate of planetary O$^+$ ions calculated in this model is 2×10^{24} sec^{-1}. This model reproduces many observed features but does not develop the tail as a domain with a distinct boundary.

Figure 12 summarizes our understanding of the sources of plasma in the near tail. This figure has been modified from that presented by *Brace et al.* [1987] by the addition of the plasma mantle/magnetic barrier region, that along with tail rays, appears to be the most important source of the tail plasma.

DISCUSSION AND CONCLUSIONS

The structure and properties of the Venus magnetic tail as an example of a mass-loaded tail of an unmagnetized body are reasonably well known. One of the important properties

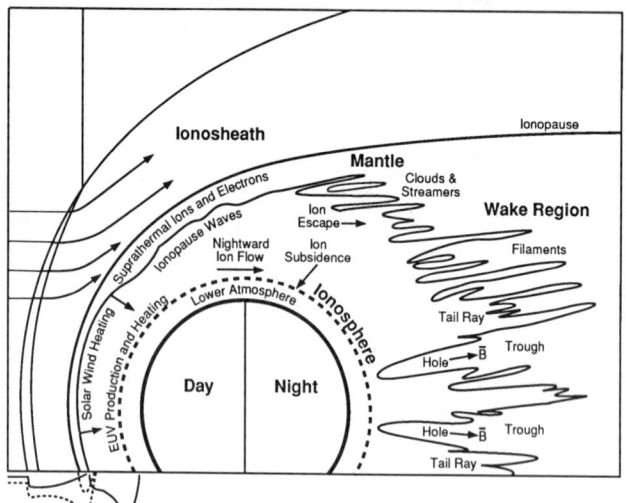

Fig. 12. Cartoon of the near-Venus environment adapted from *Brace et al.* [1987] and modified to emphasize the role of magnetic barrier and sources of the tail plasma. Schematic profiles of plasma velocity, magnetic field strength, ionospheric number density, and pick-up ion number density in the magnetic barrier on the day side are indicated in lower left corner.

of the tail that probably was not recognized in advance is its sharp boundary, observed at least to several planetary radii downstream.

There are at least several important questions that remain unanswered. One is the time history of the tail. How do magnetic flux tubes that finally form the tail pass through the environment of Venus, and what is the topological connection of the tail regions to near-planet regions: magnetic barrier and plasma mantle, ionosphere, and planetary core? Can we separate magnetic flux tubes that are connected to or passing through these regions? Another unknown is the manner of interconnection with ionosheath. Where do particles (if any) of external (solar wind) origin and tail populations start to mix and how does the two-lobe structure become disordered by $\sim 10R_V$ downstream? How does the magnetic tail evolve with distance, and how do the magnetic and electric field continue through the tail boundary?

An additional set of remaining problems is associated with the observed complex structure of the tail at distances $\sim 10 - 12 R_V$ that is presumably associated with tail flapping. The explanation may also involve convected interplanetary current sheets, the currents associated with shear flows of tail plasma, and the different histories of magnetic flux tubes populating the Venusian tail. It is not yet resolved how we can distinguish between the central current sheet and the other current sheets that are present in the tail.

Our knowledge of the tail plasmas is much more limited than our knowledge of the magnetic field. Venera 9 and 10 provided quite limited data sets, and their plasma analyzers were not-mass-selective. The Pioneer Venus plasma analyzer provided a much more extensive data set, but is was not sufficiently sensitive for tail plasma measurements. From Venera 9 and 10 measurements we know that the tail is nearly permanently populated with outflowing plasma, that this plasma is of planetary origin, and that this plasma domain is separated from the external shocked solar wind plasma, at least to several Venusian radii downstream, by a thin tail boundary. Ion populations observed in the current sheets, and frequently confined to these current sheets, have different plasma parameters from those typical of the lobe plasma. Usually the plasma populations associated with the current sheets are more energetic, but not necessary hotter, than the lobe plasma. The origin of the two different plasma populations observed at different downstream distances is poorly known. There is a reasonable degree of confidence that at least two sources of tail plasmas are operative, namely the pick-up of planetary photoions in the magnetic barrier, and nightside ionosphere escape. The possible tail entry of ions picked up in the external flow and of the plasma in detached plasma clouds are not excluded by the available observations.

The final set of problems is associated with acceleration processes. Least explored are processes of extraction and initial acceleration of the nightside ionospheric plasma. The wide range of reported values of acceleration also indicates that the acceleration of the lobe and current sheet plasma are not well understood. Another unanswered question is the origin of accelerated ions at the tail boundary.

In conclusion, we would like to mention that existing data sets still hold the promise of new results in the studies of the processes in the Venus tail. At the same time we may expect that the limitations of the plasma experiments conducted to date may not allow us to answer some of the fundamental questions concerning the origin and dynamics of the Venus tail.

REFERENCES

Brace, L. H., R. F. Theis, W. R. Hoegy, J. H. Wolfe, J. D. Mihalov, C. T. Russell, R. C. Elphic, and A. F. Nagy, The dynamic behavior of the Venus ionosphere in resonse to solar wind interactions, *J. Geophys. Res., 85*, 7663, 1980.

Brace, L. H., R. F. Theis, W. R. Hoegy, Plasma clouds above the ionopause of Venus and their implications, *Planet. Space Sci., 30*, 29, 1982a.

Brace, L. H., R. F. Their, H. G. Mayr, S. A. Curtis, and J. G. Luhmann, Holes in the nightside ionosphere of Venus, *J. Geophys. Res., 87*, 199, 1982b.

Brace, L. H., W. T. Kasprzak, H. A. Taylor, R. F. Theis, C. T. Russell, A. Barnes, J. D. Mihalov, and D. M. Hunten, The Ionotail of Venus: Its configuration and evidence for ion escape, *J. Geophys. Res., 92*, 15, 1987.

Bridge, H. S., A. J. Lazarus, C. W. Snyder, E. J. Smith, L. Davis, P. J. Coleman, and D. E. Jones, Plasma and magnetic fields near Venus, *Science, 158*, 1669, 1967.

Cloutier, P. A., R. E. Daniel, Jr., and D. M. Butler, Atmospheric ion wakes of Venus and Mars in the solar wind, *Planet. Space Sci., 22*, 967, 1974.

Cravens, T. E., and H. Shinagawa, The ionopause current layer at Venus, *J. Geophys. Res., 96*, 11,119, 1991.

Dolginov, Sh. Sh., E. M. Dubinin, Ye. G. Yeroshenko, P. L. Israilevich, I. M. Podgorny, and S. I. Shkol'nikova, On the configuration of the field in the magnetic tail of Venus, *Kosmich. Issled. 19*, 624, 1981.

Elphic, R. C., C. T. Russell, J. A. Slavin, L. H. Brace, and A. F. Nagy, The location of the dayside ionopause of Venus: Pioneer Venus orbiter magnetometer observations, *Geophys. Res. Letters., 7*, 561, 1980.

Gringauz, K. I., V. V. Bezrukikh, G. I. Volkov, M. I. Verigin, L. N. Davitayev, V. F. Kopilov, L. S. Musatov, and G. E. Slouchenkov, Study of solar plasma near Mars and along the Earth to Mars path by means of charged particles traps aboard the Soviet spacecrafts, Launched in 1971-1973, I, Techniques and devices, *Kosmich. Issled., 12*, 430, 1974.

Gringauz, K. I., V. V. Brzrukikh, T. K. Breus, T. Gombosi, A. P. Remizov, M. I. Verigin, and G. I. Volkov, Plasma observations near Venus onboard the Venera 9 and 10 satellites by means of wide angle plasma detectors, In: *Physics of Solar Planetary Environment*, edited by D. J. Williams, AGU, Boulder, pp.918-932, 1976.

Intriligator, D. S., Observations of mass addition to the shocked solar wind of the Venusian ionosheath, *Geophys. Res. Lett., 9*, 727, 1982.

Intriligator, D. S., J. H. Wolfe, and J. D. Mihalov, The Pioneer Venus Orbiter plasma analyzer experiment, *IEEE Trans., GE-19*, 39, 1980.

Kasprzak, W. T., H. B. Niemann, and P. Mahaffy, Observations of energetic ions at the nightside of Venus, *J. Geophys. Res., 92*, 291, 1987.

Luhmann, J. G., C. T. Russell, L. H. Brace, H. A. Taylor, W. C. Knudsen, F. L. Scaarf, D. S. Colburn, and A. Barnes, Pioneer Venus observations of plasma and field structure in the near wake of Venus, *J. Geophys. Res., 87*, 9205, 1982.

Lundin, R., E. Dubinin, S. Barabash, H. Koskinen, O. Norberg, N. Pizzarenko, A. Zakharov, On the momentum transfer of the solar wind to the Martian Topside ionosphere, *Geophys. Res. Lett., 18*, 1059, 1991.

Marubashi, K., J. M. Grebowsky, H. A. Taylor, Jr., J. G. Luhmann, C. T. Russell, A. Barnes, Magnetic field in the wake of Venus and the formation of ionospheric holes, *J. Geophys. Res., 90*, 1385, 1985.

McComas, D. J., H. E. Spenner, C. T. Russell, and M. A. Saunders, The average magnetic field draping and consistent plasma properties of the Venus magnetotail, *J. Geophys. REs., 91*, No. A7, 7030-7953, 1986.

Mihalov, J. D., and A. Barnes, The distant interplanetary wake of Venus: Plasma observations from Pioneer Venus, *J. Geophys. Res., 87*, 9045, 1982.

Moore, K. R., D. J. McComas, C. T. Russell, and J. D. Mihalov, A statistical study of ions and magnetic fields in the Venus magnetotail, *J. Geophys. Res., 95*, 12,005, 1990.

Moore, K. R., D. J. McComas, C. T. Russell, S. S. Stahara, and J. R. Spreiter, Gasdynamic modeling of the Venus magnetotail, *J. Geophys. Res., 96*, 5667, 1991.

Phillips, J. L., and D. L. McComas, The magnetosheath and magnetotail of Venus, *Space Sci. Rev., 55*, 1, 1991.

Phillips, J. L., J. G. Luhmann, C. T. Russell, and K. R. Moore, Finite Larmor radius effects on ion pickup at Venus, *J. Geophys. Res., 92*, 9920, 1987.

Phillips, J. L., J. G. Luhmann, W. C. Knudsen, and L. H. Brace, Asymmetries in the location of the Venus ionopause, *J. Geophys. Res., 93*, 3927, 1988.

Romanov, S. A., V. N. Smirnov, and O. L. Vaisberg, On the nature of solar wind-Venus interaction, *Kosmich. Issled., 16*, 746, 1978.

Russell, C. T., R. C. Elphic, and J. A. Slavin, Limits on the possible intrinsic magnetic field of Venus, *J. Geophys. Res., 85*, 8319, 1980.

Russell, C. T., J. G. Luhmann, R. C. Elphic, F. L. Scarf, and L. H. Brace, Magnetic field and plasma wave observations in a plasma cloud at Venus, *Geophys. Res. Lett., 9*, 45, 1982.

Russell, C. T., and O. L. Vaisberg, The interaction of the solar wind with Venus, In: *Venus*, edited by D. M. Hunten, L. Colin, T. M. Donahue, and V. I. Moroz, The University of Arizona Press, Tucson, pp.873-940, 1983.

Saunders, M. A., and C. T. Russell, Average dimension and magnetic structure of the distant Venus magnetotail, *J. Geophys. Res., 91*, 5589, 1986.

Slavin, J. A., E. J. Smith, and D. S. Intriligator, A comparative study of the distant magnetotail structure of Venus and Earth, *Geophys. Res. Lett., 11*, 1074, 1984.

Slavin, J. A., D. S. Intriligator, and E. J. Smith, Pioneer Venus orbiter magnetic field and plasma observations in the Venus magnetotail, *J. Geophys. Res., 94*, 2383, 1989.

Spenner, K., W. C. Knudsen, K. L. Miller, V. Novak, C. T. Russell, and R. C. Elphic, Observations of the Venus mantle, the boundary between solar wond and ionosphere, *J. Geophys. Res., 85*, 7655, 1980.

Spreiter, J. R., Magnetohydrodynamic and gasdynamic aspects of solar wind flow around terrestrial planets: A critical review, In: *Solar wind interaction with the planets Mercury, Venus, and Mars*, edited by N. F. Ness, NASA Publ. NASA Sp-397, Washington, 1976.

Taylor, H. A., H. C. Brinton, S. J. Bauer, R. E. Hartle, P. A. Cloutier, and R. E. Daniell, Global observations of the composition and dynamics of the ionosphere of Venus: Implications for the solar wind interaction, *J. Geophys. Res., 85*, 7765, 1980.

Vaisberg, O. L., On the asymmetry of the internal flow in the Venus wake, *Kosmich. Issled., 18*, 809, 1980.

Vaisberg, O. L., and L. M. Zelenyi, Formation of the plasma mantle in the Venusian magnetosphere, *ICARUS, 58*, 412, 1984.

Vaisberg, O. L., L. S. Zhurina, V. G. Kovalenko, E. L. Lein, B. V. Polenov, B. I. Khazanov, and A. V. Shifrin, Multichannel modular spectrometer of low energy electrons and ions, *Pribori i Technika Experimenta, No. 6*, 42-44, 1971.

Vaisberg, O. L., V. N. Smirnov, I. P. Karpinsky, B. I. Khazanov, B. N. Polenov, A. V. Bogdanov, and N. M. Antonova, Ion flux parameters in the solar wind Venus interaction region, In: *Physics of Solar Planetary Environment*, edited by D. J. Williams, AGU, Boulder, pp. 904-917, 1976.

Vaisberg, O. L., V. N. Smirnov, G. N. Zastenker, and A. O. Fedorov, Experimental data on the plasma envelopes of Mars, Venus, and Halley's and Giacobini-Zinner's comets: Comparison of loading effects, *Cosmic Res., 27*, 638, 1989.

Verigin, M. I., K. I. Gringauz, T. Gombosi, T. K. Breus, V. V. Bezrukikh, A. P. Remizov, and G. I. Volkov, Plasma near Venus from Venera 9 and 10 wide-angle analyzers data, *J. Geophys. REs., 83*, 3721, 1978.

Yeroshenko, Ye. G., Unipolar induction effects in the magnetic tail of Venus, *Kosmich. Issled., 20*, 604, 1979.

Zwan, B. G., and R. A. Wolf, Depletion of solar wind plasma near a planetary boundary, *J. Geophys. Res., 81*, 1636, 1976.

O. Vaisberg, V. Smirnov, A. O. Fedorov, L. Avanov, and F. Dunjushkin, Space Research Institute, USSR Academy of Science, 117810 GSP 7, Profsoyuznaya Ul. 84 32, Moscow, Russia.

J. G. Luhmann and C. T. Russell, Institute of Geophysics and Planetary Physics, University of California, Los Angeles, California 90024-1567, USA.

Ion Scattering and Acceleration by Low Frequency Waves in the Cometary Environment

H. Karimabadi, N. Omidi[1],

Department of Electrical and Computer Engineering, University of California, San Diego, California

and S. P. Gary

Group SST-7, Los Alamos National Laboratory, Los Alamos, New Mexico

The interaction between the solar wind and newborn cometary ions leads to the generation of low frequency magnetic fluctuations which in turn heat and scatter the ions. The nature of this interaction and the resulting particle pitch-angle scattering are studied using one- and two-dimensional hybrid simulations. A general theory for particle orbits in an arbitrary wave spectrum is presented. It is shown that particle scattering and diffusion in velocity space are controlled by three quantities. One is a newly derived constant of motion that is a locus of all points in velocity space energetically available to the particle. In case of one wave, this constant of motion reduces to a circle in $v_\perp - v_\parallel$ plane, where $v_\perp$ and $v_\parallel$ refer to the velocity components perpendicular and parallel to the static magnetic field, respectively. A second quantity is the resonance width which determines the extent of pitch-angle scattering. The third quantity which controls the amount of velocity diffusion is an adiabatic invariant which traces an arc in $v_\perp - v_\parallel$ plane. Strong velocity diffusion requires the destruction of this adiabatic invariant. This adiabatic invariant is broken in regions where neighboring resonances overlap. It also breaks on timescales over which Arnold diffusion becomes important. This latter process is, however, typically very slow and can be neglected. Detailed comparison between this new theory and simulations is presented. Finally, several previously puzzling features of the pickup process are explained.

1. INTRODUCTION

The scattering and acceleration of charged particles by electromagnetic fluctuations are fundamental processes in space plasma physics. The field energy of low frequency fluctuations (that is, those well below the proton cyclotron frequency) is predominantly magnetic; therefore, a single wave cannot change the energy of a particle but may alter the direction of its velocity vector. Thus pitch-angle scattering and the consequent reduction of ion anisotropies is an important consequence of enhanced low-frequency fluctuations in many different space plasmas including the terrestrial magnetosheath [*Sckopke et al.*, 1990] and the outer magnetosphere [*Tanaka*, 1985; *Omura et al.*, 1985]. Pitch-angle scattering via wave-particle interactions in the solar wind also plays an important role in the pickup of newborn cometary ions [*Wu et al.*, 1986; *Terasawa*, 1991; *Gary et al.*, 1988, 1991; *Miller et al.*, 1991] and the assimilation of ionized interstellar atoms [*Lee and Ip*, 1987].

If a multi-mode spectrum of low frequency fluctuations is present, interaction with the waves can lead to charged particle acceleration as well as to pitch-angle scattering. This energization process can be important

---
[1] Also at California Space Institute, University of California, San Diego, California.

Solar System Plasmas in Space and Time
Geophysical Monograph 84
Copyright 1994 by the American Geophysical Union.

at the terrestrial bow shock [*Lee*, 1982], at other planetary bow shocks [*Smith and Lee*, 1986], and at interplanetary shocks [*Lee*, 1983] as well as at astrophysical shocks.

The studies of particle acceleration and pitch-angle scattering in magnetized plasmas generally fall into four categories: (i) Orbit theory; (ii) Quasilinear theory (Q-L); (iii) Renormalization theories; and (iv) Particle-In-Cell simulations. Each approach has its set of shortcomings. The early theories of orbits showed that the energy of a particle interacting with a parallel propagating wave is constant in the wave frame. This result, due to its simplicity, has had a widespread appeal and is quoted and/or rederived in almost all papers dealing with particle scattering in space physics even though the idealization of only one wave interacting with a particle is hardly ever realized. Extensions of the orbit theory to more complex wave spectra have only recently become available [*Karimabadi et al.*, 1992; and this paper]. A second approach which sidesteps the issue of the nonlinear orbits is the Q-L which is a statistical method based on unperturbed orbits. Since the inception of Q-L, however, limitations in the method have been recognized [e.g. *Klimas and Sandri*, 1973; *Jones et al.*, 1973; *Goldstein et al.*, 1975] and it is widely known that many of the difficulties encountered in Q-L are associated with the use of free streaming orbits. The renormalization theories are aimed at providing a more accurate statistical description of plasma dynamics and in principle do not yield the unphysical secularities predicted in Q-L [for an excellent review see *Krommes*, 1984]. However, in spite of considerable added complexity in the theory, it is not clear that the results are any better than the traditional Fokker-Planck description [e.g. *Maasjost and Elsässer*, 1982]. An entirely different approach is particle simulations which provide an accurate description of wave-particle interactions. But the fact remains that without a theory, no quantitative interpretation of the simulation results can be achieved.

In this paper, we will combine the first and the last approach. We present a theory of orbits in the presence of a general wave spectrum and use the theory to explain the simulation results. Some of the advantages that the orbit theory offers over the Q-L are as follows: (i) The theory applies to a general wave spectrum independent of how large the wave amplitude or how narrow the spectrum. (ii) Detailed knowledge about the conditions under which particles can get scattered into a shell can be obtained. (iii) Analytical conditions for the onset of diffusion can be obtained.

The outline of this paper is as follows. In section 2 we present the theory of particle orbits; first for the case of one wave, then two waves, and finally a general wave spectrum. In section 3 we consider the nature of the interaction between the solar wind and newborn cometary ions, which eventually leads to the deceleration of the solar wind (mass loading) and pickup of cometary ions. The newborn ions are created when neutral particles escape from the surface of nucleus and are ionized. We present hybrid simulations of instabilities associated with the pickup process and use our new theory to explain the particle scattering/heating observed in the simulations. Finally, the summary and conclusions follow in section 4.

2. NONLINEAR ORBIT THEORY

Since wave-particle resonances are of fundamental importance to the understanding of the interactions studied here, we start our discussion of orbits by a brief discussion of this concept. Given an electromagnetic fluctuation of frequency ω and wavevector $\mathbf{k}$, a charged particle of species j and velocity $\mathbf{v}$ is said to be in resonance with the fluctuation, if the condition

$$\omega = k_\parallel v_\parallel \pm l\Omega_j \quad l = 0, 1, 2, \ldots \quad (1)$$

is satisfied. Here Ω_j is the cyclotron frequency of the j-th species and the subscript $\parallel$ refers to the component of the vector parallel to the static field. The excursion in particle velocity and/or pitch-angle is proportional to the square root of the wave amplitude for a resonant particle and proportional to wave amplitude for a nonresonant particle [see discussion following equation (B5)]. Thus, a resonant particle interacts more strongly with a wave than a nonresonant one.

2.1. One-Wave

Early studies of particle orbits [e.g. *Kennel and Engelmann*, 1966] demonstrated that the particle energy in the field of an electromagnetic plane wave propagating parallel to a background magnetic field is constant in the frame moving with the phase velocity of the wave. In addition, it was shown that the Hamiltonian in this frame is a constant of motion given by equation (A3) and is in the shape of a (semi-circle) in $v_\perp - v_\parallel$ plane as shown in Figure 1a. In the presence of such a circularly polarized wave, there is only one resonance possible with the resonance condition given by: $\omega - k_\parallel v_\parallel = -\eta\Omega\psi$, where η is the helicity, and $\psi = 1$ if the wave is propagating in the direction of the static field and $\psi = -1$ if the wave is propagating opposite to the magnetic field. The motion is integrable in this case which is the same as saying the Hamiltonian has one degree-of-freedom. Motion of a resonant particle along the Hamiltonian is further restricted by the size of the trapping width which is proportional to the square root of the wave amplitude [equation (B5a)]. The larger the wave amplitude the larger the trapping width and hence the larger the extent of pitch-angle variation.

At oblique angles of propagation, the motion is no longer integrable and the Hamiltonian has two degrees-of-freedom. The Hamiltonian changes from a circle to a circular arc with a finite width in $v_\perp - v_\parallel$ plane [equation (A3)]. The width is a measure of the velocity diffusion and is due to the fact that at oblique angles the electric field cannot be transformed away in its entirety [see discussion immediately following equation (A3); see also *Karimabadi et al.*, 1992]. There are also many resonances possible at harmonics of the gyrofrequency $\omega - k_\parallel v_\parallel = \ell\Omega$ with $\ell = -\infty, .., 0, ...\infty$. As long as the trapping widths associated with the neighboring resonances are well separated, the particle motion is nearly integrable (i.e. periodic in time). However, if the wave amplitude is sufficiently large for the neighboring trapping widths to overlap, the motion becomes stochastic and the particle can hop from one resonance to the next, covering a larger segment of the Hamiltonian in the process. For more details we refer the reader to *Karimabadi et al.* [1990; 1992].

2.2. Two-Waves

In this section, we consider the particle motion in the presence of two parallel propagating circularly polarized electromagnetic waves. It has been suggested by *Galeev and Sagdeev* [1987] that the Hamiltonian in the presence of more than one wave is a superposition of the Hamiltonians (semi-circles) due to each wave. In the case of two parallel but oppositely propagating magnetosonic waves the Hamiltonian was expected to look like that shown in Figure 1b. A rigorous derivation of the Hamiltonian given in Appendix A, however, shows the above expectation to be false. The correct Hamiltonian [equation (A5a)] is the surface (the system has two degrees-of-freedom) bounded by the two arcs in Figure 1c. Also shown are the two resonances and their trapping widths. As in the one wave case, the particle motion in velocity space is determined by the Hamiltonian and the size of the trapping width. From the trapping width one can calculate the maximum excursion in $v_\parallel$ which in turn can be used along with equation (A5a) for the Hamiltonian to find the corresponding range in $v_\perp$ at a given $v_\parallel$ (i.e. the velocity diffusion). The spread in $v_\perp$ corresponding to the trapping width is generally less than

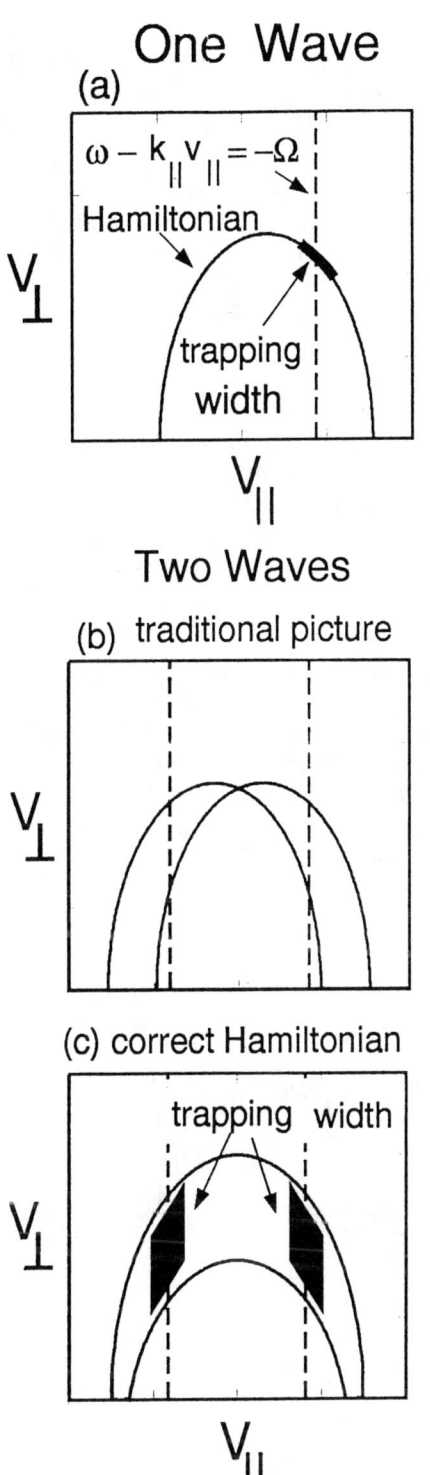

Fig. 1. Particle motion in velocity space in the presence of a parallel propagating electromagnetic wave. (a) One wave. (b) Two waves. In the traditional picture, the Hamiltonian is a superposition of Hamiltonians due to each wave. (c) The correct Hamiltonian in the presence of two parallel but oppositely propagating magnetosonic waves.

the maximum spread in $v_\perp$ allowed by the Hamiltonian (Figure 1c). However, as the wave amplitude is increased, a resonant particle can cover a larger portion of the Hamiltonian surface.

Note that the maximum width of the Hamiltonian is reached near $v_\parallel \sim 0$ where a particle would feel the force due to both waves. The finite thickness of the Hamiltonian is due to the fact that in the presence of two waves with different phase velocities there is no frame where the electric field can be eliminated. In fact, the larger the difference in the phase velocities the broader the Hamiltonian and the larger the velocity diffusion becomes. Away from $v_\parallel \sim 0$ a particle would feel the force mainly due to one of the waves, whereas near this speed both waves would contribute. Thus, the velocity diffusion is maximized at small parallel velocities and decreases as $v_\parallel$ increases (Figure 1c).

We illustrate the above points further by numerically solving the orbits of several protons. Figure 2a shows the time history of the orbits of three protons in $v_\perp - v_\parallel$ plane in the presence of two magnetosonic waves propagating in opposite directions along the static magnetic field. The boundaries of the Hamitlonian is given by the two arcs in Figure 2a. The orbits are followed in time up to $\Omega t = 2000$ using a fourth-order Runge-Kutta scheme. The wave and plasma parameters are $\omega_1/\Omega = 0.4$, $\omega_2/\Omega = 0.6$, $\alpha_1 = 0°$, $\alpha_2 = 180°$, $\omega_p/\Omega = 10000$ and $\delta B/B_o = 0.2$. Here ω_i and α_i are the frequency and propagation angle of the i-th wave respectively, ω_p is the plasma frequency, Ω is the gyrofrequency and δB is the wave magnetic field. All particles have the same Hamiltonian $H = 18mV_A{}^2$. Also shown with dashed lines are the two cyclotron resonances $\omega_i - k_i v_\parallel = -\Omega$. The particle with $v_\parallel \sim 0$ is nonresonant and thus scatters much less than the other two resonant particles. The extent of scattering and velocity diffusion (i.e. spread in $v_\perp$ at a given $v_\parallel$) are determined by the trapping width.

Figure 2b shows orbit of one particle, with initial $v_\parallel = 0$ and the same parameters as in Figure 2a except that now $\delta B/B_o = 0.7$. The two resonances have now overlapped and the particle is scattered almost randomly in phase space filling out most of the Hamiltonian surface (H-surface), which due to the larger wave amplitude has a wider thickness than that in Figure 2a. We should mention that even for $\delta B/B_o = 0.7$, there are still integrable regions in phase space that impede the random motion of particles in velocity space and temporarily trap the particles that come close to them. This peculiar behavior is a general property of Hamiltonian systems [e.g. *Karney*, 1983]. In fact, a close examination of Figure 2b reveals a U-shaped curve in an otherwise random collection of dots near $v_\parallel \sim 0$. This local trapping is only temporary and the particle eventually breaks loose.

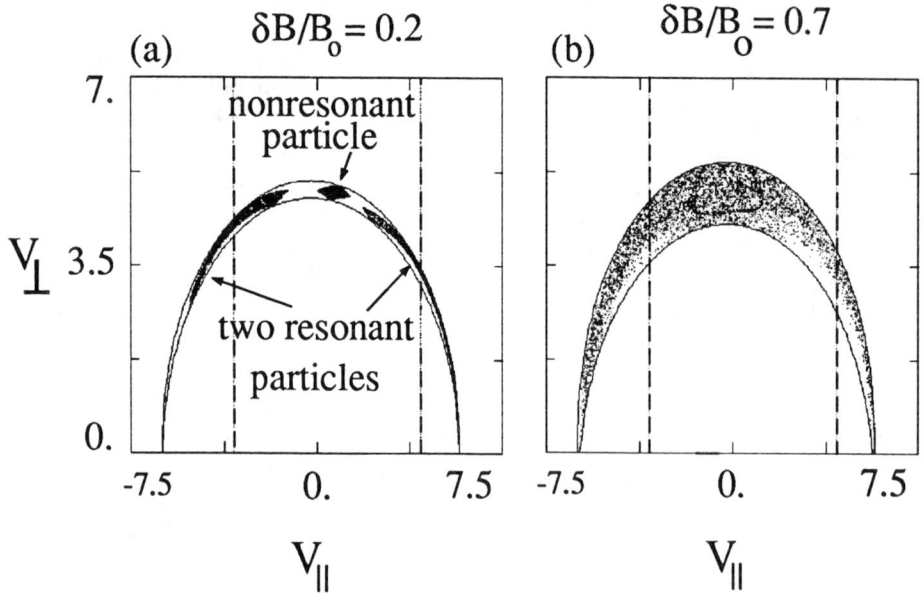

Fig. 2. Particle motion in velocity space in the presence of two electromagnetic waves propagating in opposite directions along the static magnetic field. All velocities are normalized to the Alfvén speed. (a) $\delta B/B_o = 0.2$. Orbits of three protons are shown. Also shown is the Hamiltonian which is now a surface bounded by the two arcs. The two resonances are indicated with dashed lines. The two resonant particles are scattered much more efficiently than the nonresonant particle. (b) $\delta B/B_o = 0.7$. The Orbit of one proton is shown.

2.3. Many-Waves

The expression for the Hamiltonian for a general wave spectrum is derived in Appendix A and is given by equation (A7a). As shown in Appendix A, in the presence of more than one wave, the system has three degrees-of-freedom regardless of whether the wave spectrum is continuous or consists of only two waves. The only exception is when the wave spectrum consists of two parallel propagating circularly polarized electromagnetic waves in which case the system has two degrees-of-freedom (see section 2.2).

Plate 1a shows a segment of the Hamiltonian in momentum space $(P_{z1}, P_{z2}, P_\perp)$ for the case of four waves with $\omega/\Omega = 0.1, 0.3, 0.4$ and 0.9, and propagation angles $\alpha = 0°, 180°, 0°$, and $0°$, respectively. Here P_{z1} and P_{z2} are related to $P_\parallel$ by equation (A6c), and $P_\perp$ is the component of momentum perpendicular to the static magnetic field. The system now has three degrees-of-freedom and the H-surface is a three dimensional surface in momentum space. As shown in Appendix A, in the presence of more than one wave, the system would have three degrees-of-freedom regardless of whether the wave spectrum is continuous or consists of only two waves. The only exception is when the wave spectrum consists of two parallel propagating circularly polarized electromagnetic waves in which case the system has two degrees-of-freedom (see section 2.2).

The intersection of resonance planes with the Hamiltonian surface are shown as solid curves in Plate 1a. In contrast to systems with one and two degrees-of-freedom, the H-surface intersects the resonances at infinite number of points. Thus, even for an arbitrarily small wave amplitude, a given particle can follow each resonance line to reach large energies. This diffusion along the resonances is called Arnold diffusion and is generic to many-dimensional nonlinear systems.

Due to the three dimensional nature of the H-surface, its projection onto the two-dimensional $v_\perp - v_\parallel$ plane covers the whole plane. Thus, in contrast to the case of one and two waves, the H-surface no longer restricts the motion of particles in the $v_\perp - v_\parallel$ plane. But we know from experience that particles do not in general fill up the whole velocity plane. The question then arises as to what quantities limit the diffusion of particles. In order to address this issue, we have examined the motion of a particle near a resonance. The mathematical details are given in Appendix B. Here we focus on the physics of the process. Plate 1b shows the schematics of particle motion near an isolated resonance. The resonance condition is given as before by $\omega - k_\parallel v_\parallel = \ell\Omega$ and is shown as a dashed red line in Plate 1b. Also shown is the trapping width associated with this resonance (blue curve) as well as the adiabatic invariant which is given by equation (B4). This adiabatic invariant limits the velocity diffusion of the particle whereas the extent of scattering along this arc is given by the resonance width. In order to have velocity diffusion, the adiabatic invariant has to be broken. There are two mechanisms that can lead to destruction of the adiabatic invariant. One is overlap of neighboring resonances. This leads to the usual quasilinear diffusion and occurs above a certain wave amplitude such that resonances overlap. The second is Arnold diffusion which is always present independent of wave amplitude (i.e. independent of whether resonances overlap or not) even though the rate of diffusion does depend on the wave amplitude. Arnold diffusion is typically very slow and unimportant compared to strong diffusion due to overlap of resonances.

We now demonstrate the above points by several examples. Plate 2a shows the time history of the orbits of two resonant protons in the presence of four waves with frequencies $\omega/\Omega = 0.2, 0.2, 0.4, 0.4$, propagating at $0°, 180°, 0°$ and $180°$, and a total wave amplitude $\delta B/B_o = 0.4$. The orbits are followed in time up to $\Omega t = 3000$ using a fourth-order Runge-Kutta scheme. Also plotted are the resonance widths (color coded) due to each wave; note that the trapping width becomes wider as $v_\perp$ increases. At small $v_\perp$ two resonances overlap on each side of $v_\parallel \sim 0$, whereas at larger $v_\perp$ all four resonances overlap. Even though the orbit of each particle is stochastic, their trajectories in $v_\perp - v_\parallel$ plane have only filled out a small segment of the whole available phase space. The inability of each particle to cross $v_\parallel \sim 0$ and reach the other side is due to the fact that all four resonances overlap only for $v_\perp \gtrsim 5$. It is also clear that each particle has followed the arc-shaped adiabatic invariant closely even though some velocity diffusion has occurred due to the overlap of resonances (two on each side). Plate 2b shows the orbit of one proton that started on the right-hand side but has now scattered to the other side. The increase in the wave amplitude ($\delta B/B_o = 0.7$) has brought down the overlap of all four resonances to $v_\perp \sim 3$ and the particle can scatter fully in pitch-angle. Note that at parallel propagation, $\ell = 0$ does not exist and strictly speaking there are no resonances centered at $v_\parallel = 0$. However, $\ell = -1$ resonances have widths that extend to $v_\parallel = 0$ and beyond thus allowing the penetration of particles through the 90° pitch-angle. The maximum velocity diffusion (and thus maximum deviation from the adiabatic invariant) has apparently occured at $v_\parallel \sim 0$. This is because that is where the resonance overlap is the strongest.

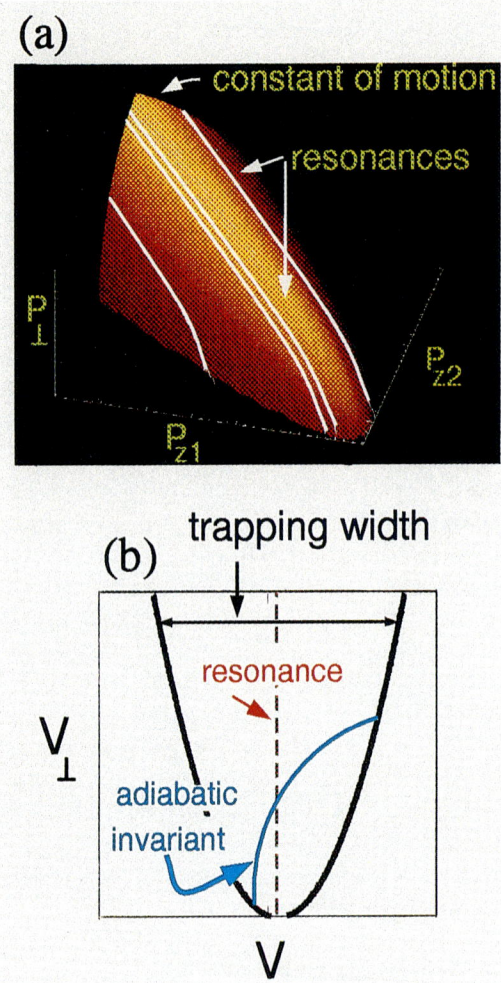

Plate 1. (a) The generic shape of the Hamiltonian in the presence of a wave spectrum. The resonance planes intersect this surface at infinite number of points (solid curves). (b) Particle motion near a resonance.

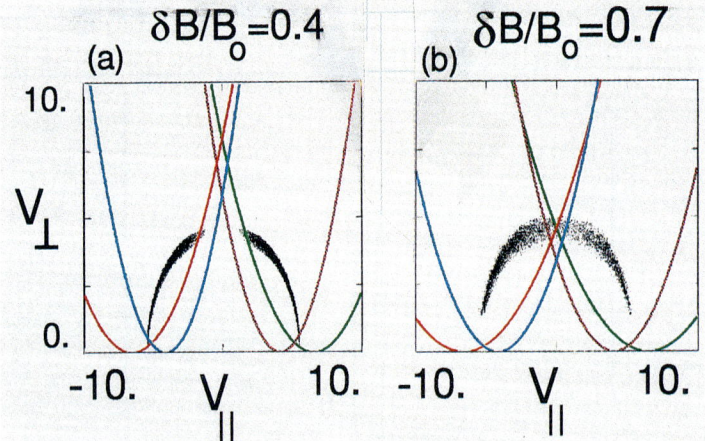

Plate 2. Particle motion in $v_\perp - v_\parallel$ plane in the presence of four parallel propagating waves with (a) $\delta B/B_o = 0.4$ and (b) $\delta B/B_o = 0.7$. The trapping width due to each of the four waves are indicated by the four colored curves. The two particles cannot scatter through pitch-angle of 90° in Plate 2a. Once the amplitude is large enough for resonances to overlap through $v_\parallel \sim 0$, the particles can scatter through pitch-angle of 90°, as in Plate 2b.

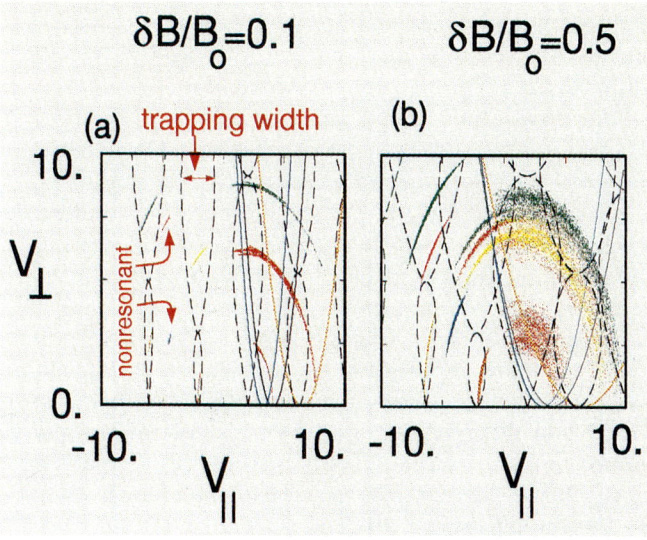

Plate 3. Particle motion in the presence of three parallel propagating waves (all in the same direction) and one obliquely propagating wave. The trapping widths due to the obliquely propagating wave are shown as dashed curves in black. The trapping widths due to the other three waves are shown as blue, purple and orange, respectively. (a) $\delta B/B_o = 0.1$. Orbits of 9 protons are shown. In all cases, the extent of scattering is determined by the resonance widths. (b) $\delta B/B_o = 0.5$. One of the nonresonant particles in Plate 3a has now become resonant due to the increase in size of the trapping widths. In the region of positive $v_{\parallel}$, many resonances overlap and particles go thorugh a strong velocity diffusion. However, their extent of scattering is again limited by the last available resonance.

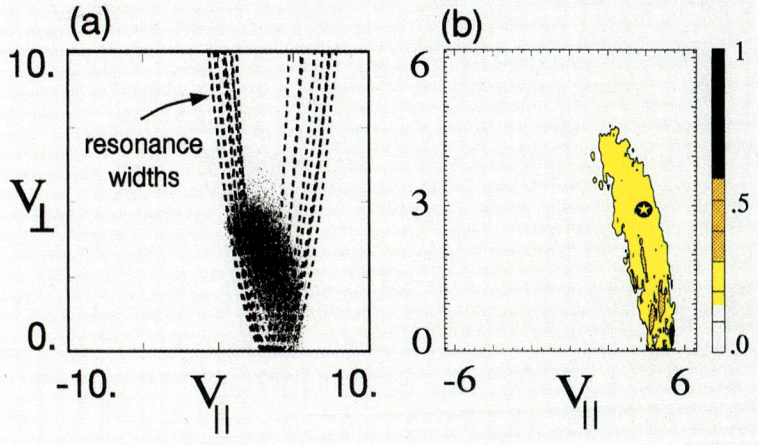

Plate 4. (a) Phase space plots of $v_{\perp} - v_{\parallel}$ at $\Omega t = 100$ for the pickup ions with 35°. The scattering is due to magnetosonic waves propagating parallel to the magnetic field and is described well by the size of trapping widths. (b) Contours of proton phase space density. The initial pickup velocity is denoted by a star.

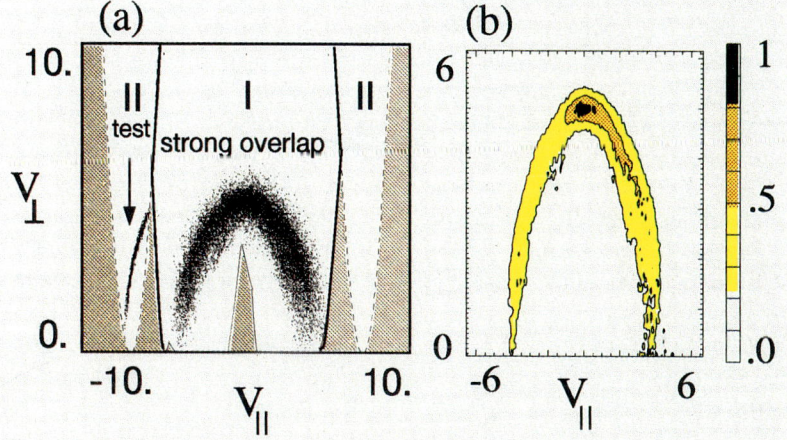

Plate 5. (a) Phase space plots of $v_{\perp} - v_{\parallel}$ at $\Omega t = 150$ for the pickup ions with $\alpha = 90°$. The scattering is due to Alfvén waves propagating in opposite directions along the magnetic field. (b) Contours of proton phase space density.

At oblique propagation angles, the number of resonances due to each wave increases from one and the trapping width becomes a complicated function of $v_\perp$. But the physics of diffusion discussed above remains essentially unchanged. Plate 3 shows time history of the orbits of several protons in the field of four waves at two wave amplitudes of 0.1 and 0.5 with $\omega/\Omega = 0.2, 0.5, 0.4, 0.6$ and $\alpha = 0°, 60°, 0°, 0°$. The orbits were followed in time up to $\Omega t = 3000$. The resonances (fundamental as well as harmonics) due to the obliquely propagating wave (dashed curves) have widths that go to zero at several $v_\perp$. Particle acceleration is severely limited at such points. The three resonances associated with the three parallel propagating waves are also shown and color coded as blue, purple and orange, respectively. In the region of negative $v_\parallel$, only resonances due to the oblique wave contribute and thus particles in this region scatter much less than the particles in the positive $v_\parallel$ region. In all cases, however, the extent of scattering is described well by the size of trapping widths. We have also plotted the orbits of two nonresonant particles in Plate 3a. As expected, nonresonant particles scatter much less efficiently than the resonant particles.

Of particular interest is the particle with an initial $v_\parallel = 4$ and $v_\perp = 0.6$ in Plate 3b. In spite of its large energy gain (the adiabatic invariant is destroyed), the absence of resonances at $v_\parallel \sim 0$ and $v_\perp \lesssim 4$ has limited its scattering into only half a shell. Another interesting point is the orbits of particles coded with red, green and yellow colors (all having positive $v_\parallel$'s initially). As is evident in Plate 3b these three particles have diffused in velocity space in regions where a number of resonances overlap. But their orbits in regions past $v_\parallel \gtrsim 0$ where there is no overlap shows little velocity diffusion and they follow the adiabatic invariant very closely. From the foregoing it is clear that diffusion in velocity space is quite nonuniform and is determined by the size and number of overlapping resonances. Thus, a correct diffusion formalism must take into account the finite trapping width of each resonance, yielding a large diffusion coefficient in regions where resonances are strongly overlapping and giving no diffusion in regions where there is no overlap. Such a formalism has been developed and tested successfully by *Karimabadi and Menyuk* [1991]. The interested reader is referred to the above reference for more detail.

3. COMPARISON OF THEORY WITH HYBRID SIMULATIONS

In this section we present results from hybrid simulations of ring-beam instabilities associated with cometary ion pickup and use the theory developed in section 2 to explain the resulting pitch-angle scattering and velocity diffusion.

The creation of ions in the neutral atmosphere of a comet leads to a strong interaction between the cometary ions and the solar wind and the growth of several different plasma instabilities. As the newborn ions undergo cyclotron motion in the magnetic field of the solar wind, they form a ring-beam distribution with perpendicular speed $v_{sw} \sin\alpha$ and parallel drift velocity $v_{sw} \cos\alpha$ in the solar wind frame, where α is the angle between the solar wind flow velocity and the ambient magnetic field.

Low-frequency electromagnetic instabilities driven by ring-beam distributions have been categorized as primarily beam-driven at $\alpha \lesssim 60°$ and primarily ring-driven at α values relatively close to perpendicular [*Gary*, 1991]. In the former regime, there is agreement among linear theory [*Winske and Gary*, 1986], computer simulations [*Gary et al.*, 1989; *Miller et al.*, 1991], and observations [*Tsurutani et al.*, 1987, 1989; *Glassmeier et al.*, 1989] that the ion/ion right-hand resonant instability, which typically has maximum growth at parallel propagation, often dominates distant cometary environments. Recently, *Omidi et al.* [1993] have shown the existence of a competing instability at small α's which is driven by the anisotropy rather than the beam component of the ring-beam and has a maximum growth at oblique angles and leads to formation of shocklets [*Omidi et al.*, 1993]. Here, we limit our study to pickup of protons and to the paramter regime where the ion/ion right-hand resonant instability is the dominant instability at small α's.

For α relatively close to 90° the ring-like character of the cometary ion distribution can drive two types of electromagnetic instability. One is the ion cyclotron instability with left-hand polarization and maximum growth rate at $k \times B_o = 0$ [*Gary et al.*, 1989 and references therein]. The other is the mirror instability, a compressional mode which propagates obliquely to the background magnetic field [*Price*, 1989]. The parameter regimes in which each of these modes should dominate have not been established; since the ion cyclotron instability typically saturates at very low amplitudes [*Gary et al.*, 1989] it is difficult to observe in the cometary environment, whereas the mirror instability usually appears at relatively large amplitudes and has been observed at Halley by *Russell et al.* [1987].

The simulation codes used in this paper are the 1-D [*Winske and Omidi*, 1991] and 2-D electromagnetic hybrid codes which treat the electrons as a fluid (adiabatic) and ions as macroparticles. We use periodic boundary conditions for both particles and fields. The simulation model consists of two ion species: solar wind protons and the newborn protons. The simulations are carried out in the solar wind rest frame and the newborn ions are assumed to have an initial ring-beam velocity distribution with a negligible thermal spread. The simulation results shown here are for $\omega_p/\Omega = 10^4$, and $V_{sw} = 5V_A$.

We first consider the case where $\alpha = 35°$ and $n_b = 0.012n_i$. By performing a 2-D simulation we found that the maximum growth occurs at parallel propagation and the particle scattering is very similar to 1-D. Thus, we performed 1-D simulations with $\vec{k} \| \vec{B}_o$ to further analyze the resulting pitch-angle scattering. One advantage of the 1-D simulations is that runs with much smaller noise level than the 2-D simulations can be made. The simulation box is 100 c/ω_p, with 128 cells, 1000 particles per cell, $\Omega \Delta t = 0.05$ and a total run time of $\Omega t = 100$. The dominant growing waves propagate in the beam direction, have positive helicity and are right-hand polarized in the center of mass frame. Thus, we have identified the growing modes as the ion/ion right-hand resonant instability in agreement with simulations of *Miller et al.* [1991] and previous linear theories [e.g. *Gary et al.*, 1989; *Winske and Gary*, 1986]. A segment of the ion/ion resonant instability branch below $k \sim 0.25\omega_p/c$ has negative frequency (and thus left-hand polarization) in the solar wind rest frame, but most of the power is at $\omega \sim 0.68\Omega$, and $k \sim 0.49\omega_p/c$.

Plate 4a shows the scattering of newborn ions at $\Omega t = 100$ which is after the saturation of the instability at $\Omega t \sim 30$. The wave energy oscillates with an ever decreasing amplitude after the saturation time. As a result there is very little change in the particle pitch-angle scattering beyond $\Omega t = 50$. In order to determine the resonant widths, detailed knowledge of the wave spectrum is required. We have performed two-dimensional fast Fourier transforms of the waves (in ω and k) starting from $\Omega t \sim 5$ up to various times. We have found no significant change in the relative power between the various modes beyond $\Omega t \sim 40$. Thus, we used the spectrum between $\Omega t = 5$ and 60 and a wave amplitude of $\delta B/B_o = 0.157$ (the wave amplitude at saturation) to calculate the resonance widths due to the 10 dominant modes shown in Plate 3a. Note that at parallel propagation, there is at most one resonance due to each mode. The 10 modes have resonance widths that overlap and thus allow the particles to diffuse both in pitch-angle and velocity. It is clear from Plate 4a that the cutoff in pitch-angle scattering is accurately described by the extent of the overlapping resonances. We have verified that the size and structure of the overlapping resonances do not change as the number of modes is increased from 10 to 400.

Contours of proton phase space density along with the color scale are shown in Plate 4b. The initial velocity of the pickup protons is marked with a star. The maximum phase space density is clearly away and at smaller pitch-angle than the initial pitch-angle. This is an important finding as it demonstrates that contrary to popular belief, it is possible to get peaks in phase space densities away from the injection angle within a simple theory of waves interacting with particles in a homogeneous medium. Note that the above finding may provide an alternative explanation for the one sided pickup proton phase space densities observed [e.g. *Neugebauer et al.*, 1989] upstream of the Halley bow shock which show peaks away from the injection angle, without having to invoke global effects. Note that simulations discussed in this section are initial value runs where all the pickup ions are initialized at the start of the run as a beaming-ring. The important point to note is that as long as the scattering rate is faster than the injection rate, the peak in the phase space density can be different than the injection angle and is determined by the size of the trapping widths and their overlap (and thus the value of the diffusion coefficient) in various parts of phase space. Particles can diffuse faster into regions where resonances are strongly overlapping. By the same token, particles can spend a long time in regions where the diffusion coefficient is small. Thus, the peak in phase space density is a time varying phenomena which is controlled by variations of diffusion coefficient in various parts of phase space.

We next consider the case where the pickup ions have $\alpha = 90°$ and $n_b/n_i = 0.0256$. By performing 2-D simulations, we found that the wave spectrum has two peaks in propagation angle α_1, one centered at $\alpha_1 \sim 0°$ and a second peak at $\alpha_1 \sim 50°$. At small angles, the spectrum is nearly circularly polarized, and there exist waves propagating both parallel (negative helicity) and antiparallel (positive helicity) to the beam direction. Both modes lie on the Alfvén/ion-cyclotron branch in agreement with linear theory [e.g. *Gary and Madland*, 1988]. At more oblique angles, the waves are predominantly polarized along the B_o-direction and at the second peak power is concentrated at zero frequency. Note that for the Alfvén wave at large angles, the wave magnetic field lies outside of the $k - B_o$ plane whereas for the mirror mode the wave magnetic field is predominantly in the $k - B_o$ plane [e.g. *Krauss-Varban et al.*, 1993] and in the direction along B_o, leading to the compressional structure of mirror waves. We thus identify the second peak to be due to the mirror instability, in agreement with linear theory [*Price*, 1989].

In order to assess the relative importance of mirror and Alfvén/ion-cyclotron modes to ion scattering/heating, we have performed several 1-D simulations. We have found that for our choice of parameters the scattering due to the mirror mode is limited to the $\ell = 0$ resonance, whereas the Alfvén mode scatters the particles into a full shell. Details of the particle acceleration due to mirror mode will be published elsewhere. Here we will concentrate on the Alfvén mode. We have thus performed 1-D simulations with $\vec{k} \| \vec{B}_o$ and 1000 particles/cell. The large number of particles/cell was used to reduce the effect of scattering due to noise.

A very cold ring leads to the growth of the electromagnetic ion cyclotron ring instability [*Gary and Madland*, 1988]. However, a low level of field fluctuations will broaden such a cold ring, in which case the mode dispersion goes over into that of the ion cyclotron anisotropy instability. In our simulations (as in the cometary environment) the cold ring is broadened very quickly and the evolution of the system is determined by the ion cyclotron anisotropy instability which grows exponentially up to $\Omega_p t \sim 75$, after which there is a continuing small increase in wave energy through the end of the run at $\Omega_p t = 150$.

Plate 5a shows the scattering of pickup ions at $\Omega t = 150$. Because $\alpha = 90°$, Alfvén waves propagating in both directions are excited. Due to the large number of growing modes in this case, we have not shown the resonance widths due to the modes. Instead, we have divided the figure into two separate zones. The shaded areas are to make the separation between the two zones visually more clear. In the area labelled zone I, the resonances overlap strongly and the particles are scattered into a bi-spherical shell. Outside of this zone, the resonances are due to waves with amplitudes comparable to the initial noise in the system $\delta B/B_o \sim 2 \times 10^{-3}$. Thus, no significant velocity diffusion is possible outside of zone I. To demonstrate this point, we have followed the orbits of a population of 12800 test particles initialized with $V_{sw} = -7.5 V_A$ and $\alpha = 23°$ in the self-consistent fields of the simulation. Clearly, the scattering is due to resonant interaction with the 'noise' and its extent is well described by the resonant width. We thus conclude that the particle scattering and velocity diffusion is described accurately by our theory.

Contours of proton phase space density are shown in Plate 5b. The maximum phase space density is close to the initial injection angle (90°). The phase density has a higher peak on the positive $v_\|$ than the negative $v_\|$ side. Note that depending on the parameters, the peak in phase space density is a time varying quantity which may or may not be at the injection angle.

4. SUMMARY AND CONCLUSIONS

In this paper, we have examined the interaction between the solar wind and the newborn cometary ions and presented a general theory of particle scattering and velocity diffusion. The main results of this study are summarized as follows.

(1) In systems with one and two degrees-of-freedom, the motion in velocity space is limited by the Hamiltonian and the size of the trapping width. An example of a system with one degree-of-freedom is of a particle in the presence of a parallel propagating circularly polarized electromagnetic wave. If the wave is obliquely propagating, the system has two degrees-of-freedom. Another example of a system with two degrees-of-freedom is that of two circularly polarized waves propagating along the magnetic field.

(2) The Hamiltonian in a general wave spectrum is a three dimensional surface in velocity space and the system has three degrees-of-freedom. The Hamiltonian surface then intersects the resonance planes at infinite number of points. Thus, unlike systems with lower degrees-of-freedom, motion of particles along the resonances is not energetically forbidden. Thus particles can follow along the resonances to large energies. This process called Arnold diffusion is very slow and usually not important. The projection of the Hamiltonian into the $v_\perp - v_\|$ plane covers the whole plane. Thus unlike systems with lower degree-of-freedom the Hamiltonian does not restrict the motion in $v_\perp - v_\|$ plane. The extent of diffusion/scattering in this case is controlled by two quantities; one is a newly derived adiabatic invariant which forces the particles to follow an arc in velocity space. The other quantity is the trapping width which further limits the scattering of the particle along this arc.

(3) Velocity diffusion can only occur if the above adiabatic invariant is broken. This can happen in two ways. First, if the time scales are long compared to the Arnold diffusion time, which typically occurs too slow to be of practical interest. The other is diffusion due to the overlap of resonances which is generally much stronger than Arnold diffusion. Note that the so-called second order Fermi acceleration is nothing more than the velocity diffusion in the presence of a spectrum of waves due to the overlap of resonances. The many difficulties associated with the quasilinear description of this diffusion process are caused by the neglect of resonance widths in the quasilinear theory.

(4) Using our theory, we have shown that velocity diffusion is controlled by the difference in parallel phase velocity $\omega_i/k_{\|i}$ of the various modes in the spectrum in two ways. First, if all the waves have the same parallel phase velocity, no velocity diffusion is possible except for a small diffusion proportional to the wave amplitude [see equation (A3)]. Secondly, the spacing between the resonances in velocity space is determined by the paral-

lel phase velocity of the various modes in the spectrum. A larger range in $\omega_i/k_{\|i}$ allows a wider coverage of velocity space by resonances. Since pitch-angle scattering and velocity diffusion are determined by the location and width of resonances, stronger scattering and diffusion are expected for cases where the waves have a wide rather than a narrow spread in $\omega_i/k_{\|i}$. At parallel propagation, the trapping width and thus the extent of pitch-angle scattering increase as a function of $v_\perp$. Thus a wave spectrum which results in a larger velocity diffusion is also likely to result in more pitch-angle scattering. This explains why in both observations as well as simulations the pickup protons are often scattered into only half a shell at small injection angles and a full shell at large injection angles, even though the wave amplitudes are typically smaller at large α's.

At small α, only waves propagating in the direction of the magnetic field are generated in the solar wind rest frame whereas at large α waves propagating in both directions with respect to the magnetic field are generated (i.e. larger range in $\omega_i/k_{\|i}$). Thus, scattering is more effective at large injection angles. At large α the source of free energy is the ion temperature anisotropy and the saturation mechanism is scattering of pickup protons into a shell. At small α, the source of free energy is the ion beam and the saturation mechanism is the thermalization and/or reduction in the beam velocity in form of scattering. Since scattering is less efficient at small α's as compared to large α's, the saturation amplitudes tend to be larger at small α's.

(5) The peak in the pickup proton phase space density is a time varying quantity that does not have to be at the injection angle. Its exact location in velocity space is determined by the the diffusion coefficient and the ion injection rate.

Acknowledgments. Useful discussions with D. Krauss-Varban and S. Fuselier are gratefully acknowledged. This research was supported by NASA grants NAGW-1806, NAG 5-1492 and by the IGPP at Los Alamos National Laboratory. N. Omidi's research was performed under the auspices of the California Space Institute. The Los Alamos portion of this work was performed under the auspices of the U.S. Department of Energy and was supported by the DOE Office of Basic Energy Sciences, Division of Engineering and Geosciences, and by the IGPP at Los Alamos National Laboratory. Computing was performed on the CRAY Y-MP at the San Diego Supercomputer Center.

APPENDIX A: DERIVATION OF THE CONSTANT OF MOTION

The Hamiltonian formalism proves very convenient for studies of orbits and is used here to derive the constants of motion for various types of wave spectra. The Hamiltonian of a particle with charge q and rest mass m in the presence of a spectrum of M waves is given by [*Karimabadi et al.*, 1992]:

$$H = \frac{1}{2m}(\vec{P} - \frac{q\vec{A}}{c})^2 + q\sum_{i=1}^{M}\Phi_i \sin\psi_i \quad (A1a)$$

where

$$\vec{A} = \sum_{i=1}^{M}\left[A_{1i}\cos\alpha_i \sin\psi_i \hat{e}_x + A_{2i}\cos\psi_i \hat{e}_y \right.$$
$$\left. - A_{1i}\sin\alpha_i \sin\psi_i \hat{e}_z\right] + xB_o\hat{e}_y, \quad (A1b)$$

$$\psi_i = k_{\perp i}x + k_{\|i}z - \omega_i t + \delta_i, \quad (A1c)$$

and Φ_i is the amplitude of the electrostatic component of the i-th wave. In equation (A1), $\vec{P}$ is the canonical momentum, B_o is the uniform and static magnetic field and is taken to be in the z direction, $k_\perp = k\sin\alpha_i$, $k_\| = k\cos\alpha_i$, α_i is the i-th wave propagation angle with respect to the magnetic field, and δ_i is the phase difference between the various waves. The subscripts $\perp$ and $\|$ refer to directions perpendicular and parallel to the static field, respectively. Since H is independent of y, P_y is a constant of motion and can be easily eliminated from the Hamiltonian by using the transformation $F_2 = P_x{'}(x - cP_y{'}/qB_o) + yP_y{'}$. The new variables are given by $P'_x = P_x$, $P'_y = P_y$ and $X = x - cP'_y/qB_o$. It proves convenient to express the Hamiltonian in terms of the zeroth-order action angle variables (θ, J) using the generating function $F_1 = \frac{1}{2}m\Omega x^2 \cot\theta$, which gives

$$H = \frac{1}{2m}\left\{P_\perp^2 + P_\|^2 + 2\frac{|q|}{c}\left[P_\perp \sin\theta \sum_{i=1}^{M}A_{2i}\cos\psi_i\right.\right.$$
$$-\frac{q}{|q|}P_\perp \cos\theta \sum_{i=1}^{M}A_{1i}\cos\alpha_i \sin\psi_i$$
$$\left.+\frac{q}{|q|}P_\| \sum_{i=1}^{M}A_{1i}\sin\alpha_i \sin\psi_i\right]$$
$$+\frac{q^2}{c^2}\left[\left(\sum_{i=1}^{M}A_{1i}\cos\alpha_i \sin\psi_i\right)^2 + \left(\sum_{i=1}^{M}A_{2i}\cos\psi_i\right)^2\right.$$
$$\left.\left.+\left(\sum_{i=1}^{M}A_{1i}\sin\alpha_i \sin\psi_i\right)^2\right]\right\} + q\sum_{i=1}^{M}\Phi_i \sin\psi_i \quad (A2a)$$

with

$$X = (2J/m\Omega)^{1/2}\sin\theta, \quad (A2b)$$

and
$$P_x = P_\perp \cos\theta, \quad (A2c)$$

and where $P_\perp = (2m\Omega J)^{1/2}$ and $P_\parallel = P_z$. We are now ready to derive the constant of motion, which amounts to eliminating the explicit time dependence in H, for various types of wave spectra.

A.1. One-Wave

The explicit time-dependence in H is eliminated by $F_2 = [z - (\omega/k_\parallel)t]$:

$$H = \frac{1}{2}mv^2 - m(\frac{\omega}{k_\parallel})v_z$$
$$+ q\left[A_1 \frac{\omega}{k_\parallel c} \sin\alpha_1 + \Phi_1\right]\sin\psi_1. \quad (A3)$$

The presence of the phase terms (i.e. those multiplying the $\sin\psi_1$) on the right-hand side of equation (A3) are due to the fact the electric field of the wave cannot in general be eliminated in its entirety even in the wave frame. For a parallel propagating wave the phase terms drop out and we recover the well known result that the particle energy in the wave frame is constant. Since H depends only on $v_\parallel$ and $v_\perp$, the system has two degrees-of-freedom. For the special case of a parallel propagating circularly polarized wave, there exists an additional constant of motion related to the helical symmetry of the wave. This additional constant of motion reduces the degrees-of-freedom of the system to one and the system becomes integrable [e.g. *Roberts and Buchsbaum*, 1964; *Karimabadi et al.*, 1990].

A.2. Two Waves

In the case of two circularly polarized waves propagating parallel to the magnetic field, the explicit time-dependence in H is eliminated by $F_2 = (\theta - k_1\sigma_1 z + \xi_1\omega_1\sigma_1 t)I_1 + (\theta - k_2\sigma_2 z + \xi_2\omega_2\sigma_2 t)I_2$, where we have defined $\xi_i = \cos\alpha_i$ and σ_i is the relative sign between A_{1i} and A_{2i} (i.e. the helicity) and is equal to one for a magnetosonic wave propagating parallel to the static field, whereas if the magnetosonic wave is propagating antiparallel to the field $\sigma_i = -1$. Below we assume $q > 0$. Note that the last transformation is strictly valid for $k_1\sigma_1 \neq k_2\sigma_2$, since otherwise J and $P_\parallel$ would only differ by a constant and no unique solution for I_1 and I_2 would be possible. However, this is just a mathematical restriction as we can take k_1 arbitrary close to k_2.

The new variables are related to the old ones by

$$\theta_i = \theta - k_i\sigma_i z + \xi_i\omega_i\sigma_i t, \quad (A4a)$$

$$J = P_\perp^2/2m\Omega = I_1 + I_2, \quad (A4b)$$

and

$$P_\parallel = -k_1\sigma_1 I_1 - k_2\sigma_2 I_2. \quad (A4c)$$

After some algebra, we obtain the constant of motion written in terms of the ordinary velocity:

$$H = (0.5 + a)v_\perp^2 + bv_\perp + 0.5v_\parallel^2 + dv_\parallel + e \quad (A5a)$$

where

$$a = \frac{\sigma_1\sigma_2\omega_1\omega_2(N_1\xi_2 - N_2\xi_1)}{2\Omega^2(N_1\sigma_1\omega_1/\Omega - N_2\sigma_2\omega_2/\Omega)} \quad (A5b)$$

$$b = 2a\left[\sigma_1\eta_1\sin(\phi + \psi_1\xi_1/\sigma_1)\right.$$
$$\left. + \sigma_2\eta_2\sin(\phi + \psi_2\xi_2/\sigma_2)\right] \quad (A5c)$$

$$d = \frac{(c/V_A)(\sigma_2\xi_2\omega_2/\Omega - \sigma_1\xi_1\omega_1/\Omega)}{\sigma_1 N_1\omega_1/\Omega - \sigma_2 N_2\omega_2/\Omega} \quad (A5d)$$

$$e = a\left[\eta_1^2 + \eta_2^2 + 2\eta_1\eta_2\sigma_1\sigma_2\right.$$
$$\left. \cdot \cos(\psi_1 - \xi_1\xi_2\psi_2/\sigma_1\sigma_2)\right] \quad (A5e)$$

$$\eta_i = \frac{c}{V_A}\frac{|q|A_i}{mc^2}, \quad (A5f)$$

$$\psi_i = k_i\xi_i z - \omega_i t. \quad (A5g)$$

$N_i = ck_i/\omega_i$, and ϕ is the gyroangle defined as $\phi = \tan^{-1}(v_y/v_x)$. α_i is the direction of propagation of the $i-th$ wave (either $0°$ or $180°$) with respect to the static field B_o. In the above, the velocities are normalized to the Alfvén speed (V_A) and H is normalized to mV_A^2. Note that the above constant of motion is exact and is valid for all wave amplitudes. Since H depends only on $v_\parallel$ and $v_\perp$, the system has two degrees-of-freedom.

A.3. Many-Waves

The constant of motion can be obtained by making the following transformation: $F_2 = (z - \sum_{i=1}^{M}\frac{\omega_i}{k_{\parallel i}}t)P_{z1} +$

$(z + \sum_{i=1}^{M} \frac{\omega_i}{k_{\|i}} t) P_{z2}$. The new variables are related to the old ones by

$$z_1 = z - \sum_{i=1}^{M} \frac{\omega_i}{k_{\|i} t}, \quad (A6a)$$

$$z_2 = z + \sum_{i=1}^{M} \frac{\omega_i}{k_{\|i} t}, \quad (A6b)$$

$$P_\| = P_{z1} + P_{z2}, \quad (A6c)$$

$$P_\perp = P_\perp', \quad (A6d)$$

$$\theta' = \theta. \quad (A6e)$$

This transformation is strictly valid for $\sum_{i=1}^{M} \omega_i/k_{\|i} \neq 0$; otherwise the transformation becomes degenerate.

The new constant of motion is:

$$H = \frac{1}{2}[P_\perp{}^2 + (P_{z1} + P_{z2})^2] - \frac{c}{V_A} \sum_{i=1}^{N} \frac{1}{N_{\|i}}$$
$$\cdot (P_{z1} - P_{z2}) + \frac{c^2}{V_A{}^2} \sum_{i=1}^{N} \epsilon_{3i} \sin \psi_i \quad (A7a)$$

and

$$\psi_i = k_{\perp i}(\frac{P_\perp}{m\Omega} \sin\theta + \frac{cP_y}{qB_o}) + 0.5 k_{\|i}(z_1 + z_2)$$
$$- \omega_i \frac{(z_2 - z_1)}{\sum_{i=1}^{M} \omega_i/k_{\|i}} + \delta_i. \quad (A7b)$$

The Hamiltonian now depends on $P_\perp$, P_{z1} and P_{z2} and the system has thus three degrees-of-freedom. P_y is a constant of motion and its only effect on the particle motion is by introducing an additional phase shift between the waves. Note that the only equation relating P_{z1} and P_{z2} to $P_\|$ is equation (A6c). However, the transformation is still well defined and the time evolution of P_{z1} and P_{z2} can be easily obtained from the equations of motion.

APPENDIX B:
PARTICLE DYNAMICS NEAR A RESONANCE

Here we consider the particle behavior near a resonance for a general wave spectrum. We start with the Hamiltonian equation (A2a) and expand it to first order in wave amplitude, obtaining:

$$H = \frac{1}{2}(P_\perp{}^2 + P_z{}^2) + \sum_{i=1}^{M} \sum_{\ell=-\infty}^{+\infty} Z_{\ell_i}$$
$$\cdot \sin(\ell\theta + k_{\|i}z - \omega_i t), \quad (B1a)$$

where

$$Z_{\ell_i} = mV_A^2 \Big\{ \frac{q}{|q|} \Big(\frac{P_\|}{mV_A} \eta_{1i} \sin\alpha_i + \eta_{3i} \Big) J_{\ell_i}(k_{\perp_i}\rho)$$
$$+ \frac{1}{2} \frac{P_\perp}{mV_A} \Big[(\eta_{2i} - \frac{q}{|q|}\eta_{1i}\cos\alpha_i) J_{\ell_i-1}(k_{\perp_i}\rho)$$
$$- (\eta_{2i} + \frac{q}{|q|}\eta_{1i}\cos\alpha_i) J_{\ell_i+1}(k_{\perp_i}\rho) \Big] \Big\}, \quad (B1b)$$

where

$$\eta_{1i} = \frac{c}{V_A} \frac{|q|A_{1i}}{mc^2}, \quad (B1c)$$

$$\eta_{2i} = \frac{c}{V_A} \frac{|q|A_{2i}}{mc^2}, \quad (B1d)$$

$$\eta_{3i} = \frac{c^2}{V_A^2} \frac{|q|\Phi_{oi}}{mc^2}, \quad (B1e)$$

$\rho = P_\perp/m\Omega$ and J_{ℓ_i} is the Bessel function of the first kind. We have made use of standard Bessel function identities [Abramowitz and Stegun, 1964] in deriving (B1). In what follows, we assume that only one wave contributes to a given resonance. In order to derive the Hamiltonian near a given resonance ℓ_o, we first make a canonical transformation to a frame that rotates with the resonant frequency. This is accomplished by means of the generating function $F_2 = (z + \ell_o\theta/k_{\|i} - \pi/2k_{\|i} - \omega_i t/k_{\|i})\hat{P}_\| + \theta\hat{J}$. The new variables are related to the old variables by

$$\hat{z} = z + \ell_o\theta/k_{\|i} - \pi/2k_{\|i} - \omega_i t/k_{\|i}, \quad (B2a)$$

$$\hat{\theta} = \theta, \quad (B2b)$$

$$\hat{P}_\| = P_\|, \quad (B2c)$$

and

$$\hat{J} = J - \ell_o P_\|/k_{\|i}, \quad (B2d)$$

Averaging over the fast angle $\hat{\theta}$, we finally obtain

$$H = \frac{1}{2m}(P_\perp{}^2 + P_\|{}^2) - \frac{\omega_i}{k_{\|i}}P_\| + Z_{\ell_o} \cos(k_{\|i}\hat{z}). \quad \text{(B3)}$$

Since H is now independent of $\hat{\theta}$, $\hat{J}$ = constant, or

$$\hat{J} = \hat{P}_\perp^2 = P_\perp{}^2 - 2m\Omega\ell_o P_\|/k_{\|i} = \text{constant}. \quad \text{(B4)}$$

The trapping width at resonance ℓ_o can be easily found as in the one wave case [*Karimabadi et al.*, 1992] and is given by

$$\Delta p_\| = 2|\mathcal{M} Z_{\ell_o}|^{1/2}, \quad \text{(B5a)}$$

with

$$\mathcal{M}^{-1} = (1 - \frac{1}{N_\|^2})/m, \quad \text{(B5b)}$$

and $Z_{\ell_o} = Z_{\ell_o}(P_{\|o}, P_{\perp o})$ where $P_{\|o}$ and $P_{\perp o}$ are obtained from the resonance condition $\omega - k_{\|i} v_\| = \ell_o \Omega$ and equation (B4), respectively. Here, $N_i = ck_i/\omega_i$ is the index of refraction of the ith-wave and $N_{\|i} = N_i \cos \alpha_i$.

The equations in this section apply both to the one wave case as well as an arbitrary wave spectra. There is, however, a significant difference between the one wave case and the more general case. In the case of one wave, the Hamiltonian has the shape of a semi-circle in $v_\perp - v_\|$ plane and intersects a given resonance line at a single point $(P_{\|o}, P_{\perp o})$. Given the Hamiltonian and the resonance condition one can then calculate the trapping width using equation (B5). In the general case, however, the Hamiltonian covers the entire $v_\perp - v_\|$ plane and intersects the resonance planes not at a unique point but at infinite number of points. Now $\hat{J}$ plays the role that the Hamiltonian plays in the one wave case; i.e. each particle has associated with it a given value of the adiabatic invariant $\hat{J}$. Once the value of $\hat{J}$ is known, one can use the resonance condition along with equation (A4) to determine $P_{\|o}, P_{\perp o}$ and hence the trapping width.

REFERENCES

Abramowitz, M., and I. A. Stegun, *Handbook of Mathematical Functions*, U.S. Government Printing Office, Washington, D.C., 1964.

Galeev, A. A., and R. Z. Sagdeev, Alfvén waves in space plasma and its role in the solar wind interaction with comets, *Astro. Space Sci., 144*, 427, 1987.

Gary, S. P., Electromagnetic ion/ion instabilities and their consequences in space plasmas: A Review, *Space Science Reviews, 56*, 373, 1991.

Gary, S. P., et al., Computer simulations of two-pickup-ion instabilities in a cometary environment, *J. Geophys. Res., 93*, 9584, 1988.

Gary, S. P., K. Akimoto, and D. Winske, Computer simulations of cometary ion/ion instabilities and wave growth, *J. Geophys. Res., 94*, 3513, 1989.

Gary, S. P., R. H. Miller, and D. Winske, Pitch-angle scattering of cometary ions: computer simulations, *Geophys. Res. Lett., 18*, 1067, 1991.

Gary, S. P., and C. D. Madland, Electromagnetic ion instabilities in a cometary environment, *J. Geophys. Res., 93*, 235, 1988.

Glassmeier, K. H., et al., Spectral characteristics of low frequency plasma turbulence upstream of comet P/Halley, *J. Geophys. Res., 94*, 37, 1989.

Goldstein, M. L., A. J. Klimas, and G. Sandri, Mirroring in the Fokker-Planck coefficient for cosmic-ray pitch-angle scattering in homogenous magnetic turbulence, *Astrophys. Journal, 195*, 787, 1975.

Jones, F. C., T. J. Birmingham, and T. B. Kaiser, Investigation of resonance integrals occuring in cosmic-ray diffusion theory, *Astrophys. J., 180*, L139, 1973.

Karimabadi, H., et al., Particle acceleration by a wave in a strong magnetic field: regular and stochastic motion, *Phys. Fluids, B2*, 606, 1990.

Karimabadi, H., and C. R. Menyuk, A fast and accurate method of calculating particle diffusion: Application to the ionosphere, *J. Geophys. Res., 96*, 9669, 1991.

Karimabadi, H., D. Krauss-Varban, and T. Terasawa, Physics of pitch angle scattering and velocity diffusion 1. Theory *J. Geophys. Res., 97*, 13,853, 1992.

Karney, C. F. F., Long-time correlations in the stochastic regime, *Physica, 8D*, 360, 1983.

Kennel, C. F., and F. Engelmann, Velocity space diffusion from weak plasma turbulence in a magnetic field, *Phys. Fluids, 9*, 2377, 1966.

Klimas, A. J., and G. Sandri, A rigorous cosmic-ray transport equation with no restrictions on particle energy, *Astrophys. J., 180*, 937, 1973.

Krauss-Varban, D., N. Omidi, and K. B. Quest, Mode properties of low-frequency waves: kinetic theory versus Hall-MHD, *J. Geophys. Res.,* , submitted, 1993.

Krommes, J. A., Statistical descriptions and plasma physics, in *Basic Plasma Physics, vol. II*, edited by A. A. Galeev, and R. N. Sudan, North-Holland, New York, 1984.

Lee, M. A., Coupled hydromagnetic wave excitation and ion acceleration upstream of the Earth's bow shock, *J. Geophys. Res., 87*, 5063, 1982.

Lee, M. A., Coupled hydromagnetic wave excitation and ion acceleration at interplanetary traveling shocks, *J. Geophys. Res., 87*, 5063, 1983.

Lee, M. A., and W.-H. Ip, Hydromagnetic wave excitation by ionized interstellar hydrogen and helium in the solar wind, *J. Geophys. Res., 92*, 11,041, 1987.

Maasjost, W., and K. Elsässer, *J. Stat. Phys., 28*, 793, 1982.

Miller, R. H., et al., Pitch-angle scattering of cometary ions into monospherical and bispherical distributions, *Geophys. Res. Lett., 18*, 1063, 1991.

Neugebauer, M., et al., The velocity distributions of cometary protons picked up by the solar wind, *J. Geophys. Res., 94*, 5227, 1989.

Omidi, N., et al., Generation and nonlinear evolution of oblique magnetosonic waves: application to foreshock and comets, *Proceedings of Yosemite meeting*, submitted, 1993.

Omura, Y., M. Ashour-Abdalla, R. Gendrin, and K. Quest,

Heating of thermal helium in the equatorial magnetosphere, *J. Geophys. Res.*, *90*, 8281, 1985.

Price, C. P., Mirror waves driven by newborn ion distributions, *J. Geophys. Res.*, *94*, 15,001, 1989.

Roberts, C. S., and S. J. Buchsbaum, Motion of a charged particle in a constant magnetic field and a transverse electromagnetic wave propagating along the field, *Physical Review*, *135*, A381, 1964.

Russell, C. T., et al., Mirror instability in the magnetosphere of comet Halley, *Geophys. Res. Lett.*, *14*, 644, 1987.

Sckopke, N., et al., Ion thermalization in quasi-perpendicular shocks involving reflected ions, *J. Geophys. Res.*, *95*, 6337, 1990.

Smith, C. W., and M. A. Lee, Coupled hydromagnetic wave excitation and ion acceleration upstream of the Jovain bow shock, *J. Geophys. Res.*, *91*, 81, 1986.

Smith, E. J., et al., Waves in the Giacobini-Zinner magnetosheath: ICE observations, in *Proc. of Chapman Conference, Plasma Waves and Instabilities in Magnetospheres and at Comets*, edited by H. Oya and B. T. Tsurutani, Sohbon Insatu, Sendai, Japan, 1987.

Tanaka, M., Simulations of heavy ion heating by electromagnetic ion cyclotron waves driven by proton temperature anisotropies, *J. Geophys. Res.*, *90*, 6459, 1985.

Terasawa, T., Acceleration mechanisms for cometary ions, in Cometary Plasma Processes, *Geophys. Monograph*, *61*, p. 277, 1991.

Tsurutani, B. T., et al., Steepened magnetosonic waves at comet Giacobini-Zinner, *J. Geophys. Res.*, *92*, 11,074, 1987.

Tsurutani, B. T., et al., Low frequency plasma waves and ion pitch angle scattering at large distances ($\xi\ 3.5 \times 10^5$ km) from Giacobini-Zinner: Interplanetary magnetic field α dependencies, *J. Geophys. Res.*, *94*, 18, 1989.

Winske, D., and S. P. Gary, Electromagnetic instabilities driven by cool heavy ion beams, *J. Geophys. Res.*, *91*, 6825, 1986.

Winske, D., and N. Omidi, Hybrid Codes: methods and application, in *Computer Simulation of Space Plasmas-Selected Lectures from ISSS-4*, Nara, Japan, 1992.

Wu, C. S., and R. C. Davidson, Electromagnetic instabilities produced by neutral-particle ionization in interplanetary space, *J. Geophys. Res.*, *77*, 5399, 1972.

Wu, C. S., D. Winske, and J.D. Gaffey, Jr., Rapid pickup of cometary ions due to strong magnetic turbulence, *Geophys. Res. Lett.*, *13*, 865, 1986.

H. Karimabadi, Department of Electrical and Computer Engineering, University of California, San Diego, La Jolla, CA 92093-0407.

N. Omidi, Department of Electrical and Computer Engineering and California Space Institute, University of California, San Diego, La Jolla, CA 92093-0407.

S. P. Gary, Group SST-7, Los Alamos National Laboratory, Los Alamos, NM 87545.

Axisymmetric Modeling of Cometary Mass Loading on an Adaptively Refined Grid: Hydrodynamic Results

Tamas I. Gombosi

Space Physics Research Laboratory, Department of Atmospheric, Oceanic and Space Sciences, University of Michigan, Ann Arbor

and

Kenneth G. Powell

Department of Aerospace Engineering, The University of Michigan, Ann Arbor

The first results of axisymmetric model of the interaction of an expanding cometary atmosphere with the solar wind are presented. The governing equations are solved on an adaptively refined unstructured grid using a Godunov-type numerical technique. The combination of the adaptive refinement with the Godunov scheme allows the entire cometary atmosphere to be modeled, while still resolving the shock and the comet nucleus. It was found that (i) a shock is formed at about 0.4 million km upstream of the comet; (ii) a pressure maximum is formed about halfway between the shock and the comet; and (iii) a contact surface is formed at about 15 comet radii upstream of the nucleus separating an outward expanding cometary ionosphere from the nearly stagnating solar wind flow.

1. INTRODUCTION

A well developed cometary atmosphere extends to distances some six orders of magnitude larger than the size of the nucleus. It is the mass loading of the solar wind with newly created cometary ions, from this extended exosphere, that is responsible for the interaction with the solar wind.

Mass loading occurs when a magnetized plasma moves through a continously-ionized background of neutral particles. Photoionization and electron impact ionization result in the addition of plasma to the plasma flow, while charge exchange replaces fast ions with almost stationary ones. "Ion pickup" (or ion implantation) is the process of accommodation (but not thermalization) of a single newborn ion to the plasma flow. It should be noted that photo- and impact-ionization add new mass to the plasma flow. Charge exchange can either add mass (such as the $H^+ + O \rightarrow H + O^+$ reaction) to the plasma or can act as a mass loss mechanism (for example $O^+ + H \rightarrow O + H^+$). The combined effect of the various ionization processes is usually net mass addition. Conservation of momentum and energy requires that the plasma flow be decelerated as newly born charged particles are "picked up". The process of continuous ion pickup and its feedback to the plasma flow is called "mass loading". The physics of mass loading in space plasmas has been extensively studied in the last quarter century (for details see the recent review by *Gombosi* [1991]).

Neutral atoms and molecules of cometary origin move along ballistic trajectories (with velocities ranging from ~0.5 km/s to a few tens of km/s) and become ionized with a characteristic ionization lifetime of $10^5 - 10^7$ seconds (cf. *Gombosi* [1986], *Mendis et al.*, [1985]). In the first study of the mass loading process *Biermann et al.*, [1967] assumed that the plasma flow rapidly accommodates the new ions, i.e. the entire plasma population can be described by a single drifting Maxwellian velocity distribution. This assumption made it possible to apply a one dimensional single-fluid hydrodynamic treatment for the continuously mass loaded supersonic and superalfvenic plasma flow and to use the conservation equations to describe the deceleration of the contaminated solar wind flow. *Biermann et al.*, [1967] had also shown that continuous deceleration of the solar wind flow by mass loading is possible only up to a certain point at which the mean molecular weight of the plasma particles reaches a critical value. At this point a weak shock forms and impulsively decelerates the flow to subsonic velocities.

In the last decade several large scale, multidimensional models have been developed to describe the solar wind interaction with an expanding cometary atmosphere (cf. *Schmidt and Wegmann* [1982], *Fedder et al.*, [1986], *Schmidt-Voigt* [1987], *Schmidt et al.*, [1988], *Huebner et al.*, [1991]). These models significantly advanced our understanding of the global nature of the interaction. However, these large scale MHD models suffer from three general limitations typical of the present generation of global MHD calculations: *(i)* the modeled volume is not large enough to include the entire upstream mass loading region, the flaring shock and the long antisunward plasma tail; *(ii)* the resolution is not fine enough to resolve many important plasma regions (such as the diamagnetic cavity, the shock structure, or the cometopause); and *(iii)* it is difficult to choose appropriate boundary conditions.

This paper presents the first results of a new generation of global models which solves the governing equations of solar wind mass loading on an adaptively refined unstructured grid. Our new model successfully addresses all three major weaknesses of earlier calculations: *(i)* it models a 0.1 AU × 0.1 AU × 0.1 AU volume, which is large enough to encompass all physically important regions; *(ii)* it uses very high resolution in high gradient regions, such as the shock, the stagnation region, and near the nucleus (as small as ~1 km near the nucleus) and is capable of resolving the shock and the inner cometary ionosphere; and *(iii)* it uses finite volume Godunov-type schemes which are robust and accurate and minimize numerical dissipation and dispersion, faithfully represent shocks and other discontinuities, and, most importantly, provide a framework in which boundary conditions can be applied in a systematic and appealing way. The adaptive model presented in this paper is the first step in developing a new generation of global modeling tools for cometary (and space) plasma physics: presently it is only 2 dimensional, steady-state and neglects the magnetic field. In the future we are planning to develop a 3D full MHD version of the present model.

2. MODEL EQUATIONS

Implanted ions in the cometary environment are mainly water group ions of cometary origin. Before being ionized these particles were escaping the comet with velocities significantly smaller than the solar wind speed. Freshly born ions are accelerated by the motional electric field of the high-speed solar wind flow. The ion trajectory is cycloidal, resulting from the superposition of gyration and E×B drift. The resulting velocity-space distribution is a ring-beam distribution, where the gyration speed of the ring is $v_\perp = u \sin \alpha$, (where u is the bulk plasma speed and α is the angle between the solar wind velocity and magnetic field vectors) and the beam velocity (along the magnetic field line) is $v_\parallel = u \cos \alpha$. The ring beam distribution has large velocity space gradients and it is unstable to the generation of low frequency transverse waves. In the supersonic, superalfvenic solar wind flow the frequency of the excited waves is much lower than the cyclotron frequency of implanted cometary ions, which in a first approximation interact with these low frequency waves without significantly changing their energy (in the plasma frame of reference). As a result of this wave-particle interaction process the pitch angles of the pickup-ring particles become randomized in the solar wind frame of reference and they become distributed on a velocity space shell around the local solar wind velocity.

Neglecting the velocity of the outflowing cometary neutrals the governing equation for the phase-space distribution function of implanted ions can be written in the following form (cf. *Gombosi* [1988]):

$$\frac{\partial f}{\partial t} + (\mathbf{u} \cdot \nabla)f + (\mathbf{v} \cdot \nabla)f - \left[\frac{\partial \mathbf{u}}{\partial t} + (\mathbf{u} \cdot \nabla)\mathbf{u} + (\mathbf{v} \cdot \nabla)\mathbf{u} - \frac{e}{m}(\mathbf{v} \times \mathbf{B})\right] \cdot \nabla_v f$$
$$= \frac{Q \exp\left(-\frac{r_c}{\lambda}\right)}{4\pi \lambda r_c^2} \delta(\mathbf{u} + \mathbf{v}) \qquad (1)$$

where t is time, $\mathbf{u}$ is the plasma bulk velocity, $\mathbf{v}$ is the implanted ion random velocity, $f(t,\mathbf{r},\mathbf{v})$ is the implanted ion phase space distribution function in the plasma frame, e and m are the charge and mass of implanted ions, $\mathbf{B}$ is the magnetic field vector, Q is the cometary neutral gas production rate, r_c is cometocentric distance, and λ is the ionization scale length of cometary neutrals. In equation (1) the wave-particle interaction effects were also neglected.

The transport model to be used in this paper is a modified version of the model of *Gombosi et al.*, [1988; 1991], which assumes that the freshly ionized cometary particles form a pickup shell distribution in the magnetized solar wind plasma. Even though this assumption is not quite justified, it represents a reasonable first approximation and greatly simplifies the mathematical problem (see the discussion by *Gombosi et al.*, [1991]). This assumption means that the implanted ion distribution is considered to be isotropic in the random velocity, $f=f(t,\mathbf{r},v)$. In this case we obtain

$$\nabla_v f = \frac{\mathbf{v}}{v}\frac{\partial f}{\partial v} \quad (2)$$

Substituting Equation (2) into (1) one obtains the following transport equation for the implanted ion distribution function:

$$\frac{\partial f}{\partial t} + (\mathbf{u}\cdot\nabla)f + (\mathbf{v}\cdot\nabla)f - \left[\frac{\partial \mathbf{u}}{\partial t} + (\mathbf{u}\cdot\nabla)\mathbf{u} + (\mathbf{v}\cdot\nabla)\mathbf{u}\right]\cdot\frac{\mathbf{v}}{v}\frac{\partial f}{\partial v}$$

$$= \frac{Q\exp\left(-\frac{r_c}{\lambda}\right)}{4\pi\lambda r_c^2}\delta(\mathbf{u}+\mathbf{v}) \quad (3)$$

It is important to note that Equation (3) is independent of the magnetic field: this feature is a direct consequence of the directional isotropy of the velocity distribution.

In the next step we multiply equation (3) by m, $m\mathbf{v}$, and mv^2, respectively, and integrate the resulting equations over the entire velocity space. This operation results in the following set of moment equations:

$$\frac{\partial \rho_c}{\partial t} + (\mathbf{u}\cdot\nabla)\rho_c + \rho_c(\nabla\cdot\mathbf{u}) = \frac{mQ}{4\pi\lambda r_c^2}\exp\left(-\frac{r_c}{\lambda}\right) \quad (4)$$

$$\rho_c\frac{\partial \mathbf{u}}{\partial t} + \rho_c(\mathbf{u}\cdot\nabla)\mathbf{u} + \nabla p_c = -\mathbf{u}\frac{mQ}{4\pi\lambda r_c^2}\exp\left(-\frac{r_c}{\lambda}\right) \quad (5)$$

$$\frac{\partial p_c}{\partial t} + (\mathbf{u}\cdot\nabla)p_c + \frac{5}{3}p_c(\nabla\cdot\mathbf{u}) = \frac{u^2}{3}\frac{mQ}{4\pi\lambda r_c^2}\exp\left(-\frac{r_c}{\lambda}\right) \quad (6)$$

where ρ_c and p_c are the mass density and pressure of implanted cometary ions.

Assuming temperature isotropies and neglecting losses, the macroscopic parameters of the ambient solar wind are determined by the following transport equations:

$$\frac{\partial \rho_{sw}}{\partial t} + (\mathbf{u}\cdot\nabla)\rho_{sw} + \rho_{sw}(\nabla\cdot\mathbf{u}) = 0 \quad (7)$$

$$\rho_{sw}\frac{\partial \mathbf{u}}{\partial t} + \rho_{sw}(\mathbf{u}\cdot\nabla)\mathbf{u} + \nabla p_{sw} = 0 \quad (8)$$

$$\frac{\partial p_{sw}}{\partial t} + (\mathbf{u}\cdot\nabla)p_{sw} + \frac{5}{3}p_{sw}(\nabla\cdot\mathbf{u}) = 0 \quad (9)$$

where ρ_{sw} and p_{sw} are the mass density and pressure of the solar wind. It should be noted that in the present model the flow velocities of the solar wind plasma and of the implanted cometary ions are assumed to be the same. This assumption makes it possible to combine the two sets of transport equations and obtain a full set of flow equations for the mass loaded plasma flow. After some manipulation the combined equations can be written in the following conservative form:

$$\frac{\partial \underline{w}}{\partial t} + \left(\nabla\cdot\underline{\underline{F}}\right)^T = \underline{P} \quad (10)$$

where $\underline{w}$ and $\underline{P}$ are the five dimensional state and source vectors, and $\underline{\underline{F}}$ is a 3×5 dimensional flux diad. These quantities are defined as follows:

$$\underline{w} = \begin{bmatrix} \rho \\ \rho u_1 \\ \rho u_2 \\ \rho u_3 \\ E \end{bmatrix} \qquad \underline{P} = \begin{bmatrix} S \\ 0 \\ 0 \\ 0 \\ 0 \end{bmatrix}$$

$$\underline{\underline{F}} = \begin{bmatrix} \rho u_1 & \rho u_1^2 + p & \rho u_2 u_1 & \rho u_3 u_1 & u_1(E+p) \\ \rho u_2 & \rho u_1 u_2 & \rho u_2^2 + p & \rho u_3 u_2 & u_2(E+p) \\ \rho u_3 & \rho u_1 u_3 & \rho u_2 u_3 & \rho u_3^2 + p & u_3(E+p) \end{bmatrix} \quad (11)$$

where the mass density, pressure, and total energy density are $\rho = \rho_c + \rho_{sw}$, $p = p_c + p_{sw}$, and $E = 3p/2 + \rho\mathbf{u}\cdot\mathbf{u}/2$, respectively, while the mass addition rate is given by

$$S = \frac{mQ}{4\pi\lambda r_c^2}\exp\left(-\frac{r_c}{\lambda}\right) \quad (12)$$

Equation (10) represents five differential equations describing the plasma flow near active comets. It is important to note that even though formally Equation (10) looks like a set of single–fluid Euler equations, we never assumed that implanted ion become thermalized (i.e., that the velocity distribution of the mass loaded solar wind plasma can be described by a single Maxwellian). The only assumption we made was that the velocity distribution of the implanted ions is isotropic in the plasma frame.

3. SOLUTION ON A 2D ADAPTIVE GRID

The numerical procedure used in solving Equation (10) is built from two pieces that are extremely well suited to the cometary mass–loading problem. The first is a data structure that allows for adaptive refinement of the mesh in regions of interest, and the second is a second–order MUSCL–type scheme [*van Leer*, 1979] based on an approximate Riemann solver [*Roe*, 1981]. The scheme is described in detail elsewhere [*De Zeeuw and Powell*,

1992]; a brief justification and description of the approach is provided here.

Adaptive refinement and coarsening of a mesh is a very attractive way to make optimal use of computational resources. It becomes particularly attractive for problems in which there are disparate spatial scales. For problems like the cometary mass-loading problem, in which the ionization length scale and the radius of the comet differ by several orders of magnitude, an adaptive mesh is a virtual necessity. The primary difficulty in implementing an adaptive scheme is related to the way in which the solution data are stored; the data structure must be much more flexible than the simple array-type storage used in the majority of scientific computations.

The basic data structure used is a hierarchical cell-based quadtree structure where "parent" cells are refined by division into four "children" cells. This concept is illustrated in Figure 1. Each cell has a pointer to its parent cell (if one exists) and to its four children cells (if they exist). The cells farthest down the hierarchy, that is, the ones with no children, are the cells on which the calculation takes place. This tree structure contains all the connectivity information necessary to carry out the flow calculations; no other connectivity information is stored. While it involves a more sophisticated implementation than a simple array-type data structure, it is an efficient and natural way to implement adaptive refinement and coarsening.

The mesh is generated in such a way that important geometric and flow features are resolved. The geometry is resolved by recursively dividing cells in the vicinity of the comet until a specified cell-size is obtained near the nucleus. A larger cell-size is specified for the remainder of the mesh. Flow features are resolved by obtaining a solution on this original mesh, and then automatically refining cells in which the flow gradients are appreciable, and coarsening cells in which the flow gradients are negligible. This is done by evaluating $\nabla \cdot \mathbf{u}$ and $\nabla \times \mathbf{u}$ in each cell, and refining cells in which either value is large, and coarsening cells in which both values are small. This automatic procedure leads, after several applications, to the grid shown in Figure 1.b.

The resulting grid has small cells next to large cells, and non-quadrilateral cells in the vicinity of the comet. A numerical procedure that is both accurate and stable on a mesh of this type cannot be developed from simple finite-difference ideas, but must be based on techniques developed for unstructured meshes [*Barth*, 1990]. The gradient calculation procedure that makes the scheme second order, and the limiting procedure that renders it monotonicity-preserving, are described elsewhere [*De Zeeuw and Powell*, 1992]. Basically, the scheme is a finite-volume scheme, in which the governing equations $\partial \underline{w}/\partial t = -\nabla \cdot \underline{\underline{F}} + \underline{P}$ are integrated over a cell of the mesh, and Gauss' theorem is invoked, yielding

$$\frac{dw_i}{dt} = -\frac{1}{\mathcal{V}_i} \sum_{j=\text{faces}} F_{ij}^T \cdot \mathbf{n}_j \, dS + P_i \quad (13)$$

where the subscript "i" denotes the i-th cell of the mesh, and $\mathcal{V}_i$ is the cell's volume and dS is the area of the cell boundary. Equation (13) represents a set of coupled ordinary differential equations in time, which can be solved by a suitable numerical integration procedure. In this work, an optimally-smoothing multi-stage scheme [*van Leer et al.*, 1989] is used. To carry out the multi-stage time-stepping procedure for Equation (13), an evaluation of the flux tensor at the interfaces between cells of the mesh is required. This evaluation is done by means of an "approximate Riemann solver," particularly, that of Roe [*Roe*, 1981]. This flux function is evaluated based on the states in the two cells that share the interface.

The advantages of the flux-based approach described above are:
- The flux function is based on the eigenvalues and eigenvectors of the Jacobian of $\underline{\underline{F}}$ with respect to $\underline{w}$, leading to an upwind-differencing scheme which respects the physics of the problem being solved;
- The scheme provides capturing of shocks and other high-gradient regions without oscillations in the flow variables;
- The scheme has just enough dissipation to provide a non-oscillatory solution, and no more;
- The scheme provides a physically consistent way to implement boundary conditions that are stable and accurate – the comet simply becomes a no-flux surface, and a physically consistent flux can be calculated at the far-field boundaries by use of the approximate Riemann solver.

These advantages are apparent in the solution shown in Figures 1 through 7: the shock is captured over a small number of points and without any oscillations, and there are no discontinuities in the solution near the boundaries of the computational domain

4. RESULTS AND DISCUSSION

Equation (10) was solved in dimensionless quantities for a Halley class comet. Normalization was done with the help of the ionization scale length, λ, the upstream mass density, ρ_∞, and the upstream acoustic speed, $a_\infty = (5 p_\infty / 3 \rho_\infty)^{1/2}$: we used $u' = u/a_\infty$, $\rho' = \rho/\rho_\infty$, $p' = p/\rho_\infty a_\infty^2$. Time and distance were normalized to $t' = t a_\infty / \lambda$ and $r' = r/\lambda$, respectively, while the normalized source term was given by

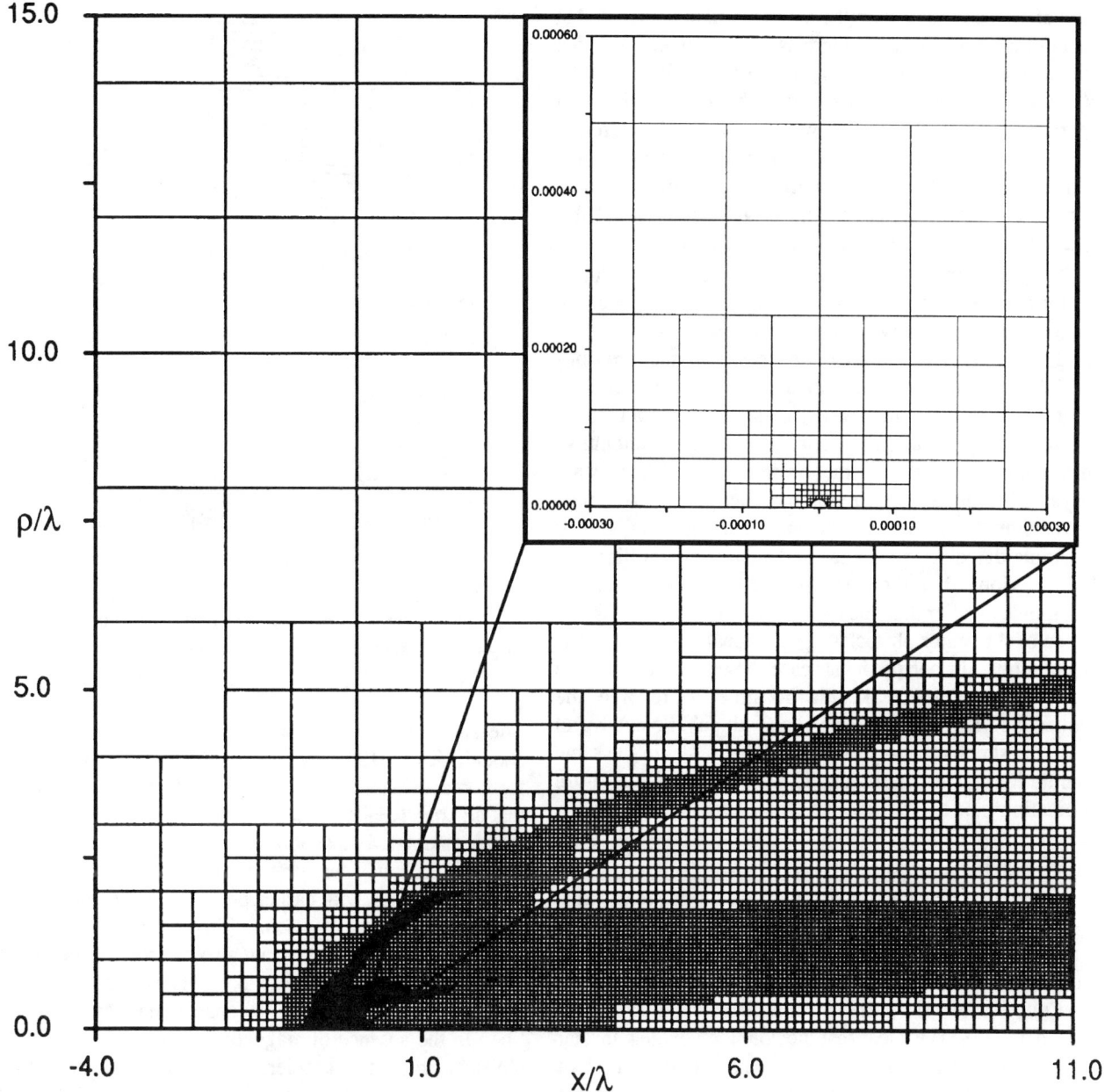

Fig. 1: The grid structure under steady-state conditions, which already incorporates a large amount of information about the interaction itself. All distances are measured in units of ionization scale length, λ. A simple conversion to a Halley type case can be obtained by assuming that distance is measured in units of 1 million km. About half of the entire simulation box of 15×30 units is shown in Fig. 1, while the insert shows the blow up of the near nucleus grid structure (the radius is assumed to be $10^{-5}\lambda$).

$$S' = \frac{mQ}{4\pi\rho_\infty a_\infty \lambda^2} \frac{e^{-r_c'}}{r_c'^2} = S_0 \frac{e^{-r_c'}}{r_c'^2} \qquad (14)$$

Assuming water group ions (mass 16), a gas production rate of $10^{30}\,s^{-1}$, ionization scale length of 10^6 km, upstream solar wind density and temperature of 10 cm^{-3} and 10^5 K, respectively, the dimensionless source coefficient turns out to have a value around $S_0 = 5$. A value of $S_0 = 5$ was used in the present calculation.

As a first step to apply adaptive grid methodology to the problem of cometary mass loading we considered a rotationally symmetric scenario in which explicit magnetic field effects are neglected (the magnetic field is included in an implicit way, because there is no ion pickup without magnetic field). This is clearly a significant simplification of the real scenario, but it seems to be a reasonable first approximation of the problem which realistically addresses the problems of solar wind deceleration before the shock, and the shock itself. Magnetic field effects become important close to the comet in the stagnation region: in this region our results must be considered with care.

It is assumed that far upstream the solar wind velocity vector is parallel to the sun–comet axis and that the entire solution is rotationally symmetric around this axis. We used the numerical method described in the previous section to obtain a solution to the set of equations (10). Figure 1 shows the final grid structure under steady-state conditions, which already incorporates a large amount of information about the interaction itself. All distances are measured in units of ionization scale length, λ. A simple conversion to a Halley type case can be obtained by assuming that distance is measured in units of 1 million km. It is obvious that the nucleus size is many orders of magnitude smaller than the ionization scale length; this fact is one of the major obstacles in traditional cometary MHD calculations. In our particular case, however, the adaptive grid methodology makes it possible to resolve the immediate vicinity of the nucleus (with subkilometer grid size), the shock structure, and the global scale, in the same calculation. The figure shows about half of the entire simulation box of 15×30 units (for Halley conditions this would correspond to a simulation volume of 0.1 AU $\times$ 0.2 AU), while the insert shows the blow up of the near nucleus grid structure. One can see that we have a quite reasonable resolution even right at the surface of the nucleus (the radius is assumed to be $10^{-5}\lambda$). It should be noted that as a result of the careful use of grid coarsening and refining, the total number of grid points in this 2D calculation is only around 17,000. It is clear that the solution contains far, far more detail than a solution from a non-adapted grid of 200x100, which would require approximately the same amount of memory and CPU to compute. (It is interesting to note that the present model is running on an IBM RISC workstation and not on a supercomputer.) This is the advantage of the adaptive-mesh approach: the ability, with modest computation resources, to resolve physics that could not be resolved even with substantial supercomputer resources on a non-adapted mesh.

The flow vector field in the coma is shown in Figure 2. The arrows represent the flow velocity vectors: their direction and length characterizes the direction and magnitude of the plasma velocity. Inspection of Figure 2 reveals that a shock wave is formed upstream of the comet.

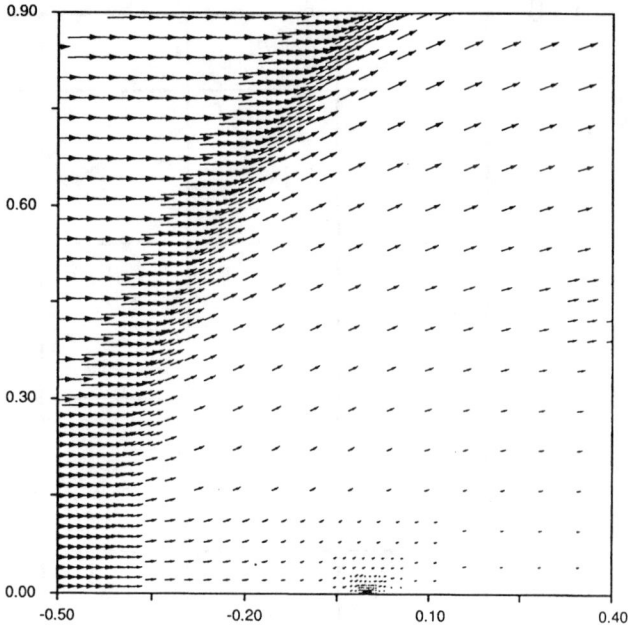

Fig. 2: The flow vector field in the coma. The arrows represent the flow velocity vectors: their direction and length characterizes the direction and magnitude of the plasma velocity.

The subsolar and terminator distances of the shock are about 0.4λ and 0.9λ, respectively (0.4 Mkm and 0.9 Mkm for comet Halley conditions). It is interesting to mention that all comet Halley shock crossings are surprisingly close to our calculated mass loading shock. It can be seen that the plasma flow is decelerated and diverted at the shock, in good accordance with earlier predictions [*Biermann et al.*, 1967; *Schmidt and Wegmann*, 1982; *Schmidt-Voigt*, 1987; *Wegmann et al.*, 1987]. It can also be seen that behind the shock the continuously mass loaded plasma flow gradually decelerates and eventually stagnates near the sun–comet line. In the absence of magnetic field there is no efficient physical process to accelerate the plasma in the tail (the pressure gradient force can not accelerate to solar wind velocities) and therefore a practically stagnating plasma tail is formed. This phenomenon is a direct consequence of the neglect of the solar wind magnetic field: in magnetized cases one can expect the tail plasma to be accelerated to near solar wind velocities. This process will be realistically modeled in future calculations.

A contour plot of the magnitude of the plasma velocity is shown in Figure 3. The plot shows the shock, the nearly stagnating inner coma and the tail, discussed above. Another interesting feature seen in Figure 3 are the nearly concentric contour lines upstream of the shock. These lines show that the plasma starts decelerating due to mass

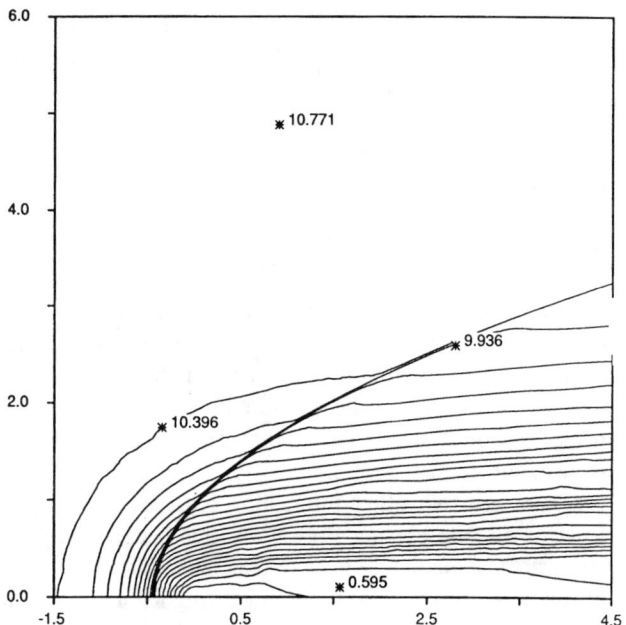

Fig. 3: Contour plot of the normalized magnitude of the plasma velocity.

loading well ahead of the shock. In effect, the shock occurs where the total amount of mass loading of a given flow line exceeds a small critical value, as discussed by *Biermann et al.*, [1967] in their classical paper some 25 years ago.

Figure 4 shows contours of normalized temperature near the comet. The temperature increase upstream of the shock is due to the continuous addition of hot cometary ions. The temperature also increases behind the shock, because some of the flow energy is converted to thermal energy. In the magnetosheath the plasma gradually cools because of expansion. It is important to note the formation of a long, cold plasma tail behind the comet.

Figure 5 presents contour plots of the normalized pressure in the cometary interaction region. It can be seen that the pressure starts to increase far upstream of the shock due to the mass loading by heavy cometary ions. The rate of pressure increase is directly proportional to the square of the local plasma velocity, u^2, because the radius of the pickup shell is u, and therefore each newly ionized particle contributes $mu^2/3$ to the total kinetic pressure. It should be mentioned at this point that particles picked up in nearly stagnating plasma regions slightly increase the total kinetic pressure but cool the kinetic temperature (because their random speed, u, is smaller than the average thermal speed in the plasma).

The pressure further increases behind the shock, because a large fraction of the bulk energy is converted to random energy. It can be seen that there is a pressure maximum near the sun–comet axis between the shock and the comet. This phenomenon can be clearly seen in Figure 6 which shows the variation of pressure (Fig. 6.a) and flow velocity (Fig. 6.b) along the sun–comet line. Inspection of Figure 6 reveals the global structure of the shock and the stagnation

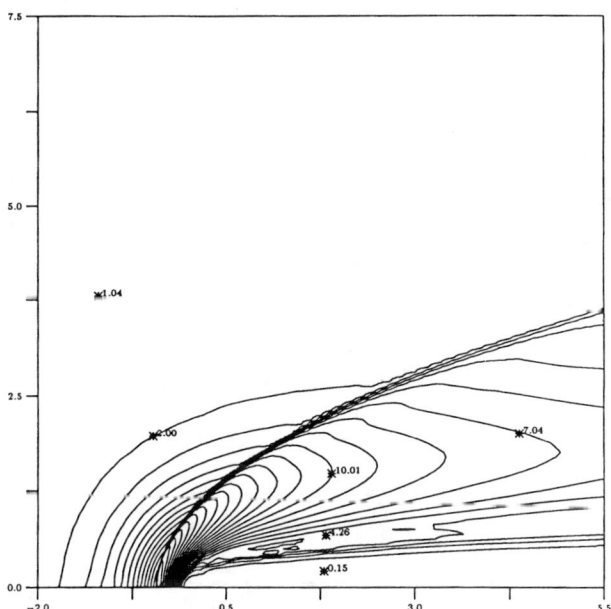

Fig. 4: Contour plots of the normalized temperature in the cometary interaction region.

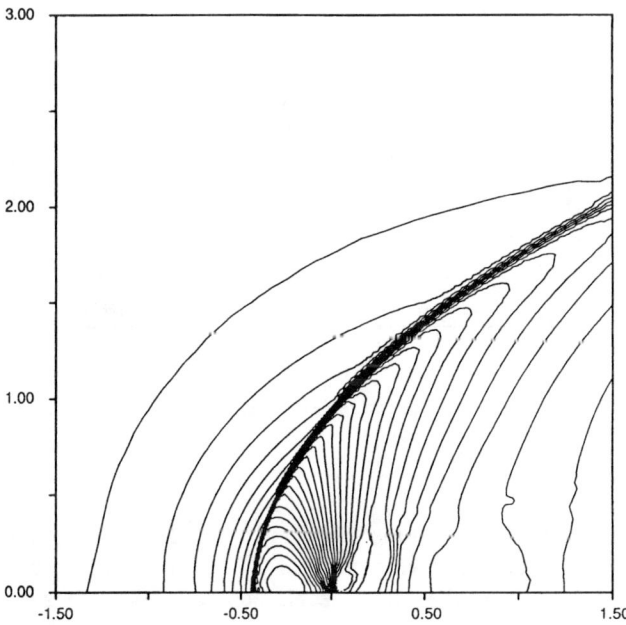

Fig. 5: Contour plots of the normalized pressure in the cometary interaction region.

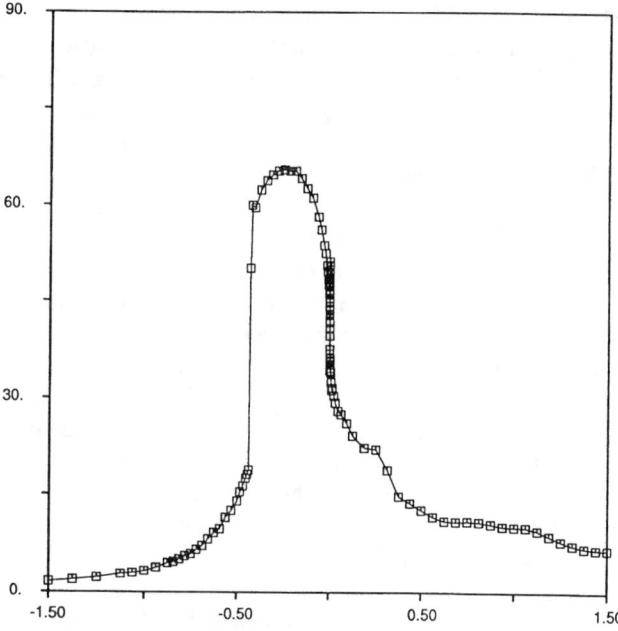

increases and the flow decelerates. In turn, ions picked up in the slower plasma flow represent a diminishing source term for the pressure and consequently the growth of pressure decreases significantly. The cumulative mass addition along plasma stream lines decreases as one moves away from the sun–comet line: this effect results in a pressure gradient force perpendicular to the axis. As a result, the plasma is expanding in the perpendicular direction, thus decreasing the pressure along the subsolar stream line. This effect, combined with the decreasing contribution of mass loading to the pressure results in a decrease of the kinetic pressure. This can be seen in Figure 6.a, which shows that the pressure along the subsolar line peaks at around the halfway point between the shock and the nucleus.

There are two other interesting features in Figure 6. First of all, the plasma flow velocities become negative (sunward) immediately near the nucleus. This effect will be discussed below. The other interesting feature is that the plasma pressure is continuously decreasing toward the tail resulting in plasma acceleration in the downstream direction.

Figure 7 shows a blow-up of the subsolar–line pressure distribution near the nucleus. A well pronounced pressure peak can be seen on both sides of the nucleus. These pressure peaks are due to very large mass loading rates in the immediate vicinity of the nucleus. The density of the

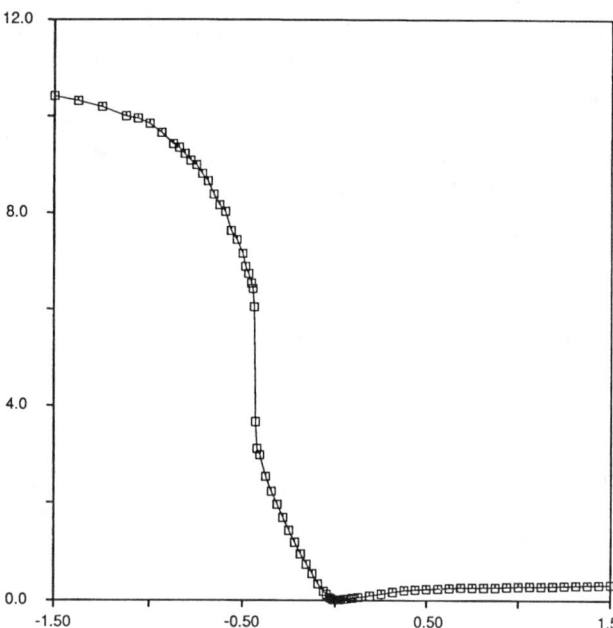

Fig. 6: Variation of pressure (Fig. 6.a) and flow velocity (Fig. 6.b) along the sun–comet line.

region behind it. As the solar wind approaches the shock it absorbs an increasing number of hot cometary ions: consequently the flow decelerates and the pressure increases. At the shock itself the velocity decreases and the pressure increases satisfying the Rankine–Hugoniot jump conditions. Behind the shock the mass loading rapidly

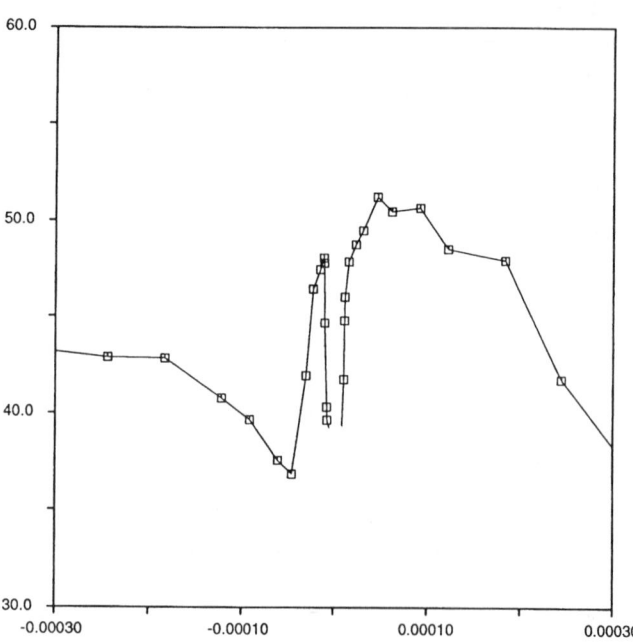

Fig. 7: Blow-up of the pressure distribution along the sun–comet line near the nucleus.

outflowing gas (and therefore the mass loading rate, too) decreases as the square of the cometocentric distance, r_c, therefore the pressure buildup is largest within a few cometary radii. It can be seen in Figure 7 that there is pronounced difference in the pressure profile upstream and downstream of the nucleus. In the downstream region the plasma flow velocity and the pressure gradient force point in the same direction, therefore the decreasing pressure buildup is gradually overcompensated by plasma expansion into the low pressure tail. This expansion results in a pressure peak at about 5 nucleus radii, followed by a gradual decrease in plasma pressure.

The situation is much more interesting and complicated upstream of the nucleus. First of all, the shocked plasma arriving from the upstream direction is very heavily mass loaded by cometary ions and, consequently, its flow velocity is quite small. The large-scale pressure profile is decreasing towards the nucleus due to the flow divergence discussed above (see Figure 6). On the other hand, as a result of the large density of outflowing cometary gas, there is a significant pressure buildup immediately upstream of the nucleus. This means that a well pronounced pressure minimum develops at about 3 nucleus radii upstream of the comet. The negative pressure gradient accelerates the plasma towards the sun (these negative velocities can be seen in Figure 6.b): this flow is gradually decelerated and diverted in the region of positive pressure gradient. The result is two colliding plasma streams which are separated by a contact surface. The diverted plasma is flowing along this contact discontinuity on both sides of the separator. This phenomenon can be clearly seen in Figure 8 which shows plasma stream lines in the immediate vicinity of the nucleus.

Inspection of Figure 8 reveals the flow pattern near the nucleus, showing sunward moving flow lines along the sun-comet axis. These flow lines eventually become diverted at the contact surface and the plasma turns towards the tail. The characteristic size of the contact surface is about 15 nucleus radii (150 km), much smaller that the observed diamagnetic cavity at comet Halley. The main reason for this discrepancy is twofold: we neglected the outflow velocity of cometary gas particles (and consequently the neutral drag on the ions) and there are no explicit magnetic field effects included in the calculation.

Finally, we would like emphasize again the advantage of using adaptive grid methodology which enabled us to achieve very high resolution in critical regions, while keeping the total number of grid points managable.

5. SUMMARY

This paper presents the first results of an axisymmetric model describing the effects of cometary mass loading on the solar wind. The model equations were numerically solved for a Halley class comet on an adaptively refined unstructured grid using ~17,000 grid points. The adaptive grid methodology made it possible to achieve very high resolution in the most interesting regions. The main findings of the present calculation are:

(i) A shock is formed at about 0.4 million km upstream of the comet. The location and the shape of the shock are in good agreement with observations at comet Halley.

(ii) A pressure maximum is formed about halfway between the shock and the comet. Upstream of this maximum mass loading continuously increases the plasma pressure, while downstream of the maximum the divergence of the plasma flow adiabatically cools the plasma.

(iii) A contact surface is formed at about 15 comet radii upstream of the nucleus separating an outward expanding cometary ionosphere from the nearly stagnating solar wind flow. This location of the contact surface is not realistic due to the neglect of the magnetic field pile-up effect.

This calculation represents a very promising first step towards the development of a new generation of cometary models. Next we are planning to develop a genuinly 3D MHD version of the model, which will account for magnetic field effects as well as include the effects of finite outflow velocity.

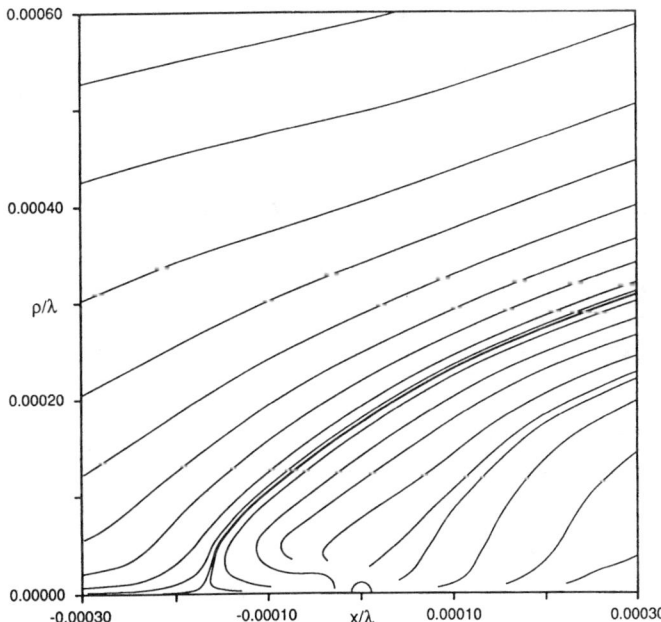

Fig. 8: Plasma flow lines near the nucleus.

Acknowwledgments. This work was supported by the NASA Planetary Atmospheres Program under grant number NAGW-1366.

REFERENCES

Barth, T.J., On unstructured grids and solvers, in *"Computational Fluid Dynamics", von Kármán Lecture Series 1990-03*, edited by H. Deconinck, pp. Rhode-St.-Genese, Belgium, 1990.

Biermann, L., Brosowski, B., and Schmidt, H.U., The interaction of the solar wind with a comet, *Solar Phys., 1*, 254, 1967.

De Zeeuw, D., and Powell, K.G., An adaptively-refined Cartesian mesh solver for the Euler equations, *J. Comp. Phys., 104*, 55, 1992.

Fedder, J.A., Lyon, J.G., and Giuliani, J., J.L., Numerical simulations of comets: Predictions for comet Giacobini-Zinner, *EOS, Trans. AGU, 67*, 17, 1986.

Gombosi, T.I., Preshock region acceleration of implanted cometary H^+ and O^+, *J. Geophys. Res., 93*, 35, 1988.

Gombosi, T.I., The plasma environment of comets, *Rev. Geophys. Suppl., 29*, 976, 1991.

Gombosi, T.I., Nagy, A.F., and Cravens, T.E., Dust and neutral gas modeling of the inner atmospheres of comets, *Rev. of Geophys., 24*, 667, 1986.

Gombosi, T.I., Neugebauer, M., Johnstone, A.D., Coates, A.J., and Huddleston, D.E., Comparison of observed and calculated implanted ion distributions outside comet Halley's bow shock, *J. Geophys. Res., 96*, 9467, 1991.

Huebner, W.F., Boice, D.C., Schmidt, H.U., and Wegmann, R., Structure of the coma: Chemistry and solar wind interaction, in *Comets in the Post-Halley Era*, edited by R.L. Newburn, M. Neugebauer and J. Rahe, pp. 907, Kluwer, Dordrecht, 1991.

Mendis, D.A., Houpis, H.L.F., and Marconi, M.L., The physics of comets, *Fund. Cosmic Phys., 10*, 1, 1985.

Roe, P.L., Approximate Riemann solvers, parameter vectors, and difference schemes, *J. Comput. Phys., 43*, 357, 1981.

Schmidt, H.U., Wegamnn, R., Huebner, W.F., and Boice, D.C., Cometary gas and plasma flow with detailed chemistry, *Comp. Phys. Comm., 49*, 17, 1988.

Schmidt, H.U., and Wegmann, R., Plasma flow and magnetic fields in comets, in *Comets*, edited by L.L. Wilkening, pp. 538, The University of Arizona Press, Tucson, Arizona, 1982.

Schmidt-Voigt, M., *Time dependent MHD models for the cometary magnetosphere*, in: Proc. of Symposium on the Diversity and Similarity of Comets, edited by E.J. Rolfe and B. Battrick, Vol. pp. ESA SP-278, Brussels, Belgium, 1987.

van Leer, B., Towards the ultimate conservative difference scheme, V. A second-order sequel to Godunov's method, *J. Comput. Phys., 32*, 101, 1979.

van Leer, B., Tai, C.H., and Powell, K.G., *Design of optimally-smoothing multi-stage schemes for the Euler equations*, in: Proc. of AIAA 9th Computational Fluid Dynamics Conference, edited by Vol. pp. 1989.

Wegmann, R., Schmidt, H.U., Huebner, W.F., and Boice, D.C., Cometary MHD and chemistry, *Astron. Astrophys., 187*, 339, 1987.

T. I. Gombosi, Space Research Laboratory, Department of Atmospheric, Oceanic and Space Sciences, University of Michigan, Ann Arbor, MI 48109-2143.

K. G. Powell, Department of Aerospace Engineering, University of Michigan, Ann Arbor, MI 48109-2140.

First High-Resolution Measurements by the Freja Satellite

R. LUNDIN[1], L. ELIASSON[1], O. NORBERG[1], G. MARKLUND[2], L. R. ZANETTI[3], B. A. WHALEN[4],
B. HOLBACK[5], J. S. MURPHREE[6], G. HAERENDEL[7], M. BOEHM[7], G. PASCHMANN[7]

The joint Swedish-German Freja satellite was successfully launched on October 6, 1992. Freja contains an extensive set of plasma and field instruments for high time-resolution measurements of the auroral plasma in the altitude range 600 - 1750 km. The satellite also contain two UV imagers to monitor the auroral activity at the geomagnetic footprint of the satellite. Freja is a low-cost project characterized by a minimum of management and operational costs. An interactive system is implemented enabling experimenters to have close contact with their experiments in-flight and to perform extensive uploading of flight software. After approximately 1.5 months testing of the scientific instruments normal data taking with all on-board instruments working simultaneously commenced.

In this paper we review the first few months of operation and the first scientific data from the Freja experiments. Of particular interest are the density cavities observed within the auroral energization region and the rather unique perspective obtained from the Freja orbit. We also discuss the great potentials of the high-time resolution measurements by Freja.

1. THE FREJA SATELLITE PROJECT

Freja is a continuation of the magnetospheric research that commenced with Viking, the first Swedish satellite launched in 1986. Like on Viking, participation in the instrumentation for Freja is spread over a number of countries, but the satellite is mainly funded by Sweden ($\approx$75%) and Germany ($\approx$25%). Operations are funded by Sweden and Canada. The extreme low cost approach on Freja follows a highly successful tradition, such as Viking and AMPTE (Germany). A piggyback launch (on the Chinese Long March II) and a very streamlined project organization was used to keep the total costs of the project to within 100 million Swedish kronor (corresponding to $\approx$15 million US dollars). Freja is equipped with 8 PI-instruments for particles, fields and auroral imaging that are capable of providing fine-structure plasma measurements with an hitherto unprecedented temporal/spatial resolution from satellites.

The scientific objective of Freja is to explore fine structure plasma properties within the low-altitude portion of the auroral acceleration region and to study the physical processes whereby ionospheric plasma is being heated/accelerated and subsequently ejected out into the magnetosphere. The altitude range traversed by Freja ($\approx$600 km - 1750 km), constitute the topside ionosphere and the low-altitude part of the auroral energization region. This altitude range is in general considered well explored by numerous polar orbiting satellites. However, a high telemetry rate and state-of-the-art design of the instrumentation, enables more than an order of magnitude increased temporal/spatial resolution compared to its

[1]Swedish Inst. of Space Physics, Kiruna, Sweden
[2]Royal Inst. of Technology, Alfvén Laboratory, Stockholm Sweden
[3]Applied Physics Lab., Johns Hopkins Univ., Laurel MD, USA
[4]National Res. Council of Canada, Ottawa, Ontario, Canada
[5]Swedish Inst. of Space Physics, Uppsala Division, Uppsala, Sweden
[6]Dep. of Physics and Astronomy, Univ. of Calgary, Calgary, Alberta, Canada
[7]Max-Planck-Institute für extraterrestrische Physik, Garching, Germany

TABLE 1. The FREJA Scientific Payload

Experiment	Measurement Technique	Principal Investigator
F1 Electric Fields	3 pairs of Wire booms, 20 tip-to-tip 2 comp. of E up to 6000 samples/s	Göran Marklund, Alfvén Lab., Royal Inst. of Technology, Stockholm, Sweden
F2 Magnetic Fields	Triaxial flux-gate on 2 m boom 3 comp. of B, 128 samples/s	Lawrence Zanetti, Johns Hopkins Univ., Applied Physics Lab., Laurel, MD, USA
F3H Particles, Hot Plasma:	2D Magnetic electron spectrometer 2D distr. 0.1-115 keV, 100 samples/s 2D Ion composition spectrometer => 3D distr. 0.001 - 10 keV in 3 s.	Lars Eliasson Swedish Institute of Space Physics, Kiruna, Sweden
F3C Particles, Cold Plasma:	2D ion/electrons on 2 m boom 3D distr. of cold plasma (<300 eV) >100 samples/s	Brian Whalen, National Research Council, Ottawa, Canada
F4 Waves	Wire booms + 3 axis search coil E, B, Δn Waves, 1 Hz - 4 MHz. Up to 8 MHz sampling frequency	Bengt Holback Swedish Institute of Space Physics Uppsala -Div., Uppsala Sweden
F5 Auroral Imager	2 UV CCD cameras Auroral images every 6 s.	John S Murphree, Univ., of Calgary, Calgary, Canada
F6 Electron Beam	Three electron guns 3 components of E, 100 samples/s	Götz Paschmann, Max Planck Institut für extraterr. Physik, Garching, Germany
F7 Correlator	2D electron spectrometer 0.01 - 20 keV, correlation with F4	Manfred Boehm, Max Planck Institut für extraterr. Physik, Garching, Germany

predecessors, which makes Freja even exploratory in some aspects.

Freja was launched as a piggy-back satellite on a Chinese Long-March II rocket in October 6, 1992. After launch into a 63° inclination, Freja was lifted to a higher perigee (≈600 km) and apogee (1750 km) by means of two separate solid fuel boost motors. The relatively low inclination means that good auroral oval coverage is obtained mainly over the American continent. Thus the Canadian Prince Albert ground station, has turned out to be an important ground segment for the direct transmission of data. However, Esrange in Sweden is the main operations centre where all uplink communications are performed. Measurements over the southern polar region will also be made for certain time periods over the Japanese Syowa station.

2. PAYLOAD AND OPERATIONS

The Freja payload comprises a full complement of high-resolution plasma diagnostic instruments and a fast auroral imager. The instruments represents the state-of-the-art in space plasma instrumentation. A number of instruments are the first of its kind. For instance, the ion and electron imaging spectrometers of F3H and F3C represent completely new spectrographic concepts. F7 combines a wave-particle correlator with a special two-dimensional geometry of sensors, fast sweeps and data packing. Altogether the particle instruments provide an hitherto unprecedented spatial-temporal resolution from satellites. The wave and field instruments are due to the extended frequency coverage of all seven wave components and a high data rate unique in many respects. The special combinations of wave form sampling are expected to bring new insight into wave coupling phenomena. Table 1 lists the individual instruments, their key elements, performance parameters, and Principal Investigators.

Freja is a relatively small spacecraft with total mass 258 kg, including the two boost motors for raising the apogee and perigee. The shape is that of a spinning disk (Fig. 1) with a diameter of 2.2 meters. The spin period is 10 rpm and the spin axis is always pointing towards the sun ($\pm 30°$). The continuous sun pointing philosophy, with solar panels mounted on the sun-facing side, gives a relatively high peak power, up to 168 W. The telemetry rate, using S-band telemetry, is 256 - 510 kbps. A total of 15 MB on-board burst memory within the experiments, gives an effective data rate of >2 Mbps for limited time periods (about 60 s). These burst periods are mainly triggered off by measurement conditions set by the magnetic field, electric field or wave experiment. The burst memory may also be used for measuring over regions not covered by the ground stations. For instance, the magnetic and electric field vector experiments are regularly providing overview measurements along the entire orbit. The Freja data taking strategy is summarized in Figure 2.

All Freja operations are controlled from Esrange in Sweden, which means that switch-on of experiments and transponders, mode control etc over other ground stations (Prince Albert and Syowa) are preprogrammed by time-tagged commands. The on-board memory for time-tagged commands enables a reduced tending of the satellite with uplink commanding concentrated to office hours, thus also reducing the costs for operations. An important part of operations is also the production of master (DAT) tapes for PI:s and the Freja Summary Plots (FSPs). FSPs are produced on a routine basis at the Freja Operations Centre and then transferred by e-mail to a common data base with access by the PI-institutions. The FSPs has turned out to be an extremely valuable element for orbit selection. Booklets

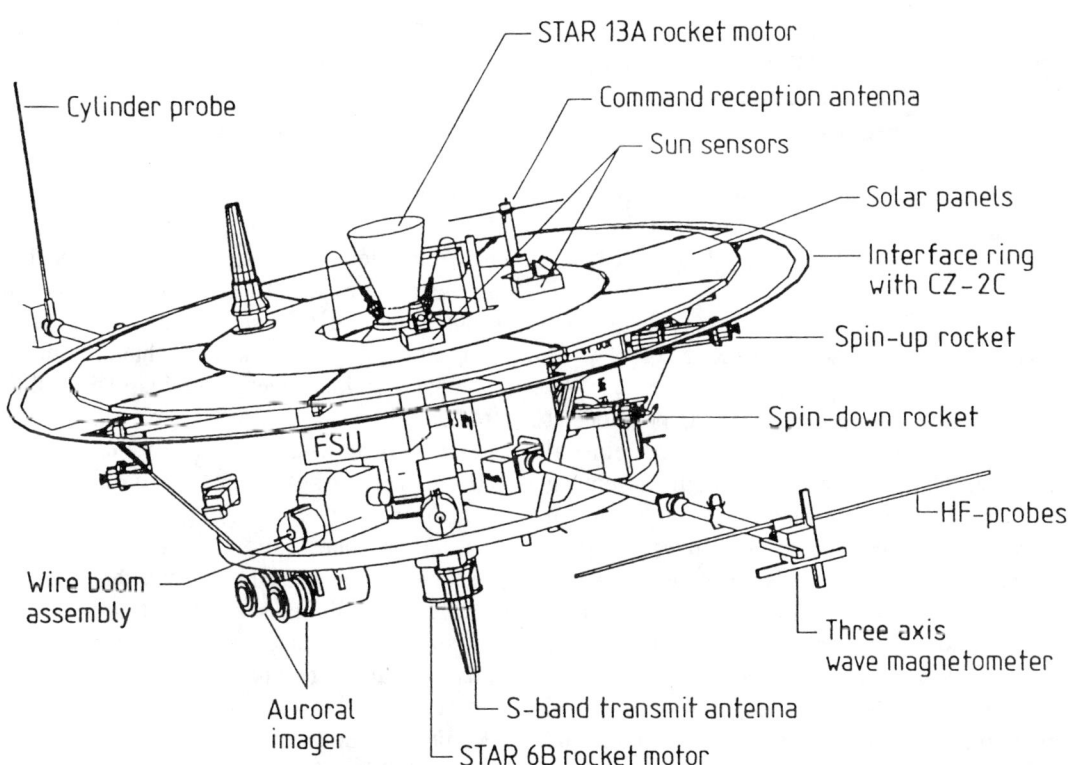

Fig. 1. View of the Freja spacecraft with some of the equipment and experiments.

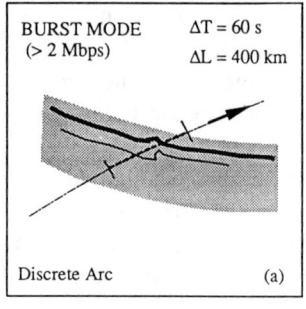
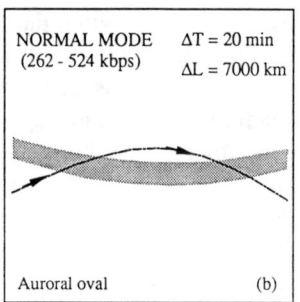
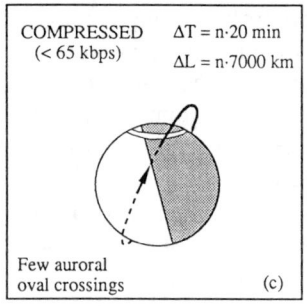
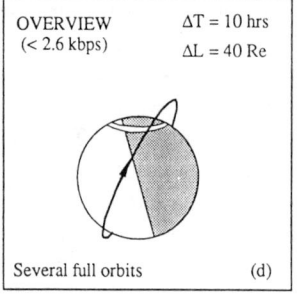

Fig. 2. Freja data taking strategy.

with FSPs are also made available for a wider distribution within the science community.

The Freja scientists are involved in the operations and orbit planning to reduce the costs for operations, but also to enhance the scientific return of the mission. Orbit planning is managed by the Freja scientist in charge, selected on a monthly basis among the PI-institutions. The orbit planning involves e.g. burst-mode orbit selection, special science topic orbits, and campaign coordination. The Freja operations finally allows for a very close contact with the experiments. Remote access to the operations centre for command uploading can be made by all PIs from their home institutions.

3. SCIENTIFIC CONTRIBUTIONS FROM FREJA

The apogee of Freja, at ≈1750 km, implies operations at the lower edge, or below, the mid-altitude auroral energization region. Thus, the Freja mission differs from e.g Viking and other mid-altitude mission that emphasizes in situ measurements within the field-aligned acceleration region. The capability of Freja lies elsewhere. It will be uniquely suited to study the impact of the auroral acceleration process on the structure of the topside ionosphere. Large holes in plasma density, created by cavitation processes in the lower edge of the auroral energization region [e.g. Haerendel, 1989; Lundin and Hultqvist, 1989], are expected to be seen regularly along Freja's trajectory at auroral latitudes. It appears likely that at least some of these cavitations are due to the heating and upward loss of electrons and ions simultaneously, possibly due to the strong low-frequency electric turbulence observed within these cavities [Hultqvist et al., 1988; Lundin et al., 1990]. Such heating processes have been known to extend very deep down into the ionosphere [e.g. Whalen et al., 1978; Wahlund and Opgenoorth, 1989]. The general characteristics and height distribution of such plasma density cavities will be explored by Freja.

Large-scale plasma density cavities have strong effects on the auroral processes. They establish and maintain the conditions for current instability, they constitute an imprint on the magnetosphere-ionosphere interface of the recent occurrence of accelerations, they modify the transmission of energy carried by e.g. Alfvén waves towards the ionosphere, and they are the source region of auroral kilometric radiation. The high-density walls surrounding the cavities may efficiently trap plasma waves and lead to the build-up of high wave intensities and non-linear decay instabilities. Micro-scale plasma density cavities detected from both satellites [Temerin et al., 1982; Boström et al, 1988] and rockets [LaBelle et al., 1986; Kintner et al., 1992] also constitute a new, still poorly known, source of wave-particle interaction. Thus, Freja will scan a region of utmost importance for the production and modification of electrostatic and electromagnetic plasma waves. Particularly interesting will be the analysis of reflection, transmission, and damping of Alfvén waves incident from above [e.g. Lysak and Dum, 1983].

The perhaps most significant advance in auroral physics achievable by Freja, will be made possible by the combination of greatly enhanced temporal/spatial resolution of the in-situ measurements. This will enable, for the first time from satellites, the study of the plasmaphysical reason for the omnipresent fine-structure of auroral arcs. For instance, auroral rays and vortex structures have typical dimensions ranging from a few hundred meters to a few kilometers at ionospheric heights. The stability of such structures in time is of the order ≈0.5 - 5 seconds. The satellite moving with a speed of ≈5 km/s (projected down to 100 km) will pass through such auroral fine structures in ≈0.1 - 1 s. With a sampling time for two-dimensional electron distributions down to 10 ms and fields and cold plasma sampling rates much faster than that, Freja is quite capable of resolving such fine-structures. Similarly, the fine-structure of discrete auroral arcs, frequently with individual arc segments in the hundred meter range, are hitherto only resolved by a relatively limited amount of sounding rocket measurements at altitudes generally much below 1000 km. Freja is not only radically improving the data base for such fine-structure measurements, it also accesses an altitude range hitherto not attended by instruments utilizing the novel fine-structure measurement technique.

4. OVERVIEW DATA AND MACROSCOPIC FIELD ALIGNED CURRENTS

The Freja auroral UV imager provides the most important overview data. An example of such an overview is shown in Fig. 3 for a Freja pass over the early morning auroral oval (Orbit 303, Oct. 29, 1992). The pass occurred during the recovery phase of a minor substorm. A striking feature of this pass is the extensive fine-structure of the aurora and the fact that many of the auroral arcs were oriented poleward, or in a north-south orientation often observed during recovery phase [Rostoker, et al., 1987]. Freja, traversing the field lines of these structures also detected strong north-south (Z) deviations of the magnetic field (Fig. 4), suggesting the traversal of north-south oriented Birkeland current sheets collocated with the north-south oriented auroral arc structures. Notice that these Z-component deviations are as large as, and superimposed on the traditional L-shell aligned deviations, characteristic of the Region 1 field-aligned current system. The longitude separation of these multiple north-south oriented current sheets, embedded within the large-scale Region 1 field aligned current, is of the order 300 km. Such north-south multiple auroral arcs, and their associated current sheet, must of course map to a similarly striated morphological dynamo structure in the tail. Thus, these structures raises a number of interesting questions about the morphology of the near-Earth tail during substorms, with large intrusions of plasma and associated

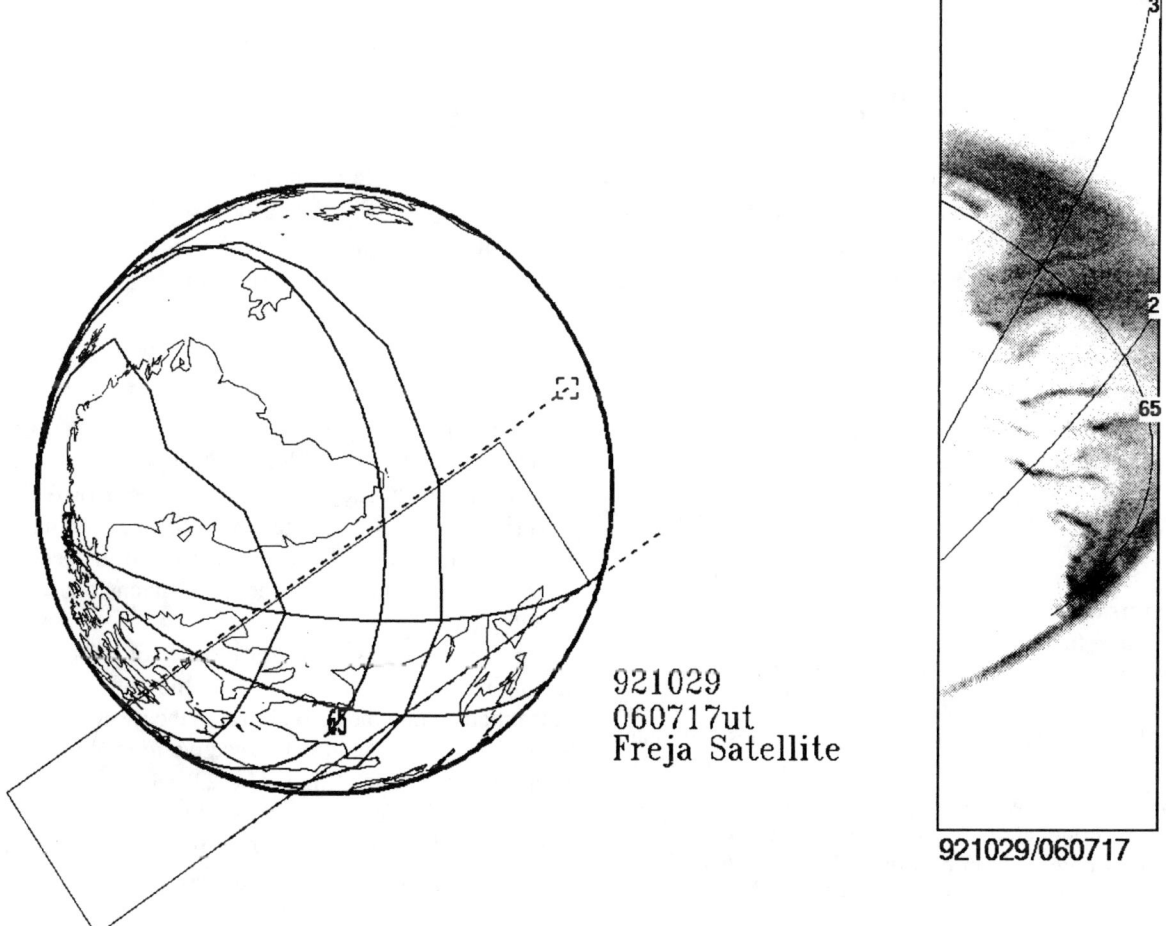

Fig. 3. Freja auroral image of an expanded auroral oval (substorm expansion/recovery) with multiple north-south oriented arc segments extending from the oval. Each arc should be associated with individual upward Birkeland currents.(Orbit 303, Oct 29, 1992).

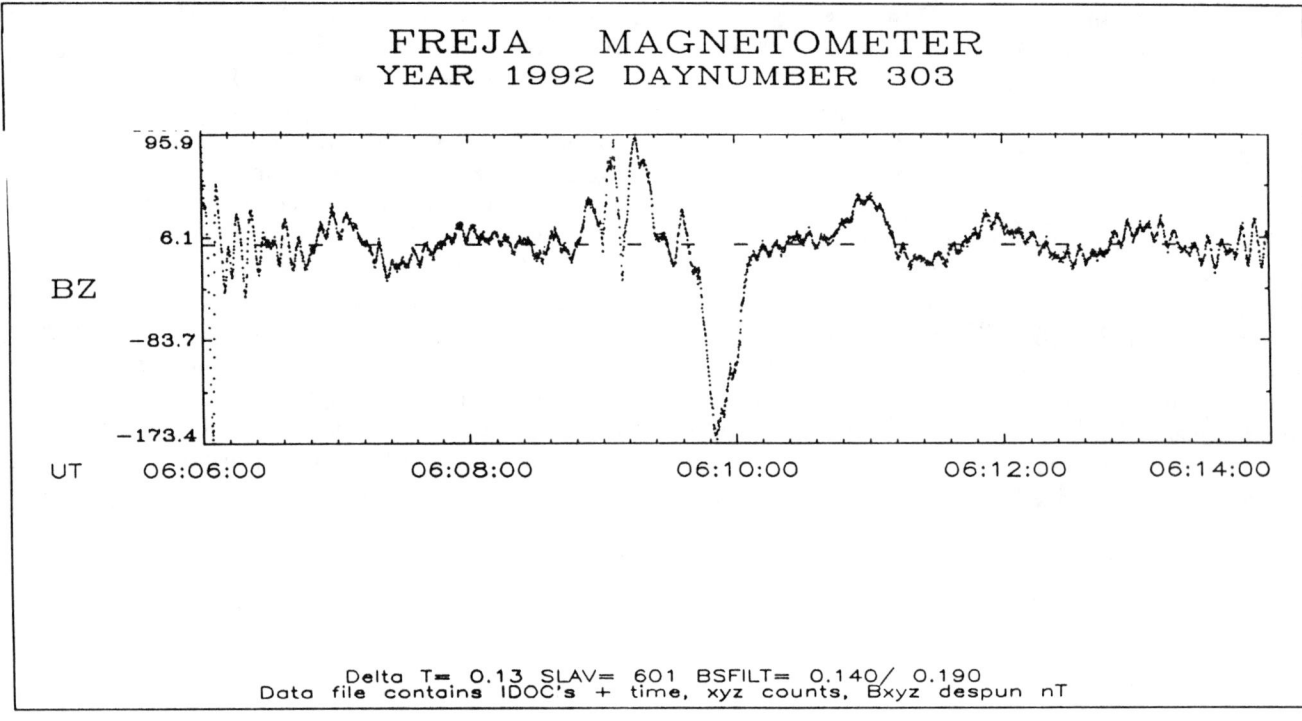

Fig. 4. Freja magnetometer data displaying traversals of multiple north-south oriented current filaments (Bz component, spin axis, approximately sunward) embedded in the large-scale downward Region 1 current structure. Although slightly out of the image frame, it is believed to be associated with similar arc structures as those observed in Fig 3.

currents from high latitude boundaries directly into the magnetosphere and to low latitudes.

5. AURORAL ACCELERATION OBSERVATIONS

Traversing the altitude range 600 - 1750 km means that Freja covers the lower part of the auroral acceleration region. This implies that most of the field-aligned acceleration, usually considered to be associated with field-aligned electric potentials, lies above the satellite. On the other hand, Freja frequently accesses the plasma density cavities associated with the heating and transverse energization of ionospheric plasma in the auroral oval. An example of a pass through field lines connected to the acceleration region with accelerating potentials in excess of 30 kV is displayed in Fig. 5 (Orbit 790). The panels of Fig. 5 shows ion and electron energy-time spectra taken from the Freja hot plasma experiment (F3H and F7). The uppermost three panels gives ion spectra in the energy range 1 - 4500 eV for O^+, He^+ and H^+ taken from one out of the 32 sectors of the three-dimensional ion composition spectrograph (TICS). The fifth panel from the top gives electron energy spectra from two-dimensional magnetic electron spectrometer (MATE) here covering the energy range ≈3 - 120 keV. Finally, the bottom panel shows 0° pitch angle electron spectra from 0.02 - 20 keV taken by the two-dimensional electron spectrometer (TESP).

Fig 5 exemplifies the characteristics of a Freja traversal through the auroral oval, i.e. the satellite moving tangentially along the auroral oval in the west-east direction. This gives quite a unique perspective in terms of both the local time coverage (MLT 17 - 22) and the time spent in the auroral zone and its corresponding acceleration region. Notice here that Freja spent over 10 minutes in this large-scale "inverted V" structure. The maximum electron acceleration is unusually high here, in fact exceeding 30 keV. The electron energy-time spectrogram plot shows that the maximum energization was achieved close to the poleward boundary, decreasing equatorward in a manner which is more or less symmetric with local time.

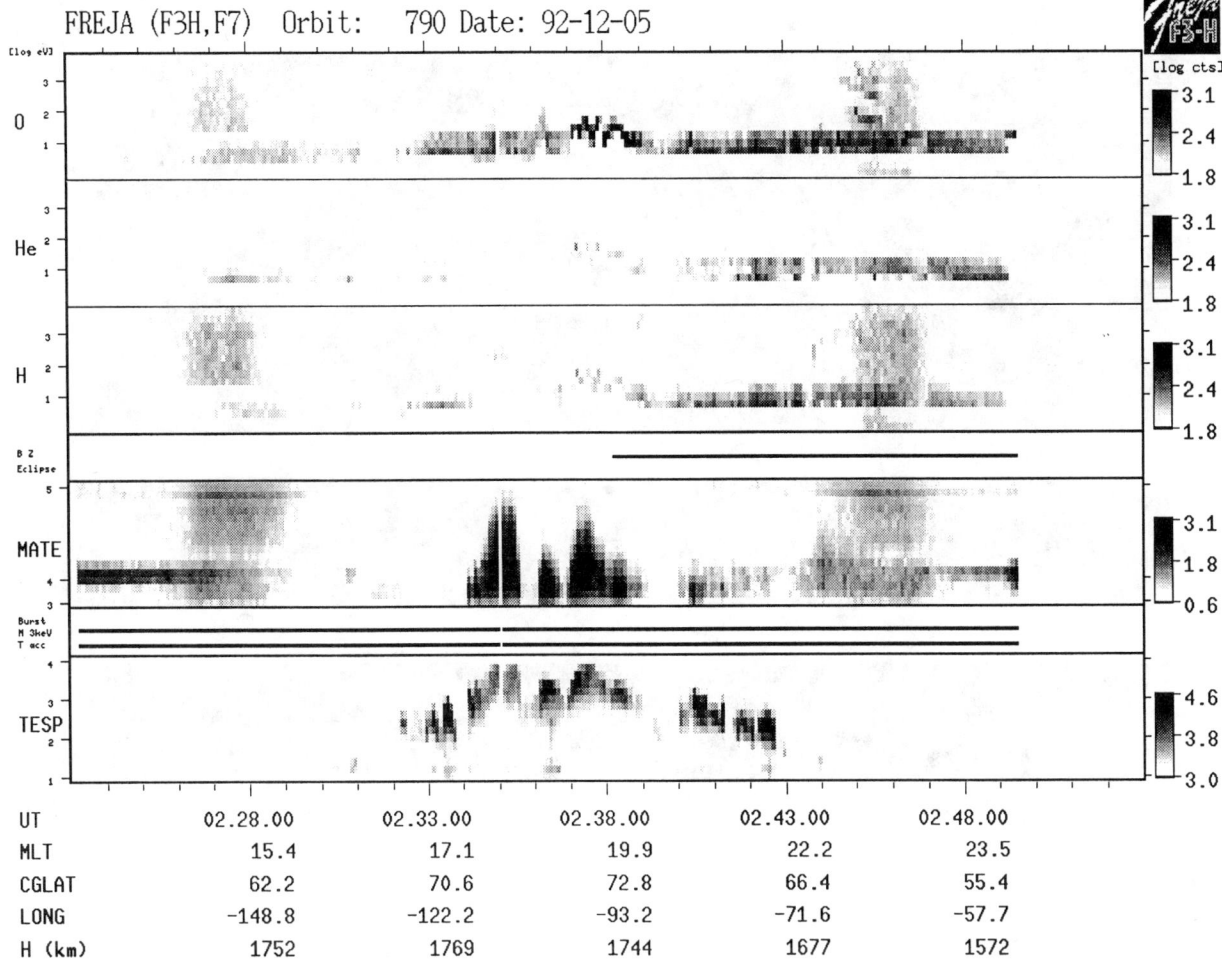

Fig. 5. Overview particle data (F3H + F7) of Freja "skimming" the oval with a large-scale "inverted V" with maximum acceleration exceeding 30 kV (Orbit 790).

Fig. 6 represents a blow-up of the time period when Freja enters the central portion of the acceleration region where the ionospheric plasma becomes heated and starts eroding upward. This can be recognized in Fig. 6 as relatively narrow distributions of energized ions around 90° in pitch angle reaching energies up to 1 keV. The time period before entering the region of transversely energized ions (TAIs) is also characterized by upgoing cold ions coincident with downward accelerated electrons (02:32 - 02:34). Notice also by the end of this 2 min period that the acceleration voltage exceeds the ≈20 keV upper energy limit of the TESP electron spectrometer. The strong, O^+ dominated, ion outflow observed within the acceleration region raises the question about the ionospheric plasma content within the energization region. Fig. 7 shows data from the Langmuir probe of the electron density, prior to and within the energization region. Clearly, there is a dramatic decrease of the local plasma density, from about 1000 cm^{-3} outside to 50 - 200 cm^{-3} within the TAI region. The ion energization region is thus associated with a large plasma density cavity in the topside ionosphere. This large scale plasma density cavity is thus caused by a massive erosion of plasma upward, most likely by a combined effect of upflowing cold ions and transverse ion energization.

Previous satellite measurements [e.g. Hultqvist et al., 1988; Lundin and Hultqvist, 1989) have established a good correlation between low-frequency, broad band, electrostatic noise (LEF) and ion energization. Fig. 8, displaying electric field data from two of the E-field probes, shows that LEFs indeed occur here too. Fig. 6 and Fig 8 demonstrates an almost perfect match between the amplitude of the LEFs and the maximum ion energization energy, quite in agreement

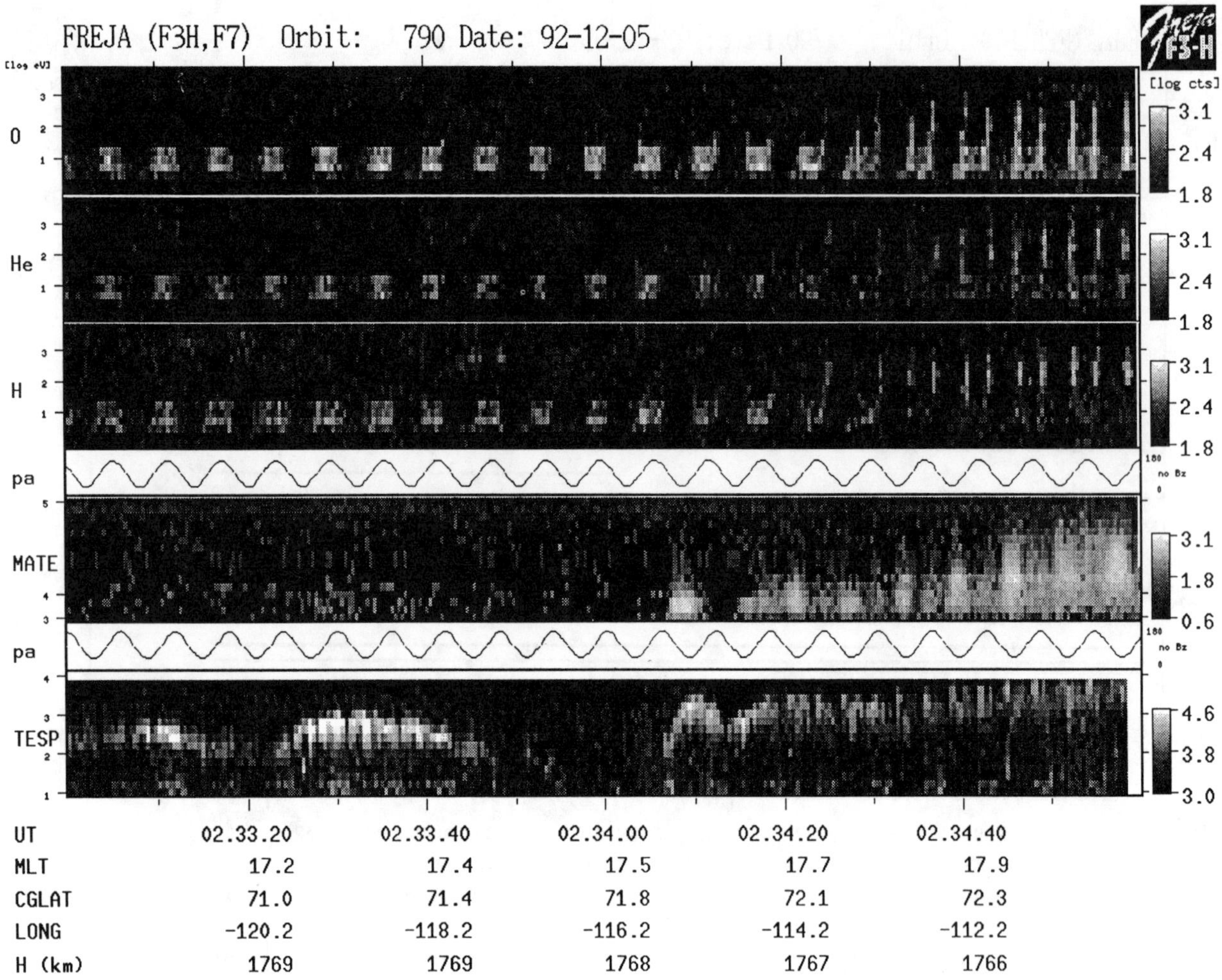

254 FIRST FREJA MEASUREMENTS

Fig. 6. Blowup of the previous pass (790) showing the appearence of transversely accelerated ions (O+, He+, H+) within the central portion of the acceleration region.

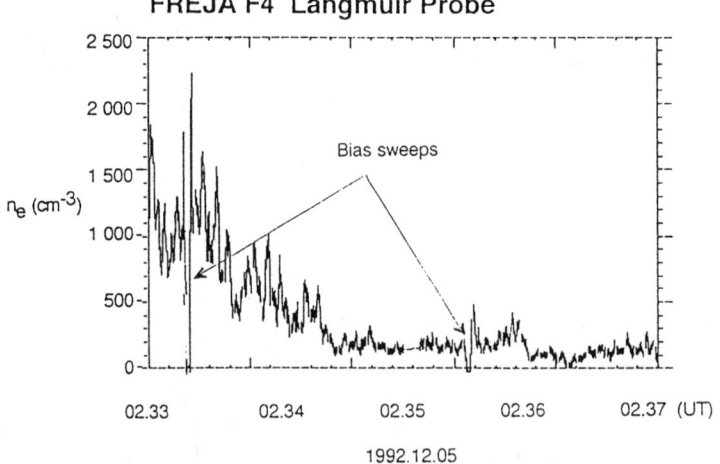

Fig. 7. Langmuir probe data illustrating the strong evaccuation of ionospheric plasma within the acceleration region in orbit 790.

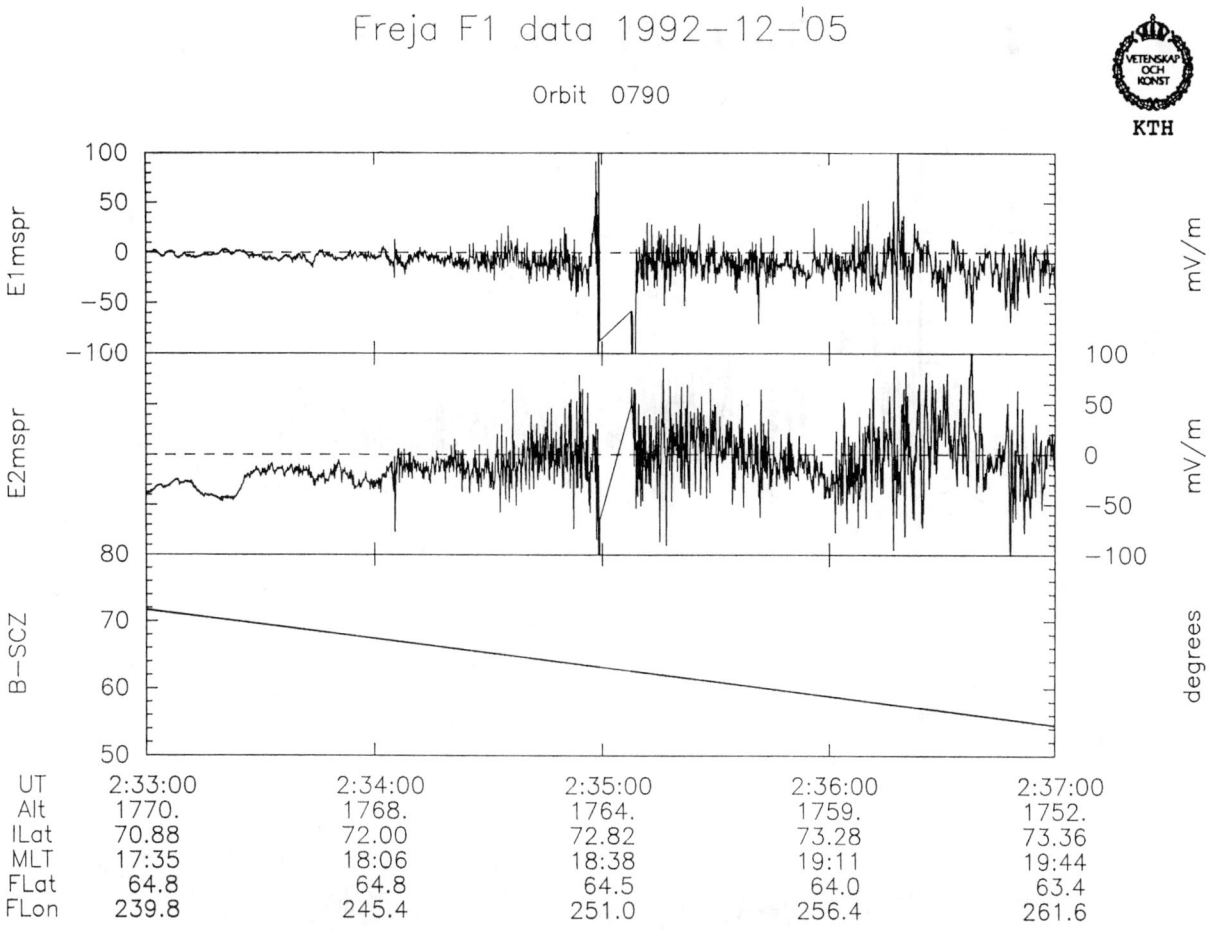

Fig. 8. Electric field data showing strong low-frequency electrostatic noise (LEFs) usually observed together with ion acceleration.

with the Viking results [Lundin et al., 1990]. The interesting new aspect of these observations is that LEFs may frequently be observed at Freja altitudes with a similar relation between LEFs and the ion energization.

Another important characteristics within the the energization region can be found in the Freja wave data between ≈0.1 Hz - 4 MHz. Fig. 9 demonstrates the capability of the wave instrument to sample the wave form for short time intervals (lower panel). In this case one may clearly resolve wave packets of what appears to be Langmuir waves with a frequency of 300 kHz modulated at whistler wave frequencies in the 10 kHz range. Notice that this data was taken much before the satellite entered the large-scale plasma density cavity. The Langmuir wave peak at ≈300 kHz corresponds to an electron density of ≈1100 cm^{-3}. The importance of measuring both the magnetic and the electric component of waves is exemplified in Fig. 10, showing the wave power spectra outside (left) as well as inside (right) the region of ion energization. The local proton cyclotron frequency (≈400 Hz) is marked by the dashed vertical bar. Clearly both the electric and magnetic field high frequency noise enhances within the ion energization region. However, it is important to note that while the noise is more electrostatic outside- is appears to be electromagnetic inside the ion energization region. Although it may be too early to speculate on this fact it is yet tempting to here conclude that electrostatic lower hybrid waves are more dominant within the region of "cold" ion upflow than it is within the region of TAIs, where they appear more electromagnetic and "whistler-like".

6. FINE-STRUCTURE OBSERVATIONS

One of the main scientific objectives of Freja is to resolve the plasma fine-structure of physical processes in the auroral

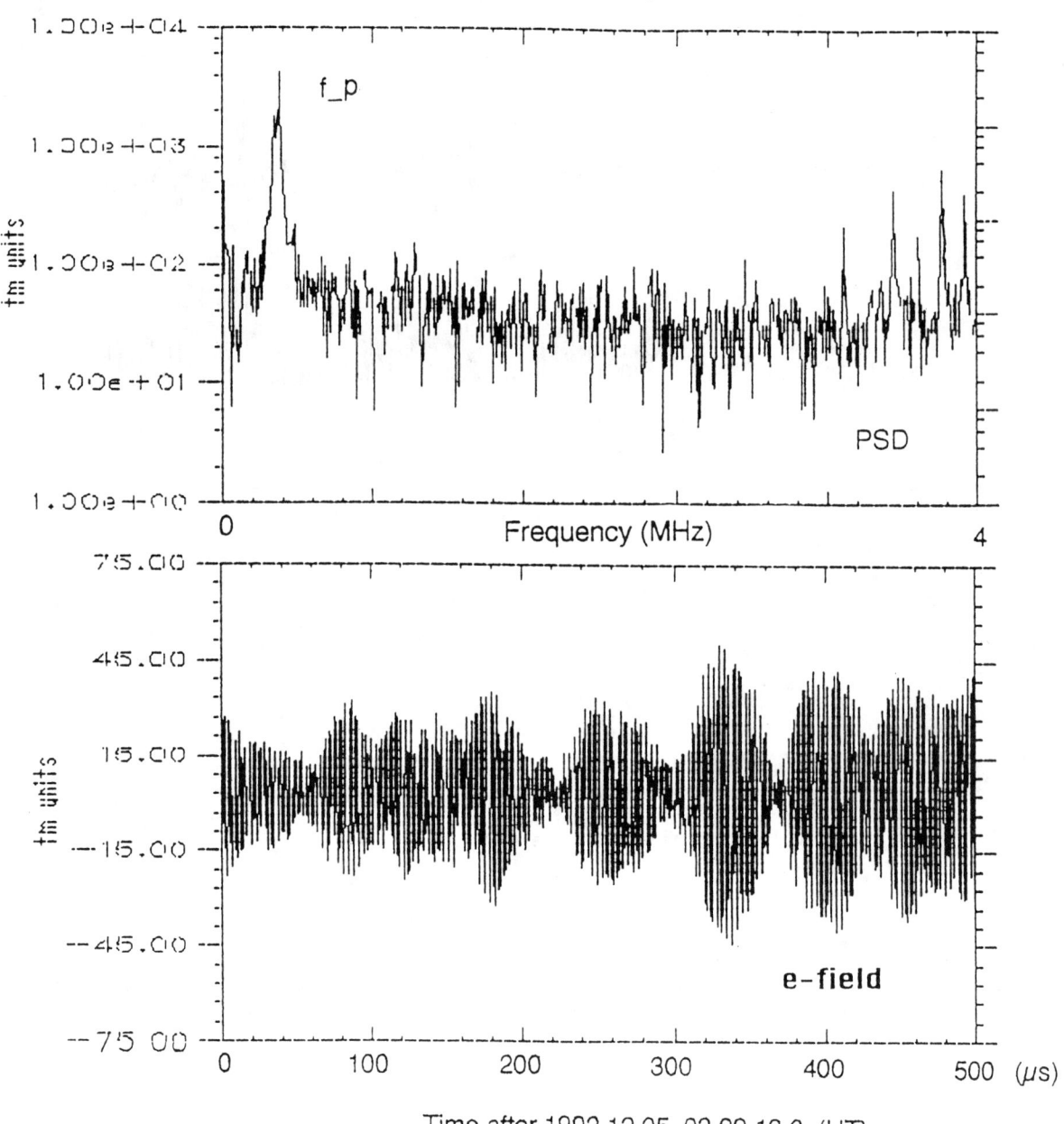

Fig. 9. Wave data of electric fields 0 - 4 MHz showing Langmuir waves and wave form data during 500 μs.

energization region. This is achieved partly by the inherent resolution of the on-board plasma instruments and partly by the data rate and storage capability of the satellite. Even the lowest TM transmission rate (256 kbps) is substantially more than that achieved by previous auroral satellites. The burst mode data rate (>2 Mbps) gives an order of magnitude improvement which so far has only been achieved from sounding rockets.

Indeed, observations from sounding rockets [e.g. McFadden et al., 1987, 1990] have stressed the necessity of high-time resolution measurements during structured auroral precipitation. Similarly, fine-structure measurements from

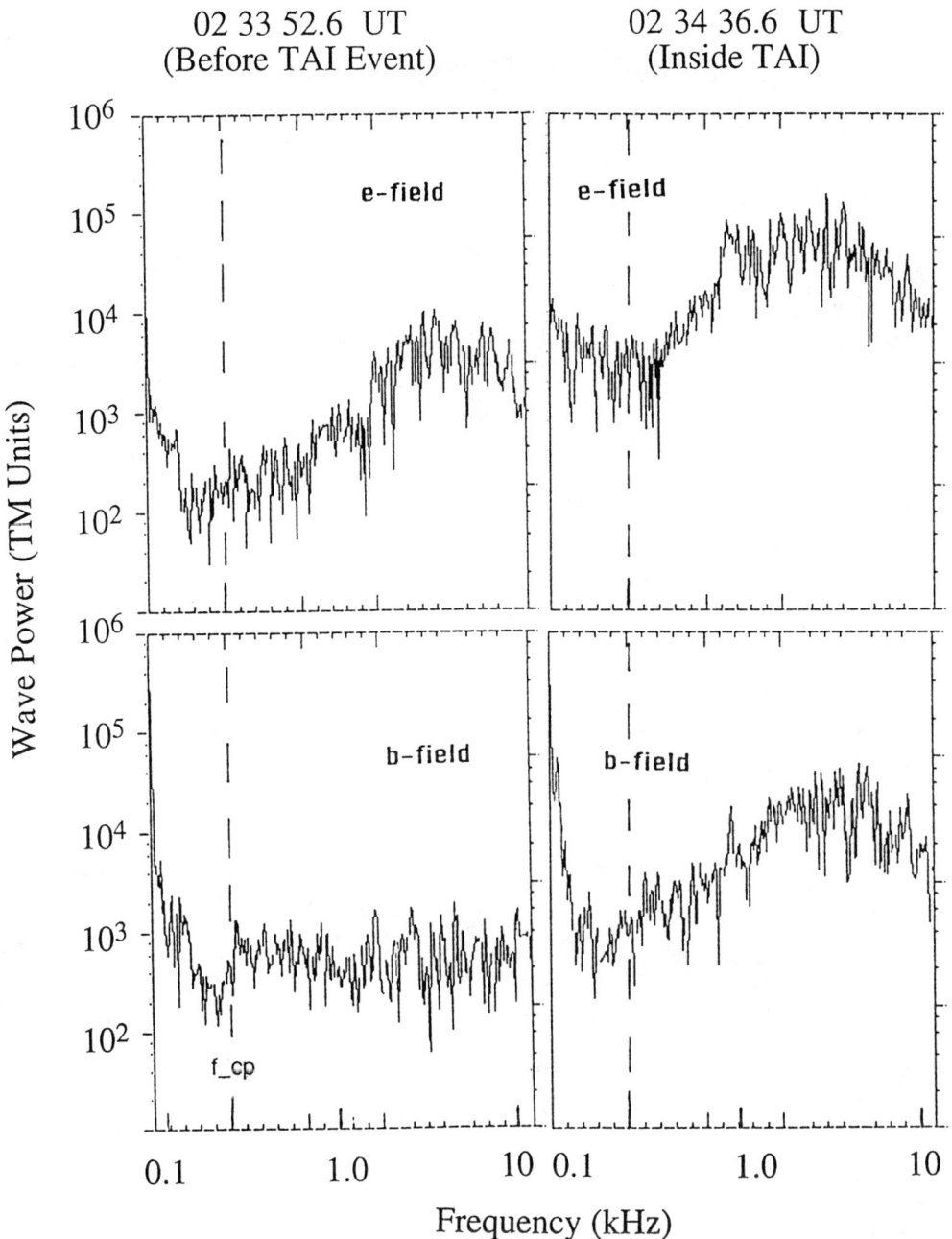

Fig. 10. Wave power spectra (0.1 - 10 kHz) within and outside the region of transverse acceleration of ions (TAI).

sounding rockets have also shown new wave aspects of the transverse ion acceleration process [e.g. Vago et al., 1992] which further stresses the need for improved wave measurements. For instance, the wave active filamentary density cavities, first denoted "spikelets" by LaBelle et al. (1986), is suggested to be of great significance for localized ion heating in the high altitude ionosphere [postulated by Chang and Coppi, 1981, verified by Kintner et al., 1992].

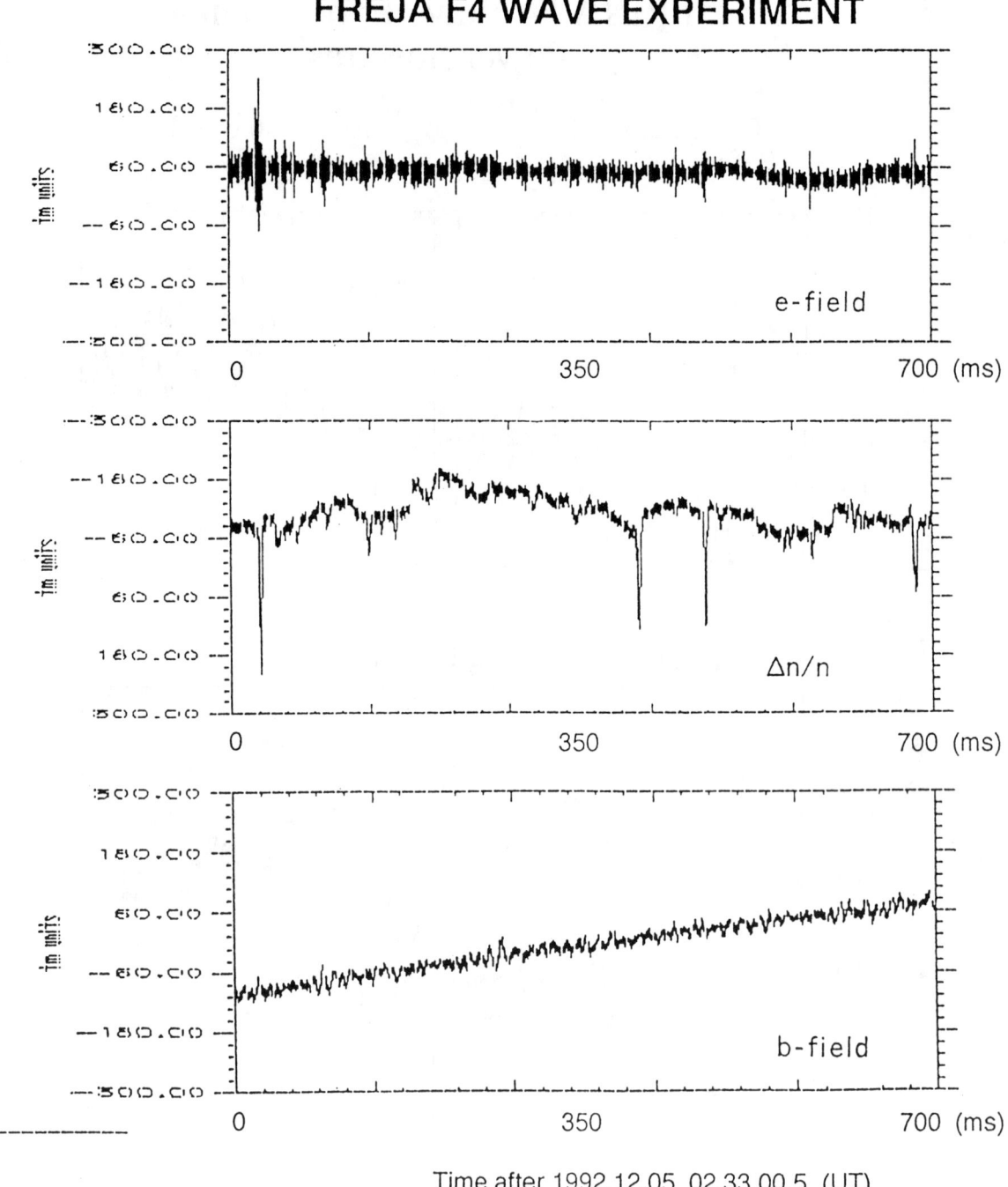

Fig. 11. Example of plasma density cavities observed outside the large-scale ion acceleration region.

Although similar filamentary density cavities have been reported at higher altitudes as well [Boström et al, 1987], it is not yet clear whether these two reported phenomena are related or represents different classes of phenomena. The first observations of similar "spikelets" from Freja does therefore represent an important step forward to their understanding. Fig. 11 display a 0.7 s time interval with wave form data of the electric field, density variation and magnetic field, taken

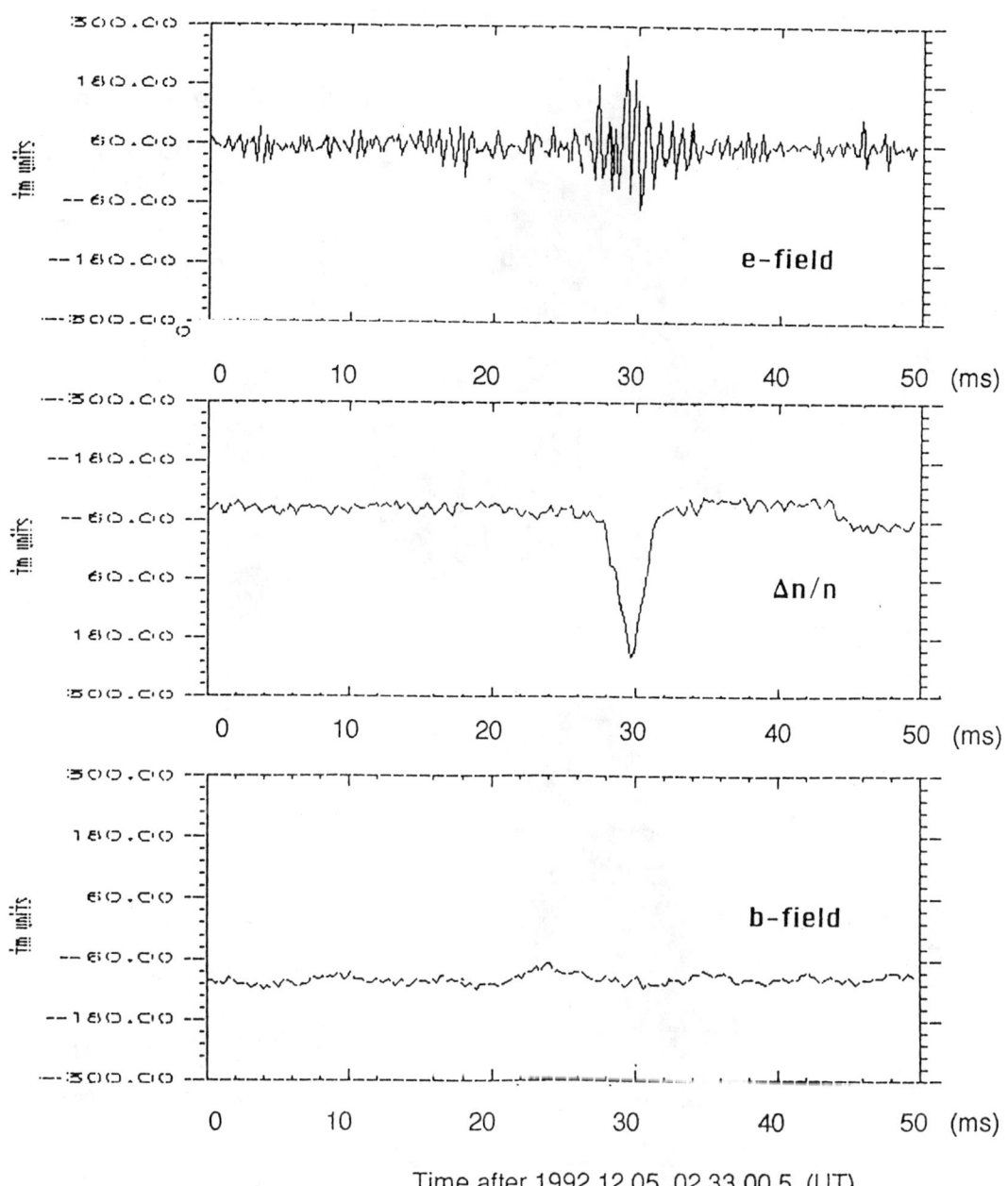

Fig. 12. Blowup of density cavity with electric field fluctuations inside the cavity (lower hybrid envelope soliton).

during the same orbit as that contained in Fig. 5 (Orbit 790) which display at least four filamentary density cavities. A blowup of the first structure, illustrating the wave envelope contained in the density cavity, is shown in Fig. 12. Because the wave frequency is close to the lower hybrid frequency (in this case around 1 kHz), and because the waves are purely electrostatic (no wave signature in the B-field) they may be termed lower hybrid envelope solitons.

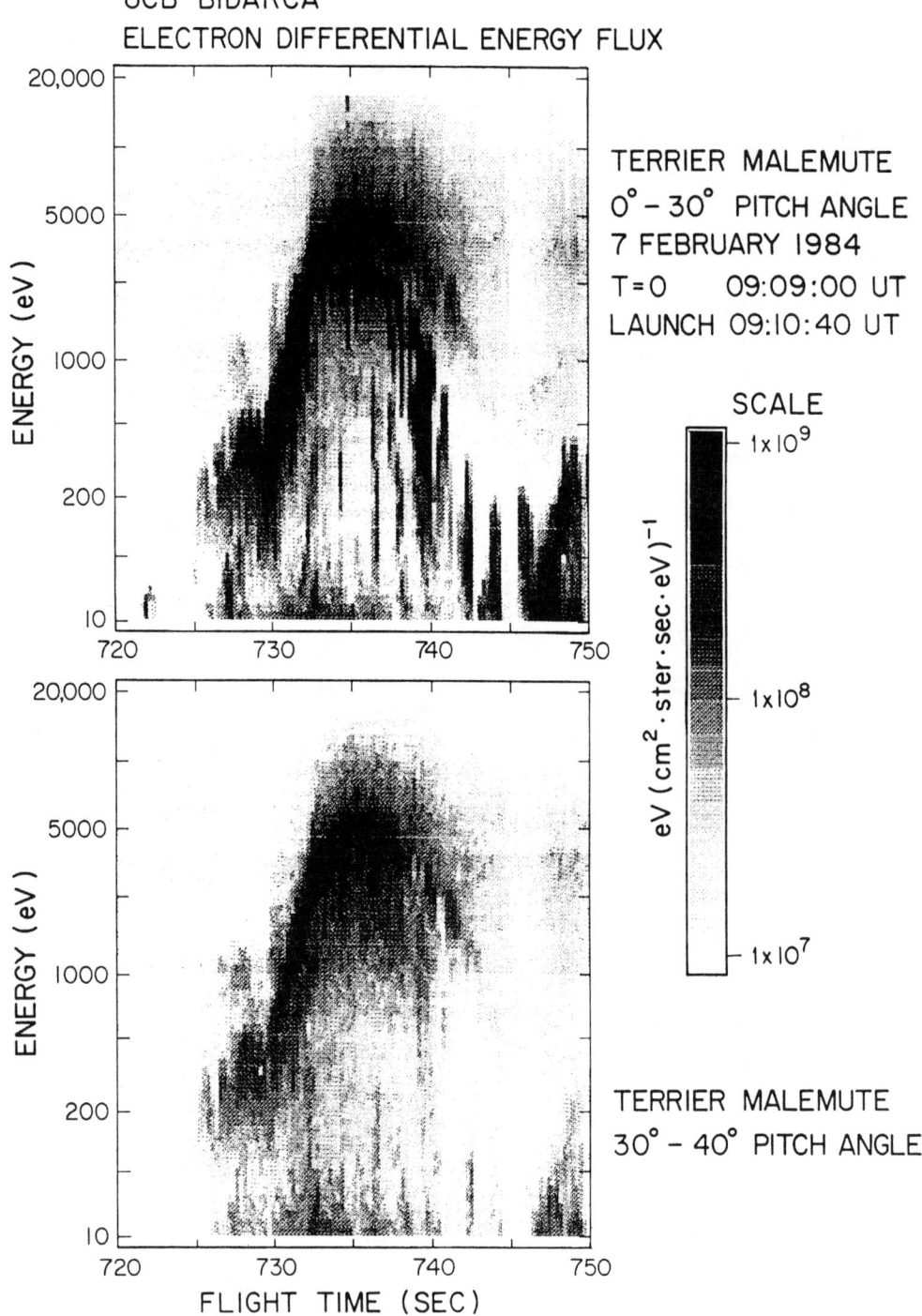

Fig. 13. Sounding rocket data of auroral arc traversal (10 km - ≈10 s traversal time) (McFadden et al., 1990).

It is important to note that these solitary structures were only observed in the region with sufficiently high plasma densities (of the order 1000 cm^{-3}) before accessing the main region of TAIs. Thus, if these solitary structures are important for the transverse ion heating, they are likely to represent a preheating phase - possibly related with the cold ion upward flow observed prior to the main TAI-region (see Fig. 6). Such a limitation of the lower hybrid ion heating is in fact also consistent with observations [Vago et al., 1992].

Freja has also sufficient temporal spatial resolution to resolve the fine-structure of auroral arcs. Fig. 13 shows an energy-time spectrogram plot for electrons from a sounding rocket pass over an auroral arc, a small scale "inverted V" with spatial dimension about 10 km [McFadden et al., 1990]. Notice the rich structure within both the electron energy peak (modulations near 0.5 - 1 Hz) as well as the presence of strong field-aligned low-energy electron fluxes below the peak energy. This suggests strong dynamics of the acceleration process with cold electrons being injected within the acceleration region in a modular way.

Fig 14 gives an example of a Freja pass over the auroral oval containing multiple auroral arcs. The total traversal time, about 300 s, corresponds to an ionospheric footprint distance of about 1500 km. As evidenced by the panel, no single structure is wider than about 50 km, i.e. of the order the size of the arc displayed in Fig. 13. Larger structures (e.g. at 180 - 220 s) are here clustering of smaller structures, of which a great variety can be found from this orbit.

Fig 15 shows a blow-up of 0° pitch angle spectra for the first 150 s (top) and a further 3 s blow-up of one of the smallest arcs segments around 110 s, appearing as just a spike in Fig. 14. The 3 s blow-up represents the highest time resolution achievable with 32 ms between each energy spectra. This small scale arc structure is in fact not wider than ≈10 km, i.e. similar to the arc width displayed in Fig.

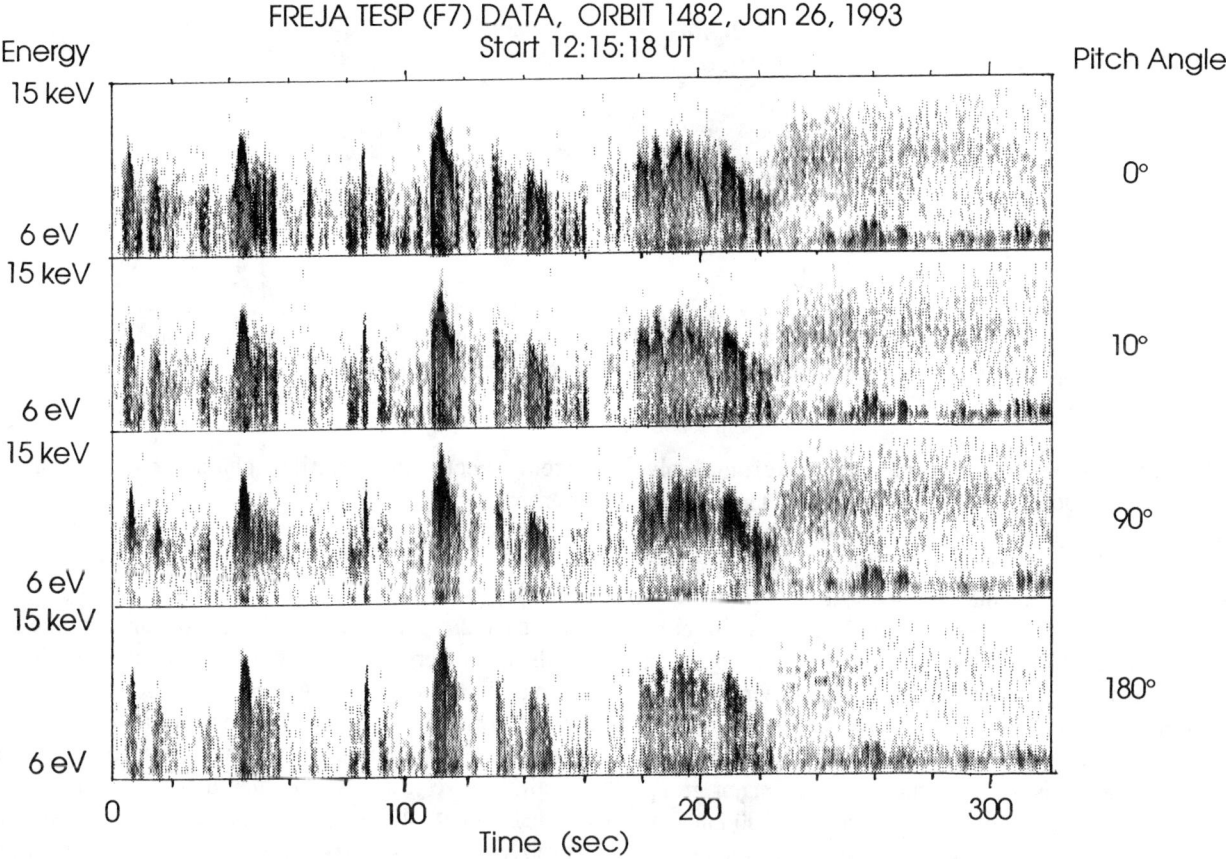

Fig. 14. Freja F7 electron energy-time spectra of multiple arc traversals with dimensions < 50 km (Orbit 1482, 930126). Notice the extreme richness of narrow arcs filaments. Notice also that each "main group" of arcs are separated by ≈300 km, thus suggesting a case of north-south oriented arcs similar to that displayed in Fig. 3 and Fig. 4.

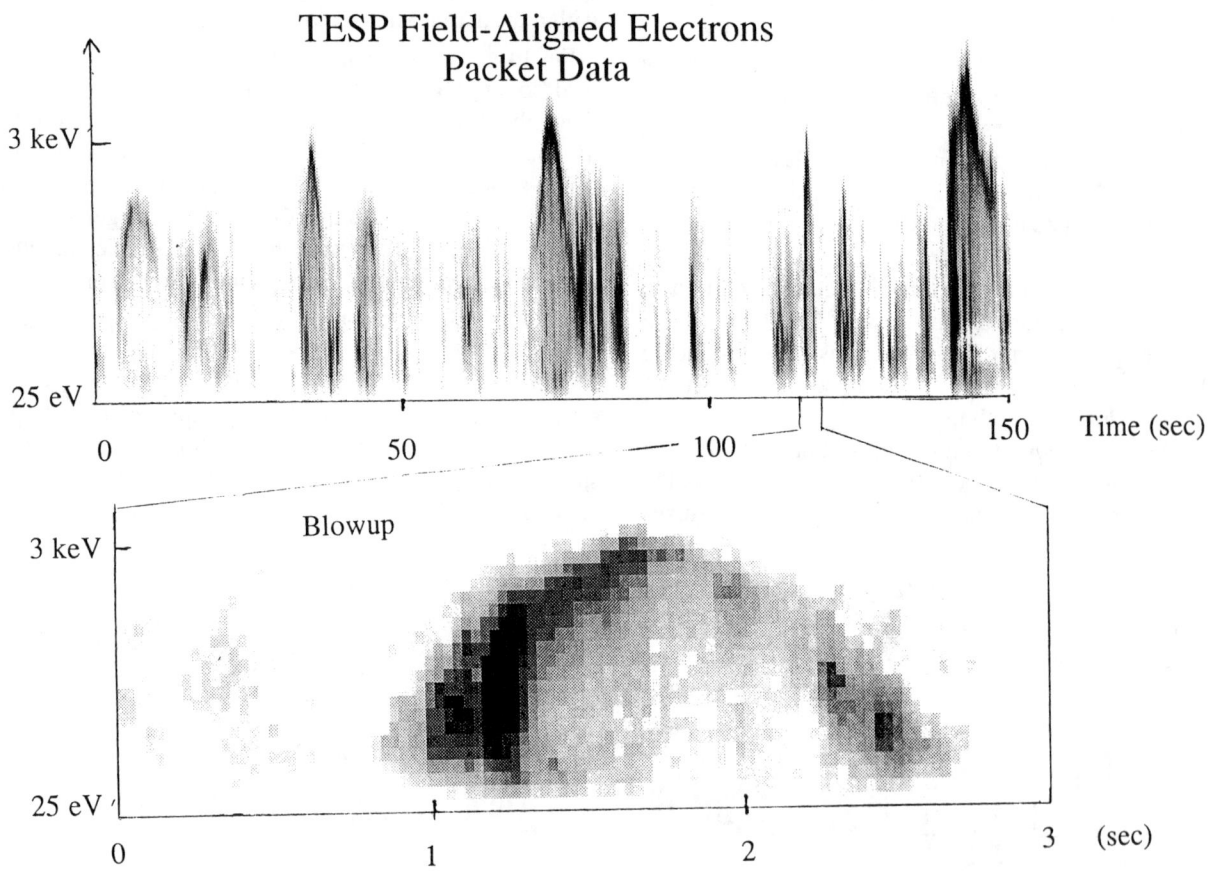

Fig. 15. Blowup of the first 150 s of electron energy-time spectra at 0° pitch angle (upper panel) and a further blowup of 3 s containing the traversal of a 10 km arc.

13. Besides demonstrating the capability of Freja to resolve small scale arc structures in the few hundred meter range, the figure also demonstrates the importance of rapid arc traversals. For instance, much of the substructure below the peak energy in Fig. 13 may be due to fluctuations of the accelerating voltage in the 1-2 Hz range [e.g. Andre et al., 1992]. Such fluctuations are not evident in this Freja high-resolution example, but they are evident in wider arcs, such as the ≈30 km arc traversed at 75 s in Fig 15.

A final peculiarity with the multiple arc structure shown in Figs. 14 and 15 is the separation between major groups of arcs, the distance corresponding to around 300 km. This is similar to the distance between north-south oriented arc segments in Fig. 2 and the distance between magnetic field Z-deviations in Fig. 3. Although this may be pure coincidence it is yet an interesting fact that the east-west grouping of arc filaments and the poleward directed Birkeland current sheets have similar characteristic separation distances.

DISCUSSIONS

We have presented some early data obtained by the joint Swedish and German Freja satellite, data taken during the first months of operations. Already in the very beginning of data analysis it is clear that the Freja mission have potentials of becoming very successful. It is difficult, though, to give justice to the measurement capabilities of all instruments in a brief overview article like this. Moreover, at this early stage the software for many instruments is more adapted for instrument surveillance than science. For instance, the cold plasma instrument (F3C), located on a 2 m boom, has demonstrated unique potentials of detailed cold plasma measurements. The dc electric and magnetic field experiments are both capable of providing waveform data up to their sampling frequency (table 1). The electron beam experiment will provide rapid 3-D electric field measurements using an independent measurement technique, which besides its own merits is an important complement to the probe measurements, in particular during satellite eclipse and within extreme plasma density cavities.

In this report we have presented examples related with three interesting topics which is expected to become important in the future analysis. These are: (1) The morphology and distribution of large-scale and small-scale Birkeland currents, their charge carriers and relation to acceleration processes. (2) The characteristics of the auroral energization process in the altitude range 600 - 1750 km and its relation to the formation of meso-scale plasma density cavities and the heating of ionospheric plasma. (3) The fine-structure of auroral forms (rays, curls etc) - its relation to the temporal/spatial characteristics of the particle precipitation, and the micro-structure of the auroral acceleration process and the associated wave-particle interaction.

The high temporal/spatial resolution of plasma measurements from Freja, as evidenced in the data presented here, represents a step towards a resolution hitherto only aquired from sounding rockets. The amount of data achieved from several months in orbit, passing the auroral oval, means that Freja already has provided a unique data base for auroral studies.

REFERENCES

André, M. and L. Eliasson, Electron acceleration by low-frequency electric field fluctuations: Electron conics, *Geophys. Res. Lett.*, *19*, 1073, 1992.

Boström, R., H. Koskinen, and B. Holback, Low frequency waves and solitary structures observed by Viking, in Proceedings of the 21st ESLAB Symposium on Small-Scale Plasma Processes, *ESA SP-291*, 185, 1987.

Chang, T., and B. Coppi, Lower hybrid acceleration and ion evolution in the subauroral region, *Geophys. Res. Lett.*, *8*, 1253, 1981.

Haerendel, G., Auroral particle acceleration - An example of a universal plasma process, *ESA Journal*, *4*, 197, 1980.

Haerendel, G., An Alfvén wave model of auroral arcs, in *High-Latitude Space Plasma Physics*, Edited by B. Hultqvist and T. Hagfors, Plenum Press, New York, p. 515, 1983.

Haerendel, G., Cosmic linear accelerators, in *Proc. of the international school and workshop on plasma astrophysics*, pp 37-44, ESA SP-285, ESA, Noordwijk, 1989.

Hultqvist, B., On the acceleration of electrons and positive ions in the same direction along magnetic field lines by parallel electric fields, *J. Geophys. Res.*, *93*, 9777, 1988.

Kintner, P.M. J.L. Vago, S.W. Chesney, R.L. Arnoldy, K.A. Lynch, C.J. Pollock, and T.E. Moore, *Phys. Rev. Lett.*, 68(16), 2448-2451, 1992.

LaBelle, J., P.M. Kintner, A.W. Yau, and B.A. Whalen, Large amplitude wave packets observed in the ionosphere in association with transverse ion acceleration, *J. Geophys. Res.*, *91*, 7113-7118, 1986.

Lundin, R. and B. Hultqvist, Ionospheric plasma escape by high-altitude electric fields: Magnetic moment pumping, *J. Geophys. Res.*, *94*, 6665-6680, 1989.

Lundin, R., G. Gustafsson, A. I. Eriksson, and G. Marklund, On the importance of high-altitude low-frequency electric fluctuations for the escape of ionospheric ions, *J. Geophys. Res.*, *95*, 5905, 1990.

Lysak, R.L., M.K. Hudson, and M. Temerin, Ion heating by strong electrostatic ion cyclotron turbulence, *J. Geophys. Res.*, *85*, 678, 1980.

Lysak, R.L. and C.T. Dum, Dynamics of magnetospheric-ionospheric coupling including turbulent transport, *J. Geophys. Res.*, *88*, 365, 1983.

McFadden J. P., C. W. Carlson, M. H. Boehm, and T. J. Hallinan, Field-aligned electron flux oscillations that produce flickering aurora, *J. Geophys. Res.*, *92*, 11133, 1987.

McFadden, J.P., C.W. Carlson, and M. Boehm, Structure of an energetic narrow discrete arc, *J. Geophys. Res.*, *95*, 6533-6547, 1990.

Rostoker, G., A.T.Y. Lui, C.D. Anger, and T.S. Murphree, "North-South stuctures in the midnight Sector Aurora as viewed by Viking Imager". *Geophys. Res. Lett.*, *14*, 407, 1987.

Temerin, M., K. Cerny, W. Lotko, and F. S. Mozer, Observations of double layers and solitary waves in auroral plasma, *Phys. Rev. Lett.*, *48*, 1175, 1982.

Temerin, M., Evidence for a large bulk ion conic heating region, *Geophys. Res. Lett.*, *13*, 1059, 1986.

Temerin, M., J. McFadden, M. Boehm, C. W. Carlson, and W. Lotko, Production of flickering aurora and field aligned electron flux by electromagnetic ion cyclotron waves, *J. Geophys. Res.*, *91*, 5769, 1986.

Vago, J.L., P.M. Kintner, S.W. Chesney, R.L. Arnoldy, K.A. Lynch, T.E. Moore, and C.J. Pollock, Transverse ion acceleration by localized lower hybrid waves in the topside auroral ionosphere, J. Geophys. Res., *97*, 16935-16957, 1992.

Whalen, B.A., W. Bernstein, and P.W. Daley, Low altitude acceleration of ionospheric ions, *Geophys. Res. Lett.*, *5*, 55, 1978.

Wahlund J.E. and H. Opgenoorth, EISCAT observations of strong ion outflows from the F-region ionosphere during auroral activity: preliminary results, *Geophys. Res. Lett.*, *727*, 1989.

R. Lundin, L. Eliasson, O. Norberg, Swedish Institute of Space Physics, Box 812, S-981 28 Kiruna, Sweden.

G. Marklund, Royal Institute of Technology, Alfvén Laboratory, S-100 44 Stockholm, Sweden.

L.R. Zanetti, Applied Physics Laboratory, The Johns Hopkins University, Johns Hopkins Road, Laurel, MD 20723-6099, USA.

B. A. Whalen, National Research Council of Canada, Ottawa, Ontario, Canada K1A 0R6.

B. Holback, Swedish Institute of Space Physics, Uppsala Division, S-755 91 Uppsala, Sweden.

J. S. Murphree, Department of Physics and Astronomy, University of Calgary, Calgary, Alberta, Canada T2N 1N4.

G. Haerendel, M. Boehm, G. Paschmann, Max-Planck-Institut für extraterrestrische Physik, D-8046 Garching, b. München, Germany.

The Inner Magnetosphere Imager Mission

D. L. Gallagher[*]

Solar Terrestrial Physics Division, NASA Marshall Space Flight Center, Huntsville, Alabama

The Inner Magnetosphere Imager (IMI) mission will carry instruments to globally image energetic neutral atoms, far and extreme ultraviolet light, and X rays. These imagers will see the ring current, inner plasmasheet, plasmasphere, aurora, and geocorona. With these observations it will be possible, for the first time, to develop an understanding of the global shape of the inner magnetosphere and the interrelationships between its parts. Seven instruments are currently envisioned on a single spinning spacecraft with a despun platform. IMI will be launched into an elliptical, polar orbit with an apogee of approximately 7 Earth radii altitude and perigee of 4800 km altitude.

1. IMI SCIENCE

After 30 years of observing the magnetospheric environment, one of our greatest challenges remains the synthesis of individual measurements in time and space into a global picture. Many of the questions that remain pertain to the scale of physical processes and the relationships between large-scale systems. What is the global extent of the substorm injection boundary? Is the formation of such boundaries the result of very localized or spatially distributed processes? How and where is the plasmapause formed and eroded during changing geophysical conditions? What are the global electric fields? What are the phenomenological connections between the dynamic processes occurring in these major magnetospheric systems and in the auroral zone?

To date the emphasis of experimental magnetospheric studies has been the development of ground truth knowledge of fields and particles through in situ measurements. In contrast, solar and astrophysics researchers have used imaging to establish the global morphologies of distant plasma systems. With the advent of magnetospheric imaging through the IMI mission, the opportunity now exists to combine our detailed in situ knowledge of magnetospheric plasmas with a global perspective of the large-scale dynamics and interactions of plasma systems. Through the imaging of photons and neutral atoms, the IMI mission will provide us global, simultaneous images of the ring current, near-Earth injection boundary, plasmasphere, auroral precipitating electrons and protons, geocorona, and outflowing ionospheric oxygen.

The IMI mission science objectives are:

1. To understand the global shape of the inner magnetosphere using simultaneously obtained images of the Earth's magnetosphere and its components: the ring current, inner plasmasheet, plasmasphere and aurora, and geocorona

2. To learn how magnetospheric current systems, field configurations, and conductivities derived from images respond on a global scale to internal and external influences

3. To visualize and identify the connections of various magnetospheric components to each other, especially as these connections act to change the components during substorms and solar wind variations

4. To relate global images of the magnetosphere to local observations in order to (a) learn how local processes combine to form the whole, (b) provide a global framework within which to place local observations, and (c) provide a "ground-truth" for the global observations.

---

[*]*For the IMI Science Definition and Preliminary Design Teams*

Solar System Plasmas in Space and Time
Geophysical Monograph 84
This paper is not subject to U.S. copyright. Published in 1994 by the American Geophysical Union.

The science rationale and mission details have been forged through a close working relationship between a science working group (Table 1) and an engineering program development team (Table 2). The science working group participants represent universities, corporations, and the government. A review and evaluation of global magnetospheric imaging techniques and scientific expectations can be found in Williams et al. [1992].

2. IMI STRAWMAN INSTRUMENTS

The IMI measurement objective is to obtain the first simultaneous images of component regions of the inner magnetosphere; i.e., the ring current and inner plasmasheet using energetic neutral atoms (ENA), the plasmasphere using extreme ultraviolet light (EUV), the electron and proton aurorae using far ultraviolet light (FUV), and the geocorona using FUV. The strawman instrument complement is shown in Tables 3 and 4. Seven instruments are described in these tables. Table 3 shows the spectral range and resolution, spin integrated field-of-view, angular resolution, and anticipated data rate for each instrument. Table 4 shows estimates of the requirements for accommodating each instrument on the IMI spacecraft. Volume, mass, power, pointing requirements, and operating temperatures are given.

Energetic neutral atoms are produced as a result of charge exchange between geomagnetically trapped, singly ionized ions, e.g., H^+, He^+, O^+, and geocorona hydrogen atoms. Since little energy is exchanged during the interaction, these neutral energetic atoms carry with them information about the original ion's spatial and energy distribution and composition. The resulting luminosity of the ring current, near-Earth plasmasheet, and heliospheric sources is sufficient to be seen by the anticipated IMI ENA imager. The detection of ENA below a few ten's keV requires a different approach than for ENA at higher energies. The use of ultra-thin foils [*McComas et al.*, 1991, 1992] and glancing angle scattering of atoms [*Herrero and Smith*, 1992] are the two techniques currently being pursued for the detection of low-energy ENA. A more detailed description of these techniques and of all IMI-related imaging techniques may be found in Wilson [1993].

The first crude images of the ring current in ENA were obtained fortuitously using the medium energy particle detector on International Sun Earth Explorer 1 (ISEE 1) [*Roelof et al.*, 1985; *Roelof*, 1987]. Instruments to be carried on the International Solar Terrestrial Physics (ISTP) Polar spacecraft [*Voss et al.*, 1992] and the SAC-B satellite [*Orsini et al.*, 1992] will be capable of imaging ENA and are expected to fly in 1994. High energy ENA imagers make use of a combination of thin foils, time-of-flight measurements, and anticoincidence discrimination to measure ENA energies and mass, while rejecting incident photons, ions, and electrons. A discussion of ENA imaging requirements can be found in Roelof et al. [1992].

Resonant scattering of solar ultraviolet light will be used by IMI to detect plasmaspheric helium, oxygen, and the geocorona. He^+ ions in the plasmasphere are luminous in extreme EUV light at 304 Å. Flux levels and, therefore, required sensitivities for an IMI 304 Å imager have been established by both spacecraft and rocket experiments [*Meier and Weller*, 1972; *Weller and Meier*, 1974; *Parsece et al.*, 1974; *Chakrabarti et al.*, 1982]. He^+ is typically found to comprise about 20% of plasmaspheric plasmas and closely follow variations in hydrogen and, therefore, total densities in the plasmasphere [*Newberry et al.*, 1989]. A further discussion of magnetospheric helium imaging can be found in Roelof et al. [1992].

TABLE 1. Science Definition Team

Team Members	Institution
T.P. Armstrong (Chairman)	University of Kansas
D.L. Gallagher (Study Scientist)	NASA Marshall Space Flight Center
A.L. Broadfoot	University of Arizona
S. Chakrabarti	Boston University
L.A. Frank	University of Iowa
K.C. Hsieh	University of Arizona
B.H. Mauk	Johns Hopkins University
D.J. McComas	Los Alamos National Laboratory
R.R. Meier	Naval Research Laboratory
S.B. Mende	Lockheed Missiles and Space Corp.
T.E. Moore	NASA Marshall Space Flight Center
G.K. Parks	University of Washington
E.C. Roelof	Johns Hopkins University
M.F. Smith	NASA Goddard Space Flight Center
J.J. Sojka	Utah State University
D.J. Williams	Johns Hopkins University

TABLE 2. MSFC Preliminary Design Team

Team Members	Area of Responsibility
C.E. DeSanctis	Chief, Space Science and Applications Group
C.L. Johnson	Study Manager
M.C. Herrmann	Lead Engineer
R.A. Alexander	Thermal Control
H.R. Blevins	Communications
T.R. Buzbee	Conceptual Art
C.K. Carrington	Guidance and Control
H.P. Chandler	Mass Budget
G.A. Hajos	Configuration
G.B. Kearns	Propulsion
J.H. Kim	Communications
L.D. Kos	Orbit Analysis
L.C. Maus	Power
F.A. Prince	Cost Analysis
T.L. Schmitt	Launch Vehicle
S.H. Spencer	Structures

TABLE 3. Inner Magnetosphere Imager (IMI) Strawman Instrument Data Table Measurements

No.	Instrument Name	Spectral Range	Spectral Resolution	Total Field of View	Angular Resolution	Data Rate
1	Hot Plasma Imager (ENA - High)	20-1000 keV	$\Delta E/\langle E_{chan}\rangle = 0.4$	~9 str	2° x 2°	12
	Hot Plasma Imager (ENA - Low)	1-50 keV	$\Delta E/\langle E_{chan}\rangle = 0.2$	4π str	4° x 4°	6
2	Plasmasphere Imager (He+304)	304 Å	50-100 Å	135° x 160°	0.5°	7
3	Plasmasphere Imager (O+834)	834 Å	50-100 Å	135° x 160°	0.5°	7
4	Geocoronal Imager	1216 Å	30 Å	4π str.	1° x 1°	2
5	Auroral Imager (FUV)	1304 Å, 1356 Å, LBH	20 Å	30° x 30°	0.03° x 0.03°	15
6	Proton Aurora Imager	1216 Å ± 40 Å	~1 Å	30° x 30°	0.06° x 0.06°	8
7	Electron Precipitation Imager	~0.3 - 10 keV	$\Delta E/E = 0.3$	3° x 3°	0.02° x 0.02°	2

Oxygen ions will be observed by IMI in scattered EUV radiation near 834 Å. A blend of three O^+ and one O^{++} solar emission lines is scattered by terrestrial oxygen [*Meier*, 1990]. Scattering efficiencies are reported by Meier [1990] to result from a complex convolution of solar emission lines with Doppler-shifted oxygen absorption lines. The observation of magnetospheric O^+ and O^{++} in scattered solar emission will be highly dependent upon ion bulk flow and thermal velocities. In addition, the concentrations of O^+ and O^{++} in the magnetosphere will be strongly dependent upon geophysical conditions and location [*Roberts et al.*, 1987]. Discussions of magnetospheric imaging can be found in Swift et al. [1989], Chiu et al. [1990], and Garrido et al. [1991]. Oxygen imaging is further distinguished from helium 304 Å imaging due to intense background ionospheric brightness near 834 Å. It is estimated that the ionosphere will be 2 to 4 orders of magnitude brighter than magnetospheric sources [*Williams et al.*, 1992] in this wavelength range. The oxygen ultraviolet imaging anticipated on IMI depends upon the use of multilayer mirrors as filters. Multilayer mirror design and fabrication is an active area of research [*Zukic and Torr*, 1991; *Chakrabarti and Edelstein*, 1992].

The interpretation of ENA images depends, in part, upon our knowledge of the geocorona. Variations in geocoronal densities with solar conditions will result in changes in observed ENA intensities that are generally independent of ring current and inner plasmasheet properties. The geocorona can be seen in scattered ultraviolet light from strong solar emission in the hydrogen Lyman-α line at 1216 Å. Geocoronal Lyman-α observations have been made from rocket [*Kupperian et al.*, 1959; *Purcell and Tousey*, 1960] and satellite [*Mange and Meier*, 1970; *Meier and Mange*, 1973] experiments. Although the geocorona is generally spherical in distribu-

TABLE 4. Inner Magnetosphere Imager (IMI) Strawman Instrument Data Accommodations

No.	Instrument Name	Volume (w × d × h) (m)	Mass (kg)	Avg. Power (W)	Pointing Stab. (deg/time)	Pointing Know. (deg)	Pointing Acc. (deg)	Detector Temp. (°C)
1	Hot Plasma Imager (ENA - High)	0.51×0.35×0.51	14.0	4.0	0.50/min	0.50	5.00	-23 to +30
	Hot Plasma Imager (ENA - Low)	0.30×0.30×0.25	7.0	7.0				-30 to +40
	Electronics Box	0.30×0.30×0.30	8.0	12.0				-30 to +40
2	Plasmasphere Imager (He+304)	0.48×0.16×0.20	7.2	4.5	0.25/min	0.25	0.50	-30 to +40
	Electronics Box	0.23×0.18×0.20	11.8	16.5				
3	Plasmasphere Imager (O+834)	0.48×0.16×0.20	7.2	4.5	0.25/min	0.25	0.50	-30 to +40
	Electronics Box	0.23×0.18×0.20	11.8	16.5				
4	Geocoronal Imager	0.30×0.60×0.30	15.0	15.0	0.50/min	0.50	0.50	≤ -100
	Electronics Box	0.30×0.60×0.30	12.0	15.0				-20 to +40
5	Auroral Imager (FUV)	0.30×0.70×0.40	18.0	20.0	0.06/min	0.03	0.20	≤ -100
	Electronics Box	0.20×0.70×0.40	12.0	15.0				-20 to +40
6	Proton Aurora Imager	0.60×0.80×0.20	20.0	25.0	0.06/min	0.03	0.20	≤ -100
	Electronics Box	0.30×0.80×0.20	10.0	15.0				-20 to +40
7	Electron Precipitation Imager	0.20×0.20×0.60	24.5	11.0	0.30/300 sec	0.30	0.30	Uncooled
	Electronics Box	0.25×0.18×0.18	3.0	9.0				

tion with a power law dependence on radial distance [*Wallace et al.*, 1970], a tailward extension [*Thomas and Bohlin*, 1972], dayside high-altitude depletion [*Bertaux and Blamont*, 1973], dawn-dusk asymmetry [*Meier and Mange*, 1973], and high-latitude depletion [*Thomas and Vidal-Madjar*, 1978] have all been observed. A source of error in observing the geocorona is from background extraterrestrial Lyman-α sources. Full sky maps of interplanetary and galactic Lyman-α intensities have been constructed by Bertaux and Blamont [1971], and by Thomas and Krassa [1971]. Long-term variation in exospheric hydrogen distributions have been found to be limited in a study by Rairden et al. [1986], using 4 years of Dynamics Explorer 1 (DE 1) observations.

Beginning with the Air Force DMSP weather service satellites [*Rodgers et al.*, 1974], the imaging of aurora from an orbiting spacecraft has been conducted for over 20 years. During this time, several low-altitude spacecraft have observed limited portions of the aurora in visible [*Anger et al.*, 1973] and ultraviolet [*Hirao and Itoh*, 1978] light. The first complete viewing of the auroral oval was accomplished by the SAI imager on the DE 1 spacecraft [*Frank et al.*, 1981, 1982]. This instrument viewed aurora in both visible (3175 to 6300 Å) and ultraviolet (1200 to 1800 Å) wavelengths, producing a complete image every 12 min. Global auroral imaging has largely been used to derive a qualitative view of global magnetospheric dynamics and topology. FUV imaging on IMI will seek to obtain quantitative measures of electron precipitation energies and fluxes through the discrimination of key spectral features in auroral emission. Meier et al. [1982] determines the characteristic electron energies and energy deposition rates from rocket observations, using the relative strengths of several N_2 LBH emission lines. In the same work, ionospheric atomic oxygen density is derived from O^+ emissions at 1356 Å, 1304 Å, and 989 Å. Strickland et al. [1983] demonstrated the importance of satellite FUV

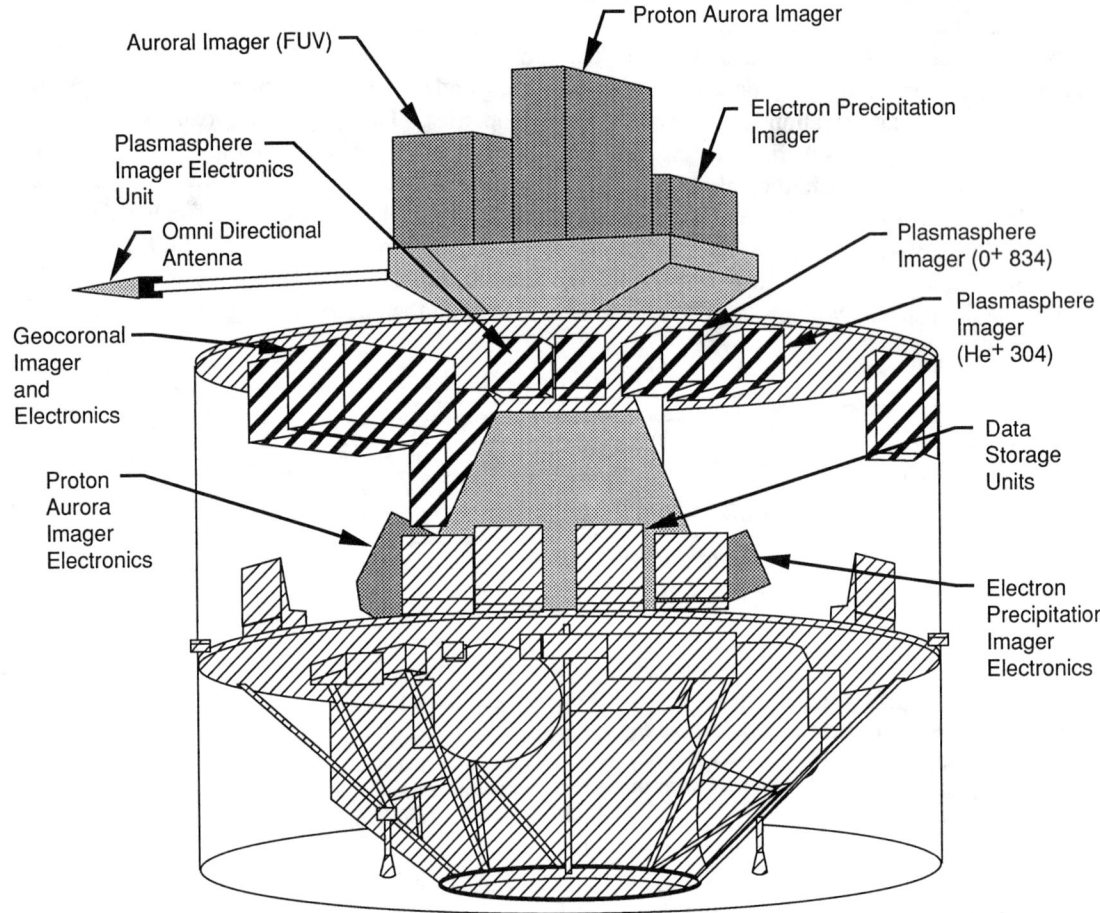

Fig. 1. Hughs HS-376 concept for IMI.

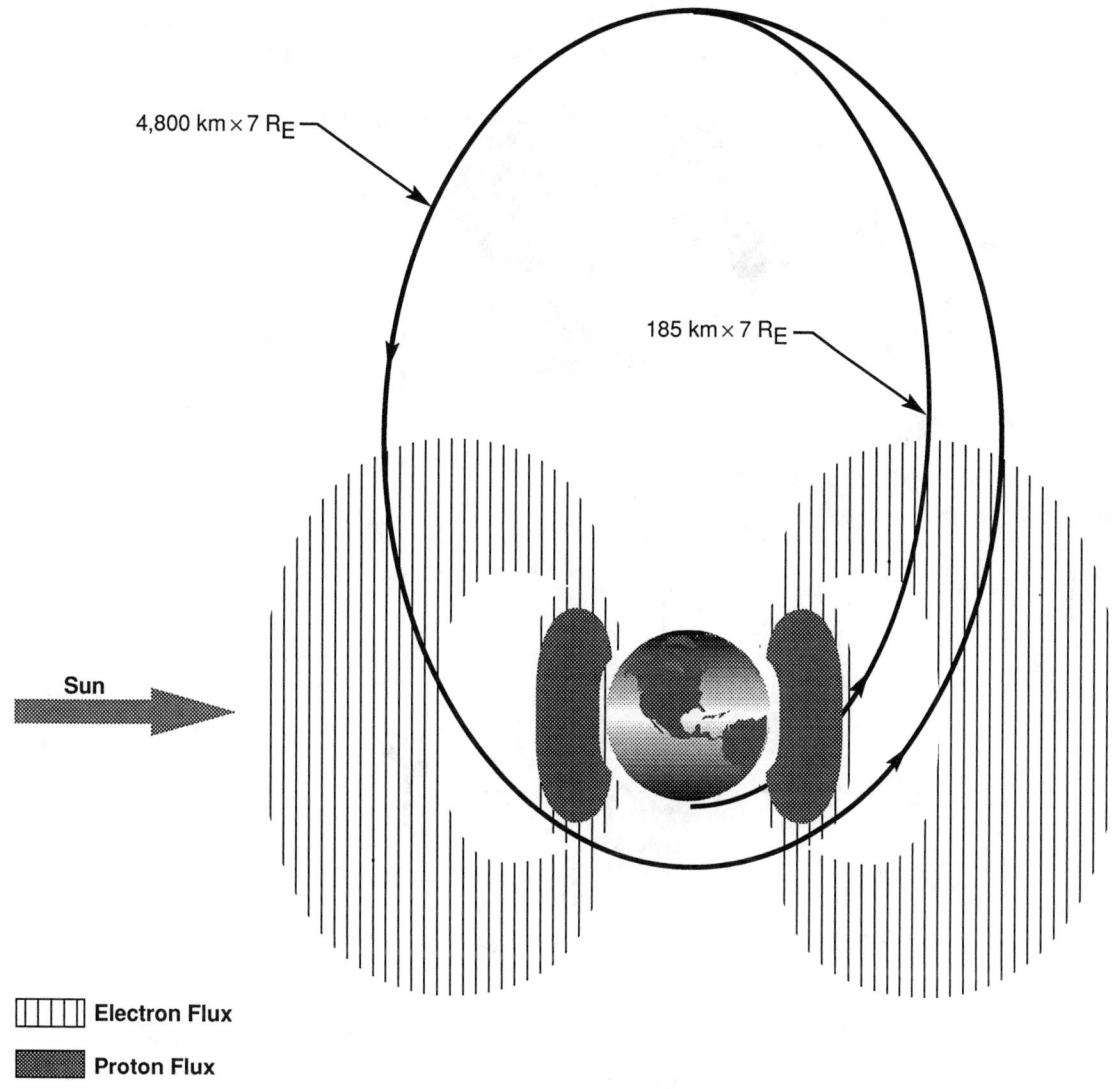

Fig. 2. lanned IMI orbit injection and perigee boost.

observations by deriving the incident auroral electron spectrum from auroral emissions of N_2^+ at 3914 Å, OI at 1356 Å, and several N_2 LBH bands.

IMI will also seek to quantitatively compare auroral electron and proton precipitation through the imaging of proton aurora. It has been shown by Lyons and Evans [1984] and Lyons et al. [1988] that discrete aurora, in general, occur in regions of nonadiabatic ion behavior on the nightside, rather than in regions of nonadiabatic electron behavior. In contrast, dayside nonadiabatic ion behavior may be associated with electron aurora [Lyons et al., 1987]. Ion precipitation may also play an important role in establishing ionospheric conductivity profiles during active periods [Senior et al., 1987]. Proton auroral emission results from excited hydrogen atoms produced in the ionosphere by charge exchange with precipitating energetic hydrogen ions. The anticipated IMI proton auroral imager will observe Doppler-shifted Lyman-α radiation in an 80 Å-wide band, centered at 1216 Å and with a resolution of 2 Å. Although the geocorona Lyman-α emission is a relatively strong source at 1216 Å, the high spectral resolution of the proton imager allows the removal of the cold geocoronal background from auroral Lyman-α emission. The high spectral resolution also allows a determination of precipitating proton energies. Images of proton aurora have been

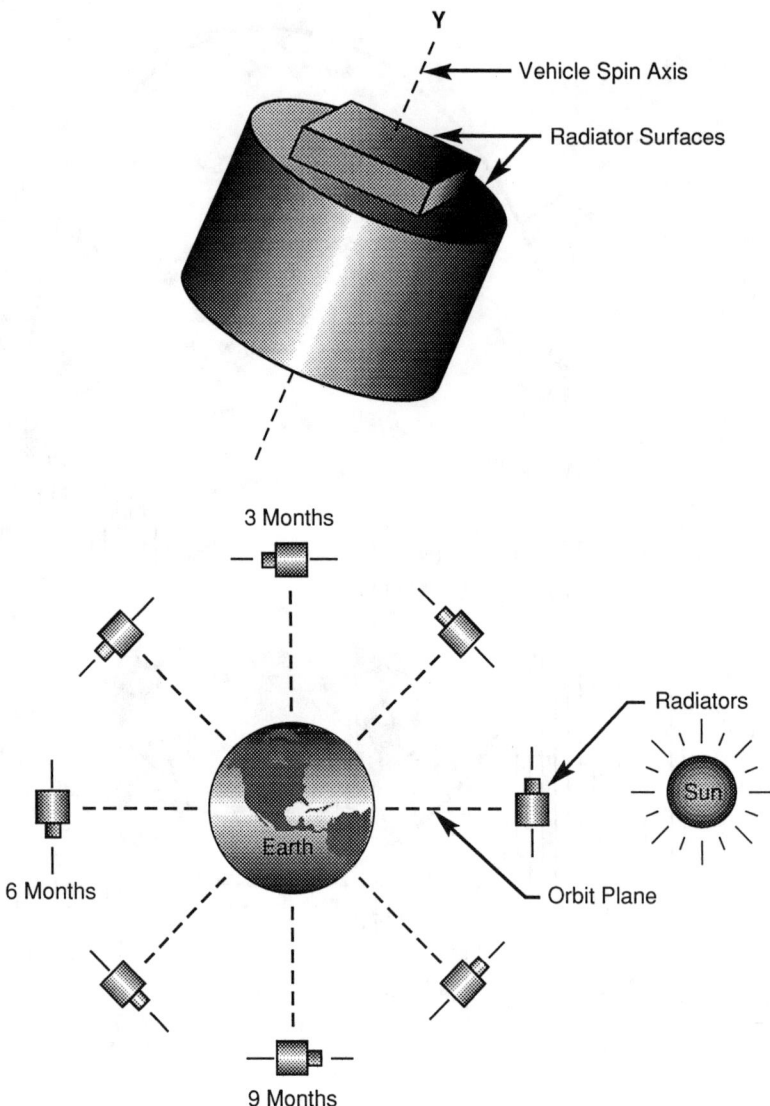

Fig. 3. Thermal control surfaces and orientations during a year.

obtained from the ground [*Ono and Hirasawa*, 1987] and are planned from space on the ISTP Polar spacecraft [*Torr et al.*, 1992].

Also planned for the IMI mission is the imaging of bremsstrahlung radiation produced by auroral-precipitating electrons. The IMI electron precipitation imager will be sensitive to X rays in the energy range from 0.3 eV to 10 keV, with a spectral resolution of $\Delta E/E=0.3$. Estimates of precipitating electron flux and energy spectrum have been obtained from satellite-borne [*Imhof et al.*, 1974; *Mizera et al.*, 1984] and rocket-borne [*Kremser et al.*, 1986] X ray experiments. Bremsstrahlung X ray measurements have been established as a remote proxy for determining precipitating energetic electron characteristics through comparisons between direct electron measurement and remote X ray observations [*Imhof et al.*, 1974; *Mizera et al.*, 1978]. The first systematic use of X ray measurements for deriving the spectra of precipitating auroral electrons was accomplished by Datlowe et al. [1988]. The ISTP Polar spacecraft will also carry an X ray imaging spectrometer instrument [*Imhof et al.*, 1992; *McKenzie et al.*, 1992], which is similar to that anticipated for IMI. This instrument will use multiple apertures at the larger distances in order to overcome low X ray luminocities. The IMI X ray imager will need to be sensitive to lower energy X rays, have a higher angular resolution, and be more sensitive than the Polar instrument.

3. PRELIMINARY IMI SPACECRAFT DESIGN

The preliminary design of the IMI spacecraft systems has been performed by a Marshall Space Flight Center (MSFC) Program Development team. The team members and areas of specialization are shown in Table 2. To date, two lightsat and one traditional design concepts have been studied. The more traditional design concept is based on the ISTP Polar spacecraft, built by Martin Marietta Corporation (formerly General Electric). The lightsat concepts are based on the Hughs HS-376 spacecraft, which is a spinner with a despun platform like Polar, and on a dual-spacecraft approach. Figure 1 shows a possible instrument configuration on the Hughs spacecraft. The auroral instruments, including the proton and electron auroral imagers, are mounted on a despun platform and pointed toward the Earth. The remaining instruments are mounted on the body of the spinning spacecraft. A total initial mass of 900.5 kg is estimated for this configuration. Of that mass the strawman instruments comprise 181 kg, the propellant needed to raise the initial perigee is 134 kg, and the on-orbit reaction control (RCS) system propellant is 32 kg. The spacecraft propulsion system uses a monopropellant hydrazine in a blowdown pressurization operation.

IMI is expected to be launched from the Western Test Range. The DELTA II, Titan IIS, Taurus 120XL/S, and Conestoga launch vehicles are being considered. The initial launch will place IMI in a polar orbit with a 185 km altitude perigee and 7 R_E altitude apogee (Figure 2). At apogee on the first orbit, a kick motor will raise the IMI perigee to an altitude of 4800 km and an orbital

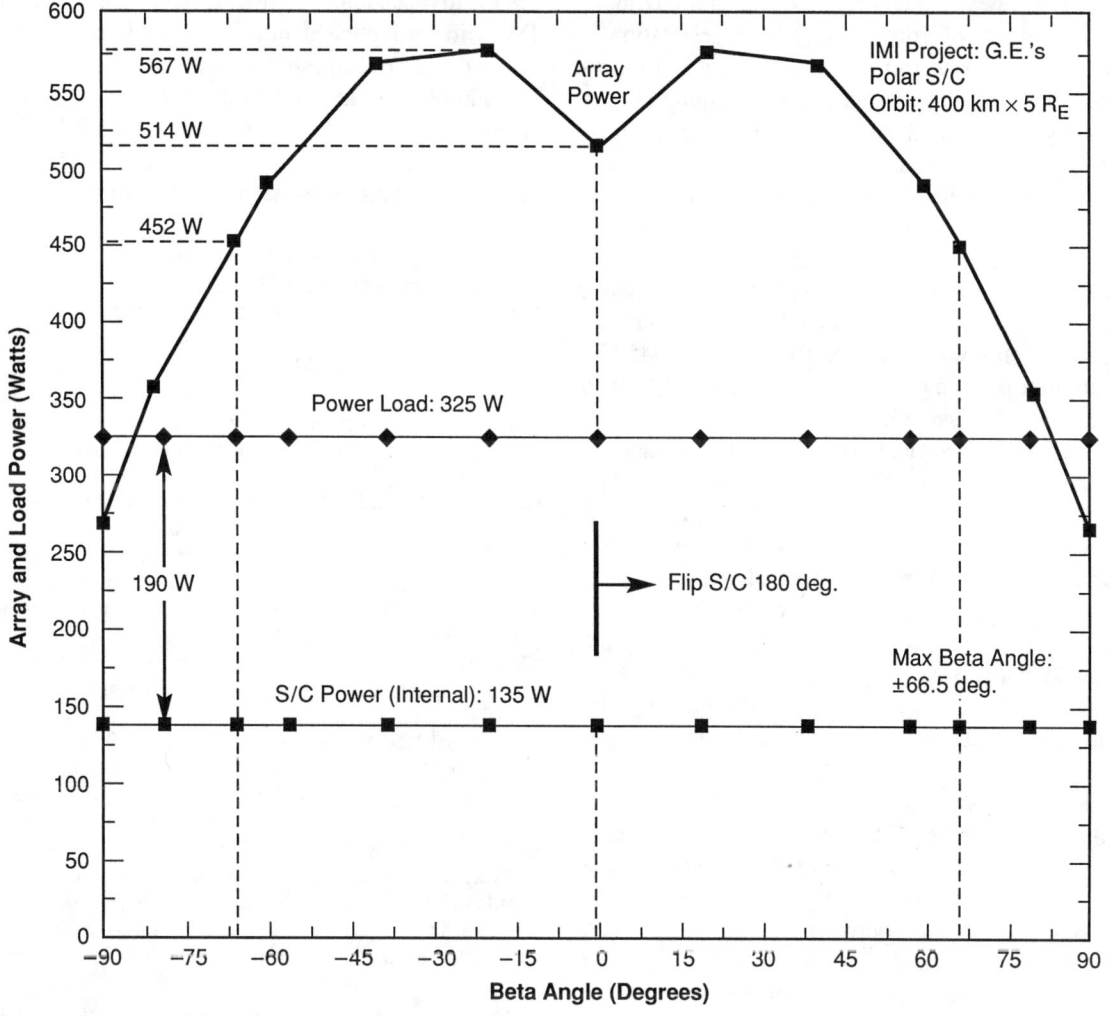

Fig. 4. Solar array and load power as a function of the angle of incidence of the Sun's light on the arrays (beta angle).

period of about 15 hours. Over the planned 2-year lifetime of the mission, the 90° inclined orbit will precess from an apogee over the northern pole to an apogee near the equator. For thermal control, a reorientation of the spacecraft is anticipated twice each year. Figure 3 shows the spacecraft orientation during the year and thermal radiating surface of the spacecraft. The reorientation is designed to keep the radiating surface pointed away from the Sun.

Electrical power is derived from approximately 14.9 m^2 of body-mounted, silicon solar cells. Figure 4 shows a plot of solar array and load power as a function of beta angle for the Polar spacecraft concept. Beta angle is the angle between the normal to the spacecraft cylindrical surface and the spacecraft-Sun direction. At large angles, the solar cells receive only glancing illumination from the Sun and can provide limited power to the spacecraft. Vertical dashed lines in Figure 4 indicate the maximum estimated beta angle of ±66°. Spacecraft reorientation is performed as the spacecraft goes through a zero beta angle. The total spacecraft systems power requirement is approximately 135 W and the total required instrument power is 190 W, for a total of 325 W. It is further estimated that cutouts for instrument viewing and thermal control will reduce the available solar cell surface area by 8%. There will also be a degradation in solar cell performance over the 2-year mission lifetime of about 25%. With these losses, it is projected that the minimum available solar cell power will drop from a high of 452 W at the maximum beta angle, as shown in Figure 4, to 355 W at the end of the mission life.

NASA's deep space network (DSN) on S-band is expected to be used for the IMI mission. Instrument data rates total 59 kbps. At full instrument data rates, 1.98 gigabits of data will be generated during each orbit. Two 1.29 gigabit capacity tape recorders are planned in one of the optional mission configurations. At a playback rate of 512 kbps, a full recorder will be able to downlink its data in 42 min. A backup real-time telemetry system at 56 kbps is also being considered. This system would be limited to ground station availability and line-of-sight viewing.

Evaluation of spacecraft systems and optional configurations continues. The dual spacecraft approach would involve the use of two small spacecraft, one spinning and one three-axis stabilized. Although the relative cost advantage between single and dual spacecraft approaches has not yet been determined, it is expected that the single spacecraft option will be the better option. Spacecraft systems complexity and use of off-the-shelf elements favor the single spacecraft approach.

4. SUMMARY

IMI will break new ground in magnetospheric research. We will, for the first time, be able to see the global morphology of major plasma systems and the dynamic relationships between them. Old and new in situ measurements will yield new insight as a result of their placement in the global framework of magnetospheric processes provided by IMI. Ring current and near-Earth plasmasheet dynamics, including storm-time injection and depletion, will be studied by the innovative observation of oxygen and hydrogen ENA. The dynamics of plasmaspheric filling and loss due to convection will be seen through the imaging of helium scattering of EUV solar light. Storm-time ionospheric outflow of oxygen will also be seen in EUV. Polar aurora will be seen in FUV and X rays. The geocorona will be seen in resonantly scattered Lyman-α radiation from the Sun. IMI will be launched into a 90° inclined polar orbit as either one or two spacecraft. Orbital perigee will be at an altitude of 4800 km and an apogee altitude of 7 R_E. IMI is expected to be a lightsat design and take the fullest advantage of technological advancements in solar cell, structure, satellite subsystems, and instrument designs.

Acknowledgments. The IMI studies are supported by the NASA Office of Space Science and Applications, Space Physics Division, Magnetospheric Physics Branch.

REFERENCES

Anger, C. D., T. Fancott, J. McNally, and H. S. Kerr, ISIS-II Scanning Auroral Photometer, *Applied Opitcs*, *12*, 1753, 1973.

Bertaux, J. L., and J. E. Blamont, Evidence for a source of an extraterrestrial hydrogen Lyman-alpha emission: The interstellar wind, *Astron. Astrophys.*, *11*, 200, 1971.

Bertaux, J. L., and J. E. Blamont, Interpretation of OGO 5 Lyman alpha measurements in the upper geocorona, *J. Geophys. Res.*, *78*, 80, 1973.

Chakrabarti, S., and J. Edelstein, An 834 Å reflective coating for magnetospheric imagery applications, *SPIE*, *1744*, 208, 1992.

Chakrabarti, S., F. Paresce, S. Bowyer, Y. T. Chiu, and A. Atkin, Plasmaspheric helium ion distribution from satellite observations of He II 304 Å, *J. Geophys. Res. Lett.*, *9*, 151, 1982.

Chiu, Y. T., R. M. Collin, S. Chakrabarti, and G. R. Gladstone, Magnetospheric and exospheric imaging in the extreme ultraviolet, *Geophys. Res. Lett.*, *17*, 267, 1990.

Datlowe, D. W., W. L. Imhof, and H. D. Voss, X ray spectral images of energetic electrons precipitating in the auroral zone, *J. Geophys. Res.*, *93*, 8662, 1988.

Frank, L. A., J. D. Craven, K. L. Ackerson, M. R. English, R. H. Eather, and R. L. Carovillano, Global auroral imaging instrument for the Dynamics Explorer mission, *Space Sci. Instr.*, *5*, 369, 1981.

Frank, L. A., J. D. Craven, J. L. Burch, and J. D. Winningham, Polar views of the earth's aurora with Dynamics Explorer, *Geophys. Res. Lett., 9*, 1001, 1982.

Garrido, D. E., R. W. Smith, D. S. Swift, and S.-I. Akasofu, Imaging the earth's magnetosphere: Effects of plasma flow and temperature, *Planet. Space Sci., 39*, 1559, 1991.

Herrero, F., and M. F. Smith, Imager of low-energy neutral atoms (ILENA): Imaing neutrals from manetosphere at energies below 20 keV, *SPIE, 1744*, 32, 1992.

Hirao, K. and T. Itoh, Scientific satellite Kyokko (Exos-A), *Solar Terr. Env. Res. in Japan, 2*, 148, 1978.

Imhof, W. L., G. H. Nakano, R. G. Johnson, and J. B. Reagan, Satellite observations of bremsstrahlung from widespread energetic electron precipitation events, *J. Geophys. Res., 79*, 565, 1974.

Imhof, W. L., H. D. Voss, and D. W. Datlowe, The imaging of X rays for magnetospheric investigation, *SPIE, 1744*, 196, 1992.

Kremser, G., A Korth, S. Ullaland, J. Stadsnes, W. Baumjohann, L. Block, K. M. Torkar, W. Riedler, B. Aparicio, P. Tanskaren, I. B. Iversen, N. Cornilleau-Wehrlin, J. Solomon, and E. Amata, Energetic electron precipitation during a magnetospheric substorm and its relationship to wave particle interaction, *J. Geophys. Res., 91*, 5711, 1986.

Kupperian, J. E., Jr., E. T. Byram, T. A. Chubb, and H. Friedman, Far ultraviolet radiation in the night sky, *Planet. Space Sci., 1*, 3, 1959.

Lyons, L. R., and D. S. Evans, An association between discrete aurora and energetic particle boundaries, *J. Geophys. Res., 89*, 2395, 1984.

Lyons, L. R., J. F. Fennell, and A. L. Vampola, A general association between discrete auroras and ion precipitation from the tail, *J. Geophys. Res., 93*, 12,932, 1988.

Lyons, L. R., A. L. Vampola, and T. W. Speiser, Ion precipitation from the magnetopause current sheet, *J. Geophys. Res., 92*, 6147, 1987.

Mange, P., and R. R. Meier, OGO 3 observations of the Lyman alpha intensity and the hydrogen concentration beyond 5 R_E, *J. Geophys. Res., 75*, 1837, 1970.

McComas, D. J., B. L. Barraclough, R. C. Elphic, H. O. Funsten, III, and M. F. Thomsen, Magnetospheric imaging with low-energy neutral atoms, *Proc. Natl. Acad. Sci. USA, 88*, 9598, 1991.

McComas, D. J., H. O. Funsten, III, J. T. Gosling, K. R. Moore, and M. F. Thomsen, Low-energy neutral-atom imaging, *SPIE, 1744*, 40, 1992.

McKenzie, D. L., D. J. Gorney, and W. L. Imhof, Auroral X ray imaging from high- and low-earth orbit, *SPIE, 1745*, 39, 1992.

Meier, R. R., The scattering rate of solar 834 Å radiation by magnetorspheric O^+ and O^{++}, *Geophys. Res. Lett., 17*, 1613, 1990.

Meier, R. R., and P. Mange, Spatial and temporal variations of the Lyman-alpha airglow and related atomic hydrogen distributions, *Planet. Space Sci., 21*, 309, 1973.

Meier, R. R., and C. S. Weller, EUV resonant radiation from helium atoms and ions in the geocorona, *J. Geophys. Res., 77*, 1190, 1972.

Meier, R. R., R. R. Conway, P. D. Feldman, D. J. Strickland, and E. P. Gentieo, Analysis of nitrogen and oxygen far ultraviolet auroral emission, *J. Geophys. Res., 87*, 2444, 1982.

Mizera, P. F., D. J. Gornet, and J. L. Roeder, Auroral X ray images from DMSP-F6, *Geophys. Res. Lett., 11*, 255, 1984.

Mizera, P. F., J. G. Luhmann, W. A. Kolansinski, and J. B. Blake, Correlated observation of auroral arcs, electrons, and X rays from a DMSP satellite, *J. Geophys. Res., 83*, 5573, 1978.

Newberry, I. T., R. H. Comfort, P. G. Richards, and C. R. Chappell, Thermal He^+ in the plasmasphere: Comparison of observations with numerical calculations, *J. Geophys. Res., 94*, 15,265, 1989.

Ono, T., and T. Hirasawa, and C. I. Meng, Proton auroras observed at the equatorial edge of the duskside auroral oval, *Geophys. Res. Lett., 14*, 660, 1987.

Parsece, F., C. S. Bowyer, and S. Kumar, On the distribution of He^+ in the plasmasphere from observations of resonantly scattered He II 304 Å, *J. Geophys. Res., 79*, 174, 1974.

Purcell. J. D., and R. Tousey, The profile of solar hydrogen Lyman-α, *J. Geophys. Res., 65*, 370, 1960.

Rairden, R. L., L. A. Frank, and J. D. Craven, Geocoronal imaging with Dynamics Explorer, *J. Geophys. Res., 91*, 13,613, 1986.

Roberts, W. T., Jr., J. L. Horwitz, R. H. Comfort, C. R. Chappell, J. H. Waite, Jr., and J. L. Green, Heavy ion density enhancements in the outer plasmasphere, *J. Geophys. Res., 92*, 13,499, 1987.

Rodgers, E. H., D. F. Nelson, and R. C. Savage, Auroral photography from a satellite, *Science, 183*, 951, 1974.

Roelof, E. C., Energetic neutral atom images of a storm-time ring current, *Geophys. Res. Lett., 14*, 652, 1987.

Roelof, E. C., D. G. Mitchell, and D. J. Williams, Energetic neutral atoms (E~50 keV) from the ring current, IMP 7/8 and ISEE 1, *J. Geophys. Res., 90*, 10,991, 1985.

Roelof, E. C., B. H. Mauk, and R. R. Meier, Instrument requirements for imaging the magnetosphere in extreme-ultraviolet and energetic neutral atoms derived from computer-simulated images, *SPIE, 1744*, 19, 1992.

Senior, C., J. R. Sharber, O. De La Beujardiére, R. A Hellis, D. S. Evans, J. D. Winningham, M. Sugiura, and W. R. Hoegy, E and F region study of the evening sector auroral oval: A Chatanika/Dynamics Explorer 2/NOAA 6 comparison, *J. Geophys. Res., 92*, 2477, 1987.

Strickland, D. J., J. R. Jasperse, and J. A. Whalen, Dependence of auroral FUV emissions on the incident electron spectrum and neutral atmosphere, *J. Geophys. Res., 88*, 8051, 1983.

Swift, D. W., R. W. Smith, and S. I. Akasofu, Imaging the earth's magnetosphere, *Planet. Space Sci., 37*, 379, 1989.

Thomas, G.E. and R.C. Bohlin, Lyman-alpha measurements of neutral hydrogen in the outer geocorona and in interplanetary space, *J. Geophys. Res., 77*, 2752, 1972.

Thomas, G.E. and R.F. Krassa, OGO 5 measurements of the Lyman alpha sky background, *Astron. Astrophys., 11*, 218, 1971.

Thomas, G.E. and A. Vidal-Madjar, Latitude variations of exospheric hydrogen and the polar wind, *Planet. Space Sci., 26*, 873, 1978.

Torr, D. G., M. R. Torr, M. Zukic, J. Spann, and R. B. Johnson, The Ultraviolet Imager (UVI) for ISTP, *SPIE, 1745*, 61, 1992.

Wallace, L., C. A. Barth, J. B. Pearce, K. K. Kelly, D. E. Anderson, Jr., and W.G. Fastie, Mariner 5 measurements of the earth's Lyman alpha emission, *J. Geophys. Res., 75*, 3769, 1970.

Weller, C. S., and R. R. Meier, First satellite observations of the He$^+$ 304 Å radiation and its interpretation, *J. Geophys. Res.*, *79*, 1572, 1974.

Williams, D. J., E. C. Roelof, and D. G. Mitchell, Global magnetospheric imaging, *Rev. of Geophys.*, *30*, 183, 1992.

Wilson, G. R., Inner Magnetosphere Imager (IMI) instrument heritage, NASA Contractor Report 4498, 1993.

Zukic, M., and D. G. Torr, High-reflectivity multi-layers as narrowband VUV filters, *SPIE, 1485,* 216, 1991.

D. L. Gallagher, ES53, NASA Marshall Space Flight Center, Huntsville, Alabama 35812.

Imaging of Magnetospheric Dynamics Using Low Energy Neutral Atom Detection

H. O. Funsten, D. J. McComas, K. R. Moore, E. E. Scime, and M. F. Thomsen

Space and Atmospheric Sciences Group, Los Alamos National Laboratory, Los Alamos, New Mexico

Recent advances in low energy neutral atom (LENA) detection technology show that imaging of magnetospheric dynamics is achievable with 4°x4° resolution and can distinguish numerous features of the magnetosphere. A critical factor in detecting low energy neutrals is their removal from the ambient UV background. One technique to accomplish this is by ionizing the neutrals by their transmission through an ultrathin charge modification foil and subsequent electrostatic deflection. We describe results of a prototype LENA imager based on this concept using a collimated beam of 10 keV H, showing that LENAs can be imaged with a 4°x4° resolution. Additionally, we illustrate model results of anticipated LENA images of the magnetosphere based on the Rice University Magnetospheric Specification Model and show that 3 keV plasma variations resulting from geomagnetic storm disturbances can be directly observed using a LENA imager.

1. INTRODUCTION

For over 30 years, the Earth's magnetosphere has been studied and modeled using single-point measurements from a large number of space-based and earth-based instruments. These measurements have proven sufficient for characterizing the statistically average nature of the magnetosphere. However, due to the relatively short time scales associated with dynamic processes in the magnetosphere and lack of a global coverage by satellites, extrapolating single point measurements to infer global magnetosphere dynamics has invariably spawned various conflicting models of the dynamic magnetosphere.

Magnetospheric plasma is composed predominantly of H^+, O^+, and He^+, and their fluxes and energy distributions vary widely both spatially and temporally within the magnetosphere depending on its dynamic state [e.g., *Young*, 1983; *Krimigis et al.*, 1985; *Gloeckler et al.*, 1985]. An emerging technique for global magnetospheric imaging is remote detection of magnetospheric plasma ions that are neutralized by charge exchange with geocoronal neutral species [*Roelof and Williams*, 1988; *McEntire and Mitchell*, 1989; *McComas et al.*, 1991; *Williams et al.*, 1992]. Since charge exchange involves electronic interactions, the velocity of a plasma ion is not altered significantly by the exchange interaction. The neutralized plasma ions subsequently follow ballistic trajectories that are not influenced by ambient electric and magnetic fields, and they can be remotely detected. Using this technique, a source plasma can be imaged, and its properties can be derived. In fact, using ISEE-1/MEPI Roelof obtained an image of the storm-time ring current with 50 keV ions [*Roelof*, 1987], showing that neutral atom imaging can provide global images of magnetospheric plasmas. Using these results, we infer that images taken at different time intervals can provide vital information on the transport of particles and energy throughout the magnetosphere on a global scale.

Considerable neutral atom imaging development and flux simulations have focused on detection of energetic neutral atoms (ENAs) at energies greater than approximately 20 keV [*Keath et al.*, 1989; *Hsieh and Curtis*, 1989; *McEntire and Mitchell*, 1989; *Cheng et al.*, 1993]. However, the magnetosphere is predominantly composed of lower energy (roughly 0.5 keV to 10s of keV) plasmas,

Solar System Plasmas in Space and Time
Geophysical Monograph 84
Copyright 1994 by the American Geophysical Union.

resulting in a greater low energy neutral atom (LENA) flux than ENA flux through much of the magnetosphere [*Young*, 1983]. By imaging LENAs, phenomena can be observed that would address a wider variety of issues associated with the magnetosphere.

Previous LENA studies include LENA flux models using a simple magnetosphere geometry [*Moore et al.*, 1992] and possible LENA detection techniques using transmission through ultrathin carbon foils [*McComas et al.*, 1991, 1992; *Funsten et al.*, 1993] and surface reflection [*Herraro and Smith*, 1992; *Funsten et al.*, 1993; *Ghielmetti et al.*, 1993]. In this paper we report new LENA simulation results using the Rice University Magnetospheric Specification Model [*Bales et al.*, 1992] and imaging results of a prototype LENA imager based on the instrument concept described in McComas et al. [1992].

2. Imaging of Magnetospheric Dynamics

The applicability of ENA imaging for monitoring the evolution of the Earth's ring current, particularly during and after geomagnetic storms, has been addressed in considerable detail [e.g., *Roelof and Williams*, 1988; *Williams et al.*, 1992]. The LENA imager considered here would extend those observations to the lower energy portion of the ring current, allowing ring current evolution to be monitored at larger radial distances and for a longer time into the storm recovery phase [e.g., *McComas et al.*, 1991]. Transient effects due to changes in the convection electric field, such as plasmaspheric plumes, should also be visible (see, for example, Plate 2c below). Furthermore, a LENA imager in a relatively low altitude polar orbit could measure vertical profiles and temporal variations of upflowing auroral ions that are field aligned and non-field aligned (conics) [*Gorney et al.*, 1981].

In addition, LENA imaging would permit the visualization of the near-Earth plasma sheet, the probable site of magnetospheric substorm initiation. With LENA imaging one could potentially observe large-scale plasma sheet motions and the thinning and reconfiguration associated with substorms. The formation and down-tail departure of plasmoids could also be monitored. Another important objective with LENA imaging of the near-Earth plasma sheet is determining the location of substorm-associated ion energization, i.e., the substorm "injection front" [*Mauk and Meng*, 1986; *Fairfield*, 1992].

LENA images of the dayside extension of the plasma sheet and ring current should reveal a discontinuity in the flux and composition at the magnetopause, enabling the position and motion of the magnetopause to be monitored globally. The compression and erosion effects of solar wind variability could thus be studied directly. Moreover, the ability to image 1 keV LENAs might enable imaging of the magnetosheath and hence to monitor the location of the bow shock, especially under compressed conditions where sheath fluxes penetrate into regions of higher geocoronal neutral densities.

3. LENA Imaging Model

The LENA flux at a satellite is dependent on the total LENA creation and losses along a particular line-of-sight vector **r**. The variation with distance r of the differential LENA flux f_i (cm^2 s sr keV)$^{-1}$ of species i along **r** is

$$\frac{df_i(r,E)}{dr} = \sigma_{ij}(E) J_i(r,E,\alpha) n_j(r) - L(r,E) \quad (1)$$

where J_i is the differential ion flux of the plasma, n_j is the neutral density of species j, and E and α are the plasma ion energy and pitch angle, respectively. The cross section σ_{ij} for charge exchange between ion species i and neutral species j (i.e., $i^+ + j^o \rightarrow i^o + j^+$) is a function of the relative speed between the charge-exchanging atoms; however, since the speed of the plasma ions is much greater than that of the geocoronal neutrals, σ_{ij} can be evaluated at the plasma ion energy. The loss term $L(r,E)$, which represents ionization of LENAs by photoionization or ionizing interactions with plasma ions or other neutrals, is in general small compared to the first term and can be ignored.

The total differential LENA flux $F_i(E)$ at a spacecraft located at **r** = (0,0,0) from all sources along **r** equals the line-of-sight integral

$$F_i(E) = \sigma_{ij}(E) \int_0^\infty J_i(r,E,\alpha) n_j(r) \, dr \quad (2)$$

As long as the source uniformly fills the pixel, the source brightness (i.e., the LENA flux $f_i(r,E)$) is independent of the distance between the source and the observer. We note here that the exclusion of the LENA loss term $L(r,E)$ suggests that LENA sources are optically thin. This underscores the complementary aspects of global imaging using neutral atoms and single point plasma measurements: single point measurements within the LENA sources provide important information on optical depth and microscale structure.

To quantify the anticipated LENA environment for instrument definition and optimization, we have developed a numerical LENA model [*Moore et al.*, 1992] patterned after the ENA model of Roelof [1987]. The model is based on global representations of magnetospheric plasmas and geocoronal neutral densities that are derived from an enormous database of single-point observations. Any uncertainty of this type of model underscores the potential unprecedented scientific return of a LENA imager.

The spherically symmetric geocoronal neutral density n_H used here is based on DE 1 Lyman α measurements [*Rairden et al.*, 1986] that is a Chamberlain model with a 1050° temperature, a 500 km exobase, an exobase density of 4.4×10^4 cm^{-3}, and a satellite critical level of 3.0 times the exobase radius. Since H is the dominant neutral geocoronal species, we use a geocorona composition of H only, and the relevant charge exchange cross sections are based on Chebychev polynomial fits to empirical cross section data [*Barnett et al.*, 1990; see also *Moore et al.*, 1992].

We use the Rice University Magnetospheric Specification Model (MSM) [*Bales et al.*, 1992] to estimate the plasma flux $J_i(r,E,\alpha)$. The MSM results for proton fluxes at 3 keV in the magnetic equatorial plane are illustrated in Plate 1 for (a) quiet-time with a uniform ring current plasma and low plasma densities near the inner boundary (~ 2R$_E$) of the MSM and (b) the end of the main phase of the geomagnetic storm period of 21-23 April 1988. In Plate 1b the enhancement of low energy ring current plasma is caused primarily by the injection of plasma sheet ions, and the dayside depletion feature is caused by a change in the cross-magnetosphere electric field that results in loss of the previously trapped plasma through the magnetopause. The plasma flux at any point out of the equatorial plane is determined by magnetically mapping the flux value in the equatorial plane to that point. The isotropic MSM plasma distributions eliminate the pitch angle dependency of J_i.

The line-of-sight, 3 keV hydrogen LENA fluxes $F_H(E)$ computed by our numerical model are illustrated in Plate 2. The vertical and horizontal axes correspond to the spacecraft elevation and azimuthal angles, respectively, and the Earth is located at 90° elevation and 180° azimuth. All panels have 4°x4° pixels and display the color-coded logarithm of the model proton flux. Plates 2a and 2b have the spacecraft located 9R$_E$ from the Earth in the magnetic equatorial plane at the dusk terminator. Plate 2c has the spacecraft located at 9R$_E$ above the earth. Plate 2a represents a quiet magnetosphere corresponding to the MSM model magnetosphere shown in Plate 1a, while Plates 2b and 2c represent magnetospheric storm conditions corresponding to Plate 1b. The spacecraft in Plates 2a and 2b is located in a region of closed field lines, so the LENA background flux equals approximately 10 (cm^2 s sr keV)$^{-1}$ due to the relatively high ambient plasma density. In Plate 2c, however, the spacecraft is located on open field lines with a relatively low plasma density, so the LENA background is comparatively small (< 4 (cm^2 s sr keV)$^{-1}$).

Plates 2a and 2b indicate that the low energy ring current intensifies (primarily due to injection of the plasma sheet ions) and moves earthward during a geomagnetic storm. The model illustrated here assumes a spherically symmetric geocorona [*Rairden et al.*, 1986]. In reality, there exists a neutral geocoronal tail (i.e., high neutral density) induced by solar photon pressure [*Bishop*, 1985]. Therefore, the computed LENA fluxes from the plasma sheet are likely underestimated. The model results in Plate 2c indicate that storm-time depletion of dayside low energy plasma and storm-time enhancement of the low energy ring current plasma would also be observable. These results demonstrate the potential of using LENA emissions to globally image the magnetospheric structures and dynamics that have been difficult or impossible to study using existing single-point measurements.

4. LENA DETECTION

The foremost technical challenge of detecting low energy (< 30 keV) neutrals in space is their separation from the tremendous Ly-α background to which low energy detectors (e.g., microchannel plates (MCPs) and channel electron multipliers) are sensitive. With an ambient UV flux of 10^8-10^{11} cm^{-2}s^{-1} [*Hsieh et al.*, 1980] and a UV detection efficiency of MCPs of 5-10%, either direct attenuation of the UV or removal of the LENAs from the UV is crucial. Proposed ENA imagers [*Keath et al.*, 1989; *Hsieh and Curtis*, 1989; *Cheng et al.*, 1993] employ a thick composite foil that attenuates the UV [*Hsieh et al.*, 1980, 1991; *Drake et al.*, 1992] but allows ENAs to pass with minimal angular scattering. However, extending this technique to LENA energies results in resolution degradation of the imager due to significant foil-induced angular scattering and energy straggling.

Historically, the first attempt to detect neutral atoms in space utilized a simple carbon foil and electrostatic deflection apparatus to convert the neutral atoms to ions and sweep them away from the ambient UV into a detector [*Bernstein et al.*, 1969]. Later instruments for LENA detection included a slotted disk velocity analyzer [*Moore and Opal*, 1975] and secondary ion emission [*Rosenbauer et al.*, 1983]. Recently, LENA imaging using surface reflection has been proposed [*Herraro and Smith*, 1992; *Ghielmetti et al.*, 1993], although some technological issues must be addressed [*Funsten et al.*, 1993]. We have developed a 0.7 to >30 keV LENA imager [*McComas et al.*, 1991, 1992] similar in concept to the simple neutral atom detector employed by Bernstein et al. [1969].

Our basic LENA imager, designed for a spinning spacecraft and shown in Figure 1, consists of four

278 LOW ENERGY NEUTRAL ATOM IMAGING

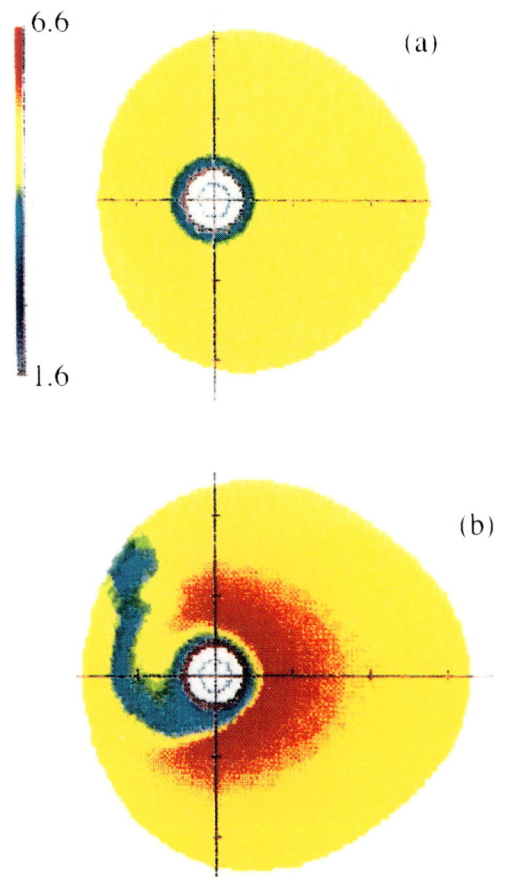

Plate 1. Equatorial fluxes of 3 keV H$^+$ (a) for a quiet-time magnetosphere and (b) at the end of the main phase of a geomagnetic storm. Fluxes were calculated using the Rice University Magnetospheric Specification Model (MSM) for the geomagnetic storm period of 21-23 April 1988 [Bales et al., 1988]. The axes tics represent 5 R$_E$. Sunward is to the left.

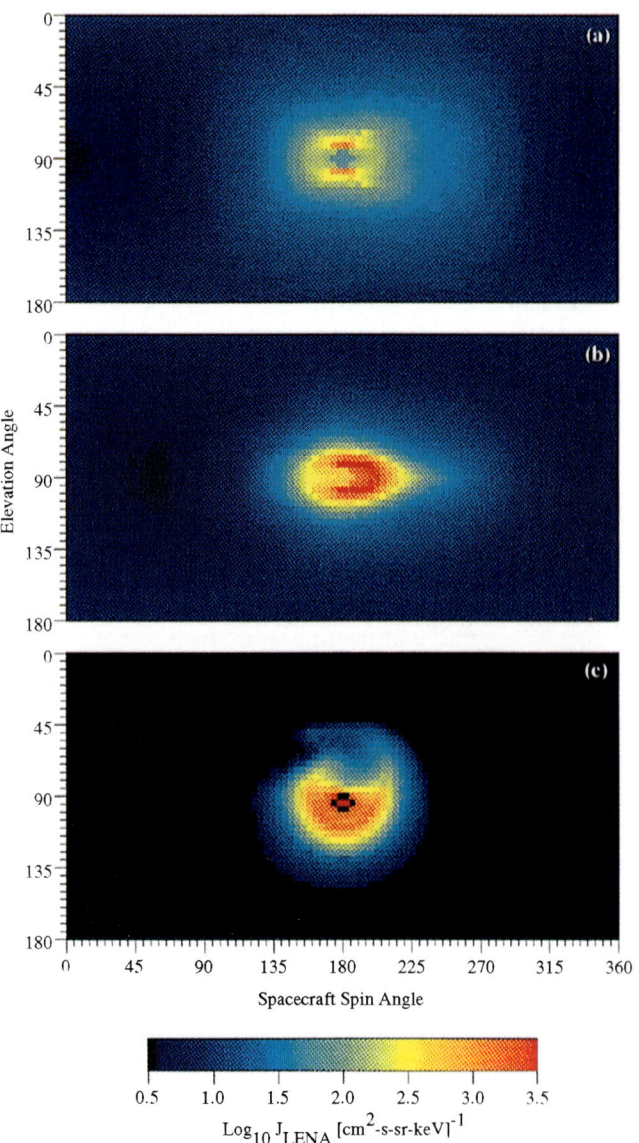

Plate 2. Computed LENA fluxes based on the MSM results shown in Plate 1: (a) quiet time ring current from a spacecraft located at 9R$_E$ from the Earth in the magnetic equator at the dusk terminator; (b) storm-time ring current from same spacecraft location, and (c) storm time magnetosphere from 9 R$_E$ above the Earth.

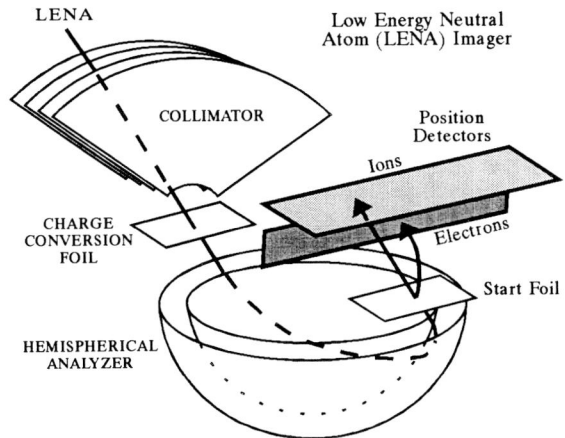

Fig. 1. Schematic of a LENA imager.

fundamental components: a collimator, a charge conversion foil, a hemispherical electrostatic analyzer, and a time-of-flight detector. In addition to high UV rejection, the LENA imager has a relatively large geometric factor, a large field-of-view, and an ability to discriminate LENAs from both ambient ions and penetrating radiation.

The collimator consists of serrated and blackened plates that are alternately biased to reject ambient plasma ions. For a spinning spacecraft, the collimator defines the azimuthal resolution. A LENA that passes through the collimator transits an ultrathin (nominal 0.5 µg cm^{-2}) carbon foil that covers a 4 cm^2 aperture. Since the exit charge state of a low energy projectile transiting a foil is independent of its incident charge state [*Oetliker*, 1989], a LENAs have a known probability of exiting the foil as a positive ion. The exit charge state distribution is clearly an important part of the instrument geometric factor. Figure 2(a) shows the measured probability that H, He, and O exit a 0.5 µg cm^{-2} foil as a positive ion [H and O data from *Funsten et al.*, 1992]. For hydrogen the probability of exiting as H$^+$ is ~ 6% at 1 keV and ~ 41% at 30 keV.

An ionized LENA enters the electrostatic spherical analyzer (ESA) and will pass through if its energy is within the energy passband. Typical values for the ESA are an inner radius of 8 cm and outer radius of 12 cm, resulting in an energy resolution of $\Delta E/E \approx 25\%$. After passing through the ESA, ionized LENAs enter the time-of-flight section that has a second ultrathin carbon foil for secondary electron emission [*Keller et al.*, 1991; *Mechbach*, 1975]. LENAs are subsequently detected on a position-sensitive (wedge anode) detector, while secondary electrons from the carbon foil are electrostatically steered toward a separate position sensitive detector. Detection of both a LENA and its correlated secondary electrons provide a coincidence scheme to reject false counts produced by penetrating radiation. Furthermore, time-of-flight analysis can be performed using the time difference between detection of a LENA and its correlated secondary electrons, and, since the approximate LENA energy is known from the ESA passband, the LENA mass can be identified. Due to the large difference in the time-of-flight of H and O, simple time-of-flight windowing can be employed to distinguish these two main magnetospheric species.

The difference between the detected position of the secondary electron(s), which reflects the point on the foil through which the LENA transited, and the detected position of the LENA itself provides information on the

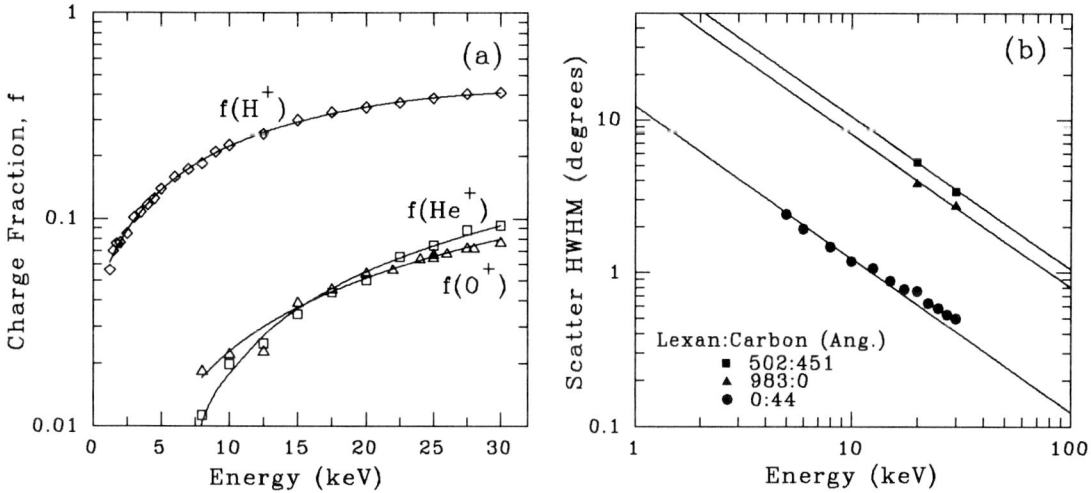

Fig. 2. (a) Exit charge state distributions H, He, and O and (b) angular scattering half width at half maximum (HWHM) of hydrogen in 44 Å carbon, 983 Å Lexan, and a 502:451 Å composite Lexan: carbon foils.

polar angle of incidence of the LENA. The polar resolution is defined by the pixel size of the anodes and the angular scattering of the foils, and at lower energies (< 4 keV) the polar resolution is dominated by the angular scattering [*Funsten et al.*, 1992]. Figure 2(b) shows measured angular scattering half-width at half maximum (HWHM) for incident H as a function of the incident ion energy for foils of 44 Å (nominal 0.5 μg cm^{-2}) carbon, 983 Å Lexan, and a composite foil of 502 Å Lexan and 451 Å carbon. The experimental apparatus used to derive these results has been previously described [*Funsten et al.*, 1992]. The solid lines are least-square fits to the data based on the theoretical result that the ratio of the HWHM to the ion energy equals a constant for a particular ion species and foil [*Meyer*, 1971].

The thick Lexan and composite foils used for the upper two curves in Figure 2(b) are approximately the thickness used for UV attenuation in ENA imagers [*Hsieh et al.*, 1980, 1991; *Drake et al.*, 1992]. They have a scatter HWHM that is almost an order of magnitude greater than that of the 0.5 μg cm^{-2} foils used in LENA imaging, illustrating the advantage of the LENA instrument approach used here.

5. PROTOTYPE LENA IMAGER RESULTS

A prototype LENA imager has been fabricated with the following characteristics: a collimator with a 50° (polar) x 4° (azimuthal) field of view and alternately biased plates for ion rejection; nominal 0.5 μg/cm^2 carbon foils for LENA ionization and secondary electron emission; a hemispherical analyzer with an analyzer constant of 18.4; and a position sensitive MCP detector with a resistive anode for polar position measurements of detected LENAs. For this prototype, the secondary electron signal was not monitored. Rather, a small (2 mm diameter) aperture was placed immediately preceding the secondary electron emission foil to identify the known transit point of an ionized LENA through the second foil. A neutral H beam was formed by inserting a nominal 0.5 μg cm^{-2} carbon foil in an incident H$^+$ beam. A significant fraction of the beam exits the foil as neutrals (see Figure 2(a) for charge conversion efficiency), and the remaining ions are electrostatically swept from the neutral beam by the collimator. The prototype imager was mounted on a stage that had independent azimuthal and polar angle control.

The polar angle response measured using the prototype is depicted in Figure 3(a) for 10 keV H. Each peak is labeled with the incident polar angle of the beam relative to the instrument. This angle was varied from 0° to 25° in increments of 5°. The abscissa in Figure 3(a) is the polar

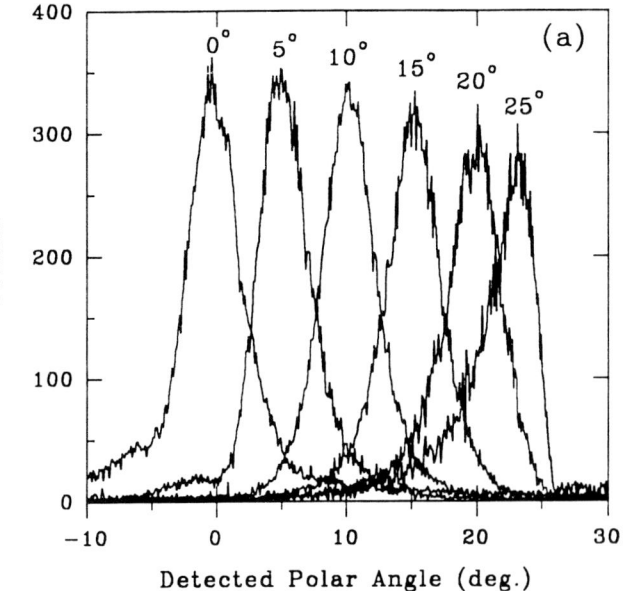

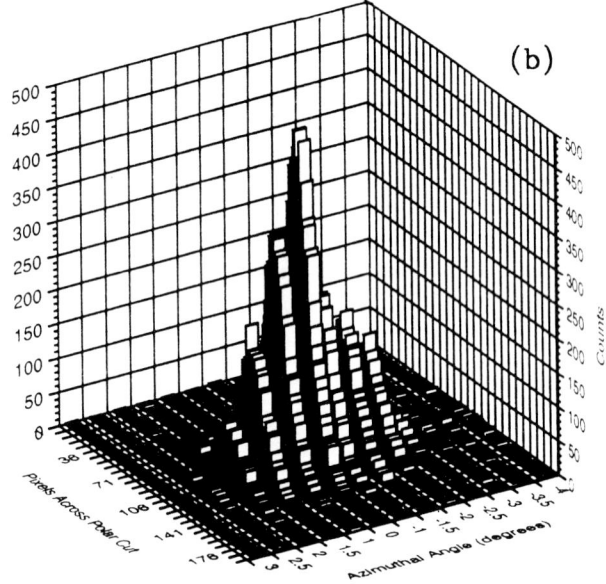

Fig. 3. Prototype LENA imager results: (a) detected polar angle distributions from a 10 keV hydrogen beam at incident polar angles ranging from 0° to 25° in increments of 5°; and (b) an image of a 10 keV H beam incident at 0° polar angle and azimuthal angles ranging from -4° to 4° in increments of 0.5°.

angle of detected atoms at the MCP detector, calculated using the detection position on the MCP detector and the distance between the secondary electron emission foil and the MCP detector. The observed polar resolution of the imager is approximately 2° HWHM for a 20 keV H beam and approximately 2.5° HWHM for a 10 keV beam which agrees with the calculated polar resolution estimated from measured angular scattering in the foil. Note that in Figure 3(a) the data for the incident polar angle of 25° are truncated at the ±25° limit of the polar field of view of the instrument.

The azimuthal response for a beam of 10 keV hydrogen is illustrated in Figure 3(b) for incident azimuthal angles of -4° to 4° in increments of 0.5° and an incident polar angle of 0°. The resulting composite image of all azimuthal look directions is that of the incident H beam. The azimuthal angular resolution, set by the collimator, is approximately 2° HWHM. The image of the beam in Figure 3(b) is representative of the base resolution of a LENA imager.

6. Discussion

The LENA imager described here has a pixel field of view of 4°x4°, an instantaneous instrument field of view is 120° (polar) x 4° (azimuthal), and a 120° x 360° field of view over one spin. A typical geometric factor for a basic imager is approximately 0.05 to 1 cm^2 sr eV/eV depending on the available spacecraft resources [*McComas et al.*, 1992]. Based on the analysis in McComas et al. [1992] for this imager, over a 2-3 minute image accumulation interval the LENA fluxes of approximately 10^3 (cm^2 s sr keV)$^{-1}$ (e.g., the red in Plates 2a-c) would produce several hundred counts per pixel and LENA fluxes on the order of 50 (cm^2 s sr keV)$^{-1}$ (e.g., the light blue in Plates 2a-c) would generated 10s of counts per pixel. This is sufficient to provide high-sensitivity, statistically significant measurements to observe macroscale magnetospheric dynamics.

Based on the LENA imager described in this paper and LENA simulation fluxes derived using MSM results (e.g., Plate 2), LENA imaging can be a viable tool to examine global magnetospheric dynamics. The simulated LENA fluxes for quiet time and storm-time magnetospheres show that the plasma sheet can be imaged and magnetospheric dynamics can be observed. The LENA imager is based on mature technologies, and foil studies have been performed to measure exit charge state fractions and scattering of H, He, and O that transit ultra-thin foils. A prototype LENA imager has been fabricated and tested, and a beam of hydrogen has been imaged with a measured resolution of 4°x4° in agreement with calculated values. We thus conclude that LENA imaging is both practical in the near-term and of immense scientific potential for global magnetospheric observations.

Acknowledgments. The authors thank J. Baldonado and D. Everett for lab support and D.G. Mitchell (APL) for providing Lexan composite foils. We also thank R.W. Spiro and R.A. Wolf for providing the MSM results. Development of the MSM at Rice University was supported by Air Force contract F19628-90-K-0012 and scientific application of the MSM by NASA grant NAGW-1655. This work was performed under the auspices of the United States Department of Energy.

References

Bales, B., J. Freeman, B. Hausman, R. Hilmer, R. Lambour, A. Nagai, R. Spiro, G.-H. Voigt, R. Wolf, W.F. Denig, D. Hardy, M. Heinemann, N. Maynard, F. Rich, R.D. Belian, and T. Cayton, Status of the development of the magnetospheric specification and forecast model, Proceedings of the Solar-Terrestrial Prediction Workshop, Ottawa, May 1992, in press.

Barnett, C.F., Collisions of H, H$_2$, He, and Li atoms and ions with atoms and molecules, Vol. 1 of Atomic Data for Fusion, *Rep. ORNL-6086*, Oak Ridge Nat. Lab., Oak Ridge, Tenn, 1990.

Bernstein, W., R.L. Wax, N.L. Sanders, and G.T. Inouye, An energy spectrometer for energetic (1-25 keV) neutral hydrogen atoms, in *Small Rocket Instrumentation Techniques*, North Holland, Amsterdam, 224-231, 1969.

Bishop, J, Geocoronal structure: The effects of solar radiation pressure and the plasmasphere interaction, *J. Geophys. Res.*, 90, 5235-5245, 1985.

Cheng, A.F., E.P. Keath, S.M. Krimigis, B.H. Mauk, R.W. McEntire, D.G. Mitchell, E.C. Roelof, and D.J. Williams, Imaging neutral particle detector, *Remote Sensing Reviews*, in press, 1993.

Drake, V.A., B.R. Sandel, D.G. Jenkins, and K.C. Hsieh, H Lyα transmission ot thin foils of C, Si/C, and Al/C for keV particle detectors, *Proc. SPIE, 1744*, 148-160, 1992.

Fairfield, D.H., Advances in magnetospheric storm and substorm research, *J. Geophys. Res.*, 97, 10865-10874, 1992.

Funsten, H.O., D.J. McComas, and B.L. Barraclough, Application of thin foils in low-energy neutral-atom detection, *Proc. SPIE, 1744*, 62-69, 1992.

Funsten, H.O., D.J. McComas, and E.E. Scime, Low-energy neutral-atom imaging techniques, *Proc. SPIE, 2008*, 93-104, 1993.

Ghielmetti, A.G., E.G. Shelley, S.A. Fuselier, F.A. Herraro, M.F. Smith, P. Wurz, P. Boschsler, and T.S. Stephen, Mass spectrograph for imaging low-energy neutral atoms, *Proc. SPIE, 2008*, 105-112, 1993.

Gloeckler, G., B. Wilken, W. Stüdemann, F.M. Ipavich, D. Hovestadt, D.C. Hamilton, and G. Kremser, First composition measurement of the bulk of the storm-time ring current (1 to 300 keV/e) with AMPTE/CCE, *Geophys. Res. Lett.*, 12, 325-328, 1985.

Gorney, D.J., A. Clarke, D. Croley, J. Fennell, J. Luhmann, and

P. Mizera, The distribution of ion beams and conics below 8000 km, *J. Geophys. Res., 86*, 83-89, 1981.

Herraro, F.A., and M.F. Smith, Imager of low-energy neutral atoms: imaging neutrals from the magnetosphere at energies below 20 keV, *Proc. SPIE, 1744*, 32-39, 1992.

Hsieh, K.C., E. Keppler, and G. Schmidtke, Extreme ultraviolet induced forward photoemission from thin carbon foils, *J. Appl. Phys., 5*, 2242-2246, 1980.

Hsieh, K.C., and C.C. Curtis, Remote sensign of planetary magnetospheres: Mass and energy analysis of energetic neutral atoms, in *Solar System Plasma Physics, Geophysical Monograph Ser.*, vol. 54, eds. J.H. Waite, J.L. Burch, and R.L. Moore, pp.159-164, AGU, Washington, D.C., 1989.

Hsieh, K.C., B.R. Sandel, V.A. Drake, and R.S. King, H Lyman α transmittance of thin C and Si/C foils for keV particle detectors, *Nucl. Instrum. and Meth., B61*, 187-193, 1991.

Keath, E.P., G.B. Andrews, A.F. Cheng, S.M. Krimigis, B.H. Mauk, D.G. Mitchell, and D.J. Williams, Instrumentation for energetic neutral atom imaging of magnetospheres, in *Solar System Plasma Physics, Geophysical Monograph Ser.*, vol. 54, eds. J.H. Waite, J.L. Burch, and R.L. Moore, pp. 165-170, AGU, Washington, D.C., 1989.

Keller, J.W., K.W. Ogilvie, J.W. Boring, and R.W. McKemie, Measurement of the secondary emission yield of electrons from thin carbon foils after passage of low energy ions, *Nucl. Instr. and Meth, B61*, 291-294, 1991.

Krimigis, S.M., G. Gloeckler, R.W. McEntire, T.A. Potemra, F.L. Scarf, and E.G. Shelley, Magnetic storm of September 4, 1984: A synthesis of ring current spectra and energy densities measured with AMPTE/CCE, *Geophys. Res. Lett, 12*, 329-332, 1985.

Mauk, B.H., and C.-I. Meng, Macroscopic ion acceleration associated with the formation of the ring current in the Earth's magnetosphere, in *Ion Acceleration in the Magnetosphere and Ionosphere, Geophysical Monograph Ser.*, vol. 38, ed. T. Cheng, pp. 351-361, AGU, Washington, D.C., 1986.

McComas, D.J., B.L. Barraclough, R.C. Elphic, H.O. Funsten, and M.F. Thomsen, Magnetospheric imaging with low-energy neutral atoms, *Proc. Natl. Acad. Sci. USA, 88*, 9598-9602, 1991.

McComas, D.J., H.O. Funsten, J.T. Gosling, K.R. Moore, and M.F. Thomsen, Low-energy neutral atom imaging, *Proc. SPIE, 1744*, 40-50, 1992.

McEntire, R.W., and D.G. Mitchell, Instrumentation for global magnetospheric imaging via energetic neutral atoms, in *Solar System Plasma Physics, Geophysical Monograph Ser.*, vol. 54, eds. J.H. Waite, J.L. Burch, and R.L. Moore, pp. 69-80, AGU, Washington, D.C., 1989.

Mechbach, W., G. Braunstein, and N. Arista, Secondary electron emission in the backward and forward directions from thin carbon foils traversed by 25-250 keV proton beams, *J. Phys., B8*, L344-L349, 1975.

Meyer, L., Plural and multiple scattering of low-energy heavy particles in solids, *Phys. Stat. Sol., 44*, 253-268, 1971.

Moore, J.H., and C.B. Opal, A slotted disk velocity analyzer for the detection of energetic atoms above the atmosphere, *Space Sci. and Instrum., 1*, 377-386, 1975.

Moore, K.R., D.J. McComas, H.O. Funsten, and M.F. Thomsen, Low-energy neutral atom imaging in the Earth's magnetosphere: Modeling, *Proc. SPIE, 1744*, 51-61, 1992.

Oetliker, M., Charge state distribution, scattering, and residual energy of ions after passing through thin carbon foils, Thesis, University of Bern, Switzerland, 1989.

Rairden, R.L., L.A. Frank, and J.D. Craven, Geocoronal imaging with dynamics explorer, *J. Geophys. Res., 91*, 13613-13630, 1986.

Roelof, E.C., and D.J. Williams, The terrestrial ring current: From in situ measurements to global images using energetic neutral atoms, *Johns Hopkins APL Tech. Dig., 9*, No. 2, 1988.

Roelof, E.C., Energetic neutral atom image of a storm-time ring current, *Geophys. Res. Lett., 14*, 652-655, 1987.

Rosenbauer, H., H.J. Fahr, E. Keppler, M. Witte, P. Hemmerich, H. Lauche, A. Loidl, and R. Zwick, The ISPM interstellar neutral gas experiment, in *The International Solar Polar Mission- Its Scientific Investigations*, Eds. K.P. Wenzel, R.G. Marsden, and B. Battrick, ESA SP-1050, 125-139, 1983.

Williams, D.J., E.C. Roeloff, and D.G. Mitchell, Global magnetospheric imaging, *Rev. Geophys., 30*, 183-208, 1992.

Young., D.T., Near-equatorial magnetopsheric particles from ~1 eV to ~1 MeV, *Rev. Geophys., 21*, 402-418, 1983.

The NASA High Energy Solar Physics (HESP) Mission for the Next Solar Maximum

R. P. Lin[1], B. R. Dennis[2], R. Ramaty[2], A. G. Emslie[3], R. Canfield[4], G. Doschek[5]

The NASA High Energy Solar Physics (HESP) mission offers the opportunity for major breakthroughs in our understanding of the fundamental energy release and particle acceleration processes at the core of the solar flare problem. HESP's primary strawman instrument, the High Energy Imaging Spectrometer (HEISPEC), will provide X-ray and gamma-ray imaging spectroscopy, i.e., high-resolution spectroscopy at each spatial point in the image. It has the following unique capabilities: (1) high-resolution (~ keV) spectroscopy from 2 keV - 20 MeV to resolve flare gamma-ray lines and sharp features in the continuum; (2) hard X-ray imaging with 2" angular resolution and tens of millisecond temporal resolution, commensurate with the travel times and stopping distances for the accelerated electrons; (3) gamma-ray imaging up to ~ 1 MeV with 4"-8" resolution with the capability of imaging in specific lines or continuum regions; (4) moderate resolution spectral measurements of energetic (20 MeV to ~1 GeV) gamma-rays and neutrons. Additional strawman instruments include an imaging Bragg crystal spectrometer for diagnostic information and a soft X-ray/XUV/UV imager to map the flare coronal magnetic field and plasma structure. The HESP mission also includes extensive ground-based observational and supporting theory programs. Recently, the HESP mission has been adapted to Lightsats, lighter, smaller, cheaper spacecraft: the baseline HESP mission now includes two Pegasus-class spacecraft. A launch by the end of the year 2000 is desirable to be in time for the next solar activity maximum.

1. INTRODUCTION

The overarching scientific objective of the High Energy Solar Physics (HESP) mission is to explore the processes of impulsive energy release and particle acceleration in magnetized plasmas. The fundamental importance of these high-energy processes transcends their significance in solar physics since they are found to play a major role throughout the universe at sites ranging from planetary and neutron star magnetospheres to active galaxies. The detailed understanding of these processes is one of the major goals of space physics and astrophysics, but in essentially all cases, we are only just beginning to perceive the relevant basic physics. Nowhere can one pursue the study of this basic physics better than in the active Sun, where solar flares are the direct result of impulsive energy release and particle acceleration. The accelerated particles, notably the electrons with energies of tens of keV, appear to contain a major fraction of the total flare energy, thus indicating the fundamental role of the high-energy processes. The acceleration of electrons is revealed by hard X-ray and gamma-ray bremsstrahlung; the acceleration of protons and nuclei is revealed by nuclear gamma-rays, pion-decay radiation, and neutrons. The proximity of the Sun means that these high-energy emissions appear orders of magnitude more intense than from any other cosmic source, plus they can be better resolved, both spatially and temporally. Consequently, the Sun is the only astrophysical object where the phenomena can be studied with the detail necessary to understand the fundamental processes.

---

[1] Physics Department and Space Sciences Laboratory, University of California, Berkeley, California
[2] Goddard Space Flight Center, Greenbelt, Maryland
[3] University of Alabama, Huntsville, Alabama
[4] Institute for Astronomy, University of Hawaii, Honolulu, Hawaii
[5] Naval Research Laboratory, Washington, DC

Solar System Plasmas in Space and Time
Geophysical Monograph 84
Copyright 1994 by the American Geophysical Union.

TABLE 1. The HESP Science Study Group

Researcher	Affiliation
Robert Lin, Chairperson	UC Berkeley
Brian Dennis, Study Scientist	GSFC
Richard Canfield	U. Hawaii
Carol Crannell	GSFC
John Davis	MSFC
George Doschek	NRL
Gordon Emslie	U. Alabama, Huntsville
Richard Fisher	GSFC
David Forrest	U. New Hampshire
Gerhard/Haerendel/Erich Reiger	Max Planck Institut, Germany
Hugh Hudson	UCSD
Gordon Hurford	Caltech
Jean-Michel Lavigne	CESR Toulouse, France
James Ling	JPL
Monique Pick/Nicole Vilmer	Observatoire de Paris, France
Reuven Ramaty	GSFC
Frank van Beek	Delft Univ., The Netherlands
William Wagner, Ex Officio	NASA Headquarters
Ken Lang, Ex Officio	NASA Headqrts (Tufts Univ.)

HESP is designed to study the high-energy processes in solar flares. These involve the rapid release of energy stored in unstable magnetic configurations, the equally rapid conversion of this energy into kinetic energy of accelerated particles and hot plasma, the transport of these particles, and the subsequent heating of the ambient solar atmosphere. Observations of hard X-rays, gamma rays, and neutrons serve as the best diagnostic of these processes by providing direct evidence for the interaction of accelerated particles in solar flares. The necessary spatial and temporal resolving powers must match the spatial and temporal scales that characterize the processes of energy release, acceleration, and transport. The sensitivity should be high enough to detect the initial energy release and particle acceleration, and also to provide observations over a wide range of intensities from microflares to large flares. Equally important, the spectral resolving power must be high enough to allow the deciphering of the rich information encoded in both the gamma-ray lines and the highly-structured photon continuum. The observations should provide imaging with spectroscopy and should cover the entire photon energy range from soft thermal X-rays to gamma rays. It is the primary goal of HESP to provide, for the first time, such comprehensive observations. Additional objectives of HESP include the study of the composition of the solar atmosphere, using gamma-ray spectroscopy, and the investigation of microflares and their contribution to the heating of the corona.

The importance of HESP was recognized by the Astronomy and Astrophysics Survey Committee [*Bahcall et al.*, 1991] and by the NASA Space Physics Strategy Implementation Study of the Space Physics Subcommittee for the Space Science and Applications Advisory Committee [1991]. In both, HESP was the highest priority intermediate space mission for solar physics. The NASA Space Sciences and Applications Advisory Committee (SSAAC) placed HESP second among all new NASA intermediate, moderate, and flagship missions in their Strategic Plan 1992. A pre-phase A study was initiated in January 1991 and an initial report was issued in July 1991 [*Lin et al.*, 1991]. Recently, studies have been initiated to accommodate the HESP mission on Lightsats: smaller, lighter, cheaper spacecraft that can be built in a shorter time. These studies have resulted in a multi-spacecraft HESP program with more schedule flexibility. To provide significant coverage of the next solar maximum a HESP launch by the end of the year 2000 is desirable. Table 1 lists the science study group members.

2. SCIENTIFIC OBJECTIVES

One of the most outstanding signatures of solar flares is the hard X-ray emission, commonly observed during the impulsive phase (Figure 1). Hard X-rays are generally attributed to bremsstrahlung produced in collisions between suprathermal electrons and the constituents of the ambient solar atmosphere. When the electron energy E_e is much larger than the kT of the ambient gas, the energy lost by the electrons to bremsstrahlung is only a small fraction (~10^{-5} in the deka-keV range) of the energy lost to Coulomb colli-

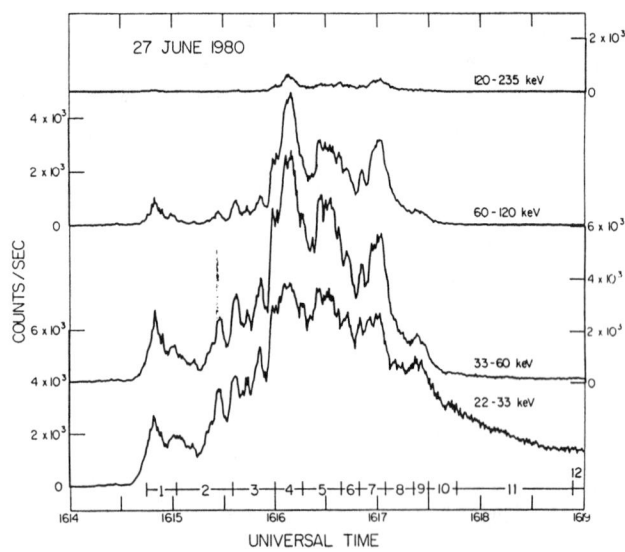

Fig. 1. Hard X-ray burst for a solar flare on 27 June 1980, showing multiple spikes [Lin et al., 1981].

sions. Thus, the fast electrons must contain many orders of magnitude more energy than the X-rays that they produce. Since observed hard X-ray spectra are typically steep power-laws produced by steep electron spectra, the total energy content of the electrons is critically dependent on the low-energy cutoff of the accelerated electron spectrum. Even for assumed cutoffs in the 20 to 30 keV range, the inferred electron energy can amount to a substantial fraction, ~10 to 50%, of the total energy released in the flare. Thus, particle acceleration must be a significant part of the flare energy release process.

The bulk of the radiative output of flares appears at lower energies, in the visible, UV, and soft X-ray bands, with a substantial fraction of the total output radiated in soft X-rays [*Canfield*, 1980]. Because so much energy is contained in the hard X-ray producing electrons, it is reasonable to assume that it is the energy deposited by these electrons into the ambient solar atmosphere which produces the observed lower-energy emissions. The thick-target model [*Brown*, 1971] assumes that the primary result of the energy release process is the impulsive acceleration of electrons, probably in the coronal part of a loop or arcade of loops. These electrons propagate along the magnetic field lines and deposit their energy predominantly in the lower corona and the chromospheric portions of the loops, heating the ambient gas. This model has been shown to be consistent with much of the observational data [*Emslie*, 1986]. The very large number of accelerated electrons required for flares and the substantial currents they carry, however, pose difficult challenges to models.

The fact that the impulsive energy release and particle acceleration are assumed as a starting point in these models illustrates how little is known about these fundamental processes. A clue to their nature may be contained in high spectral resolution (~1 keV FWHM) observations of hard X-rays (Figure 2), presently available only for a single medium-sized flare observed with a balloon-borne instrument [*Lin and Schwartz*, 1987]. These measurements show that the hard X-ray spectrum has a characteristic double power-law shape with a sharp downward break at an energy that varied during the flare from ~30 keV to greater than 100 keV. This sharp break may result from the electrons being accelerated by quasi-static electric fields parallel to the magnetic field; the break energy would then be a measure of the potential drop. This type of acceleration occurs in the Earth's aurora, but with potential drops of ~10 kV compared to the ~100 kV required in flares. The electric fields will also produce Joule heating of the thermal electrons as well as runaway acceleration of the faster electrons. Holman and Benka [1992] showed that the combination of heating of the plasma and acceleration is consistent with the overall hard

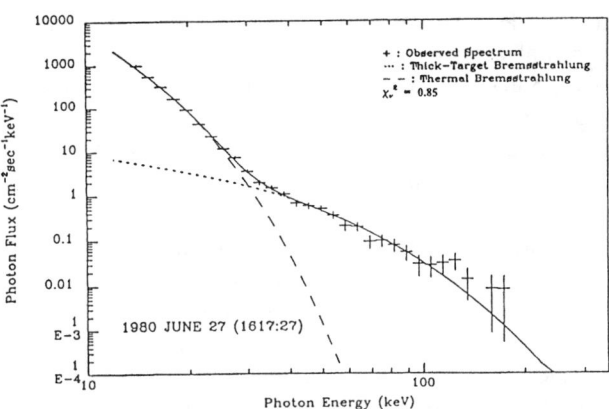

Fig. 2. High-resolution spectrum late in the hard X-ray burst of Figure 1, fitted here to a model of a D.C. electric field applied parallel to the magnetic field [Holman and Benka, 1992]. The electric field results in both Joule heating to produce the hot thermal plasma and runaway electron acceleration (thick target bremsstrahlung).

X-ray spectrum observed (Figure 2), which includes the "superhot" thermal component [*Lin et al.*, 1981].

Largely as a result of the recognition of the formidable constraints that the non-thermal models place on the primary energy release process, alternative thermal models with $E_e \simeq kT$ have been put forward. In these models [*Brown et al.*, 1979; *Smith and Harmony*, 1982, and references therein], the bremsstrahlung is produced in a quasi-thermal plasma with a temperature in excess of 100 million K located in the coronal part of the loop at or close to the initial energy release site. Since $E_e \simeq kT$, the energy lost by one electron to Coulomb collisions just goes to increase the energy of other electrons of about the same energy. Thus, primary energy losses for this plasma take the form of thermal conductive and convective losses, together with free-streaming of the very high energy tail of the electron distribution function. These losses can be considerably less than the collisional losses in models where $E_e \gg kT$. Consequently, less flare energy may be required for a given observed hard X-ray fluence.

It is quite clear from the foregoing that the fundamental questions relating to energy release and particle acceleration in solar flares remain unanswered. While very substantial amounts of energy must be contained in the hard X-ray producing electrons, this energy still cannot be reliably calculated because it is not known whether the X-ray source is predominantly thermal or nonthermal. A suitable hybrid model, involving both particle acceleration and direct heating, may be required to account adequately for all of the observed complexities and subtleties of flare emission.

The discovery of hard X-ray microflares [*Lin et al.*, 1984], impulsive bursts up to ~100 times less intense than normal flares, suggests that the flare process may be a fundamental way by which stored magnetic energy is released in the Sun's corona. During $\lesssim 2$ hours of balloon observations, one microflare was detected every ~5 minutes. These microflares last ~10 s and have power-law spectral shapes, similar to normal flares. If the spectrum extends down to ~5 keV, as appears to be the case, the rate of energy release in accelerated electrons in the observed microflares averages ~3×10^{26} ergs/sec, which is a significant fraction of the total power required to heat the active corona. Because the occurrence frequency for flare hard X-ray bursts (Figure 3) continue to increase as events get smaller, significant additional energy could be released by events below the observational threshold.

Gamma-ray and neutron emissions from solar flares are signatures of ion and relativistic electron interactions [*Ramaty and Murphy*, 1987]. Gamma-ray bremsstrahlung continuum from primary relativistic electrons has been observed up to at least 100 MeV [*Forrest*, 1985]. Gamma-ray lines are produced in nuclear reactions of accelerated protons and heavier nuclei interacting with the ambient solar atmosphere. A rich spectrum of lines, resulting from de-excitations of all the abundant constituents of the solar atmosphere up to Fe, has been observed from many flares [*Chupp*, 1984; *Murphy et al.*, 1990]. Such a spectrum is shown in Figure 4a taken from Murphy et al. [1991], with the predicted spectrum shown in Figure 4b. The nuclear reactions also lead to the production of neutrons and positrons. Capture of neutrons in the photosphere produces the 2.223-MeV line. This very narrow line is the strongest observed line from solar flares, except when the flare is close to the solar limb [*Wang and Ramaty*, 1974; *Chupp*, 1982]. Annihilation of positrons produces the 0.511-MeV line, whose shape and accompanying positronium annihilation continuum are sensitive probes of the temperature, density, and state of ionization of the ambient solar atmosphere [*Crannell et al.*, 1976]. Neutrons from flares have been detected directly [*Chupp*, 1987], as have the protons resulting from the decay of the neutrons in interplanetary space [*Evenson et al.*, 1983]. The nuclear reactions also produce pions [*Murphy et al.*, 1987; *Mandzhavidze*, 1987] and these have been detected through their decay products. The neutral pions decay into gamma rays directly. The charged pions produce secondary positrons and electrons, which produce bremsstrahlung gamma rays. In addition, the positrons also produce high-energy photons by annihilating in flight. The resultant high-energy gamma-ray emission (>10 MeV) has been observed from several flares [*Rieger*, 1989].

Before observations with the gamma-ray spectrometer on SMM, it was thought that only subrelativistic electrons are accelerated impulsively in flares. While the acceleration of ions in flares was established much earlier by direct particle observations in interplanetary space, it was thought that this acceleration was merely a secondary phenomenon that occasionally accompanied the much more frequent impulsive acceleration of electrons. This now appears not to be the case. Ion and relativistic electron acceleration, as evidenced by impulsive gamma-ray emission observed from many flares, must also be closely linked to the primary energy-release mechanisms. Relativistic neutrons and prompt gamma rays from pion decay have also been observed. Produced in GeV ion interactions, these emissions show that the acceleration of these highest-energy particles is also quite impulsive, with the lag, if any, between the acceleration of MeV and GeV protons being less than about 10 seconds. The energy contained in ions above 10 MeV, reliably determined from the gamma-ray observa-

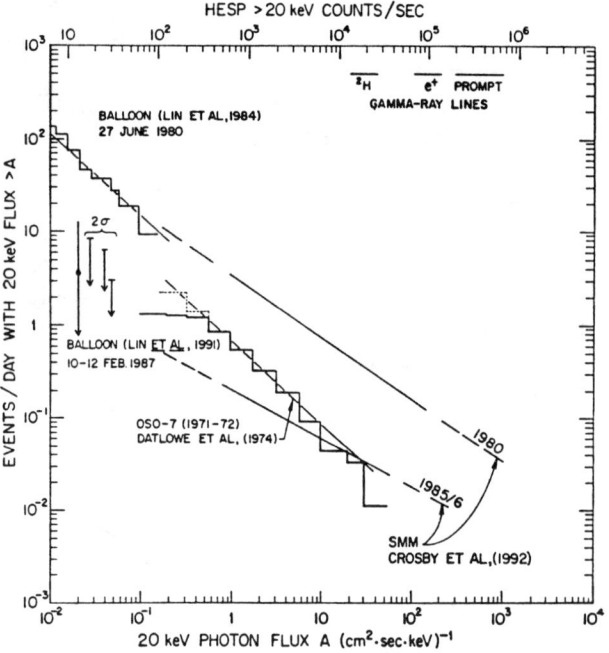

Fig. 3. The integral distribution of solar hard X-ray burst frequency per day, versus peak 20 keV photon flux. The SMM measurements [Crosby et al., 1992] in 1980 and 1985/6 are representative of the maximum and minimum activity periods, respectively, of the solar cycle. The balloon measurements [Lin et al., 1984, 1992] extend the distribution to much smaller bursts, termed microflares. The OSO-7 measurements [Datlowe, et al., 1974] show the distribution from part of the previous cycle. On top, the expected HESP >20 keV count rates are indicated, as well as 3s detection levels for gamma-ray lines of deuterium at 2.223 MeV, positron annihilation at 0.511 MeV, and prompt de-excitation lines of Ne, Mg, Si, C, N, O, Fe, etc.; computed under the assumption that the gamma-ray emission scales with the hard X-ray intensity.

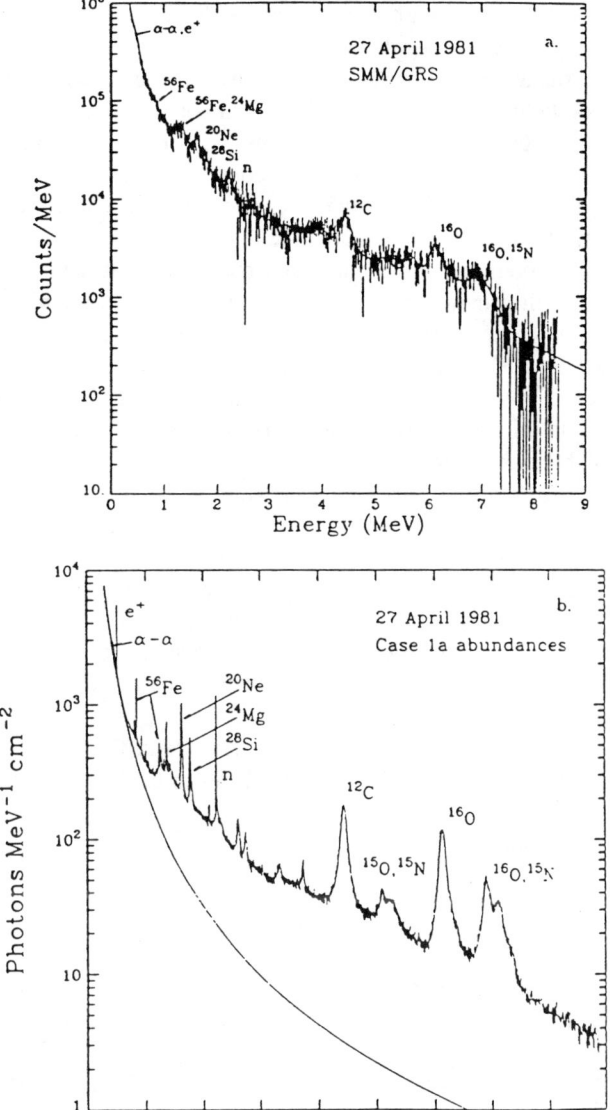

Fig. 4. (a) left Count spectrum of the 27 April 1981 flare measured with the NaI(Tl) gamma-ray spectrometer on SMM, together with a best-fit calculated spectrum [Murphy et al., 1990]. (b) right The corresponding calculated incident photon spectrum showing the ions responsible for many of the lines. (This spectrum does not reflect the statistical significance of an HESP observation of this flare.) The smooth curve represents the bremsstrahlung fluence. Clearly, much more information will be obtained with high-resolution Ge detectors since they will be able to identify and resolve the individual lines seen in the calculated spectrum.

tions, amounts to at least several percent of the flare energy. Even more energy could be contained in lower-energy ions.

It has been shown that elemental abundances can be derived from gamma-ray line observations for both the ambient gas and the accelerated particle population [Murphy et al., 1991]. The derived accelerated particle abundances for the 27 April 1981 flare resemble more closely the particle abundances measured in interplanetary space from impulsive solar flares rather than from large proton flares. The composition of the particles accelerated in impulsive flares exhibits strong enhancements of heavy elements. On the other hand, the composition of the particles seen in large proton events resembles that of the corona [Reames, 1990]. The fact that the accelerated particle composition deduced from the gamma-ray data agrees with that for particles observed escaping to interplanetary space from impulsive flares supports the result obtained from the timing observations, namely that the gamma-ray production is a direct consequence of the primary energy release.

The HESP mission addresses the following fundamental unanswered questions concerning solar flares:

What physical processes are responsible for releasing the energy stored in stressed magnetic fields to produce a solar flare?

What is the environment in which this energy release occurs?

What role do high energy particles play in the energy release process and how much of the released energy is contained in those particles?

What mechanisms accelerate both electrons and ions to high energies so rapidly and efficiently?

Is the basic flare energy release process related to coronal heating?

What mechanisms transport the flare energy, the energetic particle component in particular, away from the energy release site?

What are the characteristic radiation signatures of flares that have potentially hazardous effects, and how do these flares occur and evolve?

HESP will address these questions through the following steps:

Locate the energy release site and determine the characteristics of the energy release process.

Locate the particle acceleration and energy deposition sites during the different flare phases and determine how the particle energy is dissipated throughout the flaring region.

Determine the spectra and directivity of the accelerated electrons and ions and their evolution in time and space.

Determine the contribution of high-energy particles to flare energetics, specifically by measuring X-rays, gamma-rays, and the major contributors to the radiation energy budget, namely the soft X-ray and optical emissions.

Determine the plasma and magnetic field properties of the local environment in which the nonthermal processes occur.

Determine if all flares, large and small, have similar characteristics suggesting that they are all manifestations of the same basic processes. In particular, determine the characteristics of microflares and estimate their contribution to coronal heating.

Determine the composition of the accelerated particles and of the solar atmosphere with which they interact.

Study the long-term storage of high-energy ions at the Sun by observing and imaging pion-decay radiation and nuclear-line emission for extended time periods prior to and following the impulsive phase of flares.

3. OBSERVATIONAL APPROACH AND EXPECTED PERFORMANCE

The primary goal of HESP is to provide, for the first time, high spatial resolution and high spectral resolution imaging spectroscopy observations over the entire photon energy range from soft X-rays through hard X-rays to gamma-rays. By this we mean high-resolution spectroscopy at each point of the X-ray or gamma-ray image. This will allow the spectral evolution of the emissions to be traced in both space and time throughout a flare. It represents an important new capability not previously available in this wavelength range. Furthermore, the Sun is the only astrophysical X-ray or gamma-ray source bright enough and close enough to allow such observations to be made with present instrumentation.

In order to achieve a full understanding of the acceleration of particles and ions, and their transport through the solar atmosphere, it is essential to obtain supporting observations that can place the hard X-ray and gamma-ray events in the context of the important lower-energy (thermal) processes in flares. Such observations include measurements that must be made from space and those that can be made from the ground. Observations that require instruments in space include high-resolution spectroscopy and/or imaging at soft X-ray, XUV, and/or UV wavelengths. Sufficiently high spectral resolution is required to study Doppler shifts, line profiles, and line ratios. Such spectroscopy will provide detailed information on the plasma in which particle acceleration and transport occur, including electron temperatures and densities; turbulence, anisotropic mass motions, and atomic excitation processes; element abundances; and departures from ionization balance and Maxwellian velocity distributions. An imaging resolution commensurate with that achieved in hard X-rays would be sufficient to locate the positions of the hard X-ray and gamma-ray sources relative to magnetic field structures delineated by trapped plasma emitting in the X-ray, XUV, or UV spectral regions. Complementary direct spacecraft measurements of the energetic solar flare particles that escape to the interplanetary medium are also highly desirable.

Crucial ground-based observations include vector magnetic field measurements, and radio and optical imaging and spectroscopy. These observations will provide context information on the lower-energy components of the flaring region, including the morphology and dynamics of the thermal plasma and the magnetic field strengths and morphology in both the photosphere and the corona. In addition, they will provide complementary information on the high-energy components of flares such as the energetic electrons and ions, shocks and other plasma phenomena, and other indications of non-thermal processes.

The primary observational objectives of HESP, then, are to obtain the following:

Hard X-ray images with an angular resolution of as fine as 2 arcseconds and a temporal resolution of tens of milliseconds, commensurate with the known size scales of the flaring magnetic structures and the travel and stopping distances and times of the accelerated electrons (Figure 5). The images will be obtained with sufficient sensitivity to detect and locate the initial flare energy release and to study microflares.

High-resolution X-ray spectra with ~1-keV resolution down to energies as low as 2 keV. Since the bremsstrahl-

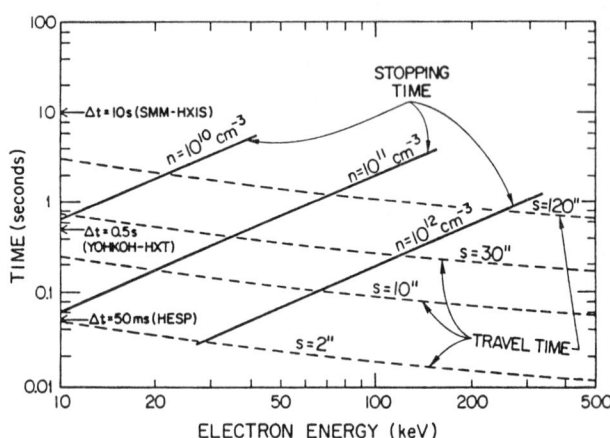

Fig. 5. Electron collisional stopping times (solid lines) for various ambient densities and travel times (dashed lines) in magnetic loops of different lengths, s, as a function of electron energy, E. The temporal resolutions (Dt) of HXIS on SMM, HXT on Yohkoh, and HESP are indicated on the vertical axis. The plot shows that the propagation of electrons along loops of typical lengths can be resolved with HESP both spatially and temporally at all energies up to >500 keV. Furthermore, for coronal densities of 10^{10} to 10^{12} cm^{-3}, the collisional degradation of the electron energy will be clearly resolved both temporally and spatially to energies of >100 keV by HESP with its 2" resolution. Neither SMM nor Yohkoh has these capabilities.

ung cross-section is well known, the hard X-ray spectrum can provide detailed quantitative information on the distribution of X-ray producing electrons at the Sun. Recently, a numerical inversion technique has been developed which derives the parent electron spectrum from the bremsstrahlung X-ray spectrum [*Johns and Lin*, 1992] for optically thin sources. The measurement of the precise shape of the X-ray continuum made possible with such fine energy resolution thus will provide unique, detailed information on the spectrum of the accelerated electrons and of the heated plasma, thus allowing the thermal and non-thermal aspects of individual flares to be clearly distinguished.

Spectrally resolved hard X-ray images. Apart from the separate scientific objectives of the imaging and spectroscopy alone, the combination of the two with sub-second time resolution will allow spectral changes to be measured as a function of space and time as the electrons propagate along the magnetic field in the flaring loop or loops, providing for the first time powerful constraints on the mechanisms of energy gain and loss.

Gamma-ray images with an angular resolution as fine as 8 arcseconds. The high-resolution gamma-ray imaging spectroscopy will allow images to be obtained in specific gamma-ray lines such as the alpha-induced lines at ~ 450 keV, and the 511-keV positron annihilation line. The intercomparison of these images from different types of particles and their comparison with the images in the electron-produced X-rays will allow the effects of differences in charge and mass on the acceleration and propagation processes to be explored for the first time.

High-resolution gamma-ray spectra with a few keV resolution to energies as high as 20 MeV. This resolution is sufficient to resolve the gamma-ray lines and to measure their shapes, thus allowing the full potential of gamma-ray line spectroscopy to be realized for the first time. Such high-resolution spectra would provide unique information on the directionality of the interacting particles, the composition of both the ambient gas and the accelerated ions, and the temperature, density, and state of ionization of the ambient gas.

Each of the above observational capabilities are new and unique; no previous solar mission has provided such capabilities. Other observational objectives, which are highly desirable but secondary to those above, are to obtain the following:

High-energy gamma-ray and neutron spectra and images for large flares at energies from 20 MeV to $\gtrsim 1$ GeV. These measurements should provide information on the acceleration of ions to the highest energies.

Soft X-ray, XUV, and/or UV spectra or images with spectroscopic resolution of about 7000 (corresponding to a velocity of 40 km/s) and/or spatial resolution as fine as 2 arcseconds, both with a temporal resolution of a few seconds. These observations will provide detailed context information on the flaring region in which the high-energy (nonthermal) processes take place.

It is important to realize that HESP will provide hard X-ray imaging spectroscopy, not just for a few flares, but for many thousands of flares in its lifetime. Figure 3 shows the integral distribution of solar flare bursts versus peak 20 keV hard X-ray flux as measured at different times in the solar cycle and with different detectors. The SMM distributions from the Hard X-Ray Burst Spectrometer (HXRBS) are indicative of the period near maximum (1980) and near minimum (1985/6). The balloon measurements on the left show the microflare distribution. The expected ≥ 20 keV count rate for the HESP instrument is shown at the top of the plot.

With the HESP background at energies between ~20 and 100 keV dominated by the diffuse sky emission, microflares would be detectable at the 3σ-level down to a 20-keV photon flux of $\sim 3 \times 10^{-3}$ (cm^2 s keV)$^{-1}$. Since a minimum of about a hundred counts are needed to form a simple image, the location and spatial size of microflares could be determined with HESP for events with a 20-keV flux as low as $\sim 2 \times 10^{-2}$ photons (cm^2 s keV)$^{-1}$.

The smallest bursts detectable by the SMM HXRBS instrument would give $\sim 10^3$ counts s^{-1} above 20 keV in HESP. In those events, rapid spatial changes in the X-ray sources could be followed on time scales of 0.1 s. Imaging spectroscopy, i.e., obtaining the spectrum as a function of spatial location, could be done with images in each of ten energy intervals with ~2-s resolution. For the once-per-day or larger flare, (i.e., $\geq 10^4$ HESP counts s^{-1} above 20 keV), imaging information could be obtained in tens of ms and imaging spectroscopy every ~100 ms.

HESP will provide, for the first time, images and high-resolution spectroscopy of gamma-ray lines and continuum for solar flares. Assuming that every flare accelerates ions with fluxes proportional to the hard X-ray flux, the 3σ-detection thresholds for the 2.223-MeV neutron-capture deuterium line, the 0.511-MeV positron annihilation line, and prompt nuclear deexcitation lines of Ne, Mg, Si, C, O, N, Fe are indicated on Figure 3. Thus the 2.223-MeV line should be detected, on average, once every ~2 to 3 days, the positron annihilation line every ~8 days, and prompt nuclear lines every ~20 days.

The narrow lines (Figure 4) are produced by protons or alpha particles colliding with the solar atmosphere. The primary background is the continuum emission from the flare itself. This consists of two components -- bremsstrahlung emission produced by relativistic electrons, and broad lines produced by the inverse process of accelerated heavy ions colliding with hydrogen and helium in the solar atmosphere. The line-to-continuum ratio is ~3 to 10 for the narrow, prompt, inelastic-scatter lines in this flare, but these ratios appear to vary from flare to flare. The line widths are typically 5 to 10 times narrower than the energy resolution of scintillation detectors (see Figure 8). The high spectral resolution of Ge detectors, thus, is essential to obtain unambiguous images of the accelerated ions.

For a large flare such as the 23 April 1981 event, it will be possible to obtain images of the energetic alphas from the 0.454-MeV Li-Be lines. At the same time, the analysis of the line shapes will provide information on the angular distribution of the energetic protons and alpha particles.

Images in the 511-keV lines will show where positrons are annihilating. Since positrons are charged, they will be guided by the magnetic structure where they are produced. Furthermore, the 0.511-MeV line width and shape will give information on the density and temperature of the annihilation region.

4. HESP STRAWMAN INSTRUMENTS

The original HESP strawman instrument payload, designed for a Delta launch vehicle, consists of a primary high energy instrument, HEISPEC, and two context instruments to provide thermal plasma diagnostic measurements. In addition, an extensive ground-based observational program (described in a later section) and supporting theory program are integral parts of the HESP mission.

The High Energy Imaging Spectrometer (HEISPEC), the main instrument of the HESP payload (Figure 6), is designed to image the range of flare energetic photons from soft X-rays (~2 keV) to ~ 1 MeV gamma-rays. Furthermore, HEISPEC has the capability to perform spatially resolved spectroscopy with high spectral resolution, to allow the full diagnostic power of hard X-rays and gamma-rays to be applied on a spatial point-by-point basis within solar flare. Table 2 summarizes the characteristics of HEISPEC.

The imaging is based on a Fourier transform technique using a set of rotating modulation collimators (RMCs), each of which is similar to those used on previous missions such as the US SAS-C and the Japanese Hinotori spacecraft. The technique is illustrated schematically in (Figure 7). Each RMC consists of two widely-spaced, fine-scale linear grids, which temporally modulate the photon signal from sources in the field of view as the RMC rotates about its long axis. The modulation can be measured with a detector having no spatial resolution placed behind the RMC. The modulation pattern over half a spin for a single RMC

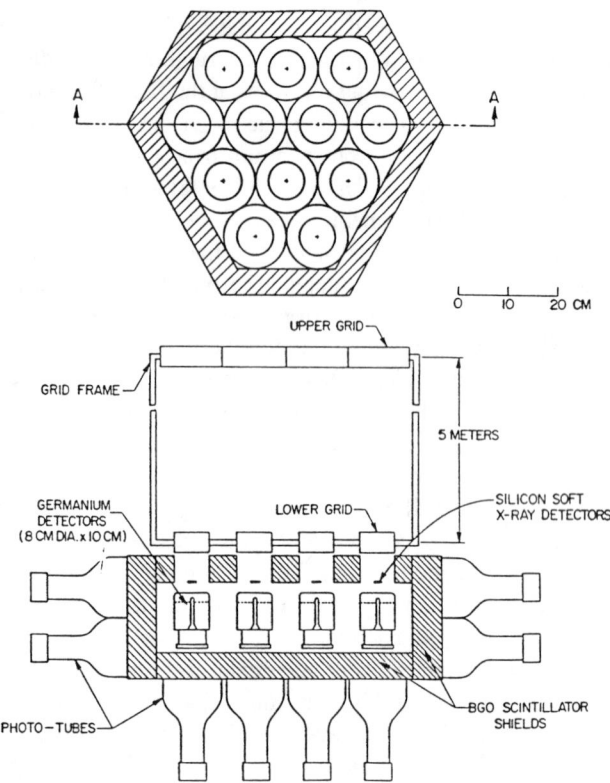

Fig. 6. Schematic cross sections of the High Energy Imaging Spectrometer (HEISPEC). The upper and lower tungsten grids, separated by 5m, form the rotating modulation collimators (RMCs). The two-segment germanium detectors provide high spectral resolution measurements from ~10 keV to 20 MeV. the combination of the Ge detectors and the bismuth germanate (BGO) shield extends the gamma-ray range to >200 MeV and provides neutron coverage from ~20 MeV to ~1 GeV. The silicon detectors cover the energy range from ~2 keV to $\gtrsim$ 20 keV. The BGO shield and collimator form an active anti-coincidence shield to reduce the background.

provides the amplitude and phase of many spatial Fourier components over a full range of angular orientations but for a small range of spatial source dimensions. Multiple RMCs, each with a different slit width, can provide coverage over a full spectrum of flare source sizes. An image is constructed from the set of measured Fourier components in exact mathematical analogy to multi-baseline radio interferometry.

The grid diameters and thicknesses are chosen to give full-Sun fields of view. Thus, upper and lower grids have diameters of 12.5 cm and 7.5 cm, respectively, and grid thicknesses range from ~2.8 mm for the finest (100-micron pitch) grids up to a maximum chosen thickness of 4 cm. With ~5-m separation between the grids, HESP will provide spatial resolution of ~2 arcseconds at hard X-ray energies

TABLE 2. HEISPEC INSTRUMENT SPECIFICATIONS

	ORIGINAL DELTA-CLASS MISSION			LIGHTSAT MISSION
Imaging				
Technique:	Fourier-transform imaging with rotating modulation collimators (RMC)			same
Angular resolution:				
~2 arcseconds	2 to 400 keV X-rays			~2 to 300 keV X-rays
~4-8 arcseconds	0.4 to ≥ 200MeV γ-rays			0.3 to 1 MeV γ-rays
~40 arcseconds	~20 MeV to 1GeV neutrons			no imaging of neutrons
Field of view:	Full Sun (0.6 degrees)			same
Temporal resolution:	~ tens of milliseconds for basic image, 1 sec for detailed image			same
Spectroscopy	Energy Range	Energy Resolution	Detector Type	
Hard X-rays & γ-rays:	10 keV to 20 MeV	0.6-5 keV FWHM	Germanium (Ge)	same
	20 MeV to ≥ 200MeV	$\Delta E/E \leq 5\%$	Ge & Bismuth Germanate (BGO)	on optional spacecraft
Soft X-rays:	2-20 keV	~1 keV FWHM	Silicon	same
Neutrons:	20 MeV to ~1 GeV	$\Delta E/E \approx 5\%$	Ge & BGO	on optional spacecraft
Instrument Characteristics				
Weight:	~800 kg			~120 kg
Power:	~300 watts (includes ~100 watts for coolers)			~100 watts
Data Rate:	~25 kbps average			~25 kbps average
Rotating Modulation Collimators				
Number of RMCs:	12			same
Length:	5 meters			1.75 meters
Grid slit spacing:	~100 microns to 6 mm (2 arcsec to 2 arcmin)			~34 microns to 2 mm
Grid thickness:	~0.3 to 4 cm (depending on slit spacing)			~0.1 to 1 cm
Material:	Tungsten (35 kg upper grids, 11 kg lower grids)			Tungsten
Germanium Detectors				
Area and volume:	600 cm^2, 6000 cm^3 (12 Ge, each 8 cm dia × 10 cm)			475 cm^2, 3800 cm^3 (12 Ge, each 7.1 cm dia × 8 cm)
Cooling:	to 85 K by mechanical coolers			same
Active anti-coincidence shield:	5-cm thick BGO well and collimator			on optional spacecraft

(below ~300 keV) and 8-16 arcseconds for gamma-ray lines and continuum up to ~ 1 MeV. The chosen rotation rate of ~15 rpm provides a complete image with the maximum number of Fourier components (~5×10^3) in 2 s, but spatial information is still available on timescales down to tens of ms, provided the count rates are sufficiently high. This high time resolution capability is important in following the propagation of electrons along typical magnetic loops as shown in Figure 5.

Behind the RMCs, dual-segment coaxial germanium detectors provide high spectral resolution from ~10 keV to ~20 MeV. The front ~2 cm thick segment measures hard X-rays up to ~200 keV, while the rear 8 cm segment provides undistorted high-resolution gamma-ray line measurements in the presence of very intense hard X-ray fluxes in large flares. The spectral resolution of Ge detectors is sufficient to resolve all of the solar gamma-ray lines with the exception of the neutron-capture deuterium line, which has an

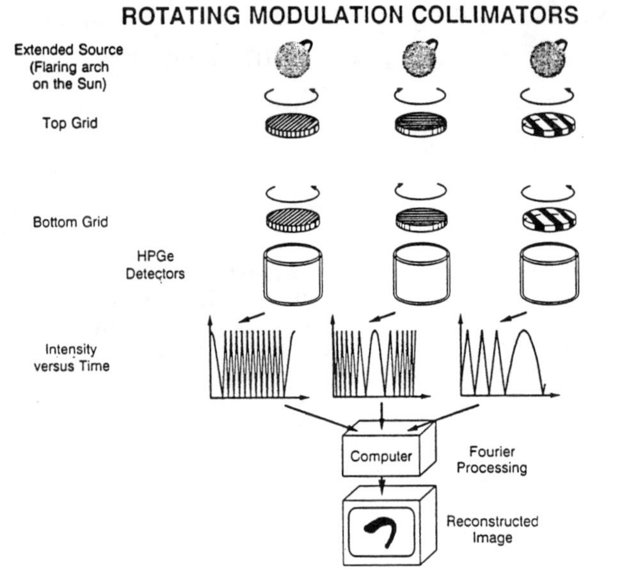

Fig. 7. Schematic representation of the Fourier-transform technique used to obtain images of solar flares with multiple rotating modulation collimators (RMCs).

expected FWHM of about 0.1 keV. This capability is illustrated in Figure 8, where the spectral resolution of the two-segment Ge detectors is compared with the typical line widths expected for gamma-rays in solar flares. It should be noted that high spectral resolution is also required to resolve sharp breaks in the non-thermal continuum and the steep super-hot thermal component of solar flares. Shown for comparison are the resolutions of the hard X-ray and gamma-ray spectrometers on SMM. The Ge detectors are cooled to their operating temperature of ~85 K by mechanical coolers.

The bismuth germanate (BGO) scintillator acts as an active anti-coincidence shield and collimator to reduce the background, provide excellent Compton rejection, and extend the energy range to ≳ 200 MeV for moderate spectral resolution gamma-ray measurements. In addition, the combination of BGO and Ge detectors provides moderate spectral resolution measurements of energetic neutrons up to ~1 GeV.

Silicon semiconductor detectors (or alternatively, proportional counters) placed in front of the germanium detectors serve to extend the imaging spectroscopy down to ~2 keV to cover the transition from non-thermal to thermal emission and to relate the high energy measurements to the thermal soft X-ray flare.

Two additional space-borne instruments are needed to determine the characteristics of the thermal, magnetic, and dynamic environment of the flaring region before, during, and after high energy flares:

A high-resolution spectrometer at soft X-ray, XUV, or UV wavelengths to study line profiles, line ratios, and Doppler shifts, and provide information on the electron temperature and densities, turbulence, anisotropic mass motions and atomic excitation processes, element abundances, and departures from ionization balance and Maxwellian velocity distributions. The strawman instrument being considered is an imaging Bragg crystal spectrometer similar to the Yohkoh BCS, but behind modulating grids to provide simple imaging.

An imager at soft X-ray, XUV, or UV wavelengths to locate the positions of the high-energy sources relative to

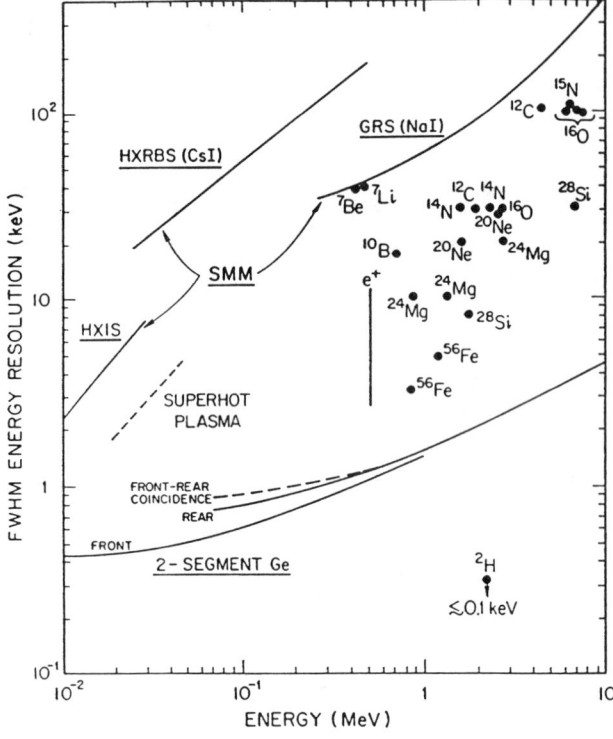

Fig. 8. The spectral resolution as a function of photon energy for a two-segment HPGe detector is compared to the resolutions of the hard X-ray and gamma-ray spectrometers on SMM. The typical widths expected for gamma-ray lines in solar flares are also shown. Note that none of these lines were resolved with the Gamma-Ray Spectrometer (GRS) on SMM but all, except the neutron-capture deuterium line at 2.223 MeV with a predicted width of <0.1 keV, can be resolved with a cooled HPGe detector. Similarly, the broken line indicating the energy resolution required to resolve the X-ray spectrum from a superhot plasma at a temperature of 30×10^6 K shows that this component was not resolved with the SMM instruments but can be clearly resolved with a cooled HPGe detector.

the magnetic field structures delineated by plasma emitting in these wavelengths. The strawman instrument is a soft X-ray telescope similar to the Yohkoh SXT.

5. HESP MISSION

The HESP Delta-class mission requirements, as driven by the science objectives and the HEISPEC instrument pointing requirements, are as follows:

Mission lifetime, 2 years minimum, ≥ 3 year goal
Spin rate, ~15 rpm
Total mass of instruments, ~800 kg
Maximum power for instrument, ~300 watts continuous.
Field of view (FOV), 36 arcminutes
Pointing accuracy, 6 arcminutes
Pointing stability, 0.1 arcseconds over 4 ms
Data collection of ~16 Gbits collected over a period of 1000 s during the most intense solar flare activity.
Dump a maximum of 16 Gbits to the ground in 24 to 48 hours.
Orbital average data rate of 20 kbps when there is no solar flare activity.

The initial HESP studies resulted in a single 4800 lb spacecraft accommodating the HEISPEC and context instruments, to be launched by a Delta 7920 rocket into a 600 km Sun-synchronous, 98 degree orbit [see Lin et al., 1991; 1993; and the following reports: (i) "Pre-Phase A Study Report on the High Energy Solar Physics Mission (HESP) - Spacecraft Study Report" by the Advance Missions Analysis Office at GSFC, and (ii) "Preliminary Mission Study for NASA Space Physics Division (Code SS) High Energy Solar Physics (HESP) Addendum" by the Ball Aerospace Systems Group].

6. HESP ON LIGHTSATS

To accommodate HESP on Lightsats, the HEISPEC instrument's weight was reduced from 800 kg to ~120 kg by removing the BGO shield and associated electronics, shortening the metering structure to 1.75 m, using slightly smaller detectors, and limiting the grids to ~1 cm maximum thickness. This modified HEISPEC, together with the Bragg crystal spectrometer, is accommodated on a spinning Lightsat spacecraft to provide imaging spectroscopy, as described in the right hand column of Table 2.. This 350 kg spacecraft can be launched by a Pegasus-class launcher. An equatorial orbit which avoids both the south Atlantic anomaly, the primary source of trapped radiation, and the polar caps, where energetic solar particles can enter, was chosen to minimize radiation damage to the Ge detectors. A second Pegasus-class non-spinning spacecraft carries the two context instruments in a Sun-synchronous polar orbit. An optional, but highly desirable third Pegasus-class spacecraft in equatorial orbit carries a compact, non-imaging germanium spectrometer with active BGO shield for high resolution gamma-ray line spectroscopy and for moderate resolution gamma-ray and neutron measurements up to ~1 GeV. Figure 9 shows the HESP spacecraft in orbit. This three-spacecraft HESP mission provides comparable imaging spectroscopy up to 20 MeV, better context measurements, and improved gamma-ray/neutron spectroscopy, compared to the original HESP, but it does not provide ≳1 MeV imaging of gamma-rays and neutrons. Alternatively, the > 20 MeV gamma-ray and neutron measurements may be made on long duration balloon flights if the required balloon technology is developed by then.

7. THE GROUND-BASED COMPONENT OF THE HESP MISSION

An integral component of the baseline HESP mission is an extensive program of crucial ground-based measurements that address the same scientific questions. The observational objectives of the ground-based program are to obtain the following information:

Vector magnetograms with approximately 2-arcseconds spatial resolution over a full active-region field of view. These will provide the structure of the magnetic fields and electric currents in the photosphere and chromosphere and will allow HESP to determine the spatial relationship between regions of electron and proton precipitation and regions of currents and magnetic shear.

Microwave imaging spectra with spatial resolution ranging from ~30 arcseconds at 1 GHz to ~1-2 arcseconds at 22

HIGH ENERGY SOLAR PHYSICS MISSIONS

Fig. 9. The baseline HESP Lightsat mission with three spacecraft, two in equatorial orbit and one in a sun-synchronous polar orbit.

GHz, spectral resolution of ~10-20%, and frequency coverage of ~1.4-22 GHz. These will allow the microwave sources to be resolved both spatially and spectrally in order to measure the strength of the coronal magnetic fields in the actual regions of electron acceleration and transport. The structure and thermodynamic conditions of these regions can be determined, not only during the energy release phase of a flare, but also prior to and following a flare. Microwave observations provide the only direct measurement of the strong magnetic fields in the coronal electron acceleration regions.

High-resolution optical images with spatial resolution of 1 to 2 arcseconds over a full active-region field of view. These images will be used to determine the preflare and flare structure and dynamics of the photosphere and chromosphere. High-resolution H-alpha images reveal the chromospheric magnetic connectivity in regions of electron and ion precipitation. Similar high resolution in white light pinpoints regions of energy precipitation to the photospheric regions accessible to only the most energetic ions. Polarimetric images allow explorations for evidence of both strong electric fields and ~100-keV nonthermal protons.

High dynamic range (10:1 to 100:1) millimeter, microwave, and meter/decimeter images of preflare and flaring conditions with sufficient spatial resolution (ranging from ~1 arcminute at dm wavelengths to ~1 arcsecond at mm wavelengths) to resolve the source structures. Such images at millimeter wavelengths provide a high quality, field-weighted observational perspective on ~1 MeV energetic electrons, responsible for the high-energy bremsstrahlung continuum. Microwave images provide a sensitive indication of the field-weighted morphology of lower-energy nonthermal electrons. Meter/decimeter images provide information on shocks, electron beams, and trapped electrons in the corona through the plasma radiation that they generate.

Optical imaging spectra in the energetically dominant lines and continua, with enough spectral and temporal resolution to relate the optical radiative energy loss to the thermalization of accelerated electrons and ions. Combined with the instruments on the HESP spacecraft, this enables the first direct measurement of the acceleration efficiency of electrons and ions in solar flares.

Coronal line and continuum images using coronagraphs and coronal polarimeters with a spatial resolution of ~2 arcseconds and full-limb field of view. Such images will be used to determine the morphology and dynamics of inner coronal structures. In combination with outer coronal images from SOHO, they relate high-energy ions and electrons to the structure and dynamics of the large-scale coronal field and to a rich variety of interplanetary wave and plasma phenomena.

8. CURRENT STATUS

Studies of the HESP spacecraft and generic instrument technology studies are proceeding. These are being managed by the HESP office at the NASA Goddard Space Flight Center. In addition, a HIgh REsolution Gamma-ray and hard X-ray Spectrometer (HIREGS) has been developed for Long Duration Balloon Flights (LDBF) in Antarctica. HIREGS consists of an array of twelve dual-segment Ge detectors surrounded by a 5-cm-thick BGO scintillation well with a 10-cm-thick drilled CsI collimator [see *Pelling et al.*, 1992 for description], very similar to the detector system for HEISPEC. HIREGS was flown successfully with four of the twelve Ge detectors on a 14-day LDBF in Antarctica in January 1992, and with the full-up twelve-detector HIREGS on a 10-day flight in January 1993.

A rotating modulation collimator imaging system, the High Energy Imaging Device (HEIDI), has been developed [see *Crannell et al.*, 1991, for description] and underwent its first flight on a conventional balloon from Palestine, Texas, in June 1993. HEIDI has a 5-meter-long boom and two RMCs with grids having slit widths of 625 and 330 microns and NaI(te) scintillation detectors. Finer grids down to ≲ 50 microns are being developed at several institutions by a variety of techniques.

Based on the flare hard X-ray burst rates measured by the SMM HXRBS in the last maximum, and current best estimates for the period of the solar activity cycle, the average flare rate should rise significantly at the beginning of 1999 and decrease substantially by the middle of 2004, with an uncertainty of ± one year. Thus, to be reasonably certain that the three-year mission life lies within the active portion of the next solar cycle, a HESP launch by the end of 2000 is desired.

Acknowledgments. We are grateful for the efforts of the other members of the HESP Science Study group, and the support of NASA Headquarter's Solar Physics discipline office. J. Phenix at Goddard Space Flight Center provided much of the technical support for spacecraft and instrument studies.

REFERENCES

Bahcall, J., et al., *The Decade of Discovery in Astronomy and Astrophysics*, National Research Council, 1991.

Brown, J. C., *Solar Phys.*, *18*, 489, 1971.

Brown, J. C., D. B. Melrose, and D. S. Spicer, *Astrophys. J.*, *228*, 592, 1979.

Canfield, R. C., in *Solar Flares*, ed. P.A. Sturrock, Colorado Assoc. Univ. Press, Boulder, 1980, p. 451.

Chupp, E. L., in *Gamma Ray Transients and Related Astrophysical Phenomena*, eds. R.E. Lingenfelter, H.S. Hudson, and D.M. Worrall, New York Assoc. Univ. Press, 1982, p. 363.

Chupp, E. L., *Ann. Rev. Astron. Astrophys.*, *22*, 359, 1984.

Chupp, E. L., *Astrophys. J.*, *318*, 913, 1987.

Crannell, C. J., G. Joyce, R. Ramaty, and C. Werntz, *Astrophys. J.*, *210*, 52, 1976.

Crannell, C. J., et al., A balloon-borne payload for imaging hard X-rays and gamma rays from solar flares, in *Proc. of AIAA International Balloon Technology Conference*, October 1991.

Crosby, N. B., M. J. Aschwanden, and B. R. Dennis, *Solar Phys.*, *143*, 275, 1993.

Datlowe, D. W., M. J. Elcan, and H.S. Hudson, *Solar Phys.*, *39*, 155, 1974.

Emslie, A. G., *Solar Phys.*, *86*, 133, 1986.

Evenson, P., P. Meyer, and K. R. Pyle, *Astrophys. J.*, *274*, 875, 1983.

Forrest, D. J., 19th Internat. Cosmic Ray Conf. Papers, 4, 146, 1985.

Holman, G. P., and S.G. Benka, *Astrophys. J. Lett.*, *400*, L79, 1992.

Johns, C. M., and R. P. Lin, *Solar Phys.*, *137*, 121, 1992.

Lin, R. P., R. A. Schwartz, R. M. Pelling, and K. C. Hurley, *Astrophys. J.*, *251*, L109, 1981.

Lin, R. P., R. A. Schwartz, S. R. Kane, R. M. Pelling, and K. C. Hurley, *Astrophys. J.*, *283*, 421, 1984.

Lin, R. P., and R. A. Schwartz, *Astrophys. J.*, *312*, 462, 1987.

Lin, R. P., et al., The High Energy Solar Physics Mission (HESP), in *Scientific Objectives and Technical Description*, 1991.

Lin, R. P., K. C. Hurley, D. M. Smith, and R. M. Pelling, *Solar Phys.*, *135*, 57, 1992.

Lin, R. P., B. R. Dennis, A. G. Emslie, R. Ramaty, R. Canfield, and G. Doschek, presented at the XXIX COSPAR, Washington, DC, 1992, in Proc. Symp. Fundamental Problems in Solar Activity (eds. M. Pick and M. E. Machaco), *Adv. Space Res.*, *13*, (9)401, 1993.

Mandzhavidze, N. Z., Thesis, Physico-Technical Inst., Leningrad, USSR, 1987.

Murphy, R. J., C. D. Dermer, and R. Ramaty, *Astrophys. J. Suppl.*, *63*, 721, 1987.

Murphy, R. J., G. H. Share, J. R. Letaw, and D. J. Forrest, *Astrophys. J.*, *358*, 290, 1990.

Murphy, R. J., R. Ramaty, B. Kozlovsky, and D. V. Reames, *Astrophys. J.*, *371*, 793, 1991.

Pelling, M., et al., A high resolution gamma-ray and hard X-ray spectrometer (HIREGS) for long duration balloon flights, in *Proc. SPIE Meeting*, San Diego, July 1992, Society of Photo-Optical Instrumentation Engineers, in press (1992).

Ramaty, R., and R. J. Murphy, *Space Sci. Rev.*, *45*, 213, 1987.

Reames, D. V., *Astrophys. J. Suppl.*, *73*, 235, 1990.

Rieger, E., *Solar Phys.*, *121*, 323, 1989.

Smith, D. F., and D. W. Harmony, *Astrophys. J.*, *252*, 800, 1982.

Wang, H. T., and R. Ramaty, *Solar Phys.*, *36*, 129, 1974.

Lin, R.P., Physics Department and Space Sciences Laboratory, University of California, Berkeley, California.

Dennis, B.R., and R. Ramaty, Goddard Space Flight Center, Greenbelt, Maryland

Emslie, A.G., University of Alabama, Huntsville

Canfield, R., Insitute for Astronomy, University of Hawaii, Honolulu, Hawaii.

Doschek, G., Naval Research Laboratory, Washington, D.C.